Francisco Pena-Pereira (Ed.)

Miniaturization in Sample Preparation

Francisco Pena-Pereira (Ed.)

Miniaturization in Sample Preparation

Managing Editor: Anna Rulka

Language Editor: Perry Mitchel

DE GRUYTER
OPEN

Published by De Gruyter Open Ltd, Warsaw/Berlin
Part of Walter de Gruyter GmbH, Berlin/Munich/Boston

ISBN: 978-3-11-041017-4
e-ISBN: 978-3-11-041018-1

Bibliographic information published by the Deutsche Nationalbibliothek. The Deutsche Nationalbibliothek lists this publication in the Deutsche Nationalbibliografie; detailed bibliographic data are available in the Internet at http://dnb.dnb.de.

Managing Editor: Anna Rulka
Language Editor: Perry Mitchel

www.degruyteropen.com

Cover illustration: © David Perez Parga & Francisco Pena-Pereira

Contents

List of Contributors — xii

Francisco Pena-Pereira

1 From Conventional to Miniaturized Analytical Systems — 1

1.1 Introduction — **1**

1.2 Miniaturizing Steps in the Analytical Process — **3**

1.2.1 A Need for Scaling Down Conventional Sample Preparation Techniques — **9**

1.2.2 Miniaturization of Analytical Separation — **11**

1.2.2.1 Gas Chromatography — **12**

1.2.2.2 Liquid Chromatography — **14**

1.2.2.3 Capillary Electrophoresis — **16**

1.2.3 Miniaturization of Detection Techniques — **17**

1.2.3.1 Molecular Spectrometry — **18**

1.2.3.2 Atomic Spectrometry — **19**

1.2.3.3 Mass Spectrometry — **20**

1.2.3.4 Electrochemical Techniques — **21**

1.3 Conclusions and Outlook — **22**

Abbreviations — **23**

Acknowledgements — **24**

References — **24**

Habib Bagheri, Hamed Piri-Moghadam, Mehrnoush Naderi, Ali Es'haghi and Ali Roostaie

2 Solid-Phase Microextraction and Related Techniques — 29

2.1 Introduction — **29**

2.2 Solid-phase Microextraction Fundamentals — **29**

2.2.1 Principle of Solid-phase Microextraction — **29**

2.2.2 Different Modes of Solid-phase Microextraction — **31**

2.2.3 Coupling to Analytical Instrumentation — **31**

2.2.3.1 Off-line Coupling — **31**

2.2.3.2 On-line Coupling — **31**

2.2.3.2.1 On-line Coupling to Gas Chromatography — **33**

2.2.3.2.2 On-line Coupling to Liquid Chromatography — **34**

2.3 Extractant Phases in Solid-phase Microextraction — **35**

2.3.1 Conventional Extractant Phases — **35**

2.3.2 Extractant Phases Based on Inorganic Polymerization — **37**

2.3.2.1 Preparation of Sorbents by Sole Precursor — **39**

2.3.2.2 Preparation of Sorbents by Precursor and Coating Polymer — **41**

2.3.2.3 Preparation of Sorbents by Precursor and a Modifier — **41**

2.3.2.4 Chemical Bonding Between Substrates and Sorbent During Sol-Gel Process — 42
2.3.2.4.1 Treatment of Fused Silica by NaOH — 42
2.3.2.4.2 Self Assembled Monolayers — 42
2.3.2.4.3 Diazonium Salts — 43
2.3.3 Conductive Polymers — 44
2.3.3.1 Structures of Some Well-known Conductive Polymers — 47
2.3.3.2 Preparation of Conductive Polymers — 48
2.3.3.2.1 Chemical Synthesis — 49
2.3.3.2.2 Electrochemical Synthesis — 49
2.3.3.3 Conductive Polymer-based Extractant Phases — 51
2.3.3.3.1 Polypyrrole-based Coatings — 51
2.3.3.3.2 Polyaniline-based Coatings — 52
2.3.3.3.3 Polythiophene-based Coatings — 53
2.3.4 Monolithic Polymers — 54
2.3.4.1 Preparation of Monolithic Polymers — 54
2.3.4.2 Monoliths for Solid-phase Microextraction — 56
2.3.4.2.1 Fiber Format — 56
2.3.4.2.2 In-tube Solid-phase Microextraction — 57
2.3.5 Composites — 58
2.3.5.1 Polymer Matrix Composites — 59
2.3.5.2 Nanocomposite-based Coatings — 61
2.3.6 Electrospun Nanofibers — 63
2.3.7 Selective Sorbents — 64
2.3.7.1 Molecularly Imprinted Polymers-based-solid-phase Microextraction — 65
2.3.7.2 Molecularly Imprinted Xerogels-based-solid-phase Microextraction — 65
2.3.8 Metal-based Coatings — 67
2.3.8.1 Metal-based Fibers Preparation by Anodization — 67
2.3.8.2 Metal-based Fibers Developed by Physical Coating — 67
2.3.8.3 Metal-based Fibers Developed by Chemical Coating — 68
2.4 Related Techniques — 70
2.4.1 Microextraction in Packed Syringe — 70
2.4.2 Stir Bar Sorptive Extraction — 74
2.4.3 Needle Trap Extraction — 75
2.5 Conclusions — 76
Abbreviations — 76
References — 78

Bin Hu, Man He and Beibei Chen
3 Novel Materials in Solid-Phase Microextraction and Related Sample Preparation Approaches — 88
3.1 Introduction — 88
3.2 Coating Preparation Techniques Applied For Solid-phase Microextraction and Related Approaches — 88
3.2.1 Sol-gel Technology — 89
3.2.2 Physical Adhesion Method — 89
3.2.3 Electrochemical Methods — 90
3.2.3.1 Electrodeposition — 90
3.2.3.2 Anodization — 90
3.2.3.3 Electrophoretic Deposition — 91
3.2.4 Polymerization — 91
3.2.5 Chemical Vapor Deposition — 92
3.2.6 Liquid Phase Deposition — 92
3.3 Commercial Solid-phase Microextraction Coatings — 92
3.4 Novel Materials for Solid-phase Microextraction and Related Approaches — 94
3.4.1 Nanostructured Materials — 94
3.4.1.1 Carbon Nanomaterials — 95
3.4.1.2 Metal Oxide Nanomaterials — 99
3.4.1.3 Mesoporous Materials — 101
3.4.1.4 Application of Nanomaterials in Solid-Phase Microextraction and Related Approaches — 105
3.4.2 Molecularly Imprinted Materials — 105
3.4.3 Ionic Liquid Coatings — 122
3.4.4 Immunosorbents — 125
3.4.5 Metal-organic Frameworks — 131
3.5 Other Novel Materials — 139
3.5.1 Monolithic Materials — 139
3.5.2 Restricted Access Materials — 142
3.6 Application of Various Materials in Solid-phase Microextraction-related Approaches — 144
3.7 Conclusions and Prospects — 157
Abbreviations — 157
Acknowledgements — 162
References — 162

Elena Fernández and Lorena Vidal
4 Liquid-phase Microextraction Techniques — 191
4.1 Introduction — 191
4.1.1 History — 191

4.1.2 Solvents — 195
4.1.3 Separation and Detection Systems — 197
4.1.4 Energy and Radiation — 200
4.1.5 Optimization Strategies — 200
4.2 Single-drop Microextraction — 201
4.2.1 Headspace Single-drop Microextraction — 203
4.2.2 Direct Immersion — 205
4.2.2.1 Direct Immersion Single-drop Microextraction — 205
4.2.2.2 Drop-in-drop and Drop-to-drop — 207
4.2.2.3 Continuous Flow Microextraction — 209
4.2.2.4 Liquid-liquid-liquid Microextraction — 210
4.2.2.5 Directly Suspended Droplet Microextraction — 212
4.2.2.6 Solidification of Floating Organic Drop Microextraction — 214
4.3 Membrane-based Liquid-phase Microextraction — 215
4.3.1 Hollow Fiber Liquid-phase Microextraction — 215
4.3.2 Electromembrane Extraction — 220
4.4 Dispersive Liquid-liquid Microextraction — 225
4.4.1 Classical Dispersive Liquid-liquid Microextraction — 225
4.4.2 Ultrasound- and Vortex-assisted Dispersive Liquid-liquid Microextraction — 230
4.4.3 Temperature-assisted Dispersive Liquid-liquid Microextraction — 231
4.4.4 In Situ Ionic Liquid Formation Dispersive Liquid-liquid Microextraction — 231
4.4.5 Supramolecular-based Dispersive Liquid-liquid Microextraction — 232
4.4.6 Air-assisted Liquid-liquid Microextraction — 234
4.5 Conclusions — 234
Abbreviations — 235
Acknowledgements — 237
References — 237

Shayessteh Dadfarnia and Ali Mohammad Haji-Shabani
5 Choice of Solvent in Liquid-Phase Microextraction — 253
5.1 Introduction — 253
5.2 Relevance of Physicochemical Properties in Extractant Phase Selection — 253
5.2.1 Solubility — 253
5.2.2 Distribution Coefficient — 254
5.2.3 Selectivity — 254
5.2.4 Immiscibility — 258
5.2.5 Density — 258
5.2.6 Interfacial Tension — 259
5.2.7 Chemical Reactivity — 259

5.2.8 Corrosiveness — 259
5.2.9 Viscosity, Boiling Point and Vapor Pressure — 259
5.2.10 Availability and Cost — 260
5.2.11 Other Criteria — 260
5.3 Extracting Solvents for Liquid-phase Microextraction — 260
5.3.1 Extractant Phases for Single-drop Microextraction — 261
5.3.2 Extractant Phases for Directly-suspended Droplet Microextraction — 264
5.3.3 Extractant Phases for Hollow Fiber Liquid-phase Microextraction — 266
5.3.4 Extractant Phases for Dispersive Liquid-liquid Microextraction — 268
5.4 Conclusions — 271
Abbreviations — 272
References — 273

Marta Costas-Rodriguez and Francisco Pena-Pereira
6 Method Development with Miniaturized Sample Preparation Techniques — 276
6.1 Introduction — 276
6.2 Evaluation of Experimental Parameters — 277
6.2.1 Type of Miniaturized Sample Preparation Technique — 278
6.2.2 Type of Extractant Phase — 279
6.2.3 Sample and Extractant Phase Volumes — 280
6.2.4 Extraction Time — 282
6.2.5 Agitation of the Sample — 284
6.2.6 pH — 285
6.2.7 Ionic Strength — 285
6.2.8 Temperature — 287
6.2.9 Derivatization — 287
6.2.10 Desorption — 289
6.3 Optimization Strategies For Analytical Method Development — 290
6.3.1 Screening of the Variables — 292
6.3.2 Optimization — 293
6.4 Valldation of Microextraction Methodologies — 298
6.5 Conclusions — 300
Abbreviations — 300
Acknowledgements — 302
References — 302

Noelia Cabaleiro and Inmaculada de la Calle
7 Miniaturized Alternatives to Conventional Sample Preparation Techniques for Solid Samples — 308
7.1 Introduction — 308

7.2 Objectives and Benefits of Miniaturized Sample Preparation Procedures — **310**
7.3 Challenges of Solid Sample Analysis — **311**
7.3.1 Types and Composition of Solid Samples — **313**
7.3.2 Pre-treatment of Solid Samples — **315**
7.3.3 Extraction Mechanisms — **316**
7.4 Sample Preparation Techniques for Solid Samples: from Conventional to Miniaturized Alternatives — **317**
7.4.1 Trace Elemental and Organometallic Analysis — **317**
7.4.1.1 Minimal Treatment-based Techniques — **326**
7.4.1.1.1 Direct Solid Sampling — **326**
7.4.1.1.2 Slurry Sampling — **327**
7.4.1.1.3 Sample Emulsification — **330**
7.4.1.2 Decomposition-based Techniques — **331**
7.4.1.2.1 Dry Ashing — **331**
7.4.1.2.2 Fusion — **333**
7.4.1.2.3 Acid Digestion and Microwave-assisted Digestion — **334**
7.4.1.2.4 Vapor-phase Acid Digestion and Vapor-phase Microwave-assisted Digestion — **337**
7.4.1.2.5 Ultrasound-assisted Digestion and Pseudodigestion — **339**
7.4.1.2.6 Enzymatic digestion — **340**
7.4.1.2.7 Tissue Solubilization — **341**
7.4.1.3 Extraction-based Techniques — **343**
7.4.1.3.1 Acid Extraction (leaching) — **343**
7.4.1.3.2 Ultrasound-assisted Extraction — **344**
7.4.1.3.3 Microwave-assisted Extraction — **346**
7.4.1.3.4 Accelerated Solvent Extraction — **348**
7.4.1.3.5 Supercritical Fluid Extraction — **349**
7.4.1.3.6 Matrix Solid-phase Dispersion — **350**
7.4.1.3.7 Solid-phase Microextraction — **352**
7.4.1.4 Combined Techniques — **353**
7.4.1.4.1 Solid-phase Extraction — **353**
7.4.1.4.2 Headspace Solid-phase Microextraction — **354**
7.4.1.4.3 Liquid-phase Microextraction — **355**
7.4.1.4.4 Fully Miniaturized Analytical Systems — **356**
7.4.2 Organic Compound Analysis — **359**
7.4.2.1 Minimal Treatment-based Techniques — **360**
7.4.2.2 Extraction-based Techniques — **366**
7.4.2.2.1 Direct Solid Treatment by Miniaturized Matrix Solid-phase Dispersion — **366**

7.4.2.2.2 Direct Solid Treatment by Miniaturized Ultrasound- and Microwave-assisted Extraction, Supercritical Fluid Extraction and Pressurized Liquid Extraction — **368**
7.4.2.2.3 Direct Solid Treatment by Microextraction Techniques — **371**
7.4.2.3 Combined Techniques — **373**
7.4.2.3.1 Solid-phase Extraction in Miniaturized Format — **373**
7.4.2.3.2 Solid-phase Microextraction — **375**
7.4.2.3.3 Liquid-phase Microextraction — **376**
7.4.2.3.4 Fully Miniaturized Analytical Systems — **377**
7.5 Future Trends — **380**
7.6 Conclusions — **381**
Abbreviations — **382**
Acknowledgements — **385**
References — **385**

Adam Kloskowski, Łukasz Marcinkowski and Jacek Namieśnik
8 Green Aspects of Miniaturized Sample Preparation Techniques — 416
8.1 Introduction — **416**
8.2 Reduction of the Amount of Organic Solvents Used — **418**
8.3 Green Extraction Phases for Microextraction Techniques — **422**
8.4 Automation in Microextraction Techniques — **430**
8.5 Chemometric Approaches for Optimization and Evaluation of Microextraction Techniques — **432**
8.6 Conclusions — **434**
Abbreviations — **436**
References — **438**

Index — 447

List of Contributors

Habib Bagueri Environmental and Bio-Analytical Laboratories, Department of Chemistry, Sharif University of Technology, P.O. Box 11365–9516, Tehran, Iran
Noelia Cabaleiro Department of Analytical and Food Chemistry, Faculty of Chemistry, University of Vigo, Campus As Lagoas-Marcosende s/n, 36310 Vigo, Spain
Inmaculada de la Calle Department of Analytical and Food Chemistry, Faculty of Chemistry, University of Vigo, Campus As Lagoas-Marcosende s/n, 36310 Vigo, Spain
Beibei Chen Key Laboratory of Analytical Chemistry for Biology and Medicine (Ministry of Education), Department of Chemistry, Wuhan University, Wuhan 430072, China
Marta Costas-Rodríguez Ghent University, Department of Analytical Chemistry, Krijgslaan 281-S12, B-9000, Ghent, Belgium
Shayessteh Dadfarnia Department of Chemistry, Faculty of Science, Yazd University, Yazd 89195–741, Iran
Ali Es'haghi Environmental and Bio-Analytical Laboratories, Department of Chemistry, Sharif University of Technology, P.O. Box 11365–9516, Tehran, Iran
Ali Mohammad Haji Shabani Department of Chemistry, Faculty of Science, Yazd University, Yazd 89195–741, Iran
Elena Fernández Martínez Departamento de Química Analítica, Nutrición y Bromatología e Instituto Universitario de Materiales, Universidad de Alicante, Apdo. 99, 03080 Alicante, Spain
Man He Key Laboratory of Analytical Chemistry for Biology and Medicine (Ministry of Education), Department of Chemistry, Wuhan University, Wuhan 430072, China
Bin Hu Key Laboratory of Analytical Chemistry for Biology and Medicine (Ministry of Education), Department of Chemistry, Wuhan University, Wuhan 430072, China
Adam Kloskowski Department of Physical Chemistry, Chemical Faculty, Gdańsk University of Technology, 11/12 G. Narutowicza St., 80–952 Gdańsk, Poland
Łukasz Marcinkowski Department of Physical Chemistry, Chemical Faculty, Gdańsk University of Technology, 11/12 G. Narutowicza St., 80–952 Gdańsk, Poland
Mehrnoush Naderi Environmental and Bio-Analytical Laboratories, Department of Chemistry, Sharif University of Technology, P.O. Box 11365–9516, Tehran, Iran
Jacek Namieśnik Department of Analytical Chemistry, Chemical Faculty, Gdańsk University of Technology, 11/12 G. Narutowicza St., 80–952 Gdańsk, Poland
Francisco Pena-Pereira Department of Analytical and Food Chemistry, Faculty of Chemistry, University of Vigo, Campus As Lagoas-Marcosende s/n, 36310 Vigo, Spain
Hadme Piri-Moghadam Environmental and Bio-Analytical Laboratories, Department of Chemistry, Sharif University of Technology, P.O. Box 11365–9516, Tehran, Iran
Ali Roostaie Environmental and Bio-Analytical Laboratories, Department of Chemistry, Sharif University of Technology, P.O. Box 11365–9516, Tehran, Iran
Lorena Vidal Martínez Departamento de Química Analítica, Nutrición y Bromatología e Instituto Universitario de Materiales, Universidad de Alicante, Apdo. 99, 03080 Alicante, Spain

1 From Conventional to Miniaturized Analytical Systems

Francisco Pena-Pereira
Analytical and Food Chemistry Department; Faculty of Chemistry;
University of Vigo, Campus As Lagoas-Marcosende s/n, 36310 Vigo, Spain
e-mail address: fjpena@uvigo.es

1.1 Introduction

Nowadays, the term miniaturization is applied to a wide spectrum of knowledge areas, including, among others, engineering, physics, medicine, materials science, computer science and chemistry. A search on the ISI Web of Knowledge provided approximately 42000 results by entering the term miniaturization, from which around 5200 results are devoted to chemistry. The number of publications concerning the miniaturization of chemical systems has experienced an important increase in the last two decades, as has the number of citations received by these publications, as shown in **Figure 1.1**. In accordance with the ISI Web of Knowledge, they currently receive around 12000 citations per year. Nevertheless, this is only the tip of the iceberg since the number of publications devoted to the development and application of miniaturized analytical systems (but not referring to miniaturization in the title or abstract sections) are not included.

In the broadest sense of the word, miniaturization can be defined as the production of novel systems that are substantially reduced in size in comparison with conventional systems. In analytical chemistry, the term miniaturization does not refer solely to the scaling-down of analytical instrumentation, apparatus and devices since it is also applicable when the components (including chemicals and solvents) needed to perform analytical operations are employed on a greatly reduced scale. In fact, size reduction is not the main driving force when shrinking analytical systems, as can be deduced from section 1.2. It is worth noting that the term miniaturization has been mainly employed in the analytical chemistry literature to refer to the micro-total analysis systems (μ-TAS) and lab-on-a-chip (LOC) devices. Even though they represent the highest degree of downsizing, the concept of miniaturization should be observed from a broader, non-exclusive perspective since this concept includes the advances achieved in every single step of the analytical process.

A recent trend in analytical chemistry is a progression towards the miniaturization of analytical systems. Different steps of the analytical process, including sample preparation, analytical separation and detection have been subjected to miniaturization, automation and portability. In addition, the integration of different analytical steps has allowed the development of fully miniaturized systems. The miniaturiza-

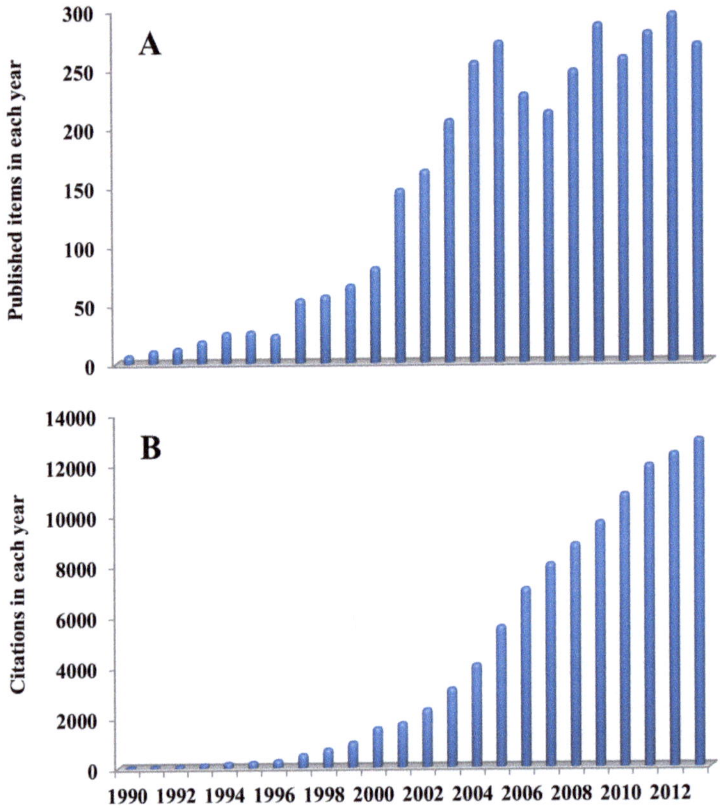

Figure 1.1 (A) Evolution of the publications devoted to miniaturization in chemistry; (B) Citations received by the publications devoted to miniaturization in chemistry (source: ISI Web of Knowledge (Web of Science) – Thomson Reuters).

tion of analytical systems has been mainly addressed as a result of the necessity to overcome problems and meet the demands from novel research areas. As a result, challenging requirements have been established for analytical microsystems, namely the analysis of highly reduced sample volumes with the highest possible sensitivity, selectivity and precision in reduced analysis times, with reduced amounts of reagents and/or organic solvents and, whenever possible, in the field.

In this chapter, an overview of the main advances towards the miniaturization of conventional systems that are commonly employed in analytical chemistry is provided. Considerations on the miniaturization of the three most developed steps of the analytical process, namely sample preparation, analytical separation and detection are provided and representative applications are briefly described.

1.2 Miniaturizing Steps in the Analytical Process

As early as 1959, Feynman introduced a field focused on the problem of manipulating and controlling things on a small scale in his inspiring talk entitled 'There is plenty of room at the bottom' (Feynman, 1992). Feynman predicted that the improvement of several scientific subjects by means of miniaturization would allow scientists to better understand fundamental problems and to overcome the limitations of full-sized devices. Miniaturization of analytical instrumentation was identified as a significant trend in the field by the journal *Analytical Chemistry* in 1970 (Senzel, 1970). Improvement and size reduction of analytical instrumentation and devices has been a challenge for several decades. In the last two decades the analytical scientific community has experienced a renewed growth of interest in miniaturized systems concerning the different analytical process steps, mainly influenced by the introduction of microextraction techniques, miniaturized separation techniques and detection systems, as well as the development of μ-TAS and the novel generations of flow injection techniques. The miniaturization of analytical systems is generally linked to other challenges in analytical chemistry, such as portability, automation and greening of analytical procedures. In addition, economy, rapidity, improved analytical performance and the size decrease of analytical systems are among the drivers for miniaturization (**Figure 1.2**).

The analytical process involves all the steps needed to obtain analytical information from a sample, namely sample collection and preservation, sample preparation, separation, detection, data processing and final decision (**Figure 1.3**). Today, almost every step of the analytical process has been subjected to miniaturization. However, the different steps of the analytical process have not been miniaturized to the same extent. For instance, sample collection and preservation is the step of the analytical process less subjected to the benefits of miniaturization, even though some autonomous and remote sensing analytical microsystems have been reported. Conversely, data acquisition and processing have achieved an excellent degree of miniaturization. Furthermore, it is generally accepted that the downscaling of sample preparation approaches has been developed after certain efforts to miniaturize both separation and detection systems.

It is worth mentioning that full miniaturization of analytical systems has also been addressed in the literature. Thus, two different systems, namely μ-TAS and lab-on-a-valve (LOV), enable the miniaturization of the different steps needed to perform a chemical analysis in a single system. The concept of μ-TAS was firstly introduced two decades ago by Manz et al. with the aim of enhancing the analytical performance of total analysis systems (TAS) rather than a simple reduction of their size (Manz et al., 1990a). In this pioneering work, the authors defined μ-TAS as "TAS systems that perform all sample handling steps extremely close to the phase of measurement". The introduction of surface techniques amenable to achieve mechanical microstructures has been critical in the development of μ-TAS systems. Nowadays, μ-TAS systems are

Figure 1.2 Drivers towards the miniaturization of analytical systems.

not a fashionable craze, but a powerful and exciting interdisciplinary research field (Marx et al., 1991). A variety of excellent books (Herold & Rasooly, 2009; Ríos et al., 2009; Lin, 2011) and reviews (Mark et al., 2010; Livak-Dahl et al., 2011; Rios et al., 2012; Nge et al., 2013) on microfluidics are available in the literature for interested readers. More recently, Ruzicka introduced the concept of LOV as a versatile methodology for downscaling reagent-based (bio)chemical assays to micro- and submicroliter level (Ruzicka, 2000). The micro-sequential injection (µSI)-LOV system, also reported as the third generation of flow injection analysis systems, enables one to carry out the unitary steps needed to perform an analysis on the basis of the use of a central sample processing unit. A number of review articles covering the evolution of flow injection techniques have been published (Hansen & Wang, 2005; Idris, 2010; Yu et al., 2011).

Size reduction is not the only reason towards miniaturization. In fact, the miniaturization of the different steps of the analytical process involves several additional benefits, as can be seen in **Figure 1.4**. The main benefits that can be obtained by downsizing the different steps of the analytical process are shown below:

1. Reduction of sample amount: The sample volume required to carry out an appropriate analysis can be highly reduced by scaling down the sample preparation,

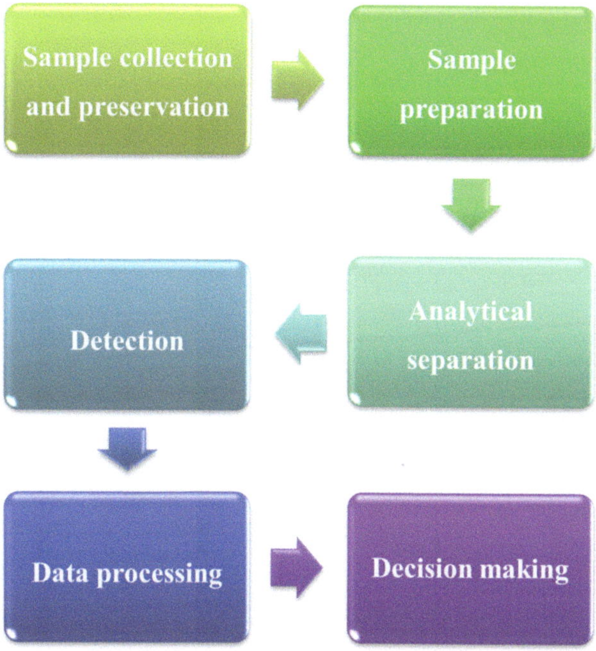

Figure 1.3 Steps of the analytical process.

separation and detection techniques. This is especially advantageous when dealing with scarce and/or precious samples.

2. Decreased consumption of chemicals and solvents: A drastic decrease in the amount of analytical reagents and organic solvents that are needed can result from the miniaturization of any analytical process step. This is especially important in the case of analytical methods involving expensive and precious reagents such as enzymes and immunochemicals, as well as in the case of analytical methodologies involving toxic reagents and/or organic solvents. Certain sample preparation strategies even allow the total removal of organic solvents and reagents, thus contributing to the environmental sustainability of analytical laboratories. The miniaturization of separation techniques enables a significant reduction of mobile phase or electrolyte, as well as the amount of stationary phase materials. As for detection techniques, reagent and gas consumption savings can be significant. In this sense, the reduction of neutral gas consumption is certainly remarkable in the case of miniaturized plasma sources.
3. Reduction of associated wastes: As a result of the above mentioned advantages, the wastes generated along the whole analytical process can be highly reduced, thus resulting in more sustainable methodologies. Recycling and recovery of chemicals and organic solvents present in wastes, as well as the on-line generation of clean wastes are important tasks aiming to be adopted in analytical laboratories (Garrigues et al., 2010).

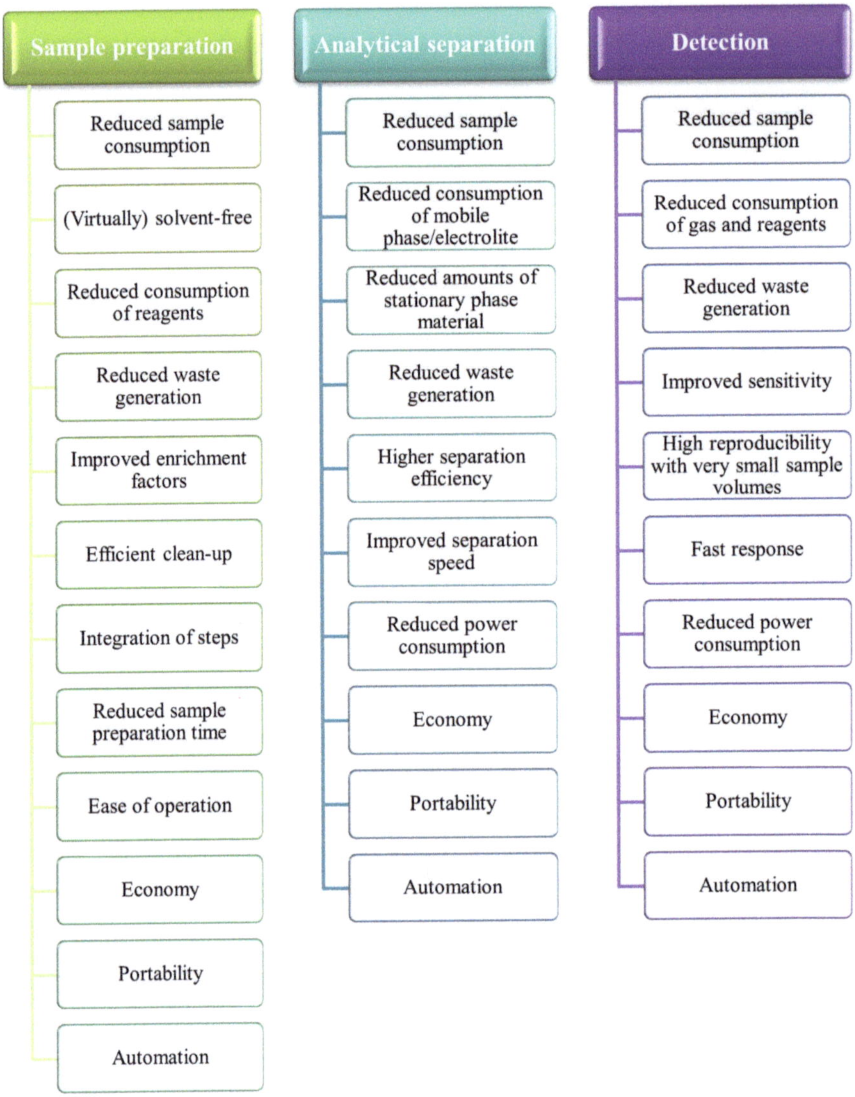

Figure 1.4 Potential benefits derived from the miniaturization of the different steps of the analytical process.

4. Improved sensitivity: Sensitivity of analytical methods can be increased by making use of an appropriate miniaturized sample preparation technique and, in certain cases, by miniaturizing detection systems. In sample preparation, high enrichment factors (EFs) can potentially be obtained as a result of the increased sample volume-to-extractant phase ratio, which can result in lower limits of detection. The improved design of recently developed analytical instrumentation can also yield increased sensitivity by using reduced sample volumes, although

in several cases the sensitivity can be significantly deteriorated when the instrumentation is miniaturized. Advances in detection systems can minimize the loss of sensitivity.

5. Rapidity: Time is a vital variable in analytical laboratories. The sample throughput is therefore an important factor in method development. The development of miniaturized sample preparation, separation and detection systems can significantly decrease the time needed to perform a single analysis. Miniaturization allows the improvement of the two most time-consuming steps, namely sample pre-treatment and analytical separation. Apart from the obvious benefit of having access to analytical data in an expeditious way, the reduction of the analysis time can provide indirect benefits, for example reduced consumption of reagents and solvents, lower energy requirements and smaller amounts of waste.
6. Portability: The development of field-portable instrumentation has been and is still a challenge in analytical chemistry. Low weight and overall dimensions, being resistant to changeable environmental conditions and efficient battery power are the requirements for portable analytical systems. The miniaturization of part or the whole of the analytical process steps contributes significantly to the portability of analytical systems to the sampling site. Furthermore, portable analytical systems deliver prompt and valuable information and reduce the risk of sample decomposition and contamination during sample storage and transportation.
7. Power consumption: The reduction of analytical systems generally involves a reduction of the power requirements. As a consequence, miniaturized instrumentation can be battery-operated, then contributing to its portability.

As can be noticed, several of the above mentioned advantages are related to each other. It is important to note, however, that the introduction of miniaturized alternatives to conventionally performed steps of the analytical process can give rise to novel challenges that need to be addressed. Therefore, the miniaturization of certain analytical processes may be not just useless but counterproductive. Examples of this fact include the development of microfluidic systems when large volumes of samples are available and/or conventional methods involve the use of small amounts of non-toxic and non-expensive reagents and solvents (Luque de Castro & Priego Capote, 2008). Besides, the use of highly reduced sample amounts can seriously affect the necessary representativeness of samples subjected to analysis. On the other hand, the fabrication of miniaturized detectors can also yield reduced resolution when compared with the corresponding full-sized counterpart (Capitan-Vallvey & Palma, 2011).

The reduced sample volume used with miniaturized analytical separation techniques can also give rise to reduced sensitivity, since the miniaturization of the chromatographic column involves the reduction of the detector volume. For instance, the use of miniaturized liquid chromatography (LC) for the analysis of easily available samples can give rise to problems of sensitivity that could be easily circumvented with

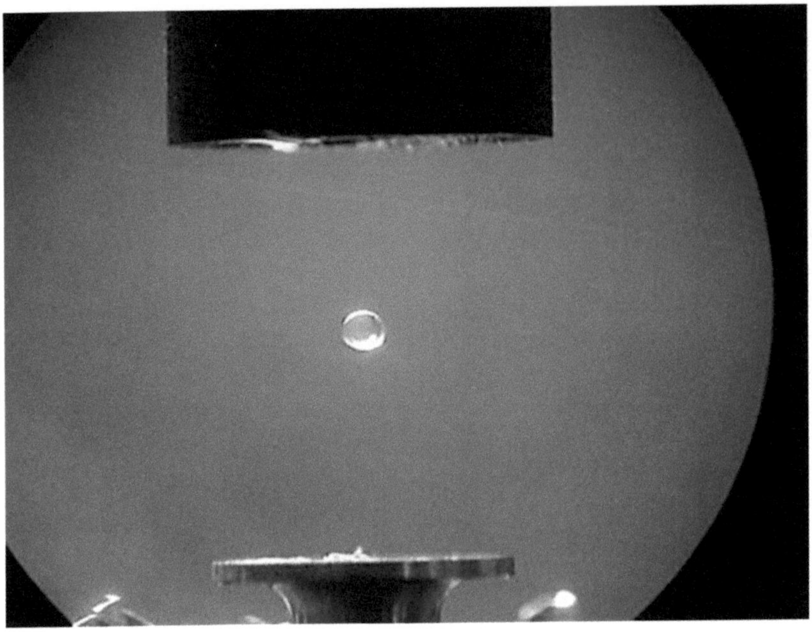

Figure 1.5 A 500-nL water drop levitated in a node in a standing wave created between an ultrasonic transducer (*bottom*) and a solid reflector (*top*). Reprinted from Santesson & Nilsson (2004) with permission from Springer.

conventional LC (Desmet & Eeltink, 2013). In addition, the development of novel analytical methods involving miniaturized sample preparation approaches introduces novel concerns derived from the limitations of the corresponding sample pre-treatment techniques. This can be the case where organic solvents with higher toxicity than in conventional analytical methods are used or when the introduction of numerous steps gives rise to tedious procedures and potential sources of contamination. On the other hand, the problem of adsorption to solid walls and interfaces when dealing with sample volumes in the picolitre to nanolitre range has been reported in the literature. The applicability of acoustic or ultrasonic levitation (**Figure 1.5**) has been proposed to avoid this problem, which can be of special concern when miniaturizing analytical and bioanalytical processes (Santesson & Nilsson, 2004; Priego-Capote & de Castro, 2006).

This technology, also known as lab-on-a-drop, is compatible with a variety of remote detection systems, and allows one to perform several analytical applications, including liquid-liquid and gas-liquid extraction, chemical and biochemical derivatization, solvent exchange, titration, crystallization, affinity two-phase separation and concentration by evaporation (Priego-Capote & de Castro, 2006). Even though the number of publications concerning levitation in chemical analysis is still relatively low in analytical chemistry, it is expected that this containerless sample handling technology will have an impact on the development of novel miniaturized systems.

1.2.1 A Need for Scaling Down Conventional Sample Preparation Techniques

Sample preparation is one of the most important steps of the analytical process, especially when dealing with the determination of trace (or ultratrace) analytes in complex samples. It is generally accepted that sample pre-treatment, together with sample collection and preservation, is the most time consuming and error-prone step of the analytical process. Sample preparation is employed in analytical chemistry to preconcentrate target analytes from samples where they are present at lower concentrations than the limit of detection of the corresponding analytical technique, to achieve a clean-up of the sample prior to instrumental analysis and/or to obtain an extract compatible with the analytical technique to be used.

A plethora of sample preparation techniques can be employed in analytical laboratories, including solid-phase extraction (SPE), conventional solvent extraction, Soxhlet extraction, pressurized solvent extraction, supercritical fluid extraction, and microwave- and ultrasound-assisted extraction. A variety of extractant phases, for example polymeric sorbents and adsorbents, organic solvents, ionic liquids, water or carbon dioxide can be used depending on the sample preparation technique. Furthermore, extraction processes can be enhanced by means of efficient selection of the experimental conditions. SPE and solvent extraction are, by far, the most commonly used sample preparation techniques in analytical laboratories. In fact, many official and standardized analytical methodologies involve their application for the extraction, preconcentration and sample clean-up prior to determination of target analytes. SPE and solvent extraction involve the partitioning of target analytes between the sample solution and a solid (adsorbent) phase or immiscible organic solvent, respectively. Both SPE and solvent extraction are exhaustive extraction techniques, so quantitative transfer of target analytes from the sample to the extractant phase is achieved under optimal conditions. However, the consumption of large amounts of organic solvents and subsequent generation of waste, the relatively low EFs that are achievable and the tediousness and significant time consumption are among the inherent drawbacks associated with these classical sample preparation approaches. These limitations led to the introduction of modern sample preparation techniques that share the common benefits of miniaturization and (virtually) solvent-free operation.

As early as 1990, Arthur and Pawliszyn presented the first miniaturized sample preparation technique, introduced under the denomination of solid-phase microextraction (SPME) (Arthur & Pawliszyn, 1990). SPME is based on the partitioning of target analytes between the sample solution (or the headspace above it) and a polymeric extractant phase coated on a fused silica fiber. SPME is a non-exhaustive solvent-free sample preparation technique that allows the extraction and preconcentration of a vast number of compounds. The wide acceptance of SPME by the scientific community is reflected by the large number of publications involving this miniaturized sample preparation technique. Since its inception, SPME has been extensively employed in a variety of research fields. Related miniaturized sample preparation techniques have been introduced in recent years as a result of the development of novel designs, as well as the implementation of novel materials. Thus, stir bar sorptive extraction (SBSE) was developed by coating a magnetic stir bar with an appropriate sorbent coating (Baltussen et al., 1999). Importantly, the surface area and sorbent coating volume are greatly improved in SBSE when compared with SPME, thus resulting in the achievement of higher extraction efficiencies (EEs) and, when thermal desorption is performed, lower limits of detection. A variety of related sample preparation techniques have also been reported in the literature, including thin film microextraction, solid-phase dynamic extraction and microextraction in a packed syringe. Relevant theoretical and experimental aspects of SPME and related sample preparation approaches are discussed in depth in chapters 2 and 3. These sample pretreatment techniques can cover most of the current requirements in terms of extractability and selectivity, with solid phase coatings showing good thermal, chemical and mechanical stabilities.

Liquid-phase microextraction (LPME) techniques have been developed recently with the aim of improving conventional solvent extraction. Specifically, the objectives required to miniaturize the solvent extraction technique were, mainly, to reduce the relatively large organic solvent volume conventionally needed to perform a single extraction process, to obtain high EFs, and, in general, to accelerate and simplify the process, thus allowing higher sample throughput. Liu and Dasgupta (1995) and Jeannot and Cantwell (1996) reported the first works concerning the miniaturization of conventional solvent extraction. The employment of a microliter-volume single drop of extractant phase exposed at the end of a capillary or, more commonly, at the tip of a microsyringe, allowed the enrichment of target analytes from both liquid and gaseous samples. This miniaturized solvent extracton technique, named as single-drop microextraction (SDME), allows the achievement of high EFs in spite of being a non-exhaustive technique. Further advances to improve the stability of the extractant phase during the extraction process involved the use of hollow fibers (Pedersen-Bjergaard & Rasmussen, 1999). Even though the use of these membranes allow the employment of experimental conditions that favorably affected the mass transfer of the analyte from the sample solution to the extractant phase (mainly, high stirring rates and extended extraction times), hollow fiber-based LPME approaches have not achieved the levels of popularity of related miniaturized sample preparation techniques, probably due to the insufficient EEs as well as its increased level of manipulation when compared with SDME. The development of dispersive liquid-liquid microextraction (DLLME) by Rezaee et al. in 2006 expanded the applicability of LPME as a result of its simplicity and the quantitative extraction recoveries achieved (Rezaee et al., 2006). DLLME is based on the use of a disperser solvent in combination with a water-immiscible extractant phase. The disperser acts as a bridge between the sample solution and the solvent. It allows the formation of tiny microdrops of the extractant phase that can disperse through the sample, then improving the mass transfer of the analyte towards the acceptor solution. Separation of the involved phases can be achieved easily by centrifugation on the basis of their different densities. Several related LPME approaches have also been reported, such as directly suspended droplet microextraction, cold induced aggregation microextraction and solidified floating organic drop microextraction. An in-depth discussion on the different LPME techniques, as well as on the different extractant phases that can be employed in LPME, is provided in chapters 4 and 5.

It is worth noting that certain sample preparation modes allow the improvement of selectivity by exploiting the physicochemical properties of target analytes making use of appropriate derivatization reactions, by exploiting kinetic discrimination, or by means of novel materials and/or separation membranes. Another advantage of miniaturized sample preparation approaches lies in the possibility of performing the extraction and derivatization of target analytes simultaneously. The integration of these unit operations reduces the number of steps that are necessary to carry out the analysis, and increases sample throughput. It should be kept in mind, however, that the possibility of integrating the extraction process and the derivatization reaction in a single step is obviously dependent on the compatibility of both processes. The advances on sample preparation towards their miniaturization are discussed in greater detail in subsequent chapters.

1.2.2 Miniaturization of Analytical Separation

Analytical separation techniques are employed in analytical chemistry for the separation of target analytes prior to their detection. A variety of separation techniques have been developed with the aim of separating and identifying a large number of compounds, with LC, gas chromatography (GC) and capillary electrophoresis (CE) being the most commonly employed analytical separation techniques.

In general, the miniaturization of the analytical separation techniques attempts to increase the separation efficiency and speed of the separation, decrease the cost and enhance portability, as well as reduce the amount of sample, solvent and reagents consumed and the wastes generated during the separation process. The first step towards the miniaturization of analytical separation techniques has been the result of the natural necessity of saving space in the laboratory. In fact, the first available analytical separation techniques were largely oversized, while much smaller analytical separation systems are currently available (Bartle & Myers, 2002). Further miniaturization has been the result of the proper search for solutions needed for challenging research activities, together with the developments on related areas, such as instrumental engineering and material sciences. The development of advanced fabrication technologies, such as micro-electromechanical systems (MEMS), has also been critical to miniaturizing analytical separation techniques.

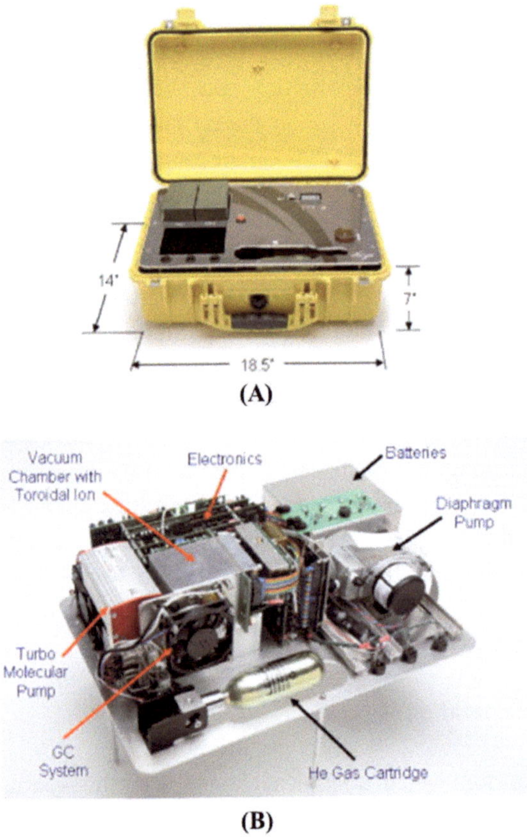

Figure 1.6 Photographs of the Guardion-7 GC-TMS showing (a) dimensions and (b) internal components. Reprinted from Contreras et al. (2008) with permission from Springer.

1.2.2.1 Gas Chromatography

GC is a standard analytical separation technique that allows the separation of complex mixtures of volatile and semi-volatile compounds. First introduced by James and Martin in the early 1950s (James & Martin, 1952), the introduction of capillary columns in GC was the first step toward the miniaturization of analytical separation techniques (Golay, 1958). GC systems have decreased in size to a great extent in recent years. For instance, a personal field portable GC system that combines a low temperature thermal mass injector, a low temperature thermal mass capillary GC and a miniature toroidal ion trap mass analyzer (TMS) has been reported. The proposed GC-TMS system (**Figure 1.6**) includes carrier gas supply and battery power source, has a relatively low weight (<13 kg, including batteries) and is totally self-contained within dimensions of 470 x 360 x 180 mm (Contreras et al., 2008).

The production of miniature gas chromatographs is of high interest not only in analytical chemistry, but also in other scientific and technological fields. The devel-

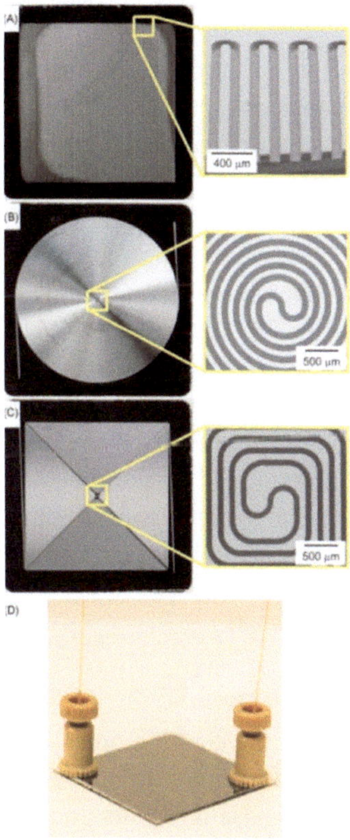

Figure 1.7 Photograph showing three different silicon-Pyrex® microcolumn configurations tested: circular-spiral with 75 µm distance between channels (A), square-spiral (B), and serpentine with 100 µm separation distance between channels (C). Each microcolumn shown is 3 m long and 100 µm × 100 µm in cross-section. Fused silica capillaries (200 µm outer diameter, 100 µm internal diameter) were attached to the microcolumn chip *via* Nanoports® (D). Reprinted from Radadia et al. (2010) with permission from Elsevier.

opment of the first chip-based GC system was reported as early as 1979, when Terry et al. fabricated a miniaturized GC on a 5 cm-diameter silicon wafer by using photolithography and chemical etching techniques (Terry et al., 1979). The system presented the components of a conventional GC, namely a carrier gas supply, sample injection system, separation column (1.5 m) and output detector (thermal conductivity detector), but in highly reduced dimensions. The preparation of the stationary phase of the proposed micro gas chromatograph (µ-GC) resulted, however, in relatively poor separation performance when compared with conventional GC systems. To improve this pitfall, the geometry of µ-GC columns is currently the focus of study (Mittermuller & Volmer, 2012). A recent article has revealed that a microfabricated serpentine geometry provides an improved separation performance when compared with circular-spiral and square spiral channel configurations in GC microcolumns (**Figure 1.7**) (Radadia et al., 2010).

μ-GC systems are known for their low size and weight. For instance, **Figure 1.8** shows a complete μ-GC system with highly reduced dimensions (300 x 170 x 80 mm) and a weight less than 3 kg, consisting of a multi-stage preconcentrator/injector, a capillary column and a photoionization detector (Jian et al., 2013).

Complete μ-GC systems have been developed on account of the advances achieved in downsizing their individual components, such as micropumps, micropreconcentrators, microcolumns and microsensors. μ-GC systems allow the separation of target compounds in highly reduced times when compared with conventional GC. The reduced power consumption is another benefit derived from the miniaturization of these systems, since the energy requirements to heat and cool smaller individual components are decreased (Dorman et al., 2010). Apart from its application in medical analysis, on-site gaseous sampling and analysis can be performed by portable and microfabricated GC, thus minimizing the problems related to analyte losses (mainly due to adsorption and reactions) that can be produced during sampling and transport (Ohira & Toda, 2008).

1.2.2.2 Liquid Chromatography

LC is an essential separation technique for a large variety of scientific areas. The miniaturization of LC has been mainly derived from the reduction of the analytical column dimensions, the development of novel stationary phase materials and the improvement of detection systems.

The decrease in the internal diameter of the analytical columns for the achievement of efficient separations was first proposed by Ishii et al. (1977) and from this innovative work, great efforts have been made to achieve a high degree of miniaturization of these analytical separation systems, yielding micro-LC, capillary LC, nano-LC and chip-based LC. The reduction of the dimensions of the LC column has an important influence on the remaining components of the chromatographic system, including the connecting tubing, injector, and detection cell volume and shape. Specifically, the tubing used to connect the LC column with the injector and the detector should be downscaled in accordance with the decrease of the chromatographic column dimensions, or even removed when the internal diameter of the LC column is lower than 100 μm. In addition, a reduction of the internal diameter of the chromatographic column can give rise to a decreased sensitivity that could be counteracted by improving the geometry of the detection system (Szumski & Buszewski, 2002).

The development of novel stationary phase materials with improved mechanical strength and chemical inertness, as well as enhanced surface area-to-volume ratio has also enabled the miniaturization of LC systems. The separation performance is, however, not just dependent on the particle size, but also on parameters such as monodispersity or porosity (Kutter, 2012). The reduction of the particle diameter of

A)

B)

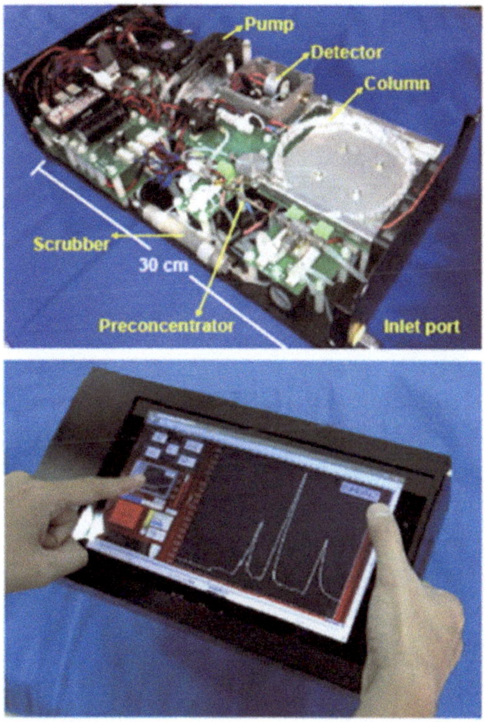

Figure 1.8 Photographs of prototype µ-GC: (A) interior structure with top lid and computer removed; and (B) complete system with built-in tablet computer. Reprinted from Jian et al. (2013) with permission from Elsevier.

the packing material of LC columns has been proposed with the aim of achieving higher resolution with reduced analysis times, even though extreme pressures are a requirement for optimal performance (Jerkovich et al., 2003). Capillary electrochromatography and ultra-high pressure LC techniques have been derived from the use of LC columns with micrometer-sized particles to overcome pressure limitations.

The sensitivity is compromised when on-column detection is performed as a result of the reduction of the analytical dimensions of the LC columns and the reduced sample volumes injected. Nevertheless, it is worth noting that electrospray ionization-mass spectrometry (ESI-MS) performs optimally at very low flow rates, thus having excellent suitability and compatibility with a miniaturized LC column (Desmet & Eeltink, 2013). Furthermore, certain strategies have been applied to enhance the sensitivity when miniaturized LC columns were used, for example selection of detection systems with improved geometry, on-column focusing, and on-line and off-line preconcentration (Zotou, 2012).

Chip-based LC systems for the separation of target analytes in a channel of a chip device were first reported in 1990 (Manz et al., 1990b). The proposed system involved a 5 x 5 mm silicon chip containing an open-tubular column of 6 µm x 2 µm x 15 cm and a conductometric detector. Since this first work, many chip-based LC systems have been reported in the literature, with applicability in environmental and clinical analysis, proteomics and genetics, among other areas. The improvement of chip-based LC systems is linked to the development of novel stationary phase materials. The separation efficiency is highly dependent on the channel size, shape and aspect ratio in microchip-based LC systems (Kutter, 2012). The improvements experienced in machining methods allow the highly accurate preparation of microchannels on the surface of silicon glass or polymeric planar substrates.

1.2.2.3 Capillary Electrophoresis

CE is a versatile analytical separation technique which provides high separation efficiency, short analysis times and reduced consumption of sample and electrolyte. The employment of 75 μm internal diameter open-tubular glass capillaries in zone electrophoresis was firstly proposed by Jorgensen and Luckacs (1981). The reduction of the inner diameter of the capillary involved an efficient dissipation of heat produced at high voltages, thus yielding highly efficient separations.

Both capillary-based CE and chip-based CE have been proposed in the literature to enable the miniaturization and portability of commercially available electrophoresis instrumentation. Capillary-based CE systems (also referred in the literature as non-chip based CE) are fabricated around a cylindrical capillary, while microfluidic channels are fabricated in chip-based CE systems by means of microfabrication techniques (Lewis et al., 2013). CE is highly amenable to portability due to it is relative simplicity as well as advantages such as the low consumption of buffer solution, few moving parts and, unlike LC systems, the absence of a pump which is not a requirement in CE.

In 1998, Kappes and Hauser (1998) presented the first portable CE instrument. The proposed instrument included, among other components, a 25 μm inner diameter fused-silica capillary, a potentiometric detector and two 12 V lead-acid batteries. The samples were electrokinetically injected. The whole CE instrument was contained in a PVC case of dimensions 340 x 175 x 175 mm and had a total weight of 7.5 kg. The applicability of the proposed field-portable CE instrument was expanded by using other two electrochemical detection techniques, namely amperometry and conductometry (Kappes et al., 2001).

Harrison et al. presented the first chip-based CE system (Harrison et al., 1992). The development of micromachining techniques allowed the formation of capillary channels in a planar glass substrate. The separation efficiency of the system was comparable to conventional open tubular capillaries when expressed as number of plates per volt. Field-portable chip-based CE systems have been reported in the literature with improved dimensions and weight, mainly due to the incorporation of small high-voltage power supplies and on-chip detectors. For instance, the first portable chip-based CE system showed highly reduced dimensions (102 x 152 x 25 mm) and weight (0.35 kg)

(Jackson et al., 2003). Two recent reviews dealing with the development of portable CE can be consulted for further details (Ryvolová et al., 2010; Lewis et al., 2013).

Novel developments towards the fabrication of autonomous chip-based CE systems for remote analysis have been recently reported. These systems allow one to perform every step of the analytical process (sampling, sample preparation, separation and detection) without the participation of the user. Culbertson et al. presented a portable, battery-operated microfluidic system capable of performing electrophoretic separations in less than 12 s under microgravity and hypergravity conditions (Culbertson et al., 2005). Skelley et al. presented a microfabricated CE instrument called the Mars Organic Analyzer (MOA) capable of determining key biomarkers present in Mars-like soil samples (Skelley et al., 2005). The MOA showed a weight of 11 kg and a peak power utilization of 15 W. The concept of lab-on-a-robot was introduced in 2008 by Berg et al. to carry out remote wireless analysis (Berg et al., 2008). This system consisted of the combination of a chip-based CE system, a mobile platform and wireless global position system (GPS) (**Figure 1.9**). The miniaturized system allows the user to select the sampling point by means of the GPS and transmitted video, followed by the collection of a gas sample, injection, separation, detection, and transfer of the data to a distant control unit.

Certain advantages are attributed to chip-based CE systems. Apart from the obvious reduction of the instrumentation size, low consumption of sample and chemicals, lower power requirements and expeditious separations are among the benefits of chip-based CE. However, it has been reported that quantitative analysis with microchip CE can be difficult for non-experienced operators, since several factors can affect the accuracy and repeatability, such as buffer electrolysis, bubble formation and clogging events, analyte-surface interactions, as well as injection and power supply-related problems (Revermann et al., 2008). Furthermore, the use of emerging materials in the fabrication of the channel surface of chip-based CE can hinder the prediction of the electroosmotic flow characteristics since the electroosmotic flow rate depends on the channel surface characteristics (Lewis et al., 2013).

It is interesting to note that the development of chip-based systems has employed mainly electrophoresis as the separation technique of choice. In general, CE is considered to be inferior to LC owing to its lack of robustness (Breadmore, 2012). Nevertheless, CE is more amenable to microchips than LC, so the current trend of miniaturization in analytical chemistry may afford new opportunities to electrophoresis.

1.2.3 Miniaturization of Detection Techniques

The advances achieved in fields such as electronics, engineering and material sciences have allowed the miniaturization of analytical detection systems. The characteristic features of MEMS fabrication techniques, namely miniaturization, multiplicity and microelectronics, have enabled the batch production of small-sized detection

Figure 1.9 Picture of the assembled *lab-on-a-robot* containing high-voltage power supply, electrochemical detector, mobile platform, field programmable gate array, radio frequency modem, compass, and GPS. Reprinted from Berg et al. (2008) with permission from John Wiley & Sons.

systems (Schuler et al., 2009). Miniaturization of analytical detection techniques requires the size reduction of different elements of conventionally-sized instruments without ignoring the performance of each component, in such a way that the overall miniaturized detection techniques yield comparable or improved analytical performance. Scale-down analytical instrumentation has been accompanied by the required advances on auxiliary devices, including micropumps (Chen et al., 2008), microvalves (Oh & Ahn, 2006), micromixers (Nguyen & Wu, 2005). It is also worth mentioning that the miniaturization of analytical instrumentation is dependent on the size of electronic systems. The advances towards the miniaturization of power sources and their integration into electronic subsystems have allowed the development of microbatteries with a size at least 500-fold smaller than commercially available batteries (Heller, 2006). In this section, a brief description of miniaturized detection systems, such as molecular and atomic spectrometry, mass spectrometry and electrochemical techniques, is discussed.

1.2.3.1 Molecular Spectrometry

Ultraviolet and visible (UV-Vis) absorption spectrometry is widely used due to its simplicity, rapidity and low cost. Decreases in the size of the sample compartment, together with improvements in radiation sources, wavelength discrimination components, fiber optics technology and the implementation of detector arrays have allowed the downscale of conventionally-sized UV-Vis spectrometers. Analysis of microliter to nanoliter sample volumes, increased optical path length-to-sample volume ratio, rapidity, low power requirements and portability are the main advantages obtained by miniaturizing UV-Vis spectrophotometers. The loss of sensitivity is, however, a drawback that can be alleviated or even resolved by increasing the optical path length by axial-direction, multireflection, liquid-core waveguides or cavity ring down spectroscopy, or by combining the different microvolume UV-Vis spectrophotometric systems available with miniaturized sample preparation techniques for appropriate preconcentration prior to the analysis. Some reviews on the miniaturization of UV-Vis spectrometric systems and accessories are available (Bacon et al., 2004; Pena-Pereira et al., 2011).

Technological improvements towards miniaturized infrared (IR) spectroscopic systems have also been reported. Specifically, the main improvements have been focused on the light sources, the interferometer and electronic circuits. A significant improvement concerning light sources for IR spectroscopy has been achieved with the development of quantum cascade lasers (QCL) (Faist et al., 1994). QCLs are characterized by their large emission power, reliability, wavelength tenability and lifetime (Kim et al., 2008). The developments achieved have enabled the hyphenation of analytical separation techniques such as LC and CE with IR spectroscopy (Kuligowski et al., 2010).

1.2.3.2 Atomic Spectrometry

Recent efforts towards shrinking plasma sources have advanced the miniaturization of analytical atomic spectrometric techniques. Plasmas consist on a highly energized mixture of positive ions and their electrons in a neutral background gas. They are initiated by supplying energy to a volume that contains a neutral gas. The miniaturization of plasma sources has emerged in response to the necessity of small and lightweight instrumentation with the potential for portability, improved cost-to-performance ratio, rapidity, reduced power and gas consumption, as well as integration with separation and detection techniques.

A variety of miniaturized plasma sources can be produced by the application of direct current (dc), alternating current (ac), radio frequency and microwave power. Those plasmas confined to dimensions of 1 mm or less are defined as "microplasmas". When coupled with optical or mass spectrometry, microplasmas are powerful tools for the detection of molecular fragments and elemental analysis. Most of the microplasmas reported in the literature have been applied to the analysis of gases or liquid samples using volatilization reactions prior to analysis. In fact, it has been stated that microplasmas lack stability in the presence of significant solvent loads more than full-sized plasmas, probably due to the combination of shorter residence times and lower gas temperatures (Webb et al., 2007). Nevertheless, some advances allow the analysis of liquid phases using microplasmas as emission sources. For instance, Webb et al. employed a small-scale (5 x 2 mm) atmospheric-pressure helium discharge that was demonstrated to be stable in the presence of significant solvent loads. The particular geometry of the source has been identified as the factor that mainly contributed to its robustness (Webb et al., 2007). Staack et al. reported the generation of a plasma discharge around electrodes with ultrasharp tips and elongated nanoparticles that enable the simultaneous determination of dissolved elements present in liquid matrices by optical emission spectroscopy (Staack et al., 2008).

The advances achieved in microfabrication techniques have enabled the integration of microplasma sources with LOC systems. The possibility of operating at atmospheric pressure affords an important advantage towards the integration of microplasmas in analytical microsystems and the development of portable devices since vacuum pumps are not required (Luo & Duan, 2012). It is also noteworthy to mention that microplasma-based atomic emission detectors have been coupled with GC for analytical purposes (Miclea et al., 2007).

The development and application of microplasmas has been discussed in a number of excellent reviews (Broekaert, 2002; Franzke et al., 2003; Karanassios, 2004; Miclea & Franzke, 2007; Luo & Duan, 2012) to which interested readers are referred for further details.

1.2.3.3 Mass Spectrometry

Mass spectrometry is a powerful analytical technique due to its high sensitivity, selectivity and speed, being widely used for both qualitative and quantitative analysis on the basis of the mass-to-charge ratio of ionized atoms and molecules. A mass spectrometer comprises three essential parts, namely the ionization source, mass selective analyzer and ion detector. Furthermore, a vacuum system and a control system are important parts of a mass spectrometer. The miniaturization of mass spectrometers has focused on downsizing mass selective analyzers. The miniaturization of a variety of mass analyzers has been reported in the literature, including miniature quadrupole, magnetic sector, time-of-flight (TOF), quadrupole ion trap and ion cyclotron resonance (Henry, 1999; Ouyang & Cooks, 2009) detectors. However, depending on the type of mass analyzer, the size reduction can be accompanied with a loss of mass resolution which therefore affects the mass analysis. Furthermore, the size reduction of the mass analyzer can also show an impact on both vacuum and control systems (Ouyang & Cooks, 2009). Developments in the design of vacuum systems and the improvement of the control system by means of modern electronics have been key milestones towards miniaturized mass spectrometers with reduced weight, size and power consumption. The vacuum system is commonly the heaviest and most power-consuming part of a mass spectrometer. Even though miniature mass spectrometers do not require pressures as low as the ones used in conventional spectrometers, the design of the vacuum system highly influences the performance of the mass spectrometer. Miniaturized high vacuum pumps have been reported in the literature, including turbo-molecular pumps, cryogenic pumps and ion-getter pumps, the former being the only reliable option when dealing with high vacuum pumping. Ion-getter pumps only operate below 10^{-3} torr, in such a way that a pre-vacuum pump is required with portable systems that operate with ion-getter pumps when starting from atmospheric pressure (Ouyang & Cooks, 2009). Nevertheless, the use of ion-getter pumps in combination with an auxiliary pump has been a common strategy that allowed the development of portable mass spectrometers. Yang et al. reported the development of a palm portable mass spectrometer for the determination of chemical warfare agents in the field (Yang et al., 2008). The miniaturized instrument, with a volume of 1.54 L and a weight of 1.48 kg, employed a miniaturized ion trap as the mass analyzer, an ion-getter pump and an embedded microcomputer for system control, all operating on a battery power of 5 W.

The miniaturization of the ion source has been mainly focused on ESI, even though other atmospheric pressure ion sources such as atmospheric pressure chemical ionization, atmospheric pressure photoionization and electron ionization have been reported in the literature (Sikanen et al., 2010). The miniaturized ESI, named as nano-electrospray ion source, was first reported by Wilm and Mann (1996). ESI is easily downscaled and offers high sensitivity at low flow rates (nL min^{-1}) using extremely small sample volumes. Moreover, ESI-MS can be easily interfaced with microfluidic systems, LC and CE.

1.2.3.4 Electrochemical Techniques

Electrochemical techniques have been employed in several scientific areas due to their high sensitivity, ease of operation, portability and low cost. Several miniaturized electrochemical techniques are commonly employed. The most representative example is the hand-held glucose meter for insulin-dependent diabetes which operates using amperometric enzyme electrodes (Wang, 2002). Furthermore, a variety of sensors have been reported for the determination of a wide range of target compounds, as has been outlined in two excellent review articles (Stetter, 2008; Kimmel et al., 2012).

Electrochemical detection techniques are well-suited to miniaturization, since they can be produced by conventional microfabrication methods and their analytical performance is not affected by the decrease in size, enabling sensitive detection of electroactive target compounds in picoliter to nanoliter sample volumes. Thus, electrochemical detectors, including voltammetric, conductometric and potentiometric detectors, have been implemented in LOC systems (Nyholm, 2005).

The development of small-size voltammetric working electrodes was reported for the first time in 1981 by Wightman who was interested in identifying changes in the chemical concentrations of neurotransmitters inside the mammalian brain (Wightman, 1981). In 1990, Penner et al. reported the fabrication of nanometer-scale electrodes that enabled the determination of electron transfer rate constants two orders of magnitude faster than with conventional electrochemical methods (Penner et al., 1990). Voltammetric electrodes with at least one dimension below 10 μm are known as "microelectrodes" (also reported as ultramicroelectrodes), while the term "nanoelectrodes" typically refers to voltammetric electrodes with at least one dimension below 100 nm. Thus, nanoelectrodes can be considered as a special type of microelectrode. Both microelectrodes and nanoelectrodes have been extensively employed in the literature, enabling remarkable advances in both fundamental and applied electrochemical research, such as electroanalysis in single cells and biomolecules, as well as electrocatalysis at nano-scales (Li & Hu, 2013; Oja et al., 2013).

Among the advantages of reducing the geometrical dimensions of voltammetric electrodes are the decrease of the ohmic drop potential, the fast establishment of a steady-state signal, current increase due to the enhanced mass transport, and the increase of the signal-to-noise ratio achieved in small size electrodes (Štulík et al., 2000). Furthermore, unlike conventional-sized electrodes (of few mm or larger), ultra small electrodes can be employed on samples of very high resistance, such as solutions without added supporting electrolyte. This is a consequence of the extremely small currents at these electrodes. Moreover, the reduced size of the electrodes enables the analysis of extremely small volumes of sample, that can be down to the attoliter level. This fact is especially relevant in areas of bioanalysis and medicine, for example, in the detection and identification of neurotransmitters and other types of chemical messengers released from certain types of cells (Li & Hu, 2011).

Small size electrodes with several geometries have been reported in the literature, including micro- and nanodisks, cylinders, arrays, bands, rings, spheres, hemispheres, and more recently, nanopores. Among the different geometries, the microdisk electrode is the most commonly employed (Li & Hu, 2013).

A variety of challenging areas have progressed substantially in recent years thanks to the development of voltammetric micro and nanoelectrodes, for example single molecule detection, single nanoparticle electrochemistry and the determination of neurotransmitters from single neuronal cells (Cox & Zhang, 2012).

1.3 Conclusions and Outlook

The motivating factors behind the miniaturization of analytical systems are diverse. First of all, we can consider the downscale of analytical instrumentation as a natural means to get more space in the laboratory. The size decrease on analytical instrumentation has been evident during the last few decades as a result of technological improvements. Nevertheless, the miniaturization of analytical systems today is more focused on the improvement of their analytical performance, rapidity, environmental sustainability and economic benefits than simply on reducing their size. In these terms, the miniaturization of the different steps of the analytical process proved to be advantageous in comparison with conventional analytical systems. However, it should not be forgotten that significant drawbacks can be produced when dealing with samples below microliter volume scales. Thus, in certain cases, the development of a miniaturized alternative to conventional analytical methods shows certain limitations that result in novel problems to solve.

The development of portable and eventually handheld instrumentation for field analysis has also been boosted by the improvements in miniaturization. Furthermore, the novel requirements of challenging research areas and applications, including but not limited to proteomics, metabolomics, forensics or remote sampling have allowed further improvements on miniaturized analytical systems. Shrinking analytical systems are continuously being improved by scientific and technological developments, including novel materials, fabrication techniques, advanced auxiliary components and improved designs. Therefore, further efforts can be envisaged on miniaturization, portability and intelligentization of analytical systems in the near future.

Abbreviations

μ-GC	micro gas chromatograph
μSI	micro-sequential injection
μ-TAS	micro-total analysis systems
CE	capillary electrophoresis
DLLME	dispersive liquid-liquid microextraction
EE	extraction efficiency
EF	enrichment factor
ESI	electrospray ionization
GC	gas chromatography
GPS	global position system
IR	infrared
LC	liquid chromatography
LOC	lab-on-a-chip
LOV	lab-on-a-valve
LPME	liquid-phase microextraction
MEMS	micro-electromechanical systems
MOA	Mars Organic Analyzer
MS	mass spectrometry
QCL	quantum cascade laser
SBSE	stir bar sorptive extraction
SDME	single-drop microextraction
SPE	solid-phase extraction
SPME	solid-phase microextraction
TAS	total analysis systems
TMS	toroidal ion trap mass spectrometer
TOF	time of flight
UV-Vis	ultraviolet and visible.

Acknowledgements

F. Pena-Pereira thanks Xunta de Galicia for financial support as a post-doctoral researcher of the I2C program.

References

Arthur, C. L., & Pawliszyn, J. (1990). Solid phase microextraction with thermal desorption using fused silica optical fibers. *Analytical Chemistry, 62*, 2145–2148.

Bacon, C. P., Mattley, Y., & DeFrece, R. (2004). Miniature spectroscopic instrumentation: applications to biology and chemistry. *Review of Scientific Instruments, 75*, 1–16.

Baltussen, E., Sandra, P., David, F., & Cramers, C. (1999). Stir bar sorptive extraction (SBSE), a novel extraction technique for aqueous samples: theory and principles. *Journal of Microcolumn Separations, 11*, 737–747.

Bartle, K. D., & Myers, P. (2002). History of gas chromatography. *TrAC Trends in Analytical Chemistry, 21*, 547–557.

Berg, C., Valdez, D. C., Bergeron, P., Mora, M. F., Garcia, C. D., & Ayon, A. (2008). Lab-on-a-robot: integrated microchip CE, power supply, electrochemical detector, wireless unit, and mobile platform. *Electrophoresis, 29*, 4914–4921.

Breadmore, M. C. (2012). Capillary and microchip electrophoresis: challenging the common conceptions. *Journal of Chromatography A, 1221*, 42–55.

Broekaert, J. A. (2002). The development of microplasmas for spectrochemical analysis. *Analytical and Bioanalytical Chemistry, 374*, 182–187.

Capitan-Vallvey, L. F., & Palma, A. J. (2011). Recent developments in handheld and portable optosensing-a review. *Analytica Chimica Acta, 696*, 27–46.

Chen, L., Lee, S., Choo, J., & Lee, E. K. (2008). Continuous dynamic flow micropumps for microfluid manipulation. *Journal of Micromechanics and Microengineering, 18*, 013001.

Contreras, J. A., Murray, J. A., Tolley, S. E., Oliphant, J. L., Tolley, H. D., Lammert, S. A., et al. (2008). Hand-portable gas chromatograph-toroidal ion trap mass spectrometer (GC-TMS) for detection of hazardous compounds. *Journal of the American Society for Mass Spectrometry, 19*, 1425–1434.

Cox, J. T., & Zhang, B. (2012). Nanoelectrodes: recent advances and new directions. *Annual Review of Analytical Chemistry, 5*, 253–272.

Culbertson, C. T., Tugnawat, Y., Meyer, A. R., Roman, G. T., Ramsey, J. M., & Gonda, S. R. (2005). Microchip separations in reduced-gravity and hypergravity environments. *Analytical Chemistry, 77*, 7933–7940.

Desmet, G., & Eeltink, S. (2013). Fundamentals for LC miniaturization. *Analytical Chemistry, 85*, 543–556.

Dorman, F. L., Whiting, J. J., Cochran, J. W., & Gardea-Torresdey, J. (2010). Gas chromatography. *Analytical Chemistry, 82*, 4775–4785.

Faist, J., Capasso, F., Sivco, D. L., Sirtori, C., Hutchinson, A. L., & Cho, A. Y. (1994). Quantum cascade laser. *Science, 264*, 553–556.

Feynman, R. P. (1992). There's plenty of room at the bottom. *Journal of Microelectromechanical Systems, 1*, 60–66.

Franzke, J., Kunze, K., Miclea, M., & Niemax, K. (2003). Microplasmas for analytical spectrometry. *Journal of Analytical Atomic Spectrometry, 18*, 802–807.

Garrigues, S., Armenta, S., & de la Guardia, M. (2010). Green strategies for decontamination of analytical wastes. *TrAC Trends in Analytical Chemistry, 29*, 592–601.
Golay, M. J. E. (1958). In D. H. Desty (Ed.), *Gas Chromatography* (pp. 36–55). London: Butterworths.
Hansen, E. H., & Wang, J. (2005). The three generations of flow injection analysis. *Analytical Letters, 37*, 345–359.
Harrison, D. J., Manz, A., Fan, Z., Lüdi, H., & Widmer, M. (1992). Capillary electrophoresis and sample injection systems integrated on a planar glass chip. *Analytical Chemistry, 64*, 1928–1932.
Heller, A. (2006). Potentially implantable miniature batteries. *Analytical and Bioanalytical Chemistry, 385*, 469–473.
Henry, C. M. (1999). The incredible shrinking mass spectrometers. *Analytical Chemistry, 71*, 264A-268A.
Herold, K. E., & Rasooly, A. (2009). *Lab on a Chip Technology (Vol. 1): Fabrication and Microfluidics*. Norfolk, UK.: Caister Academic Press.
Idris, A. M. (2010). An overview of the generations and recent versions of flow injection techniques. *Critical Reviews in Analytical Chemistry, 40*, 150–158.
Ishii, D., Asai, K., Hibi, K., Onokuchi, T., & Nagaya, M. (1977). A study of micro-high-performance liquid chromatography: I. Development of technique for miniaturization of high-performance liquid chromatography. *Journal of Chromatography A, 144*, 157–168.
Jackson, D. J., Naber, J. F., Roussel Jr., T. J., Crain, M. M., Walsh, K. M., Keynton, R. S., et al. (2003). Portable high-voltage power supply and electrochemical detection circuits for microchip capillary electrophoresis. *Analytical Chemistry, 75*, 3643–33649.
James, A., & Martin, A. (1952). Gas-liquid partition chromatography: the separation and micro-estimation of volatile fatty acids from formic acid to dodecanoic acid. *Biochemical Journal, 50*, 679–690.
Jeannot, M. A., & Cantwell, F. F. (1996). Solvent microextraction into a single drop. *Analytical Chemistry, 68*, 2236–2240.
Jerkovich, A. D., Mellors, J. S., & Jorgenson, J. W. (2003). The use of micrometer-sized particles in ultrahigh pressure liquid chromatography. *LCGC North America, 21*, 600–610.
Jian, R.-S., Huang, Y.-S., Lai, S.-L., Sung, L.-Y., & Lu, C.-J. (2013). Compact instrumentation of a μ-GC for real time analysis of sub-ppb VOC mixtures. *Microchemical Journal, 108*, 161–167.
Jorgenson, J. W., & Lukacs, K. D. (1981). Zone electrophoresis in open-tubular glass capillaries. *Analytical Chemistry, 53*, 1298–1302.
Kappes, T., Galliker, B., Schwarz, M. A., & Hauser, P. C. (2001). Portable capillary electrophoresis instrument with amperometric, potentiometric and conductometric detection. *TrAC Trends in Analytical Chemistry, 20*, 133–139.
Kappes, T., & Hauser, P. C. (1998). Portable capillary electrophoresis instrument with potentiometric detection. *Analytical Communications, 35*, 325–329.
Karanassios, V. (2004). Microplasmas for chemical analysis: analytical tools or research toys? *Spectrochimica Acta Part B: Atomic Spectroscopy, 59*(7), 909–928.
Kim, S. S., Young, C., & Mizaikoff, B. (2008). Miniaturized mid-infrared sensor technologies. *Analytical and Bioanalytical Chemistry, 390*, 231–237.
Kimmel, D. W., LeBlanc, G., Meschievitz, M. E., & Cliffel, D. E. (2012). Electrochemical sensors and biosensors. *Analytical Chemistry, 84*, 685–707.
Kuligowski, J., Quintas, G., de la Guardia, M., & Lendl, B. (2010). Analytical potential of mid-infrared detection in capillary electrophoresis and liquid chromatography: a review. *Analytica Chimica Acta, 679*, 31–42.
Kutter, J. P. (2012). Liquid phase chromatography on microchips. *Journal of Chromatography A, 1221*, 72–82.

Lewis, A. P., Cranny, A., Harris, N. R., Green, N. G., Wharton, J. A., Wood, R. J. K., et al. (2013). Review on the development of truly portable and *in-situ* capillary electrophoresis systems. *Measurement Science and Technology, 24*, 042001.

Li, C. M., & Hu, W. (2013). Electroanalysis in micro- and nano-scales. *Journal of Electroanalytical Chemistry, 688*, 20–31.

Li, T., & Hu, W. (2011). Electrochemistry in nanoscopic volumes. *Nanoscale, 3*, 166–176.

Lin, B. (2011). *Microfluidics: Techologies and applications*. Heidelberg, Germany: Springer.

Liu, S., & Dasgupta, P. K. (1995). Liquid droplet. A renewable gas sampling interface. *Analytical Chemistry, 67*, 2042–2049.

Livak-Dahl, E., Sinn, I., & Burns, M. (2011). Microfluidic chemical analysis systems. *Annual Review of Chemical and Biomolecular Engineering, 2*, 325–353.

Luo, D., & Duan, Y. (2012). Microplasmas for analytical applications of lab-on-a-chip. *TrAC Trends in Analytical Chemistry, 39*, 254–266.

Luque de Castro, M. D., & Priego Capote, F. (2008). Miniaturisation of analytical steps: necessity and snobbism. *Analytical and Bioanalytical Chemistry, 390*, 67–69.

Manz, A., Graber, N., & Widmer, H. M. (1990a). Miniaturized total chemical analysis systems: a novel concept for chemical sensing. *Sensors and Actuators B, 1*, 244–248.

Manz, A., Miyahara, Y., Miura, J., Watanabe, Y., Miyagi, H., & Sato, K. (1990b). Design of an open-tubular column liquid chromatograph using silicon chip technology. *Sensors and Actuators B, 1*, 249–255.

Mark, D., Haeberle, S., Roth, G., von Stetten, F., & Zengerle, R. (2010). Microfluidic lab-on-a-chip platforms: requirements, characteristics and applications. *Chemical Society Reviews, 39*, 1153–1182.

Marx, A., Fettinger, J., Verpoorte, E., Liidi, H., Widmer, H., & Harrison, D. (1991). Micromachining of monocrystalline silicon and glass for chemical analysis systems A look into next century's technology or just a fashionable craze? *TrAC Trends in Analytical Chemistry, 10*, 144–149.

Miclea, M., & Franzke, J. (2007). Analytical detectors based on microplasma spectrometry. *Plasma Chemistry and Plasma Processing, 27*, 205–224.

Miclea, M., Okruss, M., Kunze, K., Ahlman, N., & Franzke, J. (2007). Microplasma-based atomic emission detectors for gas chromatography. *Analytical and Bioanalytical Chemistry, 388*, 1565–1572.

Mittermuller, M., & Volmer, D. A. (2012). Micro- and nanostructures and their application in gas chromatography. *The Analyst, 137*, 3195–3201.

Nge, P. N., Rogers, C. I., & Woolley, A. T. (2013). Advances in microfluidic materials, functions, integration, and applications. *Chemical Reviews, 113*, 2550–2583.

Nguyen, N.-T., & Wu, Z. (2005). Micromixers—a review. *Journal of Micromechanics and Microengineering, 15*, R1-R16.

Nyholm, L. (2005). Electrochemical techniques for lab-on-a-chip applications. *The Analyst, 130*, 599–605.

Oh, K. W., & Ahn, C. H. (2006). A review of microvalves. *Journal of Micromechanics and Microengineering, 16*, R13-R39.

Ohira, S., & Toda, K. (2008). Micro gas analyzers for environmental and medical applications. *Analytica Chimica Acta, 619*, 143–156.

Oja, S. M., Wood, M., & Zhang, B. (2013). Nanoscale electrochemistry. *Analytical Chemistry, 85*, 473–486.

Ouyang, Z., & Cooks, R. G. (2009). Miniature mass spectrometers. *Annual Review of Analytical Chemistry, 2*, 187–214.

Pedersen-Bjergaard, S., & Rasmussen, K. E. (1999). Liquid-liquid-liquid microextraction for sample preparation of biological fluids prior to capillary electrophoresis. *Analytical Chemistry, 71*, 2650–2656.

Pena-Pereira, F., Costas-Mora, I., Romero, V., Lavilla, I., & Bendicho, C. (2011). Advances in miniaturized UV-Vis spectrometric systems. *TrAC Trends in Analytical Chemistry, 30*, 1637–1648.

Penner, R. M., Heben, M. J., Longin, T. L., & Lewis, N. S. (1990). Fabrication and use of nanometer-sized electrodes in electrochemistry. *Science, 250*, 1118–1121.

Priego-Capote, F., & de Castro, L. (2006). Ultrasound-assisted levitation: Lab-on-a-drop. *TrAC Trends in Analytical Chemistry, 25*, 856–867.

Radadia, A., Salehi-Khojin, A., Masela, R., & Shannon, M. (2010). The effect of microcolumn geometry on the performance of micro-gas chromatography columns for chip scale gas analyzers. *Sensors and Actuators B, 150*, 456–464.

Revermann, T., Gotz, S., Kunnemeyer, J., & Karst, U. (2008). Quantitative analysis by microchip capillary electrophoresis: current limitations and problem-solving strategies. *The Analyst, 133*, 167–174.

Rezaee, M., Assadi, Y., Milani Hosseini, M.-R., Aghaee, E., Ahmadi, F., & Berijani, S. (2006). Determination of organic compounds in water using dispersive liquid-liquid microextraction. *Journal of Chromatography A, 1116*, 1–9.

Ríos, A., Escarpa, A., & Simonet, B. (2009). *Miniaturization of analytical systems: Principles, designs and applications*. Chichester, UK: Wiley.

Rios, A., Zougagh, M., & Avila, M. (2012). Miniaturization through lab-on-a-chip: utopia or reality for routine laboratories? A review. *Analytica Chimica Acta, 740*, 1–11.

Ruzicka, J. (2000). Lab-on-valve: universal microflow analyzer based on sequential and bead injection. *The Analyst, 125*, 1053–1060.

Ryvolová, M., Macka, M., & Preisler, J. (2010). Portable capillary-based (non-chip) capillary electrophoresis. *TrAC Trends in Analytical Chemistry, 29*, 339–353.

Santesson, S., & Nilsson, S. (2004). Airborne chemistry: acoustic levitation in chemical analysis. *Analytical and Bioanalytical Chemistry, 378*, 1704–1709.

Schuler, L. P., Milne, J. S., Dell, J. M., & Faraone, L. (2009). MEMS-based microspectrometer technologies for NIR and MIR wavelengths. *Journal of Physics D: Applied Physics, 42*, 133001.

Senzel, A. (1970). Miniaturization in analytical instrumentation. *Analytical Chemistry, 42*, 79A-81A.

Sikanen, T., Franssila, S., Kauppila, T. J., Kostiainen, R., Kotiaho, T., & Ketola, R. A. (2010). Microchip technology in mass spectrometry. *Mass Spectrometry Reviews, 29*, 351–391.

Skelley, A. M., Scherer, J. R., Aubrey, A. D., Grover, W. H., Ivester, R. H., Ehrenfreund, P., et al. (2005). Development and evaluation of a microdevice for amino acid biomarker detection and analysis on Mars. *Proceedings of the National Academy of Sciences of the United States of America, 102*, 1041–1046.

Staack, D., Fridman, A., Gutsol, A., Gogotsi, Y., & Friedman, G. (2008). Nanoscale corona discharge in liquids, enabling nanosecond optical emission spectroscopy. *Angewandte Chemie International Edition, 47*, 8020–8024.

Stetter, J. R. (2008). Amperometric gas sensors - A review. *Chemical Reviews, 108*, 352–366.

Štulík, K., Amatore, C., Holub, K., Mareček, V., & Kutner, W. (2000). Microelectrodes. Definitions, characterization, and applications. *Pure and Applied Chemistry, 72*, 1483–1492.

Szumski, M., & Buszewski, B. (2002). State of the art in miniaturized separation techniques. *Critical Reviews in Analytical Chemistry, 32*, 1–46.

Terry, S. C., Jerman, J. H., & Angell, J. B. (1979). A gas chromatographic air analyzer fabricated on a silicon wafer. *IEEE Transactions on Electronic devices, 26*, 1880–1886.

Wang, J. (2002). Portable electrochemical systems. *TrAC Trends in Analytical Chemistry, 21*, 226–232.

Webb, M. R., Andrade, F. J., & Hieftje, G. M. (2007). The annular glow discharge: a small-scale plasma for solution analysis. *Journal of Analytical Atomic Spectrometry, 22*, 775–782.

Wightman, R. M. (1981). Microvoltammetric electrodes. *Analytical Chemistry, 53*, 1125A-1134A.

Wilm, M., & Mann, M. (1996). Analytical properties of the nanoelectrospray ion source. *Analytical Chemistry, 68*, 1–8.

Yang, M., Kim, T. Y., Hwang, H. C., Yi, S. K., & Kim, D. H. (2008). Development of a palm portable mass spectrometer. *Journal of the American Society for Mass Spectrometry, 19*, 1442–1448.
Yu, Y.-L., Jiang, Y., Chen, M.-L., & Wang, J.-H. (2011). Lab-on-valve in the miniaturization of analytical systems and sample processing for metal analysis. *TrAC Trends in Analytical Chemistry, 30*, 1649–1658.
Zotou, A. (2012). An overview of recent advances in HPLC instrumentation. *Central European Journal of Chemistry, 10*, 554–569.

2 Solid-Phase Microextraction and Related Techniques

Habib Bagheri*, Hamed Piri-Moghadam, Mehrnoush Naderi, Ali Es'haghi and Ali Roostaie

Environmental and Bio-Analytical Laboratories, Department of Chemistry, Sharif University of Technology, P.O. Box 11365–9516, Tehran, Iran
*e-mail address: bagheri@sharif.edu

2.1 Introduction

Solid-phase microextraction (SPME) is a rapid, inexpensive and solventless extraction technique for the isolation and preconcentration of solutes from liquid or gaseous matrices. SPME has several important advantages compared to the traditional sample preparation techniques:

1. It is a rapid, simple, solvent free and sensitive method for the extraction of analytes.
2. It is a simple and effective adsorption/desorption technique.
3. It is compatible with a wide range of analytical separation and detection techniques.
4. It provides linear results for wide concentrations of analytes.
5. It has small size, which is convenient for designing portable devices for field sampling.
6. It gives highly consistent, quantifiable results from very low concentrations of analytes.

2.2 Solid-phase Microextraction Fundamentals

2.2.1 Principle of Solid-phase Microextraction

For a two-phase system, the amount absorbed by the coating at equilibrium can be calculated using equation 2.1.

$$n = V_1 V_2 K C_0 / (K V_1 + V_2) \tag{2.1}$$

where n is the mass absorbed by the coating, V_1 and V_2 are the respective volumes of the coating and the aqueous solution, K is the partition coefficient of the analyte between the coating and the sample, and C_0 is the initial concentration of the analyte in the aqueous solution (Louch et al., 1992). The mass adsorbed depends on the distribution constant and the volume of the coating phase. A linear relationship is observed between the amount of analyte adsorbed and its concentration in the solution. With a three-phase system, a small amount of analyte is present in the headspace. When its volume is insignificant or when the solutes are scarcely volatile, the mass of solute adsorbed by the coating can be expressed by equation 2.1 (Rivasseau & Caude, 1995).

In a three phase system during SPME, the rules based on the equilibrium between various phases can be employed. The equilibrium constants of analyte between each two phases can be written in equation 2.2:

$$K_{fw} = \frac{C_f}{C_w}; \quad K_{hw} = \frac{C_h}{C_w}; \quad K_{fh} = \frac{C_f}{C_h} \tag{2.2}$$

where C_f is the equilibrium concentration of the analyte in the fiber coating, C_w is the equilibrium concentration of the analyte in the aqueous medium and C_h is the equilibrium concentration of the analyte in the headspace. The molar concentrations can be replaced by the number of analyte molecules in the corresponding phase volume. Also, considering that n_f is the number of molecules in the fiber and n_0, initial number of molecules in the aqueous phase, and replacing their corresponding equilibrium constants and phase volumes, n_f can be written:

$$n_f = \frac{K_{fh}K_{hw}V_f n_0}{K_{fh}K_{hw}V_f + K_{hw}V_h V_w} \tag{2.3}$$

where V_w is the volume of aqueous phase, V_f is the volume of coating and V_h is the volume of the sample headspace. Considering the Henry's law constants, it would be possible to reach to the following equation:

$$n_f = \frac{K_{fh}V_f n_0}{K_{fh}V_f + K_{hw}V_h + V_w} \tag{2.4}$$

A linear relationship between n_f and n_0 can be derived from equation 2.4. However, due to the fact that SPME is an equilibrium-based technique, the maximum recovery of SPME in equilibrium, R_{max}, can be defined:

$$R_{max} = \frac{n_f}{n_0} = \frac{K_{fw}V_f}{K_{fw}V_f + K_{hw}V_h + V_w} \tag{2.5}$$

The increase of the ratio of liquid to headspace volumes lightly improves the efficiency of extraction with any size fiber coating (Bagheri et al., 2005a). Considering equation 2.5, an increase in the headspace volume (V_h) accompanied by a decrease of the solution volume (V_w) in a defined system results in lower sorbed mass (n) on the fiber. The headspace volume is supposed to be kept to a minimum because in case of $V_h < V_f$ sampling from the headspace does not affect the amount of analyte sorbed by the coating (Zhang & Pawliszyn, 1993). It is not necessary to increase the entire volume of sampling vials because the efficiency of the extraction is presumably not enhanced if the relative volumes of liquid and headspace remain the same

(Penton, 1997). Although the principle behind SPME is based on an equilibrated partition process, it is not essential to wait for a full equilibrium to be reached. As long as the extraction time is standardized, it is be possible to obtain reproducible and sufficiently sensitive analysis. Of course, for the sake of acceptable repeatability it is rather necessary to choose equilibrium time in the region where the least possible changes on the analytical signal could be observed (Havenga & Rohwer, 1999).

2.2.2 Different Modes of Solid-phase Microextraction

It is possible to perform the SPME procedure under different configurations (**Figure 2.1**), categorized into two basic modes of diffusion mediated by stirring and diffusion mediated by flow through. The extracting phase could be coated, either physically or chemically, on the fiber, stir bar, thin film or on the inner surface of a tip, syringe and tube.

2.2.3 Coupling to Analytical Instrumentation

SPME can be coupled to different analytical instruments depending on the mode of operation and can be performed *via* both off-line and on-line configuration. The on-line combination may include the on-line extraction-desorption processes or just on-line desorption of the target analytes. Fiber-based SPME is directly coupled to gas chromatography (GC) for on-line desorption and through an appropriate desorption chamber to high performance liquid chromatography (HPLC). In the mean time, capillary microextraction (CME) is directly coupled to HPLC for extraction-desorption and its on-line combination to GC is also feasible *via* an appropriate interface. The other SPME-related techniques are usually employed in off-line combination with GC and HPLC.

2.2.3.1 Off-line Coupling

In off-line mode, the SPME extracting probe is inserted into an appropriate solvent to desorb the trapped analytes from the relevant probe. After complete desorption of analytes, the solvent containing analytes is subsequently exposed to a stream of nitrogen gas for desolvation and followed by reconstitution in a few microliter of an appropriate solvent. Eventually, it is injected into the appropriate analytical instrument.

2.2.3.2 On-line Coupling

Off-line SPME often suffers from the loss of analytes during the evaporation step and therefore lack of accuracy. To overcome this problem, on-line configuration is a very suitable alternative. Apart from the higher accuracy and precision of the on-line SPME

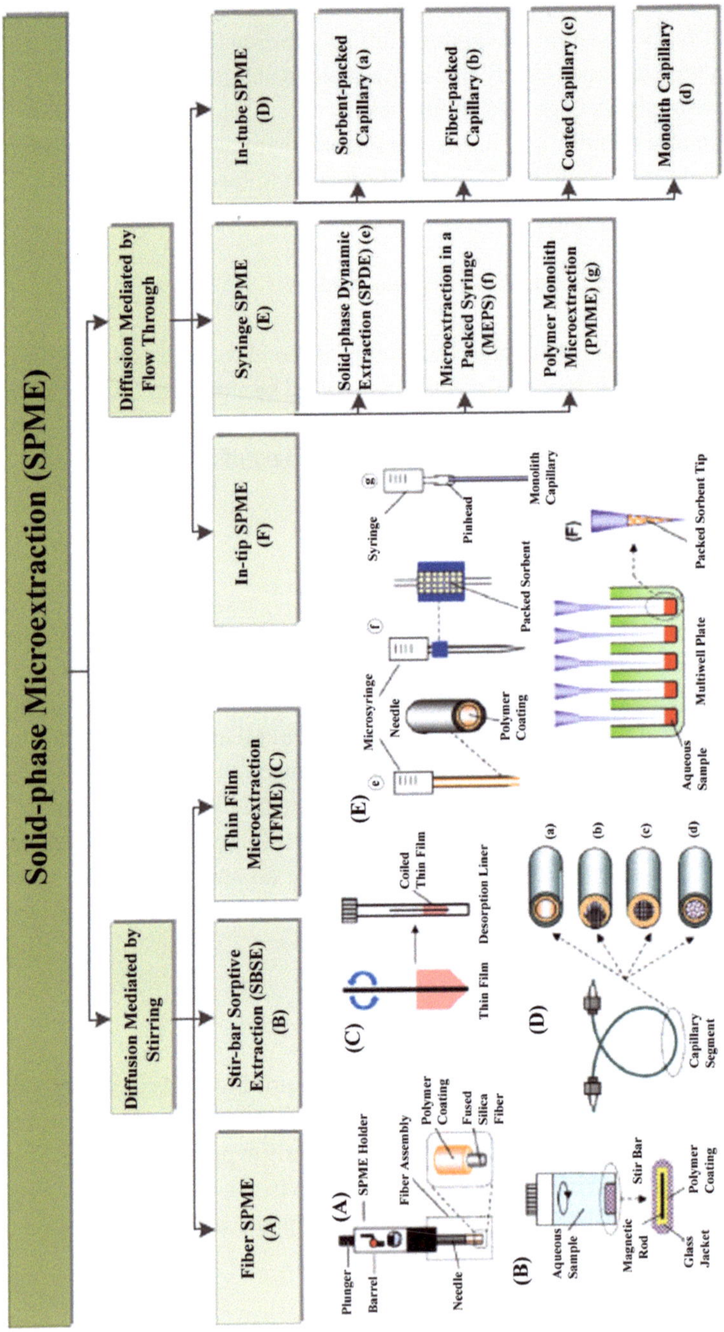

Figure 2.1 Schematic diagram for different modes of SPME. Reprinted from Kabir et al. (2013) with permission from Elsevier.

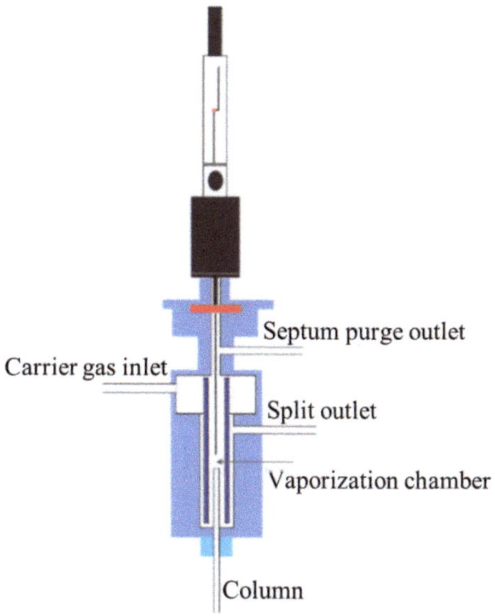

Figure 2.2 On-line desorption of fiber SPME in the injection port of GC.

set up, most of the time it is favored as far as automation and solvent consumption issues are concerned. On the other hand, CME could be directly coupled to analytical instruments using an on-line set up for extraction-desorption of the target analytes while fiber SPME needs an appropriate devoted interface containing a desorption chamber in combination with HPLC. While direct on-line coupling of CME and HPLC is mostly common, GC is also frequently used for on-line desorption of fiber SPME.

2.2.3.2.1 On-line Coupling to Gas Chromatography

Fiber SPME: In on-line SPME, desorption of analytes should be performed as rapidly as possible to avoid any possible peak broadening. In GC, the rapid desorption happens inside the high temperature injection port and the fiber SPME introduction is quite like standard injection syringe. The plunger movement allows exposure of the fiber during extraction and desorption and its protection in the needle during storage and penetration of the septum (**Figure 2.2**). When the desorption temperature of a GC is kept at high temperature and fibers with thin coatings are used, it would be expected that all the analytes are in the gas phase as soon as the fiber coating is inserted in the injection port. Under these circumstances, the sample desorption time is very much dependent on the temperature of the GC injection compartment (Lord & Pawliszyn, 2000).

CME: A schematic diagram for on-line coupling of CME to GC through an interface is shown in **Figure 2.3**. The ambient air, solution, or solution headspace is sampled by passing a gas or liquid through the open capillary. The analytes are trapped on the coating of a short piece of a capillary GC column or a prepared sorbent on the

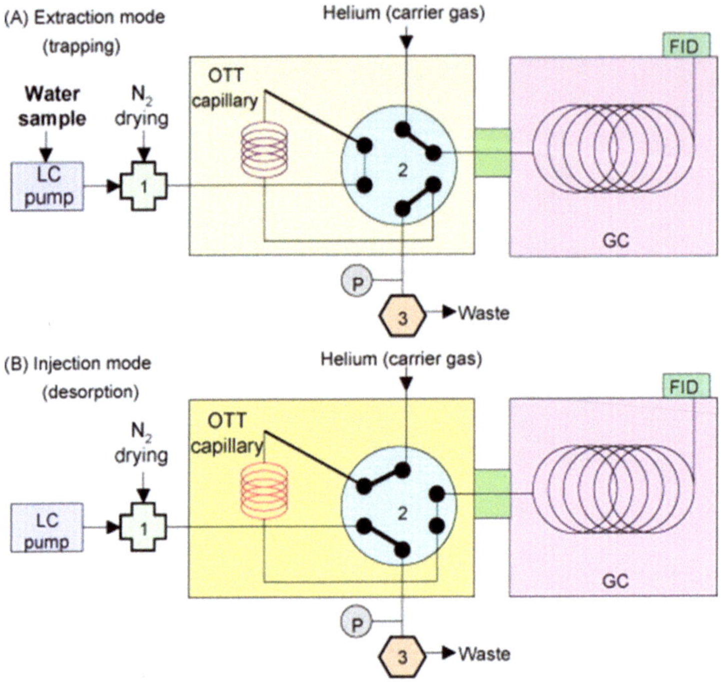

Figure 2.3 Schematic diagram of on-line CME-GC (A) Extraction and (B) Desorption. Reprinted from Kataoka et al. (2009) with permission from Elsevier.

inner surface of a capillary tube. The analytes are retained as they pass through the capillary tubing containing the stationary phase and they are subsequently desorbed, either by solvent or by thermal desorption (TD). The sample is forced to flow through the capillary and the analytes pass the trapping medium which has been coated onto the walls by diffusion. The sufficient thermal stability of GC stationary phases allows collected analytes to be thermally desorbed from the coating material after sampling. These analytes can be desorbed directly onto a GC column for analysis, avoiding dilution of the sample with solvent (Kataoka et al., 2009).

The possibility of sample cross-contamination and degradation are reduced as intermediate sample handling steps are eliminated. Another advantage is that even very volatile compounds can be enriched at ambient temperature, therefore, the need for a cryogenic refocusing step is unnecessary.

2.2.3.2.2 On-line Coupling to Liquid Chromatography

On-line combination of SPME fibers and CME to HPLC benefits from automation of the extraction. In HPLC, rapid desorption is achievable using a mobile phase with strong eluting power. Thus, the mobile phase composition should provide complete

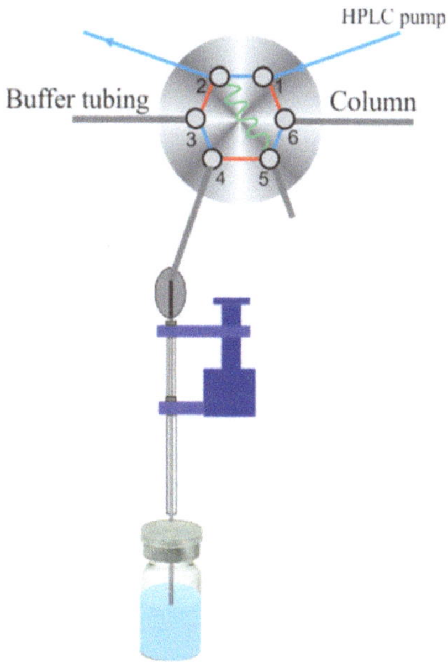

Figure 2.4 On-line extraction/desorption of fiber SPME to HPLC *via* an interface.

desorption of the extracted analytes, while the proper separation of the analytes in the analytical column can be performed.

Fiber SPME: Fiber SPME could be employed in on-line mode with HPLC in which a desorption chamber is required. **Figure 2.4** shows the appropriate configuration for this setup (Lord & Pawliszyn, 2000).

CME: According to **Figure 2.5**, extraction is performed by passing the spiked aqueous samples through the tube. After extraction, the HPLC mobile phase is used for on-line desorption and elution of the extracted analytes from the tube to the HPLC column (Bagheri et al., 2011c, 2012d).

2.3 Extractant Phases in Solid-phase Microextraction

2.3.1 Conventional Extractant Phases

Selection of the extracting phase is the most important step governing the selectivity of SPME. In the original SPME, the thermally activated polyimide film and uncoated fused silica were used as the extracting phases. The polyimide had to be treated at 350 °C prior to the extraction. Both could be applied to the determination of volatile chlorinated organic compounds in water. Benzene, toluene, ethylbenzene, and

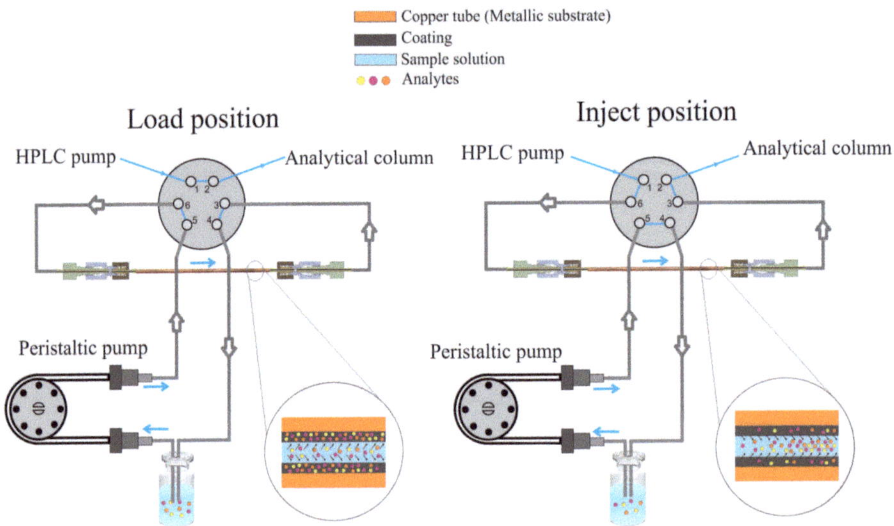

Figure 2.5 On-line extraction/desorption using CME-HPLC.

xylenes (BTEX) were also trapped on the bare silica fiber. Several types of coatings including polydimethylsiloxane (PDMS), polyacrylate (PA), polydimethylsiloxane/divinylbenzene (DVB), Carboxen/PDMS, Carbowax (CW)/template resin fibers have become commercially available. Among those, PDMS and PA coatings are the most well-studied and characterized. For a specific application, the coating is chosen based on the polarity of the analyte (Zhang et al., 1994). PDMS is less polar than PA; thus it is widely used for the extraction of non-polar compounds such as substituted benzenes (Potter & Pawliszyn, 1992; MacGillivray et al., 1994) and polyaromatic hydrocarbons (PAHs) (Potter & Pawliszyn, 1994). For polar compounds such as ketones and alcohols, polar coatings like PA and CW are preferred. The selectivity difference of polar and nonpolar coatings has been studied by analyzing organophosphorus insecticides (Magdic et al., 1996; Valor et al., 1997). Among the organophosphorus compounds tested, triazophos and methylparathion showed greater affinity towards PA than PDMS, while diazinon and prothiofos exhibited comparable or even higher affinity towards PDMS. Volatile aldehydes, pyrazines, pyridines and thiazoles were extracted by PDMS and CW/DVB. Although these fibers show different selectivity to different groups of compounds, both fibers can extract solutes with more alkyl substitutions better for all four groups of compounds. These results coincide with results obtained for some pesticides, including *N*-substituted amines, *N*-heterocyclic compounds and organophosphorus compounds. These compounds were extracted using the PA fiber coating (Eisert & Levsen, 1995). A linear correlation between extraction efficiency (relative peak response after SPME to the standard chromatogram) and the octanol-water partition coefficient for some triazine pesticides was obtained.

Despite numerous and pronounced advantages of SPME over traditional sample preparation techniques, the conventional fiber SPME suffers from significant shortcomings, which include:

1. Relatively low operating temperature due to poor thermal stability of the physically held sorbent coating;
2. Instability and swelling of the fiber coatings in organic solvents;
3. Fragility of the coated fibers;
4. Mechanical instability of the substrates;
5. Bending of the SPME syringe needle during operation;
6. Low sorbent loading that results in poor extraction sensitivity;
7. Pronounced run-to-run and batch-to-batch variability;
8. Relatively long extraction equilibrium time due to the slow diffusion of analytes into pure organic polymers used as the extracting medium;
9. Limited numbers of commercially available fiber coatings;
10. Short lifetime of the physically held sorbent coating.

Many modifications have been implemented to circumvent these problems, including the fabrication of metal-based substrates to enhance the mechanical instability of the fibers and adaption of new methodologies to improve the thermal and solvent stability of the coatings. Methodologies based on sol-gel technique, conductive polymers, nanomaterials, nanocomposites, the usage of an electrospinning system and preparation of selective sorbents by molecularly imprinted polymers (MIPs) and molecularly imprinted xerogels (MIXs) are among these efforts.

2.3.2 Extractant Phases Based on Inorganic Polymerization

Although the use of SPME fibers is increasingly popular, their relatively low recommended operating temperature (~240–280 °C), their instability and swelling in organic solvents (greatly restricting their use with HPLC), fiber breakage and stripping of coatings are obstacles to the analysis of some chemicals. Apart from the fiber breakage, other limitations are due to the physical interactions between the extracting phase and solid substrates. Low operating temperature and stripping of coatings affects the performance and lifetimes of the fibers. Preparation of chemically bonded SPME fibers has been the center of attention for many researchers to overcome some of these problems.

Inorganic polymerization, better known as the 'sol-gel process', is a general method for preparing oxides by the 'wet route' (i.e., in solution) at room temperature. The sol-gel process was discovered by Ebelmann (1846, 1847) and has benefited from developments of inorganic polymerization.

Inorganic polymerization has experienced an extraordinary revival during the last two decades, as it corresponds to a general method for the preparation of oxides

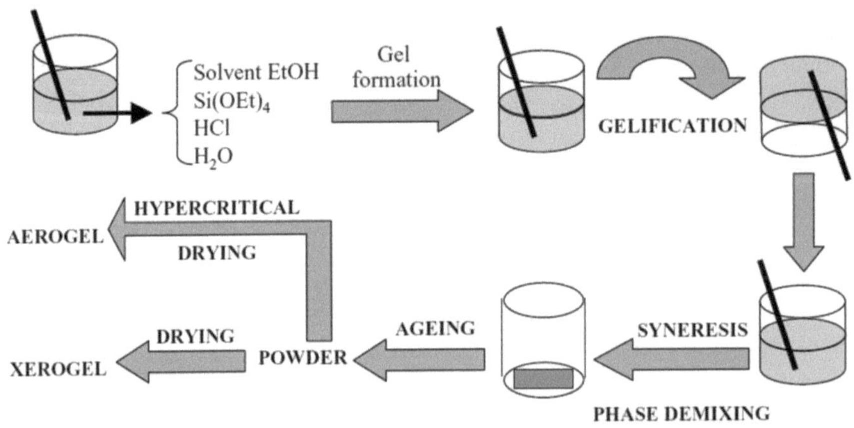

Figure 2.6 Schematic diagram of the different steps of the sol-gel process. Reprinted from Corriu & Trong Anh (2009) with permission from John Wiley & Sons.

under mild conditions. Reaction with water at room temperature and in the presence of a (acidic, basic or nucleophilic) catalyst, of a Si(OR)$_4$ (R: Me, Et, iPr, etc.) solution gives a transparent solid occupying the entire volume. This solid is not a precipitate, but a gel enclosing the solvent and sticking to the walls of the recipient. **Figure 2.6** shows the various steps of the process (Corriu & Trong Anh, 2009).

After a lapse of time which varies with the experimental conditions, a phase demixing occurs in which the solid SiO_2 expels the solvent and becomes denser. This stage is called syneresis. The solid obtained is ground, washed and vacuum dried to give a xerogel. Other treatments are possible, for instance, hypercritical drying yields aerogels, which have valuable dielectrical and insulating properties. In spite of the interesting nature of these treatments, we shall focus on the xerogels, which are mostly used in the field of nanomaterials (Corriu & Trong Anh, 2009).

In 1997, sol-gel technology was used for the preparation of sorbents in SPME by Malik and co-workers (Chong et al., 1997; Wang et al., 1997). It can be an excellent alternative to conventional coating preparation methods due to its inherent advantageous features and performances. These include single step manufacturing process, material homogeneity at the molecular level, chemical bonding between the sorbent and the fused silica surface, high thermal and solvent stability and the porous structure of the hybrid material. Moreover, sol-gel organic–inorganic hybrid materials provide desirable sorptive properties that are difficult to achieve using either purely organic or inorganic material. Precursor, coating polymer, catalyst and deactivating agent (non polar fibers) are required for preparation of sol-gel based SPME fibers. Moreover, functionality of the solid support is another major issue. The research concerning the preparation of new SPME fibers includes three sol-gel based features of selecting (i) precursors, (ii) coating polymers and (iii) modifiers to achieve fibers with specific physical capabilities and the desired polarity and selectivity (Bagheri et al., 2012e). For the preparation of a sol-gel coating in SPME, the presence of a precursor is necessary, and the addition of the coating polymer and/or modifier could be influential to improve some major interactions and available porosity.

2.3.2.1 Preparation of Sorbents by Sole Precursor

Preparation of sorbent by a sole precursor in water using an acid-catalyzed process is also possible. In this regard, a selected precursor should contain an appropriate organic group, an alkyl alkoxysilane, to prepare organically modified silica (ORMOSIL). ORMOSIL should contain appropriate groups to interact with desired analytes *via* mechanisms such as van der Waals, dipole-dipole and hydrogen bonding. A typical sol-gel processing is shown in **Figure 2.7A**.

In practice, the sol-gel process is, considering the macroscopic scale, very simple. In one step and at room temperature, it transforms a molecule into a material ready for shaping. At the nanoscopic and microscopic scales, it is in fact a very complex process consisting of several transformations of very different natures involving three distinguishable states of matter: solution, colloid and solid.

The first state occurs in solution and corresponds to the formation of the Si–O–Si bonds. At first, the precursor gives rise to nanometric linear and cyclic oligomers which, by polycondensation reactions, gradually become intermingled to yield cross-linked polymers (10–100 nm in size). These polymers aggregate into micrometer-sized colloids, which are detectable by light diffusion. This colloidal solution is a viscous sol and corresponds to a crucial phase of the process (**Figure 2.8**). It is the viscous sol which permits the shaping of the material. It can be drawn into fibers, molded as a bulk solid or used as a covering by dip or spin coating (Corriu & Trong Anh, 2009).

In the second stage, the sol is solidified into a gel, a process known as sol-gel transition. This step corresponds to a chemical reaction occurring at the surface of the colloids. The rapid gelification can be easily understood: a few chemical bonds between the voluminous colloids, whose surfaces are covered with SiOH (or SiOR) groups, are enough to create a solid wide-meshed network capable of retaining all the solvent and of occupying all the space.

The next step corresponds to a solid phase development. The number of bonds between the colloids increases, thus speeding up the densification of the solid. The initial stage, called syneresis, corresponds to the expulsion of the solvent. Continuation of this bonding process in the solid phase leads to an amorphous solid having stabilized granulometry and porosity, which depend on the experimental conditions. This is called the Oswald ageing process. It is very important because it controls the macroscopic characteristics of the solid. The final step consists of the drying of the washed solid and it should be noted that the drying conditions can significantly affect the material's texture (Corriu & Trong Anh, 2009).

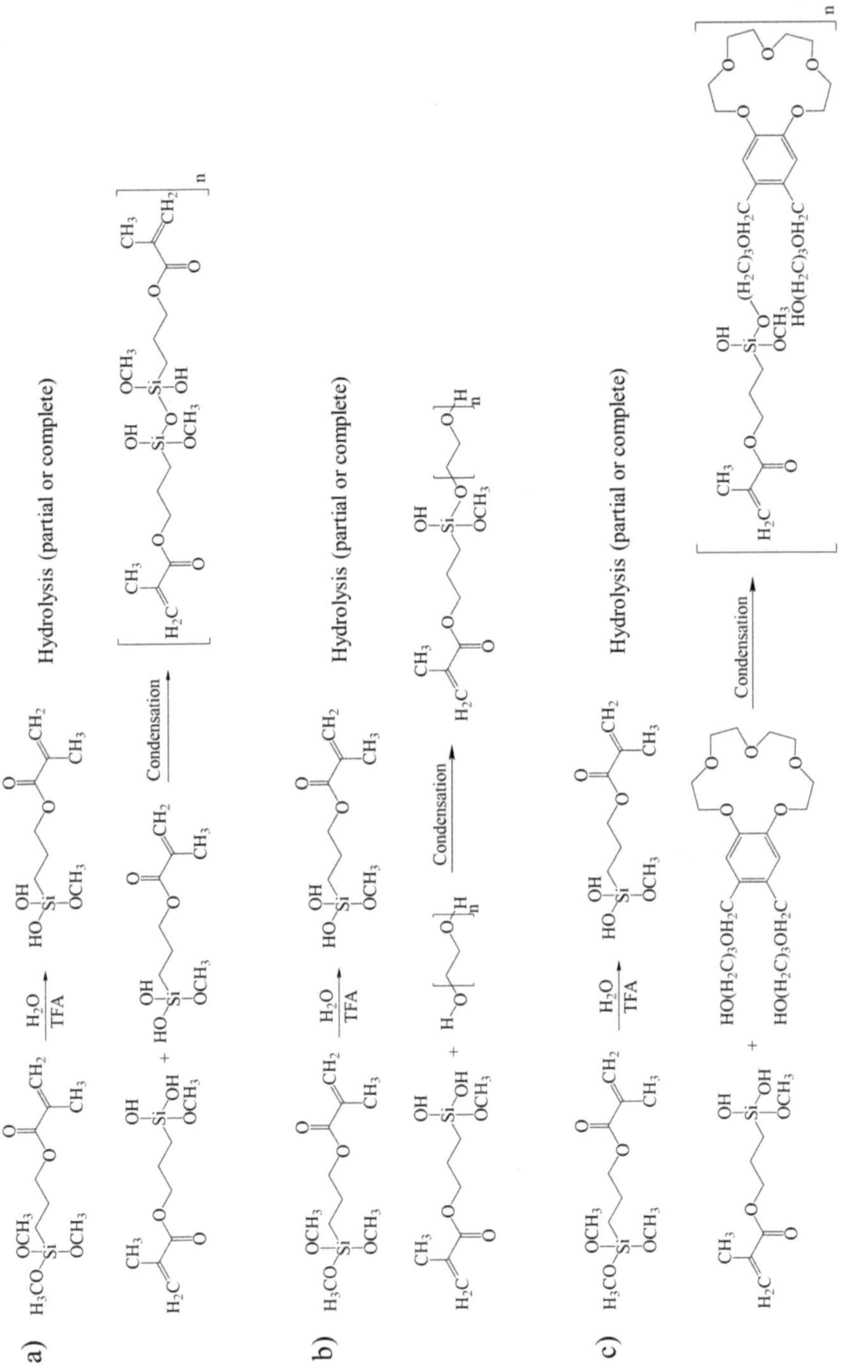

Figure 2.7 Preparation of SPME sorbents by (A) sole precursor, (B) precursor and coating polymer (C) precursor and a modifier.

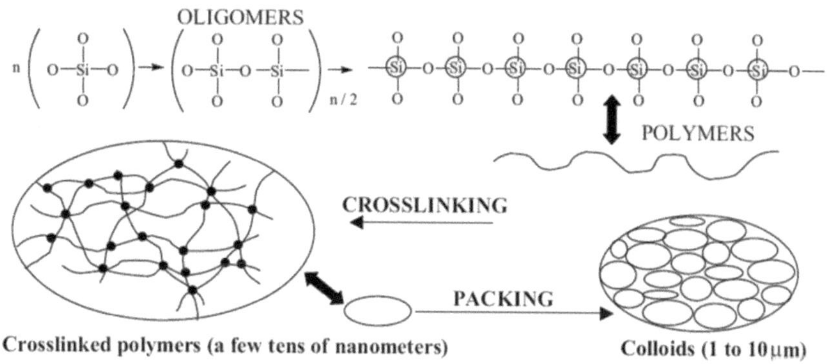

Figure 2.8 Schematic diagram for the formation of oligomers. Reprinted from Corriu & Trong Anh (2009) with permission from John Wiley & Sons.

2.3.2.2 Preparation of Sorbents by Precursor and Coating Polymer

One of the advantages of the sol-gel technique is the possibility to prepare multi-component materials which favorably affect the porosity and enhance interactions with the desired analytes through the addition of polymers into the xerogel (sorbent) structure. Addition of polyethylene glycol to a precursor is shown in **Figure 2.7B**.

2.3.2.3 Preparation of Sorbents by Precursor and a Modifier

One of the most important advantages of the sol-gel route is the possibility of designing the material structure and properties through proper selection of the sol-gel precursor and other building blocks. Different modifiers such as crown ethers (Yu et al., 2004), calixarenes (Li et al., 2004), carbon nanotubes (CNTs) (Wang et al., 2006; Sarafraz-Yazdi et al., 2010a, 2010b; Bagheri et al., 2011a), fullerene (Yu et al., 2002) and β-cyclodextrin (Zhou et al., 2007) have been used to enhance the extraction efficiency of the prepared sorbents. Sol-gel chemistry is widely used for the preparation of materials suitable for a wide range of applications. This technology has been used for preparing SPME fiber coatings and exhibit many advantages by providing efficient incorporation of organic component into the inorganic polymeric structure in solution under very mild thermal conditions. Many inherent advantages of sol-gel technology are their high thermal stability, porous structure, high degree of flexibility in coating composition, by varying the proportion of the sol solution ingredients or using a deactivation reagent. In this technique precursors are mixed at the molecular level and multicomponent materials could be formed at much lower temperature than traditional processing method. The guest molecules could be introduced by dissolving or suspending them in the hydrolyzates, without harming their physical properties. A typical reaction for this type of mechanism is shown in **Figure 2.7C**.

Calixarenes are regarded as the third generation of supramolecules next to crown ethers and cyclodextrins. They are a class of cyclic oligomers prepared from formaldehyde and para-substituted phenols *via* cyclic condensation under alkaline conditions. Considering the outstanding capacity of the calixarenes as receptors based on their variable chemical modification potential and their conformational pliability, a kind of induced fit to the shape of suitable guest molecules is allowed.

2.3.2.4 Chemical Bonding Between Substrates and Sorbent During Sol-Gel Process

One of the main advantages of the sol-gel process is the presence of chemical bonds between substrate and sorbent. The chemical bond can overcome the mentioned problems related to the conventional sorbents such as low operating temperature and swelling in organic solvents. In doing so, the substrate should be already treated and functionalized. There are three well-known methods to functionalize substrates.

2.3.2.4.1 Treatment of Fused Silica by NaOH

Prior to the coating of the SPME fiber by sol-gel technique, the fused silica, as a substrate, is dipped into 1 mol L^{-1} NaOH solution for 1 h to activate the surface of the fiber while producing silanol groups, and is then washed with water. The fiber is placed into 0.1 mol L^{-1} HCl solution for 30 min to neutralize the excess of NaOH, then washed again with water and dried in the oven. Finally, it is inserted into the prepared sol, and during the sol-gel process and preparation of gel, chemical bonds between sorbent and fused silica substrate are formed (Chong et al., 1997).

2.3.2.4.2 Self Assembled Monolayers

The main problem associated with fused silica-based sorbents is the fragility of the fibers. To overcome the mechanical instability of the fibers, metal substrates can be used. As already mentioned, functionalization of the substrate is necessary for sol-gel processing. To prepare a sorbent by sol-gel technology, the substrate must have some functional groups on its surface in order for the chemical bonding with the sorbent to occur. Metal substrates, due to the absence of functional groups, have not been assigned as suitable probes to hold the appropriate sorbents. As there is no functional group on the surface of metals, they have to be functionalized prior to sol-gel process. Self assembled monolayers (SAMs) is a technique which can be easily used to functionalize copper, gold, silver and platinum.

Self-assembled monolayers (SAMs) are ordered monomolecular films which are spontaneously formed by immersing a solid substrate into a solution containing amphifunctional molecules. The amphifunctional molecule has a head group, which usually has a high affinity towards the solid surface, a tail (typically an alkyl chain) and a terminal group that can be used to control the surface properties of the resulting monolayer. The molecular forces between the tails are chiefly responsible for the order of the monolayer. Thus, SAMs have two key features of self-assembly found in biologi-

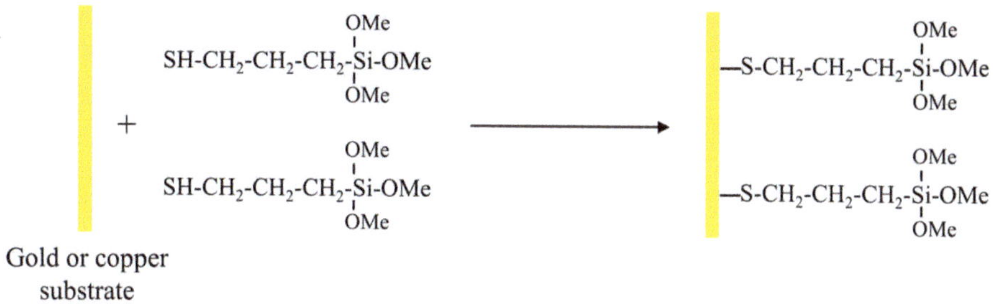

Figure 2.9 Schematic diagram of SAMs.

cal systems, namely that molecules have high affinity for each other and predictable structures are formed when the molecular units are associated. The ability to tailor both head groups and tail groups of the self assembling molecules makes control over the self-assembly behavior possible. An alkanethiol has a thiol (-SH) head group and a tail that is usually an alkyl chain. At the end of the tail is the terminal group which, in a well-packed monolayer, determines the properties of the surface of the monolayer. Alkanethiols and di-n-alkyl disulfide self-assemble on coinage metals to form well-organized monolayers, where the formation of a bond between the thiolate head group and the metal surface anchors the organosulfur molecules to the surface and interactions between the alkyl chains give the monolayer its order.

The first application of SAMs to functionalize gold fiber and copper tube was reported by Bagheri et al. (2011b, 2011c, 2011d). **Figure 2.9** shows the schematic process involved in SAMs.

Typically, alkanethiols are assembled onto gold surfaces from dilute solutions (millimolar concentrations). Common solvents are ethanol for shorter alkanethiols or hexane for longer alkanethiols. Two distinct adsorption stages are observed in the assembly –a rapid stage within the first few minutes during which time the contact angle is close to its limiting value and the thickness is 80–90% of the maximum (Bain et al., 1989). The length of this stage is dependent on the alkanethiol concentration, taking only a few minutes at a concentration of 1 mM but about 100 min at 1 μM (Pan, Durning, & Turro, 1996; Schessler, Karpovich, & Blanchard, 1996). The second slow stage occurs over several hours as the contact angle and thickness reach their final limits (Bain et al., 1989). During the latter stage, the molecules in the SAM undergo a slow reorganization equivalent to surface crystallization (Karpovich et al., 1998).

2.3.2.4.3 Diazonium Salts

Although there has been some extensive progress, SAMs are only amenable to a few metals such as Cu, Ag, Au and Pt. To overcome this limitation, aryl diazonium salts have been used to functionalize several types of metal and non-metal surfaces (Bagheri et al., 2013b). Aryl diazonium salts constitute an important series of compounds in organic chemistry. This is due to the fact that the diazonium group activates nucleophilic aromatic substitutions and provides a general pathway for introducing halogens, CN, OH and H into an aromatic ring to form a wide range of compounds. Carbon, metallic, semi-conductor and reduced polymer surfaces can be modified by the electrochemical reduction of aryl diazonium salts with the general formulae A^-, $^+N_2$–C_6H_4–R, where A^- is the anion and R stands for different functional groups (Mahouche-Chergui et al., 2011). The interest in using aryl diazonium salts lies in their ease of preparation, rapid (electrochemical) reduction, large choice of reactive functional groups, and strong aryl–surface covalent bonding (Yu et al., 2002; Boukerma et al., 2003). Aryl diazonium salts are excellent alternatives to the traditional silanes and thiols as the former bind polymers mostly to ceramics whereas the latter are mainly limited to the surface of gold, copper and platinum (Love et al., 2005).

To prepare an unbreakable sol-gel CME device using aryl diazonium salts, different metallic substrates such as copper, brass and stainless steel tubes (0.7 mm internal diameter) as well as copper, ferronickel, stainless steel wires can be functionalized. Polytetrafluoroethylene (PTFE) can also be used as a non-metal substrate. It is worth noting that the latter must be accomplished using chemical reduction. In all experiments, the capillary tube was used as the HPLC sample introduction loop. Firstly, each capillary tube was washed with acetone (three times), under sonication for 10 min, to clean its internal surfaces. The next step included functionalizing the substrate which involved the electrochemical reduction of 4-aminophenylacetic acid for *in situ* generation of the aryl diazonium cation (Breton & Bélanger, 2008). 4-aminophenylacetic acid (2 mM) was solubilized in aqueous HCl (0.5 M), and then sodium nitrite (2 mM) was added to solution under magnetic stirring at 500 rpm. The mixture was placed at room temperature for about 5 min to permit completion of the reaction, prior to the electrochemical functionalizing process. The substrates were then inserted in the electrolyte solution while acting as the cathode and a piece of aluminum foil was used as the anode. A peristaltic pump was used to deliver the electrolyte solution through the tube (wires only immersed) while a potential of –1.5 V was applied for 15 min. **Figure 2.10** shows the electroreduction of the aryl diazonium cation on the inner surface of the copper tubing. To functionalize the surface of PTFE, a representative non-metallic substrate, the PTFE was immersed into the solution of aryl diazonium cation and 150 mg of iron powder was then added. Instantaneously, small bubbles were observed in the flask corresponding to dihydrogen and dinitrogen evolution, as iron reduced protons and diazonium salts in solution (Mévellec et al., 2007). **Figure 2.10** shows the general process (Bagheri et al., 2013b).

2.3.3 Conductive Polymers

Conductive polymers (CPs) are classified as materials with a highly π-conjugated polymeric chain (MacDiarmid, 2001; Shirakawa, 2001), exhibiting the electronic properties of both metals and semiconductors and with the mechanical properties of organic

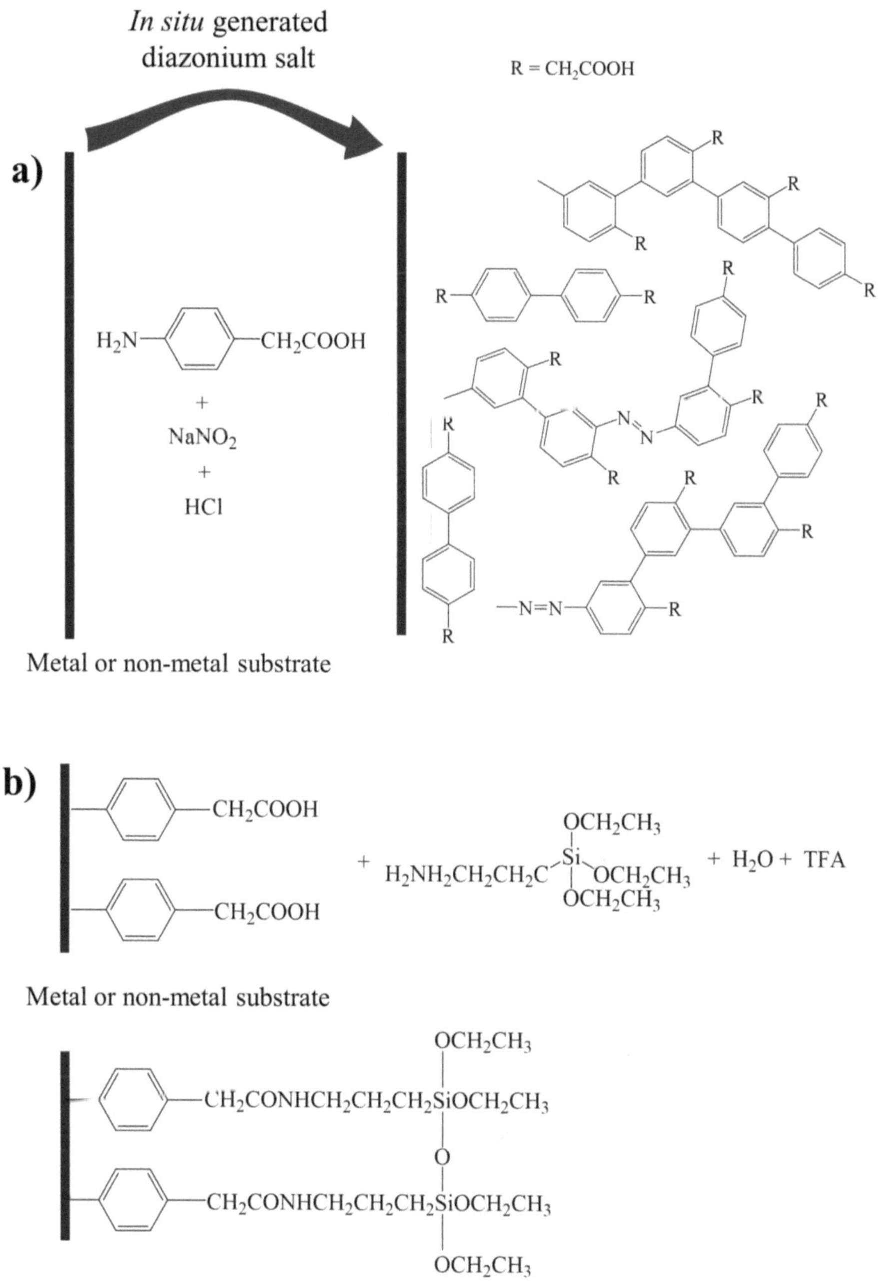

Figure 2.10 (A) Functionalization of any metal or non-metal substrates prior to the treatment and after treatment with aryl diazonium; and (B) coating of the sol-gel sorbent on the inner surface of the functionalized HPLC loop. Reprinted from Bagheri et al. (2013b) with permission from Elsevier.

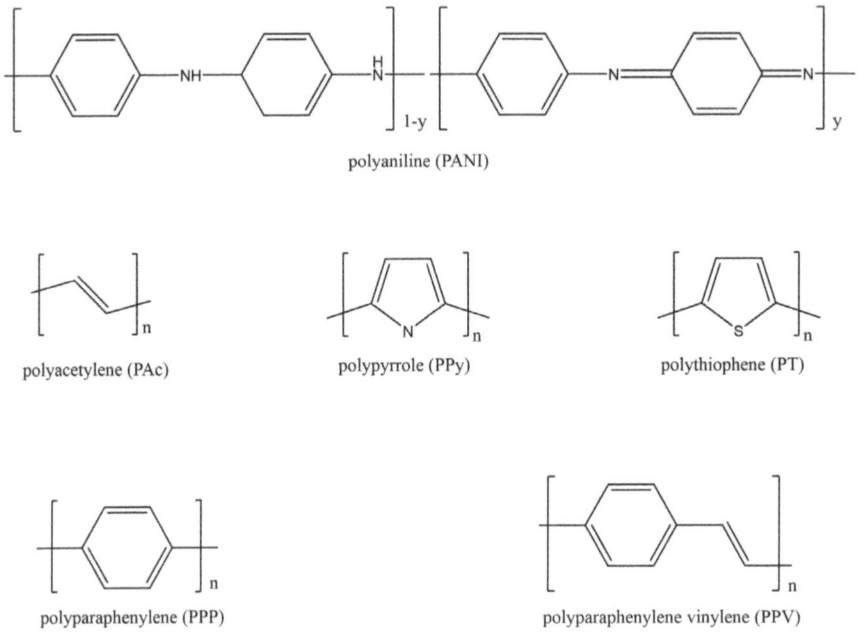

Figure 2.11 Structure of some common conductive polymers.

macromolecules. Research interest in electroactive polymers started in 1977 by preparation of the highly conducting polyacetylene (PAc) and eventually, Heeger, MacDiarmid and Shirakawa were awarded the Nobel Prize in chemistry in 2000 for the discovery and development of the so-called conducting polymers. The conductivity of PAc was demonstrated to be enhanced by several orders of magnitude upon partial oxidation or reduction, which commonly referred to as doping (Shirakawa et al., 1977). CPs have attracted more and more attention due to their electrical properties which are similar to metals but have some characteristics of organic polymers. Furthermore, they can be prepared in such a way as to be applied in different fields. This could be achieved through modifications in the polymer structure and varying the functional groups in the organic moiety. CPs with different chemical structures such as polyene-type, aromatic, heteroaromatic and mixed aromatic (or heteroaromatic) systems (Pron & Rannou, 2002), polyaniline (PANI), polypyrrole (PPy), polythiophene (PTh), poly(p-phenylene), poly(phenylene vinylene) have been studied extensively. Representative candidates of conducting polymers are given in **Figure 2.11**. Organic CPs are conformationally rigid molecules due to their extensive conjugation. Most of them are insoluble in common organic solvents while PANI displays a limited solubility in *m*-cresol, 2-pyrrolidinone and concentrated sulfuric acid, whereas no report on PPy solubility can be found (Bagheri et al., 2005b). Additionally, these polymers tend to be infusible and they are usually brittle and have poor mechanical properties. All of these properties are probably related to energetically favorable interactions of the π-systems of closely-linked chains which are stronger than those of polymer with solvent, which probably explain the insolubility and the infusibility of these polymers. Moreover, their mechanical properties derive from the anisotropic character of the intermolecular π -system interactions producing an amorphous bulk material. Nevertheless, PANI and PPy are of high interest due to their interesting properties such as tunable conductivity and stability in air (Lu et al., 2006). Due to their multi-functionality, ease of synthesis and stability, CPs are attractive and efficient sorbents for the sample preparation purposes. Among sorptive-based extraction methodologies, SPME has been used extensively in the purification and preconcentration of a vast variety of analytes and chemical species from environmental and biological samples. The microextraction techniques rely on quantitative but non-exhaustive transfer of analytes by equilibration of small portions of adsorbents or sorbents and larger amounts of samples, which are in direct or indirect contact (headspace) with each other.

The retention of target analytes in these sorptive-based techniques is due to reversible hydrophobic, polar and ionic interactions between the analytes of interest and the sorptive material. The type of coating used in SPME is therefore responsible for the efficiency of the extraction process. The availability of different materials is one of the advantages of the sorptive-based techniques. The use of CPs as efficient sorbents for sample preparation purposes is mostly due to their multifunctionality and the ability to control their structure so as to influence adsorption processes.

2.3.3.1 Structures of Some Well-known Conductive Polymers

Generally, it is assumed that CP conductivity is related to the higher degree of crystallinity and better alignment of the chains. However, this could not be confirmed for PANI, which is largely amorphous. The basic chemical backbones of conducting PAc, PPy and PANI are shown in **Figure 2.12**. PANI has different structures considering the level of reduced (amine, NH) and oxidized state (imine, N) of nitrogen atoms in its structure. Concerning emeraldine, half of the total nitrogen along the chain is in the reduced state and the remaining in the oxidized state. As a consequence, the configuration of the neighboring rings is either in a benzenic-like or quinoid-like structure, respectively.

Many other structures are known where nitrogen is in the completely reduced state (leucoemeraldine) or in the fully oxidized state (pernigraniline). Between these two limits, a very large number of intermediate states could exist. However, only a few of them, such as protoemeraldine and nigraniline, are rather stable (**Figure 2.12C**) (Snauwaert et al., 1990). The emeraldine form of PANI contains an equal number of imine and amine nitrogen atoms, the former being more basic and thus undergoing an easier protonation compared with the amine sites, which are protonated only under highly acidic conditions (Blinova et al., 2006). The secondary amine group (–NH–) in pyrrole is even less basic than the corresponding group in aliphatic compounds.

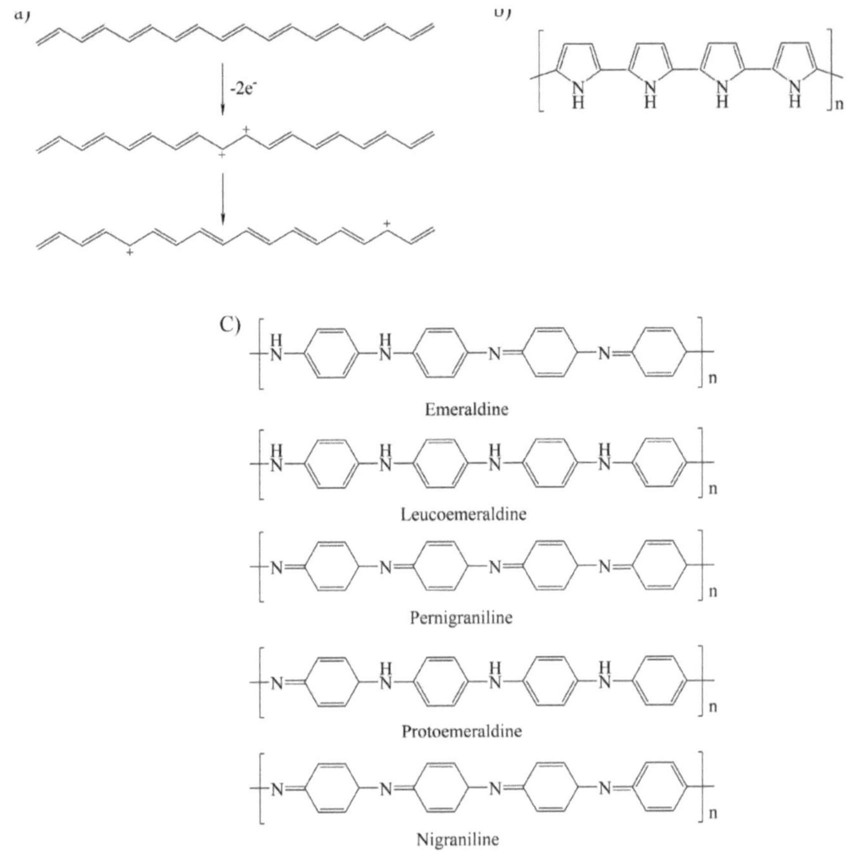

Figure 2.12 (A) Mechanism of electroconductivity in PAc, (B) Chemical structure of PPy, (C) Different forms of PANI.

The acid–base transition in the polymers is responsible for the strong change in their conductivities.

2.3.3.2 Preparation of Conductive Polymers

Organic CPs can be prepared by pyrolysis and numerous methods such as chemical, electrochemical, photochemical, concentrated emulsion, inclusion, solid-state, and plasma polymerization. Among these preparation methods, the electrochemical and chemical polymerizations have been widely applied to the preparation of CPs in industry and other fields of research. Chemical polymerization (Chen & Tsai, 1993; Wei et al., 1993) is the most useful method for preparing large quantities of CPs, since it is performed without electrodes (Ohtani & Shimadzu, 1989) but the introduction of reactants and by-products may affect the properties of the produced CPs. Electrochemical polymerization is desirable for conductive polymer thin film and nanostructure fabrication and is normally carried out in a single- or dual-compartment cell by adopting a standard three-electrode configuration in a supporting electrolyte, both dissolved in an appropriate solvent.

2.3.3.2.1 Chemical Synthesis

Chemical synthesis is the oldest and the most popular route for the preparation of bulk quantities of CPs on a batch scale. Chemical polymerization is typically carried out using relatively strong chemical oxidants such as KIO_3, $KMnO_4$, $FeCl_3$, K_2CrO_4, $KBrO_3$, $KClO_3$, $(NH_4)_2S_2O_8$ (Chen & Tsai, 1993; Wei et al., 1993). These oxidants are able to oxidize the monomers in solution, leading to the formation of cation radicals. These radicals further react with other monomers or n-mers, yielding oligomers or insoluble polymers. Concerning the chemical oxidation of aniline and pyrrole, there are no detailed mechanisms as those reported for the electrochemical polymerization. It is generally accepted that the first step of the polymerization involves the formation of the dimeric species *N*-4-aminophenylaniline, trans-azobenzene and benzidine. By adapting the chemical procedures, the structure, chemical and physical properties of the obtained polymers heavily depend on the properties of the oxidizing agent, the molar ratio of oxidant to monomer, the electrolyte and the temperature (Ghosh, 2006). Chemical oxidative polymerization is limited by the range of available chemical oxidants. The counterion of the oxidant ultimately will act as a dopant or co-dopant in the polymer. Therefore, it may be difficult to obtain CPs with different dopants. The limited range of oxidants also makes it difficult to control the oxidizing power in the reaction mixture and subsequently the degree of overoxidation during synthesis. Both the type of dopant and the level of doping are influential parameters affecting the final properties of CPs such as molecular weight, crosslinking and conductivity (Freund & Deore, 2007). Moreover, a strong drawback in using these stoichiometric oxidants is the formation of a large amount of by-products, in the case of ammonium persulfate (APS), the amount of resulting ammonium sulfate is about 1 kg per kg of organic polymer.

2.3.3.2.2 Electrochemical Synthesis

The electrochemical technique has received wider attention, because of the simplicity and the added advantage of obtaining simultaneously doped CPs. Additionally, in the electrochemical polymerization process a wider choice of cations and anions as dopant ions is available. The electrochemical method also permits the control of film thickness and morphology. In general, chemical oxidation method produces CPs as powders while electrochemical synthesis generally leads to deposition of conducting polymers films onto a supporting electrode surface by anodic oxidation (electropolymerization) of the corresponding monomer (Sadki et al., 2000). When a positive potential is applied to the working electrode, the oxidation starts with the formation of the radical cation. The delocalized radical cations induce the radical–radical

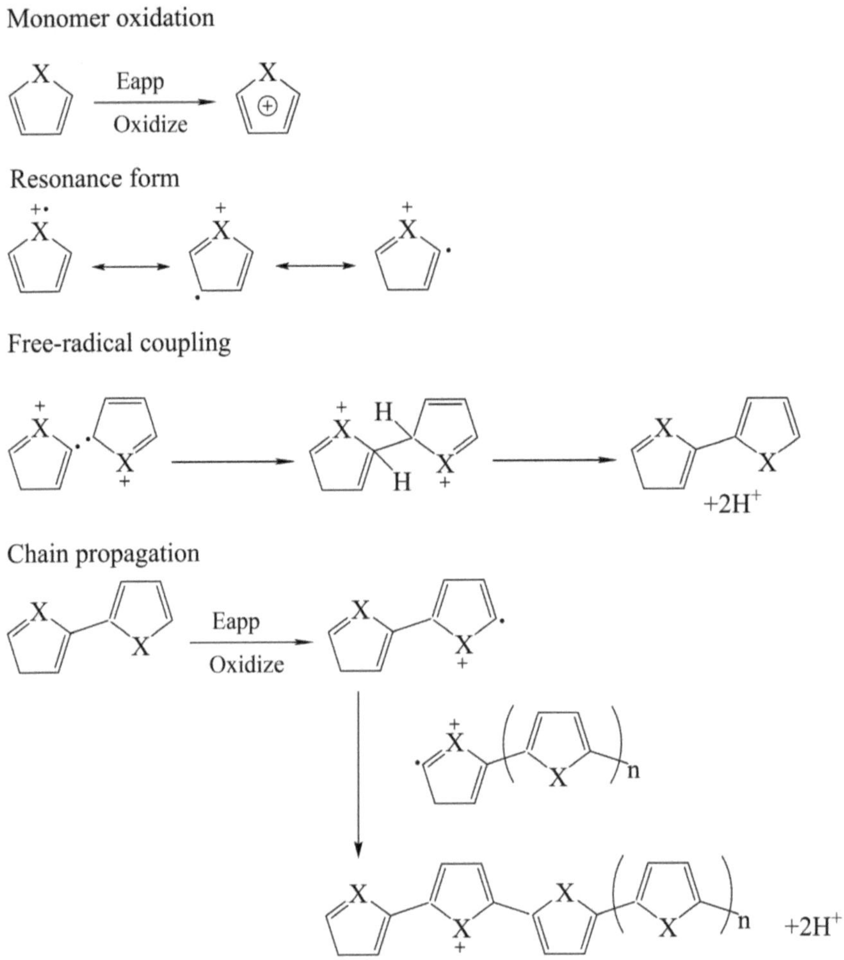

Figure 2.13 Synthesis mechanism of PPy by electrochemical route.

coupling to form dimers by the deprotonation at π-position. The extended conjugation in the polymer results in a lowering of the oxidation potential compared to the monomer. The electrochemical oxidation and radical coupling process is repeated continuously and finally the deposited CP film is produced on the working electrode. The doping of the polymer is generally performed simultaneously by incorporation of the doping anion into the polymer to ensure the electrical neutrality of the film. The mechanism of electrochemical polymerization of PPy is depicted in **Figure 2.13**, although the electropolymerization mechanism is rather unknown.

The electrochemical synthesis method has some advantages over the chemical synthesis: (i) a highly electrochemically conductive polymer film can be easily produced on an electrode surface, which is of special interest for applications such as modified electrodes, batteries or protective coatings; (ii) film thickness, morphology and conductivity can be easily controlled by monitoring the applied potential and current; (iii) they are cleaner than chemical methods because no oxidizing agent is used; (iv) they provide an *in situ* condition to investigate the polymerization process and the properties of the resulting conducting polymer.

2.3.3.3 Conductive Polymer-based Extractant Phases

Despite the high specific surface areas of the commercially available hydrophobic sorbents, they suffer from poor interaction and retention of polar compounds which is mostly due to their hydrophobic nature. One solution to this problem is to introduce a polar moiety into their networks to favor the polar interaction between the sorbent and analytes, and enhance the extraction recoveries (Fontanals et al., 2007). Combination of (i) reversed phase mechanism, rising from the polymeric skeleton and (ii) π–π interactions induced by the available functional groups, provides higher mass transfer which increases the interaction possibilities between the available sorbent sites and the analytes functional groups. All these lead to the enhanced retention of polar compounds. A hydrophilic sorbent can be prepared by copolymerizing monomers containing suitable functional groups or by introducing a functional group to the existing hydrophobic polymers.

2.3.3.3.1 Polypyrrole-based Coatings

Tailored properties of PPy coatings can be achieved by proper selection of the counter ion. This dependence has been studied for the determination of volatile organic compounds (VOCs) in headspace mode and for polar and ionic compounds in aqueous solution (Wu et al., 2000). It was reported that the thermal stability and mechanical properties of PPy films could be improved when large aromatic counter anions were incorporated into the films (Nalwa, 1997; Sadki et al., 2000). It was also found that the adhesion of the PPy film to a platinum wire surface was affected by the counter ion used. A more stable PPy coating was produced using polystyrenesulfonate (PSS) as the counterion compared with a similar film containing a chloride counterion. In other work, PPy was synthesized in the presence of different dopants such as chloride, perchlorate, acetate, sulfate, and dodecylsulfate (DS) ions in aqueous medium. Polymer films doped with chloride, perchlorate, and acetate anions showed weak adhesion on the platinum wire and were unstable at temperatures higher than 200 °C. Corresponding thermogravimetric curves revealed that the fibers coated with sulfate-doped PPy (PPy–S) and PPy–DS under argon atmosphere are thermally stable up to 300 °C. In addition, the perfect adhesion of these films to Pt wires was also achieved. The extraction properties of these two fibers toward some PAHs were examined using a HS-SPME set-up. The results showed that PPy–DS possesses up to three times higher extraction efficiency toward the PAHs compared to the PPy doped with sulfate. These results can be explained by the presence of a C_{12} chain alkyl group in the DS ion which increased the hydrophobic interaction of PPy–DS film with PAHs (Mohammadi et al., 2005). A PPy modified with tetrasulfonated nickel phthalocyanine (NiPcTS) doping anion was prepared and used for the extraction of BTEX compounds from water samples by headspace solid-phase microextraction (HS-SPME). It was shown that the PPy doped with NiPcTS could extract up to five times as much of the BTEX compounds in comparison with the PPy doped with other anions such as toluene-*p*-sulfonate (TOS), tetrafluoroborate (BF_4^-) and perchlorate (ClO_4^-). This is probably due to the presence of phenyl groups in NiPcTS incorporated into PPy as counterions (Djozan et al., 2004). The effect of different types of doping ions in PPy-based SPME fibers have been examined for sodium polyphosphate (PP), a new dopant for the electropolymerization of pyrrole on the surface of a steel fiber with the aim of achieving a polymer coating with improved qualities for SPME analysis. It was revealed that the PPy/PP coating has the highest extraction efficiency compared with the PPy polymer containing other counter-ions (perchlorate, tartarate and oxalate). In addition, thermogravimetric analysis of PPy/PP revealed that it is more thermally stable than other CPs. Also, the prepared PPy/PP coating was physically and thermally more stable and did not show any major weight loss even at 450 °C. Moreover, the PPy doped with PP could overcome the adhesion problem through incorporation of the inorganic PP into the organic polymeric structure, which led to a chemically adhered coating through Fe–P bonding to the metal surface (Rout et al., 2006). Additionally, an enhancement in the extraction efficiency of organochlorines was observed by increasing the PP concentration up to 10%, and then the extraction efficiency reached a plateau region (Mollahosseini & Noroozian, 2009). Interestingly, the successful application of PPy in SPME fiber coatings as both an extraction phase and a surface to enhance laser desorption/ionization (SELDI) of analytes was reported. The developed SPME/SELDI fiber integrated sample preparation and sample introduction on the tip of a coated optical fiber, which also acted as the transmission medium for the UV laser light. A SELDI surface could be used to transfer the laser energy to ionize the analytes instead of using a matrix (Wang et al., 2004).

2.3.3.3.2 Polyaniline-based Coatings

Along with PPy, PANI has been applied as a SPME fiber coating as well. Cyclic voltammetry (CV) has been used to deposit PANI on platinum wire for the determination of phenolic compounds (Bagheri et al., 2005b) and PAHs in water (Bagheri et al., 2007). This polymer exists in a variety of protonated and oxidized forms which can be specifically achieved by adjusting the scan rate and sweep potential (Ghassempour et al., 2005) during the electrocoating process using CV. The leucoemeraldine base form of PANI fiber coating in a setup with gas chromatography-mass spectrometry (GC–MS) provided the highest sensitivity for the determination of anatoxin-a, an alkaloid produced by cyanobacteria from cultured media and contaminated water. Also, the same coating could be prepared on steel (Minjia et al., 2004) and gold (Djozan & Bahar, 2004) supports by the application of a constant voltage. In order to improve the application of CPs as SPME fibers, a trifluoroacetic acid-doped PANI fiber was introduced for SPME (Wang et al., 2009). The extraction capacity of this coating was much greater than commonly used PANI (sulfuric acid-doped PANI) and PTh-based SPME fibers, and comparable to a commercially available 30 μm PDMS fiber. This SPME fiber coating had a thinner stationary phase but proved to have high thermal stability. Fluorine in this fiber was thought to play an important role in the explanation of its good performance. To further increase the extraction efficiency and to test the role of fluorine, a perfluorinated acid has been introduced to the PANI backbone and applied to the SPME fiber. It was reported that the perfluorooctanesulfonic acid-doped PANI (PFOS–PANI) was superhydrophobic. It was demonstrated that the PFOS–PANI coating, roughly 10 μm thick, exhibited high extraction efficiency for polybrominated diphenyl ethers (PBDEs), phenols, polychlorinated biphenyls (PCBs) and PAHs in comparison with the 100 μm PDMS, 85 μm PA and 65 μm PDMS/DVB fibers (Wang et al., 2010). In addition, the prepared PANI fiber coating showed to have high thermal (up to 350 °C) and solvent stability compared to the commercial fibers (Li et al., 2007). In other work, a poly(p-phenylenediamine-co-aniline) composite SPME coating was prepared on the surface of a stainless steel wire *via* electrochemical deposition (Rong et al., 2012). Interestingly, adapting a novel deposition approach, SAMs of polyaminothiophenol (PATP) were used as a covalently bonded coating for SPME. Thiolated aniline-analog monomers (mixture of 2- and 3-aminothiophenols, 2/3-ATP) were anchored on the gold surface and then electropolymerized. The thiol-terminated coating on the gold surface was very stable due to the strong S–Au bond. The proposed covalently bonded coating showed higher mechanical (re-usability up to 100 times) and thermal stability (up to 320 °C) than a non-covalent bonded PANI coating which could be reused up to 20 times and its thermal stability was just below 250 °C (Mehdinia et al., 2011).

2.3.3.3.3 Polythiophene-based Coatings

PTh was firstly proposed as SPME fiber by Li et al. (2008). Thiophene is a heterocyclic aromatic monomer and due to its high oxidation potential is rather difficult to be electropolymerized (Gratzl et al., 1990). In order to resolve this issue, a solution containing boron trifluoride diethyl etherate (BFEE) was used, which could act as a strong Lewis acid while possessing good ionic conductivity and electrocatalytic properties. The complexation of aromatic monomers of thiophene with BFEE can significantly lower the anodic oxidation potentials of the monomers, which is beneficial to the preparation of high quality CPs (Shi et al., 1995). Thus, the electropolymerization of thiophene can be carried out in pure BFEE. There are limited reports in the field of application of polythiophene as a SPME coating. Olszowy et al. have applied PTh and PPy as sorbent phases for SPME in order to extract multi-resistant antibiotic drugs such as linezolid and daptomycin from whole blood sample followed by HPLC-UV detection. The PTh SPME coating was shown to have better adsorption capacity compared to a PPy coating for both drugs (Olszowy et al., 2010). In other work, the PPy and PTh coatings were electropolymerized to be used in SPME analysis after which the fibers were modified by an ozone treatment in a gaseous phase. Extraction efficiencies of both kinds of fibers were compared by the microextraction of linezolid from standard solutions. In these investigations a better adsorption capacity was obtained for PPy fibers (Olszowy et al., 2011).

2.3.4 Monolithic Polymers

2.3.4.1 Preparation of Monolithic Polymers

Organic monolithic polymers are porous materials that can be synthesized in a one-step polymerization procedure. The pre-polymerization mixture consists of the monomers, cross-linkers, porogenic solvents, and initiators. **Figs. 2.14** and **2.15** show some compounds usually employed as monomers and cross-linkers, respectively. A container is filled with the pre-polymerization mixture and the polymer can be synthesized *via* a free-radical mechanism. The prepared polymer forms the container shape by *in situ* polymerization (Bagheri et al., 2013c). Polymerization can be initiated by heating, UV radiation or gamma-ray radiation in the presence of initiators. After reaction for the prescribed time at an appropriate temperature, the resulting material is washed with organic solvents to remove the unreacted components and porogenic solvents. Monolithic polymers are highly porous materials with a network of interconnected pores and have been utilized in many applications in separation science such as chromatography stationary phases and also as the extracting phase in solid phase extraction (SPE) and SPME. They are generally highly biocompatible and pH-stable, making them suitable for biological samples (Svec, 2010). However, in most cases, they suffer from shrinkage or swelling when exposed to organic solvents, leading to poor mechanical stability.

To obtain a large surface area, a large number of small pores should be incorporated into the polymeric network. A highly substantial contribution to the overall surface area originates from the micropores, with sizes smaller than 2 nm, followed by the mesopores which range from 2 to 50 nm (Svec, 2010). Influencing variables such as temperature, composition of the porogenic solvent and content of cross-linker have major roles in the tuning of the pore size over a broad range.

Temperature: The polymerization temperature, *via* its effects on the kinetics of polymerization, is particularly effective in controlling pore size distributions. The effect of temperature can be explained in terms of the nucleation rates and the shift in pore size distribution induced by changes in temperature (Svec & Frechet, 1995; Svec, 2010). The temperature mostly affects the specific surface area which is typically related to the small pores. Since the monomers are better solvating agents thermodynamically than the porogenic solvent, the precipitated insoluble gel-like nuclei swell with the available monomers in the surrounding liquid. Following the nucleation step, the polymerization continues both within the separated phase of monomer

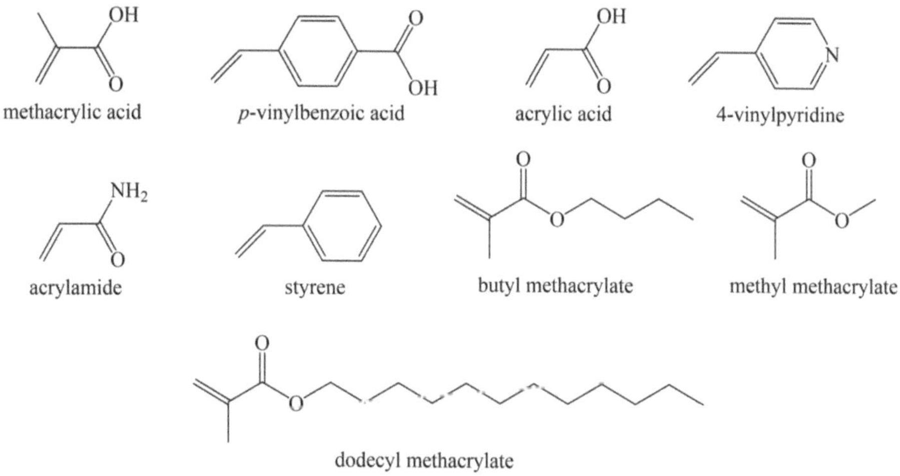

Figure 2.14 Some monomers used for monolithic polymer preparation.

swollen nuclei and in the remaining liquid polymerization mixture. At low temperatures, the reaction rate is rather slow and transfer of a substantial part of monomers from solution to the nuclei can occur. Polymerization within the nuclei is kinetically preferred because the local concentration of the monomers is higher than in the surrounding solution. By increasing the temperature, the number of polymer molecules that are formed in the solution after the original nucleation increases. These are captured by the growing nuclei and form larger clusters with less individualized texture and a reduction in the surface area.

Cross-linker: Changing the monomer to cross-linker ratio not only induces the formation of different porous structures but also leads to materials with different compositions. A higher content of cross-linker causes the formation of more cross-linked polymers in the early stages of the polymerization process and therefore earlier phase separation can occur. The nuclei are more cross-linked and since this crosslinking has inverse effects on their swelling with the monomers, their sizes remain relatively small. Since the final structure consists of smaller globules, it also has smaller voids. Thus, this approach is useful for the preparation of monoliths with very large surface areas (Santora et al., 2001). Overall, the pore size distribution is controlled by the swelling of cross-linked nuclei.

Porogenic solvent: The choice of porogenic solvent is another option that may be used to control porous properties without changing the chemical composition of the final polymer. In general, larger pores are obtained using a weaker solvent due to an earlier onset of phase separation. The porogenic solvent controls the porous properties of the monolith through the solvation of the polymer chains in the reaction medium during the early stages of the polymerization. The choice of porogenic solvent for the preparation of porous polymer monoliths mostly relies on the exper-

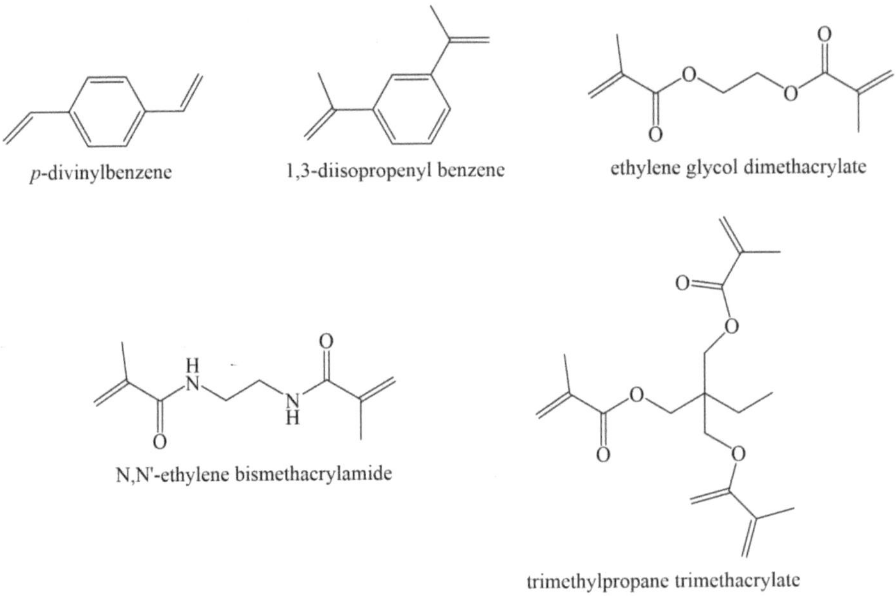

Figure 2.15 Some cross-linkers used for monolithic polymer preparation.

tise of researchers. That might be the reason why few porogenic solvents have been used and most often proven porogen mixtures are applied.

Time: In the polymerization process, relatively low duration times result in low yields for polymer reactions and most of the monomers remain non-polymerized. This initial monolith has a high pore volume and surface area. After a quantitative conversion of monomers to polymer at extended time, the surface area and pore volume are decreased.

2.3.4.2 Monoliths for Solid-phase Microextraction

Monolithic polymers have been used in SPME under two categories: fiber and in-tube formats.

2.3.4.2.1 Fiber Format

A monolithic fiber was constructed from methacrylic acid-ethylene glycol dimethacrylate (MAA-EDMA) for selective extraction of diacetylmorphine and analogous compounds (Djozan & Baheri, 2007). In another work, a monolithic fiber-based MIP-SPME was developed for selective and sensitive determination of ephedrine and pseudoephedrine (Deng et al., 2012). Polymerization of MAA, EDMA and ephedrine as a template was performed *in situ* in a silica capillary mold. In another publication, a MIP-SPME fiber was synthesized for the extraction of acetaldehyde from the headspace of beverages stored in poly(ethylene terephthalate) bottles (Rajabi Khorrami

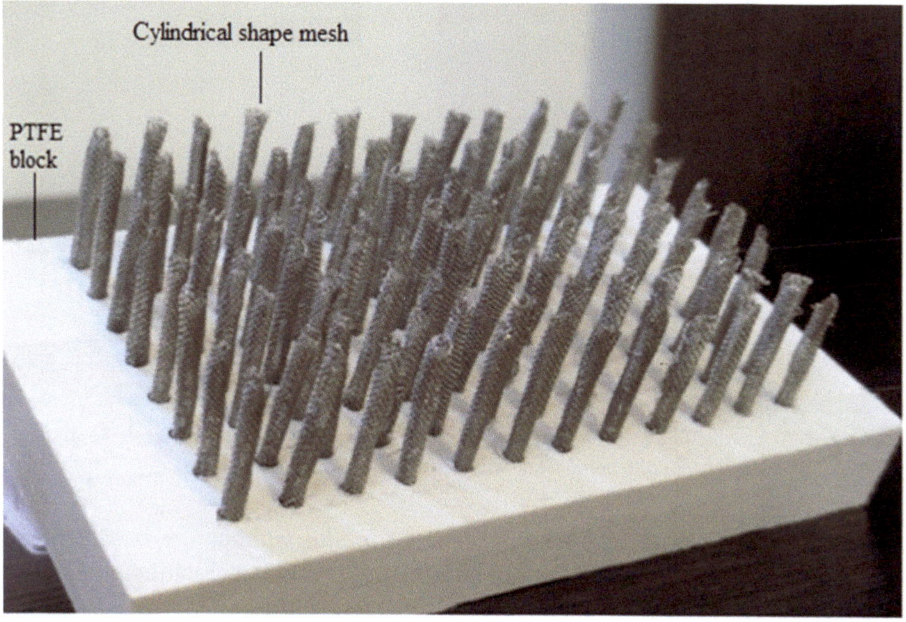

Figure 2.16 Photo describing the homemade 96 well μ-SPE device. Reprinted from Bagheri et al. (2013c) with permission from Elsevier.

& Narouenezhad, 2011). Djozan et al. reported a simple polymerization strategy to produce a monolithic SPME fiber on the basis of MIP for selective extraction of triazines from water, rice and onion samples (Djozan & Ebrahimi, 2008).

Recently, a novel high-throughput device based on a 96-micro-solid phase extraction (μ-SPE) system with monolithic sorbent was reported (Bagheri et al., 2013c). The extraction procedure was performed on a commercially available 96-well plate system. The extraction module consisted of 96 pieces of 1×3 cm of cylindrically shaped stainless steel meshes. The prepared meshes were fixed in a homemade PTFE-based block with 96 holes for possible simultaneous immersion of meshes into the center of individual wells (**Figure 2.16**). Dodecyl methacrylate and EDMA were copolymerized as a monolithic polymer with pyramid shape (**Figure 2.17**). These sorbents were placed in the cylindrically shaped stainless steel meshes. The method was successfully developed for the extraction and determination of the some selected pesticides in water samples.

2.3.4.2.2 In-tube Solid-phase Microextraction

Because of fragility, poor mechanical stability of monolithic polymers and low extraction capacity, in-tube format of SPME has gained more attention. This format increases the range of available coatings and conveniently integrates the extraction (on-line or off-line) with separation and detection systems. This setup in particular is quite promising when SPME is coupled with HPLC.

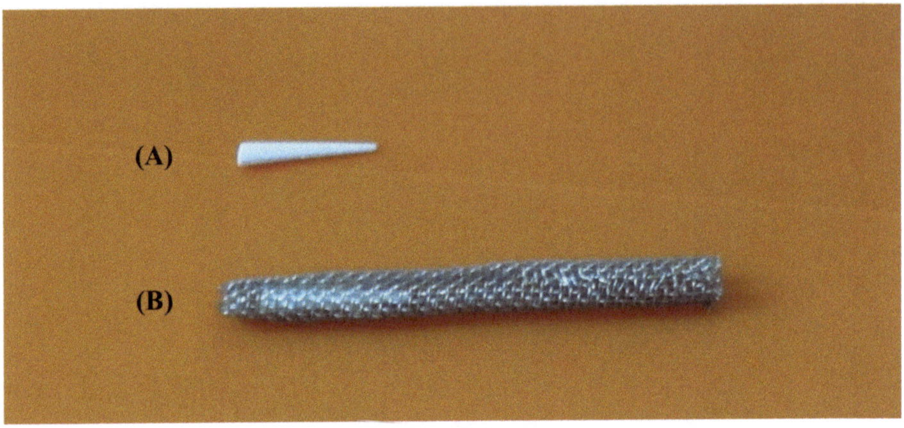

Figure 2.17 (A) A monolithic polymer with pyramid shape, (B) A cylindrical shaped mesh. Reprinted from Bagheri et al. (2013c) with permission from Elsevier.

Monolithic materials are always prepared *in situ* without the need for frits (**Figure 2.18**). Due to their satisfactory phase ratios and bimodal nature, monolithic materials result in convective mass transfer rather than diffusive mass transfer during the extraction process, which is a more desirable feature for high speed extraction.

In the early works on organic monolith, Fan et al. (2004) reported the use of a poly MAA-EDMA monolithic capillary as an in-tube microextraction. Because of polymer biocompatibility, only simple sample pre-treatment before extraction was required. Since then, monolithic polymers with different functional groups have been constructed for a variety of SPME applications (Fan et al., 2004; Zhang et al., 2006; Zheng et al., 2009, 2010; Chen et al., 2012; Wang et al., 2013; Zhang & Chen, 2013).

2.3.5 Composites

Composite materials are solids resulting from the combination of two or more simple materials that contain a continuous phase (polymer, metal, ceramic, etc.), and a dispersed phase such as glass fibers, carbon particles, silica powder, clay minerals and other relevant materials. In addition, composite materials have properties that are quite different from the components taken separately. Within the vast collection of inorganic-organic hybrid materials, composites are an emerging group that received a great deal of attention not only because of their potential in industrial applications but also from their scientific point of view (Ahlrichs et al., 1975; Akelah et al., 1994). Important advantages of composites over their constituent compounds rely on their high specific stiffness, strength, toughness, corrosion resistance, low density and thermal insulation. In most composite materials, one phase is usually continuous and called the matrix, while the other phase called the dispersed phase. On the basis of the nature of the matrices, composites can be classified into four major categories:

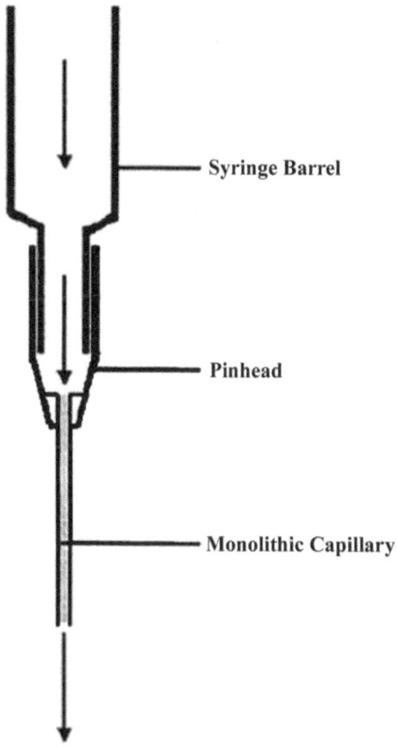

Figure 2.18 Scheme of the in-tube monolith microextraction. Reprinted from Zhang et al. (2006) with permission from Elsevier.

1. Polymer matrix composites
2. Metal matrix composites
3. Ceramic matrix composites
4. Carbon matrix composites.

Polymer matrix composites can be processed at a much lower temperature in compared with other composites. Depending on the types of polymer, polymer matrix composites are classified as thermosetting composites and thermoplastic composites.

2.3.5.1 Polymer Matrix Composites

To improve the properties of composite-based materials, the investigation of composites have been performed with lower filler sizes which led to the development of microcomposites and nanocomposites. Nanocomposites refer to composites in which one phase has nanoscale morphology such as nanoparticles, nanotubes or lamellar nanostructure (Friedrich et al., 2005). The improvement of the properties by the addition of particles can be achieved when (i) adequately good interactions between the nanoparticles and the matrix exist and (ii) the doped particles (nanoparticles) are well dispersed within the matrix.

In nanocomposites, covalent bonds, ionic bonds, van der Waals forces and hydrogen bonding interctions may exist between the matrix and filler components. Two types of the most common nanofillers including nanoparticles and nanoclays are under active investigation. The main purpose is to have a valuable nanocomposite with the largest possible surface area of nanofillers. In practice, production aims to avoid aggregation of nanoparticles and exfoliation of nanoclays. Nanoparticles are commercially available from different sources. Sols of nanosilica as colloid solutions in water or in organic solvents are used in the preparation of nanocomposites. Fused silica is available as individual particles ranging from 10–20 nm to micrometers, and can be more or less successfully dispersed in a polymer (Zhou et al., 2002). Layered aluminosilicate clays and especially montmorillonite (bentonite) are widely used in nanocomposites. Silicates have a characteristic interlayer distance of 1 nm; the basal spacing of a gallery is also ca. 1 nm. Inorganic cations like Na^+ between galleries hold negatively charged galleries together. The replacement of the inorganic cations in the galleries of the native clay by alkylammonium (onium) salts or quarternary amines with long alkyl substituents (surfactants) leads to a better compatibility between the inorganic clay and hydrophobic polymer matrix. The replacement leads to an increase in the space between galleries which facilitates the intercalation of polymer molecules into the clay.

The preparative methods of nanocomposites are divided into three main groups based on the starting materials and processing techniques. The first group is based on the intercalation of polymer or pre-polymer from solution. Here, a solvent system is chosen in which the polymer or pre-polymer is soluble and the silicate layers can be swollen. The layered silicate is first swollen in a solvent, such as water, chloroform, or toluene. When the polymer and layered silicate solutions are mixed, the polymer chains intercalate and displace the solvent within the interlayer of the silicate. After solvent removal, the intercalated structure remains, leading to the desired nanocomposite. In this method, which is also named the *in situ* intercalative polymerization method, the layered silicate is swollen within the liquid monomer or a monomer solution so that polymer formation can occur between the intercalated sheets. Polymerization can be initiated either by heat or radiation, by the diffusion of a suitable initiator or by an organic initiator or catalyst fixed through cation exchange inside the interlayer before the swelling step.

The second technique is called the melt intercalation method. This method involves annealing a mixture of the polymer and silicate above the softening point of the polymer. This method has great advantages over either *in situ* intercalative polymerization or polymer solution intercalation. First, this method is environmentally friendly as no organic solvent is needed. Second, it is compatible with current industrial processes such as extrusion and injection molding. The melt intercalation method allows the use of polymers which were previously not suitable for *in situ* polymerization or solution intercalation. In addition, polymer nanocomposites can commonly be obtained by the sol-gel method. Sol-gel nanocomposites (polymer/silica nanocomposite) are prepared by *in situ* hydrolysis and condensation of mononuclear precursors such as tetraethoxysilane (TEOS) and tetramethoxysilane (TMOS) in organic polymer matrices (Lan et al., 1995; Tong et al., 2002). Recently, the intercalation method for synthesizing polymer/clay nanocomposites has received more attention where polymer chains may penetrate into the host layers while ordered silicate registry remains (intercalated structure), or the exfoliated individual silicate layers (about 1 nm thickness) are homogeneously dispersed in the organic polymer matrix (delaminated structure) (Krishnamoorti et al., 1996). Since the successful synthesis of a nylon 6/clay nanocomposite, many polymer/clay nanocomposites have been reported (Kawasumi et al., 1999). Research indicates that, compared to the intercalated nanocomposites, the exfoliated nanocomposites have higher Young's modulus, a larger increase in elongation at the breaking point and better thermal stability and the extent of exfoliation strongly influences the improvement of the properties. Natural silicates have strong interactions between the layers which are due to negative charges and hydrogen bonding in their structures. The basal space of pristine silicate is about 1 nm, which is smaller than the radius of gyration of general polymers. This might present an obstacle for polymers to penetrate into or delaminate between the silicate layers. Thus, most hydrophobic polymers are limited in terms of penetration depth into layered regions of hydrophilic silicates. So far, the preparation of polymer/silicate nanocomposites has been mostly focused on the use of poly(ethyl acrylate)/bentonite nanocomposites.

2.3.5.2 Nanocomposite-based Coatings

Recently, advancements in nano-oriented science and the synthesis of nanomaterials, nanoparticles and nanocomposites have drawn the attention of researchers in different fields. Nanomaterials, specifically due to their high surface area and more efficient availability of the adsorption sites, have attracted more attention for use as the extracting phase for sample preparation purposes and among them, nanocomposites are quite dominant. In this regard, some reports concerning the preparation and application of polymer-based nanocomposites as SPME fiber coatings have been appeared in the literature. For instance, a novel nano-structured polyaniline-ionic liquid (1-butyl-3-methylimidazolium hexafluorophosphate, $BMIPF_6$) composite (BPANI) film coated on steel wire was prepared by electrochemical deposition (Gao et al., 2011). Scanning electron microscopy (SEM) images showed that the obtained porous BPANI coating consisted of nanofibers, with diameters ranged from 50 to 80 nm. The BPANI coating showed better analytical performance toward organochlorine pesticides (OCPs) in compared with conventional PANI and PDMS coatings. Furthermore, the novel nano-structured composite coating was very stable at relatively high temperatures (up to 350 °C) and it could be used more than 250 times without obvious reduction in the extraction efficiency. It was reported that the polymerization of aniline in the presence of $BMIPF_6$ can enhance the thermal stability of the PANI film (Gao et al., 2004) and the enhanced temperature resistance may be due to the introduction of fluorine into the polymeric network (Mecerreyes et al., 2002). Later on, a novel nano-structured copolymer of aniline and *m*-aminobenzoic acid (*m*-ABA), with a diameter lower than 75 nm, was introduced as a SPME coating for isolation of saturated fatty acids in zooplanktons (Mehdinia et al., 2010). Improved temperature resistance (up to 350 °C), relatively improved lifetime (more than 50 extractions) and satisfactory extraction efficiencies were obtained by incorporating carboxylate groups into the framework of PANI. In another study, the effects of structure and morphology of the SPME coating on the equilibrium extraction time was investigated. PANI was used as an extraction phase for the comparison of extraction capacity and equilibrium extraction time of nano- and micro-structured coatings. PCBs were used as model compounds to examine the extraction properties of nano- and micro-structured coatings. The results revealed that the nano-structured PANI coating, due to its larger surface area, exhibited a higher extraction rate and shorter desorption time in comparison with the micro-structured coating (Mehdinia & Mousavi, 2008). In another study, pristine and oxidized multi-walled carbon nanotubes–polypyrrole composite (MWCNTs–PPy) deposited electrochemically on a stainless steel wire (Asadollahzadeh et al., 2010; Chen et al., 2011) and MWCNTs–PANI film coated platinum wire (Du et al., 2009) were used for the extraction of pyrethroids, phthalate esters and phenolic compounds from aqueous samples. It was reported that the electrodeposited coating presents a porous structure with high specific surface area and adsorption capacity and thus higher extraction efficiency for the analytes. This might be due to the fact that both MWCNTs and CPs contribute to the extraction. Moreover, a novel PANI and polyamide nanocomposite based on nanosilica was synthesized and was used as SPME coating. In this study nanosilica was used as a reinforcing agent to improve the extraction capability of PANI. Evaluation of PANI polymer and aniline–silica nanocomposite fibers by SEM has shown that the synthesized nanocomposite coating possesses a non-smooth, porous structure in comparison with the PANI film. In addition, the use of the aniline–silica nanocomposite led to higher extraction results in comparison with a PANI fiber (Bagheri & Roostaie, 2012). A PPy/graphene (G) composite was prepared and applied as a novel SPME coating for determination of phenols. The PPy/G-coated fiber was synthesized by electropolymerization of pyrrole and G on a stainless-steel wire. The extraction efficiency of PPy/G-coated fiber for phenols was the highest in comparison with either PPy or PPy/graphene oxide (GO) fibers. Extraction capability of a novel PPy/G-coated fiber was evaluated by comparison with the extraction efficiency of 85 μm CAR/PDMS fiber and a 85 mm PA fiber and the results showed that the extraction efficiency of PPy/G coated fiber is better or comparable compared to the tested commercial fibers (Zou et al., 2011).

2.3.6 Electrospun Nanofibers

Today, nanomaterials are considered to be a well-established and interesting class of materials due to their unique physical properties and small sizes. These properties could improve and bring new features which are currently difficult to achieve with bulk materials (Gómez-Hens et al., 2008). Recently some research groups tried to prepare nanomaterial-based SPME fibers (Mehdinia et al., 2006, 2011; Cao et al., 2008; Rastkari et al., 2010; Zewe et al., 2010). An essential and rather important characteristic of sorbents is its surface area-to-volume ratio, with materials having long lengths and nano- to microscale diameters possessing high surface area-to-volume ratio (Huang et al., 2003). These nanomaterials can be regarded as good candidates to be used as extracting phases in analytical and environmental applications. Electrospinning is a simple and convenient method for producing nanofibers with adjustable diameters, polarities and porosities. The nanofibers have malleability to conform to a wide variety of sizes and shapes. It is possible to control the nanofiber composition to achieve the desired properties and functionality (Bhardwaj & Kundu, 2010). A variety of polymers have been used for producing nanofibers (Huang et al., 2003). The electrospun nanofibers have found a great deal of applications in medicine, enzyme immobilization, batteries, fuel cells, capacitors, filter membranes, drug delivery, tissue engineering, sensors and protective cloth for warfare agents (Ramakrishna et al., 2005). This methodology has been used in SPE (Kang et al., 2007; Xu et al., 2010), semi-micro-SPE (Chigome et al., 2010), affinity membranes (Yoshimatsu et al., 2008) and more recently as a SPME fiber (Zewe et al., 2010). The electrospinning process operates based on similar principle as electrospray ionization mass spectrometry (ESI-MS) (Rutledge & Fridrikh, 2007). In this approach, a fiber jet is produced rather than the charged droplets mist in ESI-MS. In the electrospinning process, a solution of a high molecular weight polymer with high viscosity is used. When the electrostatic repulsion overcomes the surface tension, the spherical droplets deform to Taylor conical form. By increasing the voltage across the polymer-containing syringe and the collector, the jet of the polymer ejects from the Taylor cone. The entanglement of polymer chains and the surface tension prevent the formation of polymeric droplets. As the polymer jet is flying towards the collector electrode, its diameter gets smaller. At the end of the process, a non-woven nanofiber mat is collected on the surface of the conductive collector, possessing nano- to microscale diameters. Miscellaneous materials with different shapes can be used as collector electrodes (Teo & Ramakrishna, 2006). In this regard, some reports concerning the preparation and application of polyurethane- and polyamide-based nanofibers as SPME fiber coatings have appeared in the literature. For instance, a novel polyamide nanofiber coated on a steel wire was prepared by the electrospinning technique (Bagheri & Aghakhani, 2011, 2012). SEM images showed that the obtained porous polyamide coating consisted of nanofibers, with diameters ranging from 150 to 600 nm. It was found that the electrospun coating presents a porous structure with high

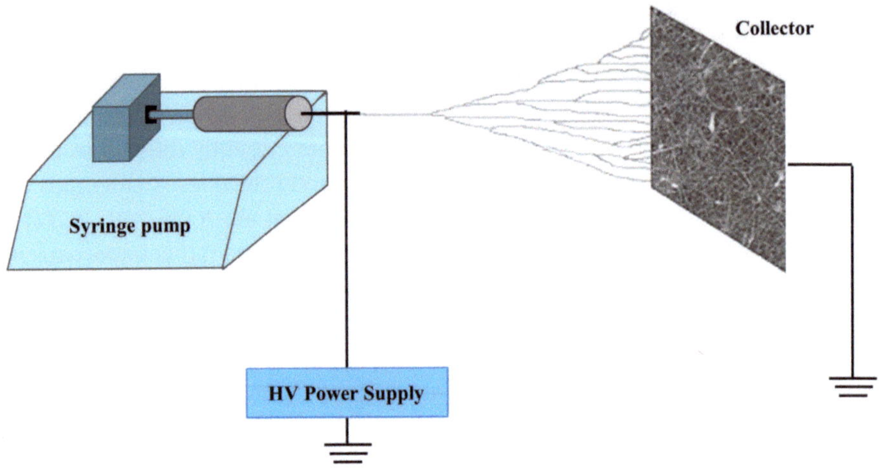

Figure 2.19 Scheme of the electrospinning setup. Reprinted from Bagheri & Aghakani (2012) with permission from Elsevier.

specific surface area and adsorption capacity, and thus high extraction efficiency for the analytes (**Figure 2.19**).

2.3.7 Selective Sorbents

Preparation of selective sorbents is achievable through MIPs. The principle of molecular imprinting was inspired by Fischer's lock-and-key metaphor. In the first step, the selected key molecule is mixed with lock building blocks. The building blocks and the key are allowed to associate with each other, firmly or loosely. The complexes that are formed between the key and the building blocks are subsequently glued together in order to fix the building block positions around the key. Removing the molecular key then leaves a construction which, if everything works properly, is selective for the original key. In molecular imprinting, molecules are used to create the marks or imprints, normally within a network polymer. In molecular imprinting, the key molecules described above can be denoted by a variety of expressions (*e.g.*, templates, template molecules, target molecules, analytes, imprint molecules, imprint antigens or print molecules), any of which is frequently encountered. The lock building blocks are normally called functional monomers, although polymers have also been used as imprinting building blocks. The molecular glue used to fix the key-building block complexes is almost always perceived as a cross-linker or a cross-linking monomer.

2.3.7.1 Molecularly Imprinted Polymers-based-solid-phase Microextraction

The combination of MIP and SPME technique provides a powerful sample preparation tool in terms of selectivity, simplicity, and flexibility. The first attempt to use MIPs in CME (in-tube SPME) was reported by Mullett et al. (2001). An automated, online MIP-SPME method was developed for the determination of propranolol in biological fluids and showed improved selectivity compared to in-tube stationary phase materials, thus overcoming the limitations of the existing SPME coating materials. Preconcentration of the sample by the MIP adsorbent increased the sensitivity, yielding low limits of detection (LODs). Koster et al. (2001) reported the first work dealing with the use of MIP coatings on SPME fibers. A silica SPME fiber was silanized, followed by *in situ* synthesis of the MIP coating on the external surface of the fiber. They prepared the MIP-based SPME fiber using clenbuterol as a template and demonstrated the possibility of selective extraction of brombuterol. This fiber was brittle and the MIP coating stripped during withdrawal of the fiber in the needle. The application of monolithic MIPs with SPME followed by GC was also reported (Djozan & Baheri, 2007; Djozan & Ebrahimi, 2008). The MIP was prepared through thermal radical copolymerization of MAA and EDMA in the presence of diacetylmorphine (DAMO) as template, and acetonitrile (ACN) and azobis(isobutyronitrile) (AIBN) were also used as a solvent and an initiator, respectively.

2.3.7.2 Molecularly Imprinted Xerogels-based-solid-phase Microextraction

Despite many advantages, MIPs still need development to overcome limitations such as template leakage, poor accessibility of the binding sites, low binding capacity and non-specific binding (Ansell et al., 1996). The applications of imprinted polymers are also restricted to the use of organic solvents for the dissolution of acrylate or acrylic type polymers which are most commonly used for the preparation of MIPs (Haupt, 2003; Gupta & Kumar, 2008). To overcome these problems, preparation of MIX by sol-gel technology has been examined by many researchers (Bagheri & Piri-Moghadam, 2012; Bagheri et al., 2013d).

Sol-gel imprinting is an emerging field, due to its straightforward synthesis path. Silica-based materials are extremely rigid because of the high degrees of cross-linking in their network. This property is very important and influential in designing and synthesizing the imprinted materials since both the size and shape of the cavities created by the template molecule should be retained after the template removal. High thermal stability of sol-gel derived materials and the possibility of using high temperatures assist the removal process. In addition, sol-gel glasses are structurally porous and can be engineered to have extremely high surface areas. These properties make silica sol-gel matrices an appropriate imprinting host. **Figure 2.20** shows a typical process for the imprinting process (Bagheri & Piri-Moghadam, 2012).

In one study, a simple route for preparation of a novel selective MIX by sol-gel technology was developed. Accordingly, 3-(trimethoxysilyl)propylmethacrylate

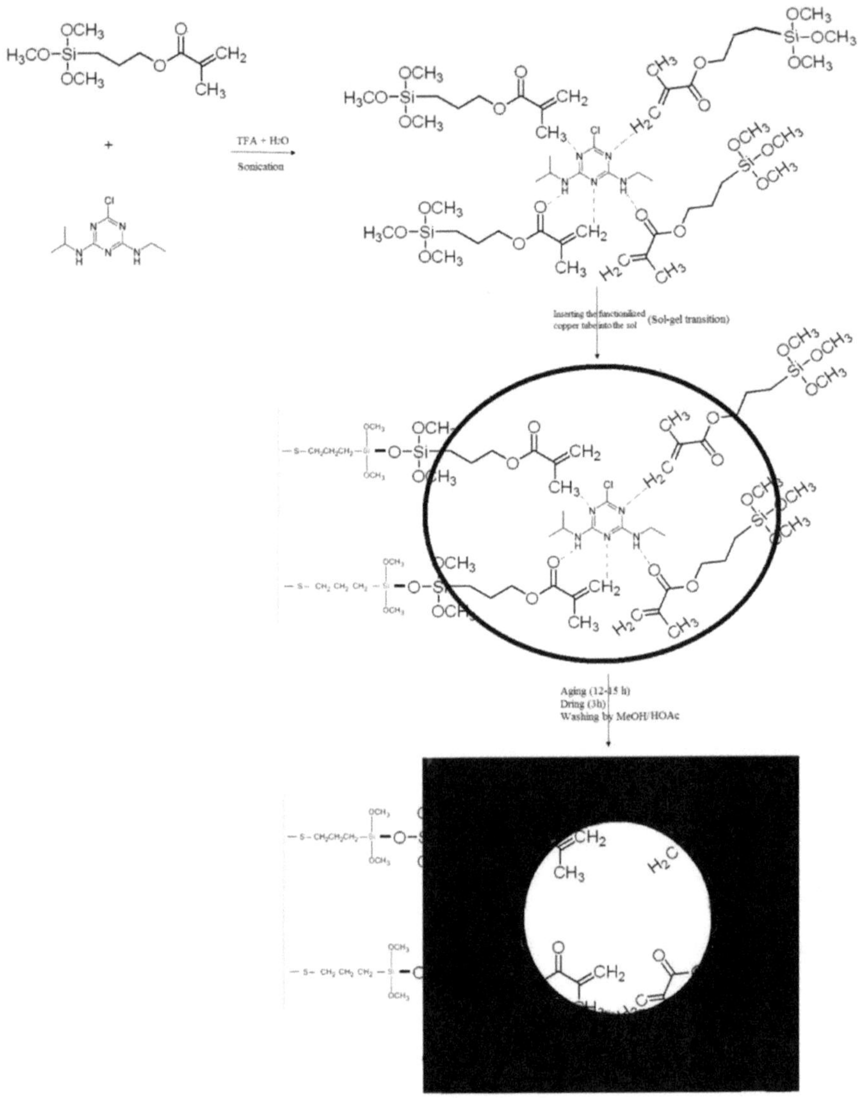

Figure 2.20 A typical process for preparation of MIX of atrazine.

(TMSPMA) was selected as only reagent in the sol-gel process and led to the preparation of a selective sorbent. In TMSPMA, the methacrylate group acts as the functional monomer and the methoxysilane acts as the cross-linker. Atrazine was selected as a model compound for evaluation of the developed method. The prepared MIX showed a high degree of thermal, chemical and mechanical stability along with extensive selectivity towards triazines while a significant reduction in the preparation time was also achieved.

2.3.8 Metal-based Coatings

Because of some shortcomings of commercialized fibers such as breakage of the fused silica fibers, stripping of the coating and damage in organic solvents, many attempts have been made to construct and evaluate fibers with more stability.

2.3.8.1 Metal-based Fibers Preparation by Anodization

It is found by Djozan and co-workers that metal-oxide fiber can be efficient for extracting some organic compounds (Djozan et al., 2001; Djozan & Abdollahi, 2003). Aluminum wires were anodized by direct current in a solution of sulfuric acid at room temperature. These fibers were used for the extraction of some aliphatic alcohols, benzene, toluene, ethylbenzene and xylenes (BTEX) and petroleum products from gaseous samples. The extraction ability is due to the porous layer of aluminum oxide, which is formed on the metal surfaces (Djozan et al., 2001).

In another work, well-aligned TiO_2 nanotubes were grown *in situ* on a titanium wire substrate by anodization of Ti wire substrates in an electrolyte containing ethylene glycol and NH_4F. The prepared fiber showed good ability to extract PAHs, anilines, phenols and alkanes from water samples (Liu et al., 2010).

2.3.8.2 Metal-based Fibers Developed by Physical Coating

Different sorbents can be coated onto metal wires and blades using appropriate glues and then used as an SPME fiber. Rastkari et al. reported attachment of single-walled carbon nanotubes (SWCNTs) onto a stainless steel wire by an organic binder and its usage as SPME fiber (Rastkari et al., 2009). Graphene (Ponnusamy & Jen, 2011) and carbon nanocones (Jiménez-Soto et al., 2010) were also coated onto metal wires. Vuckovic et al. coated silica particles onto the stainless steel wires with a thin layer of Locktite 349 glue. A total of 96 of these fibers were constructed and used for high-throughput extraction of diazepam from blood samples (Vuckovic et al., 2008). In another work, C_{18} particles were immobilized on the surface of stainless steel blades using biocompatible polyacrylonitrile glue (Mirnaghi et al., 2011). The proposed C_{18}-polyacrylonitrile blade SPME system showed the potential to be employed for high-throughput analysis in a wide variety of research areas. In a recent work, 1.0 cm pieces of silicon tubing were precisely cut and then coated on the end part of stainless steel wires (**Figure 2.21**). The prepared fibers were positioned in a homemade PTFE-based constructed 96 hole block to have the possibility of simultaneous immersion of the SPME fibers into the center of individual wells. The constructed high-throughput device was successfully applied for the determination of some selected pesticides in cucumber samples.

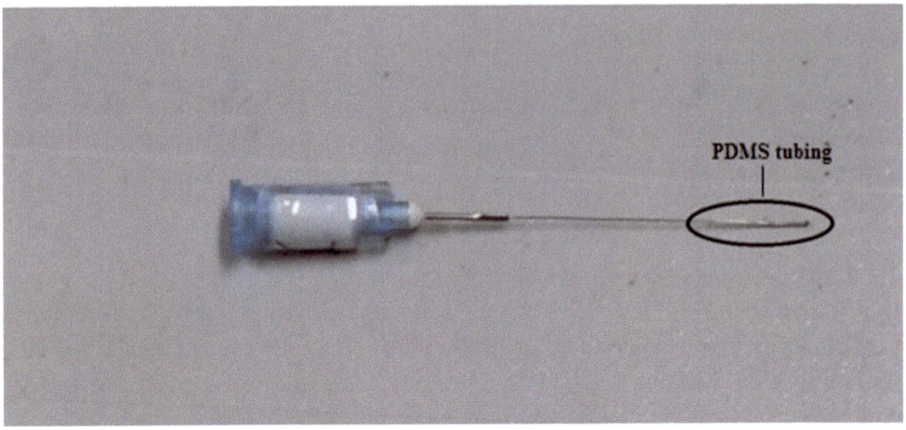

Figure 2.21 1.0 cm of silicon tubing coated on the end of stainless steel wires.

2.3.8.3 Metal-based Fibers Developed by Chemical Coating

Performing the physical coating is easy, but there are some shortcomings like decomposition of glues, fiber swelling and instability in some organic solvents because of the absence of chemical bonding between the sorbent and metal wire. To enhance the coating stability and fiber durability, some efforts were carried out to perform chemical bonding between metal substrate and coating.

Es-haghi et al. (2012) replaced a fused silica substrate by titanium wire which provided high strength and longer fiber life cycle. Titanium isopropoxide was employed as the precursor which provides a sol solution containing Ti-OH groups which are similar to the functional groups on the substrate. The similar composition of the sol-gel solution provided more links to the titanium substrate, so the stability and the uniformity of the coating could be greatly improved. The applicability of these fibers was assessed for the HS-SPME of BTEX.

In another work, a gold wire was used as a solid support onto which a film was deposited that consists of a two-dimensional polymer obtained by hydrolysis of a SAM of 3-(trimethoxysilyl)-1-propanthiol (Bagheri et al., 2011b). This first film was covered with a layer of 3-(triethoxysilyl)-propylamine. Next, a stationary phase of oxidized MWCNTs was chemically bound to the surface. The synthetic strategy was verified by Fourier transform infrared spectroscopy and SEM. The fiber was used for HS-SPME of organophosphorus pesticides in water samples.

Pang and Liu reported a novel approach for the fabrication of a SPME fiber by coating a stainless steel fiber with a polymeric ionic liquid (PIL) through covalent bonding (Pang & Liu, 2012). The stainless steel fiber was sequentially coated with a gold film by a replacement reaction between Fe and Au when immersed in chloroauric acid. A monolayer of 3-(mercaptopropyl) triethoxysilane on the gold layer through the Au–S bond was assembled. Eventually, the coating was accomplished with a silica layer by the hydrolysis and polycondensation reaction of the surface-bonded siloxane moieties and the active silicate solution. Then, 1-vinyl-3-(3-triethoxysilylpropyl)-

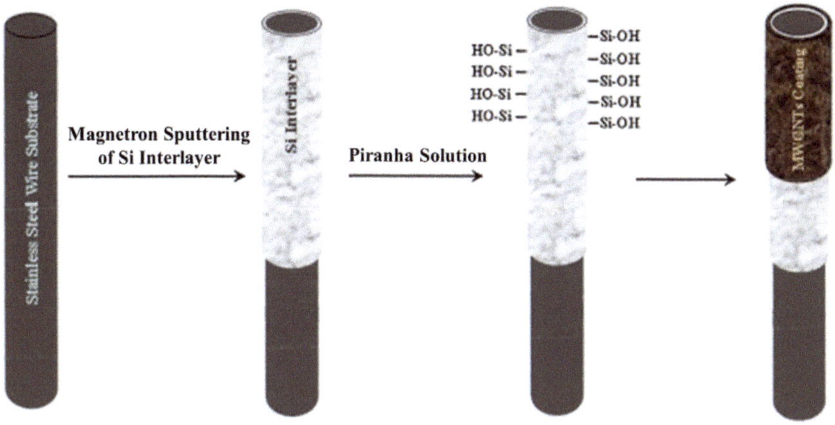

Figure 2.22 Schematic illustration of the preparation of SPME fiber. Reprinted from Liu et al. (2011) with permission from Elsevier. Firstly, a firm Si interlayer was deposited on the stainless steel wire using magnetron sputtering. Then the stainless steel wire with Si interlayer was treated with piranha solution to form –OH group on the surface. Finally, multilayer-MWCNTs coating was formed on the stainless steel wire by the covalent bond between the active groups on both sides of the substrate and MWCNTs, and the followed van der Waals force-induced spontaneous adsorption of MWCNTs.

4,5-dihydroimidazolium chloride ionic liquid was anchored on the silica layer by covalent bonding, and the PIL film was further formed by free radical copolymerization between 1-vinyl-3-(3-triethoxysilylpropyl)-4,5-dihydroimidazolium and vinyl-substituted imidazolium with AIBN as initiator. The performance of the PIL fiber was evaluated by determination of PAHs in water samples. The developed PIL fiber showed good linearity between 0.5 and 20 μg L^{-1} with regression coefficient in the range of 0.963–0.999, detection limit ranging from 0.05 to 0.25 μg L^{-1}, and relative standard deviation of 9.2–29% (n=7). The developed PIL fiber exhibited comparable analytical performance in compared with a commercial 7 μm thickness PDMS fiber in the extraction of PAHs.

In another study, the Si interlayer was prepared using a medium frequency unbalanced magnetron sputtering method in a multifunctional deposition system (Liu et al., 2011). A stainless steel wire was placed in a vacuum chamber and a base pressure of less than 4.0×10^{-3} Pa was obtained. Prior to the deposition, a 20 min sputter cleaning in argon plasma was firstly carried out so as to improve the adhesion between the film and substrate. Then, the Si deposition was performed using Ar gas as the sputtering gas, and Si as the target material. The sample holder was kept revolving during the deposition process to ensure the uniformity of the Si layer on the stainless steel wire. The prepared MWCNTs/Si/stainless steel wire fiber (**Figure 2.22**) not only preserved the excellent SPME behaviors of MWCNTs coatings, but also exhibited a number of advantages including high rigidity, long service life and good stability at high temperature and in acid and alkali solutions.

Li et al. (2009) reported the fabrication of a SPME fiber containing platinum coated with SWCNTs that was prepared by electrophoretic deposition and applied to the determination of phenols in aqueous samples. The results revealed that electrophoretic deposition was a simple and reproducible technique for the preparation of SPME fibers coated with SWCNTs without the use of adhesive. The obtained SWCNT coating did not swell in organic solvents or strip off the substrate and also possessed high mechanical strength.

2.4 Related Techniques

2.4.1 Microextraction in Packed Syringe

Microextraction in packed syringe (MEPS) is a miniaturized version of SPE in which sorbent amounts, sample volumes and desorption solvent volumes are extensively minimized. In this methodology, the tiny sorbent material is inserted either into the barrel of a liquid handling syringe as a plug with polyethylene filters on both sides or between the syringe barrel and the injection needle as a cartridge (Abdel-Rehim, 2010). On-line combination of MEPS with a LC (Abdel-Rehim et al., 2004; Altun et al., 2004) or a GC (Abdel-Rehim et al., 2006; El-Beqqali et al., 2006) can be achieved without any instrumental modification. The environmental or biological sample is drawn through the sorbent by an autosampler (draw and eject in same vial or draw and eject into waste). Similar to SPE, after the entrapment of analytes the solid phase is then washed once by water and/or an acidic solution to remove the proteins and all other possible interfering materials. The analytes are then eluted with an organic solvent such as methanol or the LC mobile phase. A key factor in MEPS is that the volume of solvent used to desorb the analytes from the extracting medium is of a suitable order of magnitude to be injected directly into an LC or GC system. For this purpose, GC systems have to be equipped with a programmed temperature vaporizing (PTV) system. When working with split–splitless injector systems, it is not possible to inject the whole desorption solvent into the injection port. Therefore, it is necessary to use MEPS in an off-line combination with GC systems.

MEPS has found more biologically-oriented applications (Abdel-Rehim, 2004, 2010; Altun et al., 2004) and reports concerning its use in environmental analysis can be found in the literature (El-Beqqali et al., 2006; Prieto et al., 2010; Bagheri, & Ayazi, 2011). The use of CPs has been implemented as the extracting phase in MEPS for environmental analysis (Bagheri et al., 2013a). For instance, PPy nanowire networks synthesized by a soft template method (Bagheri, & Ayazi, 2011) were used for MEPS to isolate triazines while PPy/PA composite nanofibers, prepared using electrospinning (Bagheri et al., 2012a), were applied as sorbents to extract organophosphorous pesticides from aqueous media. The PPy nanowire networks were prepared using a soft template technique and its characterization was studied by SEM. The use of micelles in this methodology played an important role in the shape of the growing polymer. The pyrrole monomer was introduced into cetyltrimethylammonium bromide (CTAB) micelles and this led to the formation of nanowires with diameters ranging from 30 to 60 nm. The bulk PPy prepared without CTAB contained spherical particles with diameters of 100–400 nm. The PPy nanowire network was shown to have 7–28 times higher extraction capability in compared with the bulk PPy. Moreover, a polyaniline nanowire network was also prepared by the same method. CTAB was used as structure-directing agent in PANI preparation procedure and this was led to the formation of nanowires with diameters ranging from 35 to 45 nm. The synthesized PANI nanowires network showed higher extraction capability for multiclass pesticides in comparison with the bulk PANI (Bagheri et al., 2012c). In other work, polydiphenylamine (PDPA) nanocomposite reinforced with MWCNTs was applied as a MEPS sorbent for multiresidue determination of pesticides in river waters. This nanocomposite was synthesized by oxidation of diphenylamine in 4 mol L^{-1} sulfuric acid solution containing a fixed amount of CNTs in the presence of CTAB. The effect of CNT doping level and the presence of surfactant on the extraction capability of nanocomposite was investigated and it was revealed that when 4% (w/w) of CNT in the presence of CTAB is used, the highest extraction recovery could be achieved (Bagheri et al., 2012b). A headspace adsorptive microextraction technique based on an electrospun polyaniline–nylon-6 (PANI–N6) nanofiber sheet was developed in which the nanofiber diameters were around 200 nm. The novel nanofiber sheet was successfully applied as an extracting medium to isolate some selected chlorobenzenes, as model compounds, from aqueous media (Bagheri, & Aghakhani, 2012).

For a typical application of MEPS (Bagheri, & Ayazi, 2011), 1 mL insulin injection syringes were used. 2 mg of PPy nanowires network was manually inserted inside the syringe between two polyethylene filters (SPE frits, 20 µm pore size). For this purpose the size of the SPE frits was changed to match the syringes used. Prior to the first use, the sorbent was manually conditioned with 2 mL of methanol followed by 4 mL of water. After that, the spiked sample (7 mL) was drawn through the syringe forward and backward several times using a variable speed cycling motor which was attached to a circular plate **(Figure 2.23)**. It is important that samples are drawn with sufficient speed to decrease the extraction time and to obtain good percolation between the sample and the solid support. In this study, the speed of the cycling motor was adjusted at 10 rpm (170 µL s^{-1}). After the extraction, the syringe was dried under a flow of nitrogen gas and the analytes were then desorbed with 200 mL of acetonitrile. The desorption step was performed manually by solvent aspiration into the syringe. Afterward, the desorbed analytes were transferred into a glass vial. Next, the desorbing solvent was evaporated under a nitrogen gas flow until solvent drying was complete. Finally, 10 mL of acetonitrile was added to the desorption vial and then 2 mL of desorbed solution was injected into the injection port of the GC system. After each

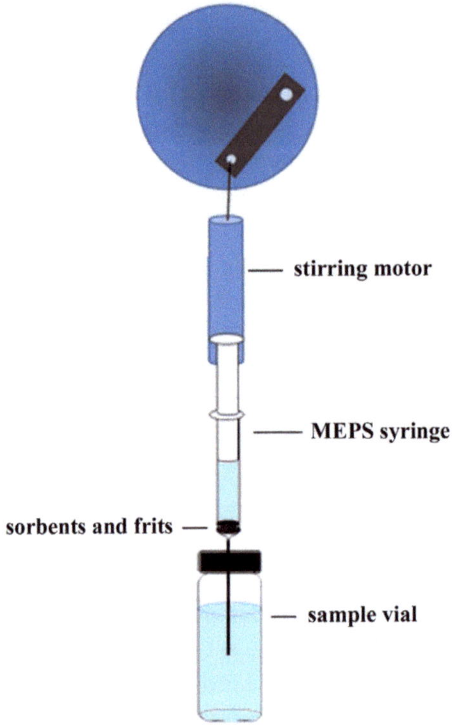

Figure 2.23 Schematic diagram for the extraction set up and MEPS syringe (Bagheri & Ayazi, 2011).

extraction the MEPS syringe was washed with approximately 200 mL of acetonitrile, 200 mL of methanol and 1 mL of water.

Preparation and characterization of PPy nanowires has also been extensively studied. Template synthesis is a chemical oxidative method which introduces structural directors into the chemical polymerization bath. These structural directors could contain soft templates such as surfactants, organic acids or polyelectrolytes that assist the self-assembly of nanostructures. When the soft-template method is considered, usually micelles or three-dimensional aggregations of micelles are the structure-directing agents. Depending on the specific conditions such as the composition of the surfactant, its concentration, ionic strength, hydrodynamics and temperature, the polymer grows inside or outside the micelles, thus replicating the morphology of the micelles. Most frequently spherical polymer particles are formed. But if the spherical micelles aggregate to form three-dimensionally organized cylindrical structures, then the conditions for the formation of rods can be overcome. In the cases of soft templates it is necessary to remove the corresponding surfactant after termination of the polymerization reaction. The mechanism of formation of the PPy nanowires network could be explained as follows: when pyrrole monomer is added into an aqueous solution containing CTAB and citric acid hydrate, it has been suggested that a three-dimensional network is formed from the assembly or aggregation of pyrrole,

Figure 2.24 SEM image of prepared (A) PPy nanowire network. T junctions are circled on the SEM image (scale bar, 500 nm), (B) bulk PPy (scale bar, 1.00 μm) (Bagheri & Ayazi, 2011).

CTAB, and citric acid before the oxidation polymerization of pyrrole occurs. Based on these results, it can be concluded that when CTAB, citric acid, and pyrrole are added to water, micelles would be formed because of the long alkyl chain of CTAB. Two micelles of CTAB might be connected by citric acid molecules to form one or more T junctions. When APS is added into the reaction solution, the three-dimensional networks of the soft templates can be elongated and form the three dimensional PPy nanowires network. Also, CI acts as a dopant for PPy in this procedure and it is not a prerequisite for wire formation, but the ability of the micelles to link together to form a micelle network is improved when the PPy is formed after addition of the oxidant APS. The surface characteristics of the prepared polypyrrole nanowires were investigated by SEM. According to **Figure 2.24A**, the formation of nanowires is confirmed and all of them are packed in rough porous structures. Also, it is shown that the nanowire diameters are in the range of 30–60 nm and there are some observable T junctions as discussed above.

The SEM image of bulk PPy (CTAB was not used in preparation of this polymer) is illustrated in **Figure 2.24B**. Clearly, the PPy prepared without CTAB are all spherical particles with diameters of 100–400 nm, indicating that these particles are linked and packed together.

Analysis of the extraction capability of the prepared PPy nanowires network was also carried out. Features such as high surface area and π-functional groups of the CP, together with polar functional groups of CI, are important characteristics of the fabricated PPy nanowire network which make it a suitable candidate for extraction purposes. To investigate the extraction capability of the prepared sorbent, the MEPS syringes were prepared using the PPy nanowires and the bulk PPy. The results revealed that the PPy nanowires network exhibits 7 to 28 times higher extraction capability in comparison with that of the bulk PPy. This enhancement in extraction capability is probably due to the increase in surface area and hydrophilicity of the PPy nanowires network. To investigate the extraction capability of the prepared PPy nanowires, the MEPS syringes were prepared using two different kinds of PPy.

These examples demonstrate that MEPS is suitable for analysis of pollutants and pesticides in water samples as well as drugs in biofluids. The matrix effect is rather insignificant as far as the water sample analysis is concerned while this strategy has already proven to be quite suitable for biofluids. Furthermore, employing such a miniaturized extraction scale meets the demands of reduced solvent consumption and waste production.

2.4.2 Stir Bar Sorptive Extraction

Stir bar sorptive extraction (SBSE) was developed in 1999 by Sandra and co-workers (Baltussen et al., 1999) and commercialized under the name "Twister". SBSE is able to extract and preconcentrate the desired compounds from liquid matrices without the use of solvents (David et al., 2003a, 2003b). SBSE has been applied successfully to trace analysis of environmental samples and has satisfactory analytical reproducibility for the determination of volatile and semi-volatile components of biological mixtures. In SBSE, organic analytes are enriched from aqueous samples by sorption onto a thick film of PDMS coated onto a glass-enveloped magnet (Mitra, 2003; Sánchez-Rojas et al., 2009). The sample extraction takes place during stirring for a fixed time. The spin bar is then removed and placed in a glass tube, which is transferred into a thermal desorption system where the analytes are thermally recovered and analyzed on-line either by GC or LC. Similar to other miniaturized sorptive extraction systems, SBSE is an equilibrium-based technique and the analyte extraction from the aqueous phase into the extracting phase is controlled by the partitioning coefficient of the analyte between PDMS and the aqueous phase. Detailed studies have correlated this partition coefficient with the octanol–water distribution coefficient (K_{OW}).

Using PDMS as the sorbent, the primary mechanism of interaction with organic analytes is *via* absorption or partitioning into the PDMS coating such that the distribution constant (equation 2.6) between PDMS and water ($K_{PDMS/W}$) is proposed to be proportional to the octanol–water partition coefficient (K_{OW}) (Baltussen et al., 1999):

$$K_D = \frac{[X]_B}{[X]_A} = K_{PDMSW} \approx K_{OW} \tag{2.6}$$

According to the theoretical development for this technique (Baltussen et al., 1999):

$$K_{PDMSW} \approx K_{OW} = \frac{[X]_{PDMS}}{[X]_W} = \frac{m_{PDMS}}{m_W} x \frac{V_W}{V_{PDMS}} \tag{2.7}$$

where $[X]_{PDMS}$ and $[X]_W$ and m_{PDMS} and m_W are the analyte concentration and the analyte mass in the PDMS and water phase, respectively, while V_{PDMS} and V_W represent the volume of the PDMS sorbent and water phase, respectively. Therefore, the parameters determining the mass of an analyte recovered by SBSE using the PDMS sorbent are the partition coefficient of the analyte (K_{OW}) and the phase ratio (V_W/V_{PDMS}) of the volume of the water phase to the volume of the PDMS coating on the stir bar.

In general, SBSE is considered to be superior to SPME in terms of sensitivity and accuracy for determining trace level quantities in difficult matrices. The amount of analyte extracted is proportional to the volume of the extraction phase, which in SPME is very small. Hence the detection limits of a method can be improved by increasing the volume of the extraction phase. This can be achieved by increasing the film thickness, either on a fiber support or by using a multifiber method. However, much longer equilibration time is required because the extraction rate is controlled by the diffusion from the sample matrix through the boundary layer into the extraction phase. SBSE is advantageous in dealing with very dilute media and trace concentration samples; compared to SPME, SBSE generally yields better detection limits. On the other hand, SPME can be fully automated which is not yet completely true for SBSE. It has been demonstrated that a wide range of volatile and semi-volatile substances such as phenylethyl alcohol, 1-ethenyl- 4-methoxybenzene, 2-phenylethyl acetate, 1-(ethylthio)-2-methylbenzene, 3,7-dimethyl-2,6-octadienoic acid, methyl ester, α-cubelene, copaene, terpenes, PCBs and pesticides can be retained on a polymer-coated magnetic bar (Baltussen et al., 1999, 2002; David et al., 2003a, 2003b; Soini et al., 2005). The amount of PDMS coating on a stir bar can be precisely controlled, increasing the reproducibility and reliability of extraction. A typical polymer phase volume of 24–100 μL in SBSE well exceeds that of SPME (typically, 0.5 μL). By rapid magnetic stirring of the aqueous medium, equilibrium on the PDMS layer is ensured during sorption so that the analyte concentration is constant and reproducible. PDMS sorption-based analyses have been described in environmental samples, essential oils, foodstuffs and beverages. Similar techniques for human biological samples have been reported. A number of new applications of SBSE and other extraction procedures have been published. These improvements involve SBSE with *in situ* derivatization, SBSE with *in situ* de-conjugation, TD in the multi-shot mode and TD with in tube derivatization method. These methods were applied successfully to the trace analysis of environmental and biological samples. (Kawaguchi et al., 2006; Quintana & Rodríguez, 2006; Buchberger & Zaborsky, 2007; David & Sandra, 2007; Picó et al., 2007).

2.4.3 Needle Trap Extraction

In the last few years, the inside needle capillary adsorption trap (INCAT) technique emerged as a solventless sample preparation approach and alternative extraction method derived from SPME. Two approaches could be associated with this technique. In the first one, the sorbent is placed as adsorbent layer, acting as non-polar or polar phase, on the interior surface of a needle (Musshoff et al., 2003). In the first approach, Murphy described a technique based on the use of an internally coated hollow needle (Murphy, 1996). Shojania et al. have used an adsorbing carbon coating on the interior surface of a hollow stainless steel needle as an INCAT device (Shojania et al., 1999). Later, a similar approach called "solid-phase dynamic extraction (SPDE)" was successfully applied to the analysis of pesticides in water (Lipinski, 2001) and amphetamines and synthetic design drugs in hair (Musshoff et al., 2002).

The other category is in-needle packing with commercially available sorbents or chemically synthesized compounds. The needle-trap devices are inexpensive, robust and reusable, and are suitable for the sampling and analysis of volatile organic compounds from many different sample matrices. Compared to the fragile silica-based SPME fiber, the needle device is rather impossible to break mechanically. Moreover, the extraction medium possesses a large extraction phase volume and a high preconcentration ability. Similar to SPME, needle-like devices are particularly convenient for automation and development of on-line procedures (Bagheri et al., 2009, 2011a).

2.5 Conclusions

SPME, since its introduction in the early 90's, has shown to have inherent potential to be adapted as a powerful technique in sample preparation methodologies. This versatile technique, together with other relevant methods, has been extended in many aspects of technical, physical and chemical points of view. Selective chemical deposition approaches have led to the emerging methodologies in which most analytical demands are almost met. The implementation of nanomaterials, sol-gel technology and metal-based coatings in SPME and other related techniques are developments which led to more stable, precise and accurate analysis while ease of operation and lower costs are also encountered. It is anticipated that SPME and other relevant techniques will play a major role in the future development of sample preparation.

Abbreviations

μ-SPE	micro-solid phase extraction
2-ATP	2-aminothiophenol
3-ATP	3-aminothiophenol
ACN	acetonitrile
AIBN	azobis(isobutyronitrile)
APS	ammonium persulfate
BFEE	boron trifluoride diethyl etherate
$BMIPF_6$	1-butyl-3-methylimidazolium hexafluorophosphate
BPANI	polyaniline-1-butyl-3-methylimidazolium hexafluorophosphate composite
BTEX	benzene, toluene, ethylbenzene and xylenes; butyronitrile
CME	capillary microextraction
CNTs	carbon nanotubes
CPs	conductive polymers
CTAB	cetyltrimethylammonium bromide

CV	cyclic voltammetry
DAMO	diacetylmorphine
DS	dodecylsulfate
EDMA	ethylene glycol dimethacrylate
ESI	electrospray ionization
G	grapheme
GC	gas chromatography
GO	graphene oxide
HPLC	high performance liquid chromatography
HS-SDME	headspace solid-phase microextraction
INCAT	inside needle capillary adsorption trap
LC	liquid chromatography
LOD	limit of detection
MAA	methacrylic acid
m-ABA	m-aminobenzoic acid
MEPS	microextraction in a packed syringe
MIP	molecular imprinted polymer
MIX	molecular imprinted xerogel
MS	mass spectrometry
MWCNTs	multi-walled carbon nanotubes
NiPcTS	tetrasulfonated nickel phthalocyanine
OCPs	organochlorine pesticides
ORMOSIL	organically modified silica
Pac	polyacetylene
PANI	polyaniline
PATP	polyaminothiophenol
PBDEs	polybrominated diphenyl ethers
PCBs	polychlorinated biphenyls
PFOS-PANI	perfluorooctanesulfonic acid-doped PANI
PDPA	polydiphenylamine
PIL	polymeric ionic liquid
PMME	polymer monolith microextraction
PP	polyphosphate
PPy	polypyrrole
PPy-S	sulfate-doped PPy
PSS	poly(styrenesulfonate)
PTFE	Polytetrafluoroethylene
PTh	polythiophene
PTV	programmed temperature vaporizing
SAMs	self assembled monolayers
SBSE	stir-bar sorptive extraction
SELDI	surface to enhance laser desorption/ionization

SEM	scanning electron microscopy
SPDE	solid-phase dynamic extraction
SPE	solid phase extraction
SPME	solid-phase microextraction
SWCNTs	single-walled carbon nanotubes
TD	thermal desorption
TEOS	tetraethoxysilane
TFME	thin film microextraction
TMOS	tetramethoxysilane
TMSPMA	3-(trimethoxysilyl)propylmethacrylate
TOS	toluene-p-sulfonate
VOCs	volatile organic compounds.

References

Abdel-Rehim, M. (2004). New trend in ample preparation: on line microextraction in packed syringe for liquid and gas chromatography applications: I. Determination of local anaesthesics in human plasma samples using gas chromatography-mass spectrometry. *Journal of Chromatography B: Analytical Technologies in the Biomedical and Life Sciences, 801*, 317–321.

Abdel-Rehim, M. (2010). Recent advances in microextraction by packed sorbent for bioanalysis. *Journal of Chromatography A, 1217*, 2569–2580.

Abdel-Rehim, M., Altun, Z., & Blomberg, L. (2004). Microextraction in packed syringe (MEPS) for liquid an gas chromatographic applications. Part II – Determination of ropivacaine and its metabolites in human plasma samples using MEPS with liquid chromatography/tandem mass spectrometry. *Journal of Mass Spectrometry, 39*, 1488–1493.

Abdel-Rehim, M., Dahlgren, M., Claude, S., Tabacchi, R., & Blomberg, L. (2006). Microextraction in packed syringe (MEPS) utilizing methylcyanopropyl-silarylene as coating polymer for extraction of drugs in biological samples. *Journal of Liquid Chromatography & Related Technology, 29*, 2537–2544.

Ahlrichs, J. L., Serna, C., & Serratosa, J. M. (1975). Structural hydroxyls in sepiolites. *Clays and Clay Minerals, 23*, 119–124.

Akelah, A., Salahuddin, N., Hiltner, A., Baer, E., & Moet, A. (1994). Morphological hierarchy of butadieneacrylonitrile/montmorillonite nanocomposite. *Nanostructured Materials, 4*, 965–978.

Altun, Z., Abdel-Rehim M., & Blomberg, L.G. (2004). New trends in sample preparation: On-line microextraction in packed syringe (MEPS) for LC and GC applications: Part III: Determination and validation of local anaesthesics in human plasma samples using a cation-exchange sorbent, and MEPS-LC-MS-MS. *Journal of Chromatography B: Analytical Technologies in the Biomedical and Life Sciences, 813*, 129–135.

Ansell, R. J., Kriz, D., & Mosbach, K. (1996). Molecularly imprinted polymers for bioanalysis: Chromatography, binding assays and biomimetic sensors. *Current Opinion in Biotechnology, 7*, 89–94.

Asadollahzadeh, H., Noroozian, E., & Maghsoudi, S. (2010). Solid-phase microextraction of phthalate esters from aqueous media by electrochemically deposited carbon nanotube/polypyrrole composite on a stainless steel fiber. *Analytica Chimica Acta, 669*, 32–38.

Bagheri, H., & Aghakhani, A. (2011). Novel nanofiber coatings prepared by electrospinning technique for headspace solid-phase microextraction of chlorobenzenes from environmental samples. *Analytical Methods, 3*, 1284–1289.

Bagheri, H., & Aghakhani, A. (2012). Polyaniline-nylon-6 electrospun nanofibers for headspace adsorptive microextraction. *Analytica Chimica Acta, 713*, 63–69.

Bagheri, H., & Ayazi, Z. (2011). Polypyrrole nanowires network for convenient and highly efficient microextraction in packed syringe. *Analytical Methods, 3,* 2630–2636.

Bagheri, H., & Piri-Moghadam, H. (2012). Sol-gel-based molecularly imprinted xerogel for capillary microextraction. *Analytical and Bioanalytical Chemistry, 404*, 1597–1602.

Bagheri, H., & Roostaie, A. (2012). Aniline-silica nanocomposite as a novel solid phase microextraction fiber coating. *Journal of Chromatography A, 1238*, 22–29.

Bagheri, H., Ayazi, Z., & Aghakhani, A. (2011a). A novel needle trap sorbent based on carbon nanotube-sol-gel for microextraction of polycyclic aromatic hydrocarbons from aquatic media. *Analytica Chimica Acta, 683*, 212–220.

Bagheri, H., Ayazi, Z., & Naderi, M. (2013a). Conductive polymer-based microextraction methods: A review. *Analytica Chimica Acta, 767,* 1–13.

Bagheri, H., Ayazi, Z., & Sistani, H. (2011b). Chemically bonded carbon nanotubes on modified gold substrate as novel unbreakable solid phase microextraction fiber. *Microchimica Acta, 174,* 295–301.

Bagheri, H., Ayazi, Z., Aghakhani, A., & Alipour, N. (2012a). Polypyrrole/polyamide electrospun-based sorbent for microextraction in packed syringe of organophosphorous pesticides from aquatic samples. *Journal of Separation Science, 35,* 114–120.

Bagheri, H., Ayazi, Z., Es'haghi, A., & Aghakhani, A. (2012b). Reinforced polydiphenylamine nanocomposite for microextraction in packed syringe of various pesticides. *Journal of Chromatography A, 1222,* 13–21.

Bagheri, H., Babanezhad, E., & Es-haghi, A. (2007). An aniline-based fiber coating for solid phase microextraction of polycyclic aromatic hydrocarbons from water followed by gas chromatography-mass spectrometry. *Journal of Chromatography A, 1152*, 168–174.

Bagheri, H., Babanezhad, E., & Khalilian, F. (2009). An interior needle electropolymerized pyrrole-based coating for headspace solid-phase dynamic extraction. *Analytica Chimica Acta, 634,* 209–214.

Bagheri, H., Bayat, P., & Piri-Moghadam, H. (2013b). Grafting the sol-gel based sorbents by diazonium salts: A novel approach toward unbreakable capillary microextraction. *Journal of Chromatography A, 1318*, 58–64.

Bagheri, H., Es-haghi, A., & Rouini, M. R. (2005a). Sol-gel-based solid-phase microextraction and gas chromatography-mass spectrometry determination of dextromethorphan and dextrorphan in human plasma. *Journal of Chromatography B: Analytical Technologies in the Biomedical and Life Sciences, 818*, 147–157.

Bagheri, H., Es'haghi, A., Es-haghi, A., & Mesbahi, N. (2012c). A high-throughput approach for the determination of pesticide residues in cucumber samples using solid-phase microextraction on 96-well plate. *Analytica Chimica Acta, 740*, 36–42.

Bagheri, H., Es'haghi, A., Es-haghi, A., & Mohammadkhani, E. (2013c). High-throughput micro-solid phase extraction on 96-well plate using dodecyl methacrylate-ethylen glycol dimethacrylate monolithic copolymer. *Analytica Chimica Acta, 792*, 59–65.

Bagheri, H., Mir, A., & Babanezhad, E. (2005b). An electropolymerized aniline-based fiber coating for solid phase microextraction of phenols from water. *Analytica Chimica Acta, 532*, 89–95.

Bagheri, H., Piri-Moghadam, H., & Ahdi, T. (2012d). Role of precursors and coating polymers in sol-gel chemistry toward enhanced selectivity and efficiency in solid phase microextraction. *Analytica Chimica Acta, 742*, 45–53.

Bagheri, H., Piri-Moghadam, H., & Es'haghi, A. (2011c). An unbreakable on-line approach towards sol-gel capillary microextraction. *Journal of Chromatography A, 1218*, 3952–3957.

Bagheri, H., Piri-Moghadam, H., & Naderi, M. (2012e). Towards greater mechanical, thermal and chemical stability in solid-phase microextraction. *TrAC Trends in Analytical Chemistry, 34*, 126–138.

Bagheri, H., Piri-Moghadam, H., Bayat, P., & Balalaie, S. (2013d). Application of sol-gel based molecularly imprinted xerogel for on-line capillary microextraction of fentanyl from urine and plasma samples. *Analytical Methods, 5*, 7096–7101.

Bagheri, H., Sistani, H., & Ayazi, Z. (2011d). Novel unbreakable solid-phase microextraction fiber by electrodeposition of silica sol-gel on gold. *Journal of Separation Science, 34*, 3246–3252.

Bain, C. D., Troughton, E. B., Tao, Y. T., Evall, J., Whitesides, G. M., & Nuzzo, R. G. (1989). Formation of monolayer films by the spontaneous assembly of organic thiols from solution onto gold. *Journal of the American Chemical Society, 111*, 321–335.

Baltussen, E., Cramers, C.A., & Sandra, P. (2002). Sorptive sample preparation – A review. *Analytical & Bioanalytical Chemistry, 373*, 3–22.

Baltussen, E., Sandra, P., David, F., & Cramers, C. (1999). Stir bar sorptive extraction (SBSE), a novel extraction technique for aqueous samples: Theory and principles. *Journal of Microcolumn Separation*, 1999, 11, 737–747.

Bhardwaj, N., & Kundu, S. C. (2010). Electrospinning: A fascinating fiber fabrication technique. *Biotechnology Advances, 28*, 325–347.

Blinova, N. V., Stejskal, J., Trchová, M., & Prokeš, J. (2006). Polyaniline prepared in solutions of phosphoric acid: Powders, thin films, and colloidal dispersions. *Polymer, 47*, 42–48.

Boukerma, K., Chehimi, M. M., Pinson, J., & Blomfield, C. (2003). X-ray photoelectron spectroscopy evidence for the covalent bond between an iron surface and aryl groups attached by the electrochemical reduction of diazonium salts. *Langmuir, 19*, 6333–6335.

Breton, T., & Bélanger, D. (2008). Modification of carbon electrode with aryl groups having an aliphatic amine by electrochemical reduction of *in situ* generated diazonium cations. *Langmuir, 24*, 8711–8718.

Buchberger, W., & Zaborsky, P. (2007). Sorptive extraction techniques for trace analysis of organic pollutants in the aquatic environment. *Acta Chimica Slovenica, 54*, 1–13.

Cao, D. d., Lü, J. x., Liu, J. f., & Jiang, G. b. (2008). *In situ* fabrication of nanostructured titania coating on the surface of titanium wire: A new approach for preparation of solid-phase microextraction fiber. *Analytica Chimica Acta, 611*, 56–61.

Chen, L., Chen, W., Ma, C., Du, D., & Chen, X. (2011). Electropolymerized multiwalled carbon nanotubes/polypyrrole fiber for solid-phase microextraction and its applications in the determination of pyrethroids. *Talanta, 84*, 104–108.

Chen, M. L., Wei, S. S., Yuan, B. F., & Feng, Y. Q. (2012). Preparation of methacrylate-based monolith for capillary hydrophilic interaction chromatography and its application in determination of nucleosides in urine. *Journal of Chromatography A, 1228*, 183–192.

Chen, S. A., & Tsai, C. C. (1993). Structure/properties of conjugated conductive polymers. 2. 3-ether-substituted polythiophenes and poly(4-methylthiophenes). *Macromolecules, 26*, 2234–2239.

Chigome, S., Darko, G., Buttner, U., & Torto, N. (2010). Semi-micro solid phase extraction with electrospun polystyrene fiber disks. *Analytical Methods, 2*, 623–626.

Chong, S. L., Wang, D., Hayes, J. D., Wilhite, B. W., & Malik, A. (1997). Sol-gel coating technology for the preparation of solid-phase microextraction fibers of enhanced thermal stability. *Analytical Chemistry, 69*, 3889–3898.

Corriu, R., & Trong Anh, H. (2009). *Molecular Chemistry of Sol-Gel Derived Nanomaterials*. UK: John Wiley & Sons.

David, F., & Sandra, P. (2007). Stir bar sorptive extraction for trace analysis. *Journal of Chromatography A, 1152*, 54–69.

David. F., Tienpont, B., & Sandra, P. (2003a). Stir-bar sorptive extraction of trace organic compounds from aqueous matrices. *LCGC North America, 21*, 108–118.

David. F., Tienpont, B., & Sandra, P. (2003b). Stir-bar sorptive extraction of trace organic compounds from aqueous matrices. *LCGC Europe, 16,* 410–417.

Deng, D. L., Zhang, J. Y., Chen, C., Hou, X. L., Su, Y. Y., & Wu, L. (2012). Monolithic molecular imprinted polymer fiber for recognition and solid phase microextraction of ephedrine and pseudoephedrine in biological samples prior to capillary electrophoresis analysis. *Journal of Chromatography A, 1219*, 195–200.

Djozan, D., & Abdollahi, L. (2003). Anodized zinc wire as a solid-phase microextraction fiber. *Chromatographia, 57*, 799–804.

Djozan, D., & Bahar, S. (2004). Solid-phase microextraction of aliphatic alcohols based on polyaniline coated fibers. *Chromatographia, 59*, 95–99.

Djozan, D., & Baheri, T. (2007). Preparation and evaluation of solid-phase microextraction fibers based on monolithic molecularly imprinted polymers for selective extraction of diacetyl-morphine and analogous compounds. *Journal of Chromatography A, 1166*, 16–23.

Djozan, D., & Ebrahimi, B. (2008). Preparation of new solid phase micro extraction fiber on the basis of atrazine-molecular imprinted polymer: Application for GC and GC/MS screening of triazine herbicides in water, rice and onion. *Analytica Chimica Acta, 616*, 152–159.

Djozan, D., Assadi, Y., & Haddadi, S. H. (2001). Anodized aluminum wire as a solid-phase microextraction fiber. *Analytical Chemistry, 73*, 4054–4058.

Djozan, D., Pournaghi-Azar, M. H., & Bahar, S. (2004). Modified polypyrrole with tetrasulfonated nickel phthalocyanine as a fiber for solid-phase microextraction. Application to the extraction of BTEX compounds from water samples. *Chromatographia, 59*, 595–599.

Du, W., Zhao, F., & Zeng, B. (2009). Novel multiwalled carbon nanotubes-polyaniline composite film coated platinum wire for headspace solid-phase microextraction and gas chromatographic determination of phenolic compounds. *Journal of Chromatography A, 1216*, 3751–3757.

Ebelmann, M. (1846). Sur les combinaisons des acides borique et silicique avec les ethers. *Annales de Chimie et de Physique, 16*, 129–166.

Ebelmann, M. (1847). Sur l'hyalite artificielle et l'hydrophane. *Les Comptes rendus de l'Académie des sciences, 25*, 854–856.

Eisert, R., & Levsen, K. (1995). Determination of pesticides in aqueous samples by solid-phase microextraction in-line coupled to gas chromatography-mass spectrometry. *Journal of the American Society for Mass Spectrometry, 6*, 1119–1130.

El-Beqqali, A., Kussak, A., & Abdel-Rehim, M. (2006). Fast and sensitive environmental analysis utilizing microextraction in packed syringe online with gas chromatography-mass spectrometry. Determination of polycyclic aromatic hydrocarbons in water. *Journal of Chromatography A, 1114*, 234–238.

Es-haghi, A., Hosseini, S. M., & Khoshhesab, Z. M. (2012). Development and application of a new solid-phase microextraction fiber by sol-gel technology on titanium wire. *Analytica Chimica Acta, 742*, 74–79.

Fan, Y., Feng, Y. Q., Da, S. L., & Shi, Z. G. (2004). Poly (methacrylic acid-ethylene glycol dimethacrylate) monolithic capillary for in-tube solid phase microextraction coupled to high performance liquid chromatography and its application to determination of basic drugs in human serum. *Analytica Chimica Acta, 523*, 251–258.

Fontanals, N., Marcé, R. M., & Borrull, F. (2007). New materials in sorptive extraction techniques for polar compounds. *Journal of Chromatography A, 1152*, 14–31.

Freund, M. S., & Deore, B. A. (2007). *Self Doped Conducting Polymers*. UK: John Wiley & Sons.

Friedrich, K., Fakirov, S., & Zhang, Z. (2005). *Polymer Composite: From Nano-to Macro-Scale*. New York: Springer.

Gao, H., Jiang, T., Han, B., Wang, Y., Du, J., Liu, Z., et al. (2004). Aqueous/ionic liquid interfacial polymerization for preparing polyaniline nanoparticles. *Polymer, 45*, 3017–3019.

Gao, Z., Li, W., Liu, B., Liang, F., He, H., Yang, S., et al. (2011). Nano-structured polyaniline-ionic liquid composite film coated steel wire for headspace solid-phase microextraction of organochlorine pesticides in water. *Journal of Chromatography A, 1218*, 6285–6291.

Ghassempour, A., Najafi, N. M., Mehdinia, A., Davarani, S. S. H., Fallahi, M., & Nakhshab, M. (2005). Analysis of anatoxin-a using polyaniline as a sorbent in solid-phase microextraction coupled to gas chromatography-mass spectrometry. *Journal of Chromatography A, 1078*, 120–127.

Ghosh, S. K. (2006). *Functional Coatings*. Weinheim: Wiley-VCH.

Gómez-Hens, A., Fernández-Romero, J. M., & Aguilar-Caballos, M. P. (2008). Nanostructures as analytical tools in bioassays. *TrAC Trends in Analytical Chemistry, 27*, 394–406.

Gratzl, M., Hsu, D. F., Riley, A. M., & Janata, J. (1990). Electrochemically deposited polythiophene. 1. Ohmic drop compensation and the polythiophene paradox. *Journal of Physical Chemistry, 94*, 5973–5981.

Gupta, R., & Kumar, A. (2008). Molecular imprinting in sol-gel matrix. *Biotechnology Advances, 26*, 533–547.

Haupt, K. (2003). Imprinted polymers - Tailor-made mimics of antibodies and receptors. *Chemical Communications, 9*, 171–177.

Havenga, W. J., & Rohwer, E. R. (1999). Chemical characterization and screening of hydrocarbon pollution in industrial soils by headspace solid-phase microextraction. *Journal of Chromatography A, 848*, 279–295.

Huang, Z. M., Zhang, Y. Z., Kotaki, M., & Ramakrishna, S. (2003). A review on polymer nanofibers by electrospinning and their applications in nanocomposites. *Composites Science and Technology, 63*, 2223–2253.

Jiménez-Soto, J. M., Cárdenas, S., & Valcárcel, M. (2010). Carbon nanocones/disks as new coating for solid-phase microextraction. *Journal of Chromatography A, 1217*, 3341–3347.

Kabir, A., Furton, K. G., & Malik, A. (2013). Innovations in sol-gel microextraction phases for solvent-free sample preparation in analytical chemistry. *TrAC Trends in Analytical Chemistry, 45*, 197–218.

Kang, X., Pan, C., Xu, Q., Yao, Y., Wang, Y., Qi, D., et al. (2007). The investigation of electrospun polymer nanofibers as a solid-phase extraction sorbent for the determination of trazodone in human plasma. *Analytica Chimica Acta, 587*, 75–81.

Karpovich, D. S., Schessler, H. M., & Blanchard, G. J. (1998). *Self-Assembled Monolayers of Thiols* (Vol. 24). New York: Academic Press.

Kataoka, H. (2011). Current developments and future trends in solid-phase microextraction techniques for pharmaceutical and biomedical analyses. *Analytical Sciences, 27*, 893–905.

Kataoka, H., Ishizaki, A., Nonaka, Y., & Saito, K. (2009). Developments and applications of capillary microextraction techniques: A review. *Analytica Chimica Acta, 655*, 8–29.

Kawaguchi, M., Ito, R., Saito, K., & Nakazawa, H. (2006). Novel stir bar sorptive extraction methods for environmental and biomedical analysis. *Journal of Pharmacutical & Biomedical Analysis, 40*, 500–508.

Kawasumi, M., Hasegawa, N., Usuki, A., & Okada, A. (1999). Liquid crystal/clay mineral composites. *Applied Clay Science, 15*, 93–108.

Koster, E. H. M., Crescenzi, C., Den Hoedt, W., Ensing, K., & De Jong, G. J. (2001). Fibers coated with molecularly imprinted polymers for solid-phase microextraction. *Analytical Chemistry, 73*, 3140–3145.

Krishnamoorti, R., Vaia, R. A., & Giannelis, E. P. (1996). Structure and dynamics of polymer-layered silicate nanocomposites. *Chemistry of Materials, 8*, 1728–1734.

Lan, T., Kaviratna, P. D., & Pinnavaia, T. J. (1995). Mechanism of clay tactoid exfoliation in epoxy-clay nanocomposites. *Chemistry of Materials, 7*, 2144–2150.

Li, X., Chen, J., & Du, L. (2007). Analysis of chloro- and nitrobenzenes in water by a simple polyaniline-based solid-phase microextraction coupled with gas chromatography. *Journal of Chromatography A, 1140*, 21–28.

Li, X., Li, C., Chen, J., & Sun, C. (2008). Polythiophene as a novel fiber coating for solid-phase microextraction. *Journal of Chromatography A, 1198–1199*, 7–13.

Li, X., Zeng, Z., Zhou, J., Gong, S., Wang, W., & Chen, Y. (2004). Novel fiber coated with amide bridged-calix[4]arene used for solid-phase microextraction of aliphatic amines. *Journal of Chromatography A, 1041*, 1–9.

Lipinski, J. (2001). Automated solid phase dynamic extraction – Extraction of organics using a wall coated syringe needle. *Fresenius' Journal of Analytical Chemistry, 369*, 57–62.

Liu, H., Ji, L., Li, J., Liu, S., Liu, X., & Jiang, S. (2011). Magnetron sputtering Si interlayer: A protocol to prepare solid phase microextraction coatings on metal-based fiber. *Journal of Chromatography A, 1218*, 2835–2840.

Liu, H., Wang, D., Ji, L., Li, J., Liu, S., Liu, X., et al. (2010). A novel TiO_2 nanotube array/Ti wire incorporated solid-phase microextraction fiber with high strength, efficiency and selectivity. *Journal of Chromatography A, 1217*, 1898–1903.

Lord, H., & Pawliszyn, J. (2000). Evolution of solid-phase microextraction technology. *Journal of Chromatography A, 885*, 153–193.

Louch, D., Motlagh, S., & Pawliszyn, J. (1992). Dynamics of organic compound extraction from water using liquid-coated fused silica fibers. *Analytical Chemistry, 64*, 1187–1199.

Love, J. C., Estroff, L. A., Kriebel, J. K., Nuzzo, R. G., & Whitesides, G. M. (2005). Self-assembled monolayers of thiolates on metals as a form of nanotechnology. *Chemical Reviews, 105*, 1103–1169.

Lu, X., Chao, D., Chen, J., Zhang, W., & Wei, Y. (2006). Preparation and characterization of inorganic/organic hybrid nanocomposites based on Au nanoparticles and polypyrrole. *Material Letters, 60* 2851–2854.

MacDiarmid, A. G. (2001). "Synthetic metals": A novel role for organic polymers (Nobel lecture). *Angewandte Chemie International Edition, 40*, 2581–2590.

MacGillivray, B., Pawliszyn, P., Fowlie, P., & Sagara, C. (1994). Headspace solid-phase microextraction versus purge and trap for the determination of substituted benzene compounds in water. *Journal of Chromatographic Science, 32*, 317–322.

Magdic, S., Boyd-Boland, A., Jinno, K., & Pawliszyn, J. B. (1996). Analysis of organophosphorus insecticides from environmental samples using solid-phase microextraction. *Journal of Chromatography A, 736*, 219–228.

Mahouche-Chergui, S., Gam-Derouich, S., Mangeney, C., & Chehimi, M. M. (2011). Aryl diazonium salts: A new class of coupling agents for bonding polymers, biomacromolecules and nanoparticles to surfaces. *Chemical Society Reviews, 40*, 4143–4166.

Mecerreyes, D., Alvaro, V., Cantero, I., Bengoetxea, M., Calvo, P. A., Grande, H., et al. (2002). Low surface energy conducting polypyrrole doped with a fluorinated counterion. *Advanced Materials, 14*, 749–752.

Mehdinia, A., & Mousavi, M. F. (2008). Enhancing extraction rate in solid-phase microextraction by using nano-structured polyaniline coating. *Journal of Separation Science, 31*, 3565–3572.

Mehdinia, A., Asiabi, M., Jabbari, A., & Kalaee, M. R. (2010). Preparation and evaluation of solid-phase microextraction fiber based on nano-structured copolymer of aniline and m-amino benzoic acid coating for the analysis of fatty acids in zooplanktons. *Journal of Chromatography A, 1217*, 7642–7647.

Mehdinia, A., Mohammadi, A. A., Davarani, S. S. H., & Banitaba, M. H. (2011). Application of self-assembled monolayers in the preparation of solid-phase microextraction coatings. *Chromatographia, 74*, 421–427.

Mehdinia, A., Mousavi, M. F., & Shamsipur, M. (2006). Nano-structured lead dioxide as a novel stationary phase for solid-phase microextraction. *Journal of Chromatography A, 1134*, 24–31.

Mévellec, V., Roussel, S., Tessier, L., Chancolon, J., Mayne-L'Hermite, M., Deniau, G., et al. (2007). Grafting polymers on surfaces: A new powerful and versatile diazonium salt-based one-step process in aqueous media. *Chemistry of Materials, 19*, 6323–6330.

Minjia, H., Chao, T., Qunfang, Z., & Guibin, J. (2004). Preparation of polyaniline coating on a stainless-steel wire using electroplating and its application to the determination of six aromatic amines using headspace solid-phase microextraction. *Journal of Chromatography A, 1048*, 257–262.

Mirnaghi, F. S., Chen, Y., Sidisky, L. M., & Pawliszyn, J. (2011). Optimization of the coating procedure for a high-throughput 96-blade solid phase microextraction system coupled with LC-MS/MS for analysis of complex samples. *Analytical Chemistry, 83*, 6018–6025.

Mitra, S. (2003). *Sample Preparation Techniques in Analytical Chemistry*. Canada: John Wiley & Sons.

Mohammadi, A., Yamini, Y., & Alizadeh, N. (2005). Dodecylsulfate-doped polypyrrole film prepared by electrochemical fiber coating technique for headspace solid-phase microextraction of polycyclic aromatic hydrocarbons. *Journal of Chromatography A, 1063*, 1–8.

Mollahosseini, A., & Noroozian, E. (2009). Polyphosphate-doped polypyrrole coated on steel fiber for the solid-phase microextraction of organochlorine pesticides in water. *Analytica Chimica Acta, 638*, 169–174.

Mullett, W. M., Martin, P., & Pawliszyn, J. (2001). In-tube molecularly imprinted polymer solid-phase microextraction for the selective determination of propranolol. *Analytical Chemistry, 73*, 2383–2389.

Murphy, G.E. US Patent 1996, 5, 565, 622.

Musshoff, F., Lachenmeier, D.W., Kroener, L., & Madea, B. (2003). Automated headspace solid-phase dynamic extraction for the determination of cannabinoids in hair samples. *Forensic Science International, 133*, 32–38.

Musshoff, F., Lachenmeier, D.W., Kroener, L., & Madea, B. (2002). Automated headspace solid-phase dynamic extraction for the determination of amphetamines and synthetic designer drugs in hair samples. *Journal of Chromatography A, 958*, 231–238.

Nalwa, H. S. (1997). *Handbook of Organic Conductive Molecules and Polymers*. New York: Wiley.

Ohtani, A., & Shimadzu, T. (1989). Effective doping of polymer anion during chemical polymerization of pyrrole using $Fe(OH)_3$ oxidant. *Bulletin of the Chemical Society of Japan, 62*, 234–238.

Olszowy, P., Szultka, M., Fuchs, P., Kegler, R., Mundkowski, R., Miekisch, W., et al. (2010). New coated SPME fibers for extraction and fast HPLC determination of selected drugs in human blood. *Journal of Pharmaceutical and Biomedical Analysis, 53*, 1022–1027.

Olszowy, P., Szultka, M., Nowaczyk, J., & Buszewski, B. (2011). A new way of solid-phase microextraction fibers preparation for selected antibiotic drug determination by HPLC-MS. *Journal of Chromatography B: Analytical Technologies in the Biomedical and Life Sciences, 879*, 2542–2548.

Pan, W., Durning, C. J., & Turro, N. J. (1996). Kinetics of alkanethiol adsorption on gold. *Langmuir, 12*, 4469–4473.

Pang, L., & Liu, J. F. (2012). Development of a solid-phase microextraction fiber by chemical binding of polymeric ionic liquid on a silica coated stainless steel wire. *Journal of Chromatography A, 1230*, 8–14.

Penton, Z. E. (1997). Sample preparation for gas chromatography with solid-phase extraction and solid-phase microextraction. *Advances in Chromatography, 37*, 205–236.

Picó. Y.; Fernández, M., Ruiz, M.J., & Font, G. (2007). Current trends in solid-phase-based extraction techniques for the determination of pesticides in food and environment. *Journal of Biochemical & Biophysical Methods, 70*, 117–131.

Ponnusamy, V. K., & Jen, J. F. (2011). A novel graphene nanosheets coated stainless steel fiber for microwave assisted headspace solid phase microextraction of organochlorine pesticides in aqueous samples followed by gas chromatography with electron capture detection. *Journal of Chromatography A, 1218*, 6861–6868.

Potter, D. W., & Pawliszyn, J. (1992). Detection of substituted benzenes in water at the pg/ml level using solid-phase microextraction and gas chromatography-ion trap mass spectrometry. *Journal of Chromatography, 625*, 247–255.

Potter, D. W., & Pawliszyn, J. (1994). Rapid determination of polyaromatic hydrocarbons and polychlorinated biphenyls in water using solid-phase microextraction and GC/MS. *Environmental Science & Technology, 28*, 298–305.

Prieto, A., Schrader, S., & Moeder, M. (2010). Determination of organic priority pollutants and emerging compounds in wastewater and snow samples using multiresidue protocols on the basis of microextraction by packed sorbents coupled to large volume injection gas chromatography-mass spectrometry analysis. *Journal of Chromatography A, 1217*, 6002–6011.

Pron, A., & Rannou, P. (2002). Processible conjugated polymers: From organic semiconductors to organic metals and superconductors. *Progress in Polymer Science (Oxford), 27*, 135–190.

Quintana, J.B., & Rodríguez, I. (2006). Strategies for the microextraction of polar organic contaminants in water samples. *Analytical & Bioanalytical Chemistry*, 2006, 384,1447–1461.

Rajabi Khorrami, A., & Narouenezhad, E. (2011). Synthesis of molecularly imprinted monolithic fibers for solid-phase microextraction of acetaldehyde from head-space of beverages stored in PET bottles. *Talanta, 86*, 58–63.

Ramakrishna, S., Fujihara, K., Teo, W. E., Lim, T. C., & Ma, Z. (2005). *An Introduction to Electrospinning and Nanofibers*. Singapore: World Scientific.

Rastkari, N., Ahmadkhaniha, R., & Yunesian, M. (2009). Single-walled carbon nanotubes as an effective adsorbent in solid-phase microextraction of low level methyl tert-butyl ether, ethyl tert-butyl ether and methyl tert-amyl ether from human urine. *Journal of Chromatography B: Analytical Technologies in the Biomedical and Life Sciences, 877*, 1568–1574.

Rastkari, N., Ahmadkhaniha, R., Samadi, N., Shafiee, A., & Yunesian, M. (2010). Single-walled carbon nanotubes as solid-phase microextraction adsorbent for the determination of low-level concentrations of butyltin compounds in seawater. *Analytica Chimica Acta, 662*, 90–96.

Rivasseau, C., & Caude, M. (1995). Comparison of on-line SPE-HPLC and SPME-GC for the analysis of microcontaminants in water. *Chromatographia, 41*, 462–470.

Rout, T. K., Bandyopadhyay, N., & Venugopalan, T. (2006). Polyphosphate coated steel sheet for superior corrosion resistance. *Surface and Coatings Technology, 201*, 1022–1030.

Rutledge, G. C., & Fridrikh, S. V. (2007). Formation of fibers by electrospinning. *Advanced Drug Delivery Reviews, 59*, 1384–1391.

Sadki, S., Schottland, P., Brodie, N., & Sabouraud, G. (2000). The mechanisms of pyrrole electropolymerization. *Chemical Society Reviews, 29*, 283–293.

Sánchez-Rojas, F., Bosch-Ojeda, C., & Cano-Pavón, J.M. (2009). A review of stir bar sorptive extraction. *Chromatographia, 69*, 79–94.

Santora, B. P., Gagné, M. R., Moloyt, K. G., & Radu, N. S. (2001). Porogen and cross-linking effects on the surface area, pore volume distribution, and morphology of macroporous polymers obtained by bulk polymerization. *Macromolecules, 34*, 658–661.

Sarafraz-Yazdi, A., Piri-Moghadam, H., Es'Haghi, Z., & Sepehr, S. (2010). Comparative study of the three sol-gel based solid phase microextraction fibers in extraction of BTEX from water samples using gas chromatography-flame ionization detection. *Analytical Methods, 2*, 746–752.

Sarafraz-Yazdi, A., Sepehr, S., Es'Haghi, Z., & Piri-Moghadam, H. (2010). Application of sol-gel based poly(ethylene glycol)/multiwalled carbon nanotubes coated fiber for SPME of methyl tert-butyl ether in environmental water samples. *Chromatographia, 72*, 923–931.

Schessler, H. M., Karpovich, D. S., & Blanchard, G. J. (1996). Quantitating the balance between enthalpic and entropic forces in alkanethiol/gold monolayer self assembly. *Journal of the American Chemical Society, 118*, 9645–9651.

Shi, G., Jin, S., Xue, G., & Li, C. (1995). A conducting polymer film stronger than aluminum. *Science, 267*, 994–996.

Shirakawa, H. (2001). The discovery of polyacetylene film: The dawning of an era of conducting polymers (Nobel lecture). *Angewandte Chemie International Edition, 40*, 2575–2580.

Shirakawa, H., Louis, E. J., MacDiarmid, A. G., Chiang, C. K., & Heeger, A. J. (1977). Synthesis of electrically conducting organic polymers: Halogen derivatives of polyacetylene, (CH)x. *Journal of the Chemical Society, Chemical Communications*, 578–580.

Shojania, S., Oleschuk, R.D., McComb, M.E., Gesser, H.D., & Chow, A. (1999). The active and passive sampling of benzene, toluene, ethyl benzene and xylenes compounds using the inside needle capillary adsorption trap device. *Talanta, 50*, 193–205.

Snauwaert, P., Lazzaroni, R., Riga, J., Verbist, J. J., & Gonbeau, D. (1990). A photoelectron spectroscopic study of the electrochemical processes in polyaniline. *The Journal of Chemical Physics, 92*, 2187–2193.

Soini, H.A., Bruce, K.E., Wiesler, D., David, F., Sandra, P., & Novotny, M.V. (2005). Stir bar sorptive extraction: A new quantitative and comprehensive sampling technique for determination of chemical signal profiles from biological media. *Journal of Chemical Ecology, 31*, 377–392.

Svec, F. (2010). Porous polymer monoliths: Amazingly wide variety of techniques enabling their preparation. *Journal of Chromatography A, 1217*, 902–924.

Svec, F., & Frechet, J. M. J. (1995). Modified poly(glycidyl methacrylate-co-ethylene dimethacrylate) continuous rod columns for preparative-scale ion-exchange chromatography of proteins. *Journal of Chromatography A, 702*, 89–95.

Teo, W. E., & Ramakrishna, S. (2006). A review on electrospinning design and nanofibre assemblies. *Nanotechnology, 17*, R89-R106.

Tong, X., Tang, T., Feng, Z., & Huang, B. (2002). Preparation of polymer/silica nanoscale hybrids through sol-gel method involving emulsion polymers. II. Poly(ethyl acrylate)/SiO_2. *Journal of Applied Polymer Science, 86*, 3532–3536.

Valor, I., Moltó, J. C., Apraiz, D., & Font, G. (1997). Matrix effects on solid-phase microextraction of organophosphorus pesticides from water. *Journal of Chromatography A, 767*, 195–203.

Vuckovic, D., Cudjoe, E., Hein, D., & Pawliszyn, J. (2008). Automation of solid-phase microextraction in high-throughput format and applications to drug analysis. *Analytical Chemistry, 80*, 6870–6880.

Wang, D., Chong, S. L., & Malik, A. (1997). Sol-gel column technology for single-step deactivation, coating, and stationary-phase immobilization in high-resolution capillary gas chromatography. *Analytical Chemistry, 69*, 4566–4576.

Wang, J. X., Jiang, D. Q., Gu, Z. Y., & Yan, X. P. (2006). Multiwalled carbon nanotubes coated fibers for solid-phase microextraction of polybrominated diphenyl ethers in water and milk samples before gas chromatography with electron-capture detection. *Journal of Chromatography A, 1137*, 8–14.

Wang, X., Ma, Q., Li, M., Chang, C., Bai, Y., Feng, Y., et al. (2013). Automated and sensitive analysis of 28-epihomobrassinolide in Arabidopsis thaliana by on-line polymer monolith microextraction coupled to liquid chromatography-mass spectrometry. *Journal of Chromatography A, 1317*, 121–128.

Wang, Y., Li, Y., Zhang, J., Xu, S., Yang, S., & Sun, C. (2009). A novel fluorinated polyaniline-based solid-phase microextraction coupled with gas chromatography for quantitative determination of polychlorinated biphenyls in water samples. *Analytica Chimica Acta, 646*, 78–84.

Wang, Y., Walles, M., Thomson, B., Nacson, S., & Pawliszyn, J. (2004). Solid-phase microextraction combined with surface-enhanced laser desorption/ionization introduction for ion mobility spectrometry and mass spectrometry using polypyrrole coatings. *Rapid Communications in Mass Spectrometry, 18*, 157–162.

Wang, Y., Zhang, J., Sheng, D., & Sun, C. (2010). Preparation and applications of perfluorinated ion doped polyaniline based solid-phase microextraction fiber. *Journal of Chromatography A, 1217*, 4523–4528.

Wei, Y., Tian, J., MacDiarmid, A. G., Masters, J. G., Smith, A. L., & Li, D. (1993). Preparation and conductivities of fullerene-doped polyanilines. *Journal of the Chemical Society, Chemical Communications*, 603–604.

Wu, J., Yu, X., Lord, H., & Pawliszyn, J. (2000). Solid phase microextraction of inorganic anions based on polypyrrole film. *The Analyst, 125*, 391–394.

Xu, Q., Wu, S. Y., Wang, M., Yin, X. Y., Wen, Z. Y., Ge, W. N., et al. (2010). Electrospun nylon6 nanofibrous membrane as SPE adsorbent for the enrichment and determination of three estrogens in environmental water samples. *Chromatographia, 71*, 487–492.

Yoshimatsu, K., Ye, L., Lindberg, J., & Chronakis, I. S. (2008). Selective molecular adsorption using electrospun nanofiber affinity membranes. *Biosensors and Bioelectronics, 23*, 1208–1215.

Yu, J., Dong, L., Wu, C., Wu, L., & Xing, J. (2002). Hydroxyfullerene as a novel coating for solid-phase microextraction fiber with sol-gel technology. *Journal of Chromatography A, 978*, 37–48.

Yu, J., Wu, C., & Xing, J. (2004). Development of new solid-phase microextraction fibers by sol-gel technology for the determination of organophosphorus pesticide multiresidues in food. *Journal of Chromatography A, 1036*, 101–111.

Zewe, J. W., Steach, J. K., & Olesik, S. V. (2010). Electrospun fibers for solid-phase microextraction. *Analytical Chemistry, 82*, 5341–5348.

Zhang, M., Wei, F., Zhang, Y.-F., Nie, J., Feng, Y. Q. (2006). Novel polymer monolith microextraction using a poly(methacrylic acid-ethylene glycol dimethacrylate) monolith and its application to simultaneous analysis of several angiotensin II receptor antagonists in human urine by capillary zone electrophoresis. *Journal of Chromatography A, 1102*, 294–301.

Zhang, W., & Chen, Z. (2013). Mussel inspired polydopamine functionalized poly(ether ether ketone) tube for online solid-phase microextraction-high performance liquid chromatography and its application in analysis of protoberberine alkaloids in rat plasma. *Journal of Chromatography A, 1278*, 29–36.

Zhang, Z., & Pawliszyn, J. (1993). Headspace solid-phase microextraction. *Analytical Chemistry, 65*, 1843–1852.

Zhang, Z., Yang, M. J., & Pawliszyn, J. (1994). Solid-phase microextraction. *Analytical Chemistry, 66*, 844A-853A.

Zheng, M. M., Ruan, G. D., & Feng, Y. Q. (2009). Evaluating polymer monolith in-tube solid-phase microextraction coupled to liquid chromatography/quadrupole time-of-flight mass spectrometry for reliable quantification and confirmation of quinolone antibacterials in edible animal food. *Journal of Chromatography A, 1216*, 7510–7519.

Zheng, M. M., Wang, S. T., Hu, W. K., & Feng, Y. Q. (2010). In-tube solid-phase microextraction based on hybrid silica monolith coupled to liquid chromatography-mass spectrometry for automated analysis of ten antidepressants in human urine and plasma. *Journal of Chromatography A, 1217*, 7493–7501.

Zhou, J., Yang, F., Cha, D., Zeng, Z., & Xu, Y. (2007). Headspace solid-phase microextraction with novel sol-gel permethylated-β-cyclodextrin/hydroxyl-termination silicone oil fiber for determination of polybrominated diphenyl ethers by gas chromatography-mass spectrometry in soil. *Talanta, 73*, 870–877.

Zhou, S., Wu, L., Sun, J., & Shen, W. (2002). The change of the properties of acrylic-based polyurethane via addition of nano-silica. *Progress in Organic Coatings, 45*, 33–42.

Zou, J., Song, X., Ji, J., Xu, W., Chen, J., Jiang, Y., et al. (2011). Polypyrrole/graphene composite-coated fiber for the solid-phase microextraction of phenols. *Journal of Separation Science, 34*, 2765–2772.

3 Novel Materials in Solid-Phase Microextraction and Related Sample Preparation Approaches

Bin Hu*, Man He and Beibei Chen
Department of Chemistry, Wuhan University, Wuhan 430072, China
*e-mail address: binhu@whu.edu.cn

3.1 Introduction

Pawliszyn and co-workers introduced solid-phase microextraction (SPME) in the early 1990s, which addressed several drawbacks of traditional sample preparation techniques and allowed rapid laboratory and on-site analysis (Arthur & Pawliszyn, 1990; Pawliszyn, 1997). In addition to ease of use, minimized organic solvent consumption and short sample preparation times, SPME features low sample amount requirements, automation capability and the ability to produce high-quality qualitative and quantitative analytical results for gaseous, aqueous and solid real-world samples of high complexity (Souza Silva et al., 2013). To meet diverse requirements in real sample analysis, SPME has been developed rapidly in terms of configuration. Many related approaches derived from conventional fiber SPME, such as in-tube SPME (also called capillary microextraction, CME), stir-bar sorptive extraction (SBSE), thin-film microextraction (TFME), dispersive SPME, needle trap device (NTD) microextraction and microextraction in a packed syringe (MEPS).

For any one of these operation modes, the extraction process is based on the partitioning equilibrium of target analytes between the coatings fixed on a SPME fiber or inside a capillary, or the adsorbents packed in needle or syringe and the sample matrix, which can be an aqueous solution or the headspace vapor above it. Similar to solid phase extraction (SPE), the extraction phase or adsorbent material plays a key role in the extraction process and determines the selectivity and sensitivity of the SPME-based analytical method. Therefore, the demand for fast and selective extraction of analytes of interest by SPME has inspired great advancements in the preparation of commercial and novel SPME coatings and adsorbents.

3.2 Coating Preparation Techniques Applied For Solid-phase Microextraction and Related Approaches

For SPME and related approaches, the extraction phases are typically immobilized on geometrically diverse surfaces except those packed inside the tubes, needles or syringe tips. The coating procedure affects the thermal and mechanical stabilities of the coating along with its porosity and is crucial to introduce commercially applicable and reproducible SPME devices. A series of preparation techniques have been applied for SPME coating procedures including sol-gel technology, monolithic technology, physical adhesion methods, chemical grafting, electrochemical methods and liquid-phase deposition.

3.2.1 Sol-gel Technology

The sol-gel method has been widely used in the synthesis of inorganic materials or organic-inorganic hybrid materials. In the sol-gel method, a precursor with high chemical activity is dissolved in water or other solvents. After mixing them together, the hydrolysis and condensation reactions occur, forming gels with a three-dimensional network structure. The gels can be transformed into powders, fibers, porous films and coating materials through evaporation, drying and aging. The sol-gel method has the advantages of mild preparation conditions (synthesis can be carried out at room temperature), simple operation, and controllable structure and morphology. In addition, various precursors are available which enable easy introduction of different functional groups into the coating materials. Chemical binding between the solid phase coating and the substrate surface created by this process facilitates stable performance of the coating under varying extraction conditions. The sol-gel method effectively integrates fiber surface treatment, deactivation, coating and stationary phase immobilization into a single step and provides a versatile way of designing organic-inorganic hybrid coatings with desirable extraction properties. Furthermore, the coating porosity provided by the sol-gel method enhances the surface area.

The sol-gel method has been widely employed for the preparation of extraction materials in SPE (Fontanals et al., 2005), SPME (Dietz et al., 2006), CME (Zheng & Hu, 2007) and SBSE (W.M. Liu et al., 2004; Yu et al., 2008), and it is also the most frequently employed methodology in SPME coating preparation. Since Chong et al. (1997) introduced the sol-gel method to the preparation of SPME coatings, many kinds of SPME coatings have been prepared, including functionalized or polymer-functionalized nanomaterials, ionic liquid-mediated SPME coatings, sol-gel-derived polymeric ionic liquid (PIL)-based SPME coatings, SPME coatings on metal wires, molecularly imprinted polymers (MIPs) coatings and immunosorbents.

3.2.2 Physical Adhesion Method

Physical coating was employed for the initial preparation of SPME coatings which were generally suitable for the extraction of materials featuring high viscosity to the substrate. Organic polymer coatings (Ligor & Buszewski, 1999; Ding et al., 2005; Yin et al., 2009), ionic liquid coatings (J. F. Liu et al., 2005; Hsieh et al., 2006) and inorganic coatings (de Oliveira et al., 2005; Wang et al., 2006) for SPME application are mainly prepared by physical coating. In this method, a fiber is placed in a concentrated organic solvent solution of the (polymeric) extraction phase for a period of time and then removed. The solvent is evaporated, resulting in a coating on the fiber (Belardi & Pawliszyn, 1989). For solid materials with large particles, the physical adhesion method can be used to adhere them to the surface of the stir bar by using appropriate adhesion glue (Hou et al., 2004; Hu et al., 2004; Yu & Hu, 2012).

3.2.3 Electrochemical Methods

Electrochemical preparation methods were introduced for SPME coatings in the last decade in which extraction films with variable thickness on unbreakable metal support can be obtained at alow cost and with a simple setup. These methods can be classified into three modes including electrodeposition, anodization and electrophoretic deposition (EPD).

3.2.3.1 Electrodeposition

Using the electrodeposition technique, metallic or conductive polymer (CP) coatings can be deposited onto a base material through electrochemical reduction of metal ions or electropolymerization of CPs from an electrolyte. This technique favors the creation of coatings with porous structure and high thermal stability on metal supports. In this regard, the electrodeposition method has been applied to coat metal supports with CPs, CP nanocomposites, PIL composites and metal oxides.

3.2.3.2 Anodization

Anodization is an electrolytic passivation process used to increase the thickness of the natural oxide layer on the surface of metal parts. Djozan and co-workers used anodized aluminum as a SPME coating for the extraction of some aliphatic alcohols, benzene, toluene, ethylbenzene, and *o*-xylene (BTEX) and petroleum products from gaseous samples (Djozan et al., 2001). Aluminum wires were anodized by direct current in a solution of sulfuric acid at room temperature and were conditioned at 300 °C for 30 min. The results obtained demonstrate the ability of anodized aluminum wire as a new fiber for sampling of organic compounds from gaseous samples. This behavior is likely due to the porous layer of aluminum oxide which is formed on the metal surfaces. This method has low cost, high thermal stability, firmness and long durability.

3.2.3.3 Electrophoretic Deposition

Electrophoretic deposition is a direct particle assembly method that utilizes an electric field to deposit charged nanoparticles from solution onto a substrate. The merit of this method includes a) good reproducibility and easily controlled thickness of coating *via* regulated voltage and electrophoretic deposition time, b) the absence of organic binder, c) low cost and simple setup, d) the high conductivity of the resulted coatings which is proper for electrosorption-enhanced (EE)-SPME applications and e) high chemical, mechanical and thermal stabilities of coating with long lifespan. This approach has been applied for the preparation of single-walled carbon nanotube (SWCNT) and multi-walled carbon nanotube (MWCNT) coatings on the surfaces of platinum and stainless steel wires, respectively.

Compared with other preparation methods, the electrochemical deposition method features the following advantages: (1) the whole coating process is controlled by the electrochemical workstation, ensuring a good reproducibility; (2) metal wires are used as the support material, avoiding the easy corruption of commonly used quartz fibers; (3) a wide range of applications is available including the sorption of organic compounds and the controllable extraction and release of inorganic ions. Currently, the SPME coatings prepared by electrochemical deposition method include the polyaniline (PANi), polythiophene (PTH) and polypyrrole (PPY) coatings (Wu et al., 2002; X. Li et al., 2008a, 2008b). Inherent characteristics of coatings prepared by these methods increase the capability of SPME to sorb polar compounds. In addition, these methods result in high thermal stability and long life span.

3.2.4 Polymerization

In this method, a chemical reaction is performed directly on the SPME substrate to generate coating materials. The molecularly imprinted SPME coatings (Koster et al., 2001; Hu et al., 2007a, 2007b) and some capillary coatings (Fan et al., 2005b; Wei et al., 2005; L. Zhang et al., 2011) represent *in situ* polymerization methods. Polymerization offers another simple *in situ* way for the preparation of SPME coatings, especially silica-based and polymer-based monolithic capillaries in the presence of a functional monomer, cross-linking agent, porogenic solvent and initiator such as heat, light, oxidation and irradiation. Typically, an organic mixture containing monovinyl and divinyl monomers, an initiator and a porogenic solvent are dispersed upon stirring in aqueous medium. Traditional porogens include inert diluents that may either be solvating or non-solvating solvents for the polymer being produced. Induced decomposition of the initiator causes polymerization to start within the individual droplets, ultimately leading to the production of polymer particles. When the polymerization process is completed, the inert porogen is removed from the monolith by leaching, leaving behind the pores.

3.2.5 Chemical Vapor Deposition

As a versatile deposition technique, chemical vapor deposition (CVD) has become one of the main processing methods for the deposition of amorphous, single-crystalline, polycrystalline thin films and coatings for a wide range of applications, with chemical reactions of gaseous reactants on or near the vicinity of a heated substrate surface. CVD can be performed in various formats including plasma-assisted CVD, combustion CVD, flame-assisted CVD, electrostatic-assisted CVD, metal organic CVD and photo-assisted CVD. CVD can provide highly dense and pure materials and has the ability to control crystal structure, surface morphology and orientation of the coating by controlling the process parameters.

3.2.6 Liquid Phase Deposition

Liquid phase deposition (LPD) method is a low cost and environmentally-friendly process for thin film preparation. LPD refers to the formation of oxide thin films from an aqueous solution of a metal-fluoro complex which is slowly hydrolyzed by adding water, boric acid (H_3BO_3) or aluminum metal. The addition of water directly forces precipitation of the oxide while H_3BO_3 or aluminum metal acts as a fluoride scavenger, destabilizing the fluoro complex and promoting precipitation of the oxide. LPD was first developed for depositing SiO_2 thin films and it was later used to prepare other metal oxide films. Lin et al. (Lin et al., 2008) deposited a thin layer of titanium dioxide (TiO_2) nanoparticles onto the surface of a capillary column by the LPD technique and applied the device to selectively concentrate phosphopeptides from protein digest products.

3.3 Commercial Solid-phase Microextraction Coatings

So far, a variety of extraction phases for SPME have been commercialized, including single-phase sorbents, such as polydimethylsiloxane (PDMS), polyacrylate (PA) and Carbowax (CW) as well as mixed-phase sorbents, such as Carboxen/polydimethylsiloxane (CAR/PDMS), polydimethylsiloxane/divinylbenzene (PDMS/DVB), divinylbenzene/Carboxen/polydimethylsiloxane (DVB/CAR/PDMS) and Carbopack Z/PDMS (Pawliszyn, 1997, 2009). **Table 3.1** lists the commercial SPME coatings generally employed in gas chromatography (GC) or high performance liquid chromatography (HPLC) analysis, along with their characteristics and recommended uses.

Most of the commercial SPME coatings involve such polymers as PDMS, PA, poly(divinylbenzene) (PDVB), CAR, polyethylene glycol (PEG) and CW.The original purpose of polymeric coatings was to protect the optical fibers from breakage. The thickness of the SPME coating generally ranges between 10 and 100 μm. Thermally stable polar or nonpolar polymeric sorbents that allow fast solute diffusion are suitable for use as the extraction phase.

PDMS is the most frequently used sorbent in SPME due to its inherent versatility, high thermal stability, and application potential for a wide range of analytes. Commercially available PDMS-coated fibers have a PDMS sorbent immobilized on the fiber using a cross-linkable functionality that is present in the polymeric structure. This cross-linking provides PDMS coatings with higher thermal stability (~340 °C) as well as solvent stability. However, due to the difficulty of stabilizing thick coating through a cross-linking reaction, the only commercially available PDMS coated fiber that can withstand solvent rinsing is the one with a 7 μm coating thickness. PDMS is a nonpolar polymer which usually extracts nonpolar analytes such as volatile organic compounds (VOCs), polycyclic aromatic hydrocarbons (PAHs), alkanes, BTEX compounds and some pesticides. PA is a highly polar sorbent immobilized by partial cross-linking. It is commercially available in an 85 μm thickness. Due to its high polarity, PA coating is frequently used to extract polar analytes such as alcohols, organic acid, aromatic amines, phenols and their derivatives and polar pesticides. PA is a rigid, low density solid polymer at room temperature, leading to slower analyte diffusion and longer extraction time.

Composite (mixed phase) coatings were introduced to enhance the selectivity toward target analytes. They are prepared by embedding porous particles (one or more types) in the partially cross-linked polymeric phase. Introduced in 1996, PDMS/PDVB (Mani, 1999), is suitable for the extraction of polar compounds such as alcohols and amines. (Werciski, 1999). CAR/PDMS employed a highly porous polymeric material named Carboxen and was used in SPME for the extraction of VOCs (Tumbiolo et al., 2004) and hydrocarbons (Pinho et al., 2006). CW/PDVB, introduced in 1997 (Hall & Brodbelt, 1997), is a blend of porous PDVB and polar CW polymeric phase. Similar to PDMS/PDVB, CW/PDVB coating is also used to extract polar compounds such as alcohols (Zuba et al., 2002). However, the swelling tendencies of CW in water and its oxygen sensitivity at temperatures above 220 °C are the major drawbacks of CW/PDVB. Like other composite coatings, Carbowax/Templated resin (CW/TR), is made by blending porous templated resin with polar CW polymer. Due to the presence of both hydrophilic (CW) and hydrophobic (Templated Resin) moieties in this polymeric blend, it provides remarkable selectivity for the extraction of surfactants from aqueous media (Aranda et al., 2000). However, compared to homogeneous polymeric coatings such as PDMS and PA, composite coatings have lower mechanical stability.

Even though the commercialization of SPME coatings greatly promoted the rapid development of SPME technology, commercial SPME fibers often feature short service-life, high cost, lack of selectivity and other shortcomings. Therefore, the preparation of novel SPME coatings with high extraction efficiency (EE), long service-life and good selectivity for target analytes is of great interest.

3.4 Novel Materials for Solid-phase Microextraction and Related Approaches

In SPME, the fiber coating is the most important factor that determines the sensitivity and selectivity, which is similar to the adsorbents used in SPE. The ideal SPME coatings are expected to have the following characteristics: (1) porosity, having a large specific surface area; (2) low background signal; (3) high chemical and mechanical stability; (4) fast kinetics of adsorption and desorption; (5) reversible adsorption; (6) high selectivity; (7) high recovery rate. Great efforts have been made in the exploration of novel materials for SPME applications. A variety of novel materials including carbonaceous materials (X. Q. Liu et al., 2004; Agrawal, 2007; Wu et al., 2013; Su et al., 2014), molecular/ion imprinted polymers (Zambrzycka et al., 2011; Zheng et al., 2011; Lv et al., 2012; Zhang & Hu 2012), nanostructured materials (Liang et al., 2001; Pu et al., 2005; Yin et al., 2005; Huang et al., 2007; Suleiman et al., 2007), restricted access materials (RAM) (Hu et al., 2009; F. Wang et al., 2012), monolithic materials (L. Zhang et al., 2011, 2013; Choi et al., 2013; Y. Zhang et al., 2013) and metal-organic frameworks (MOFs) (Zhou et al., 2006; Zheng et al., 2007; Gu et al., 2011a; Hu et al., 2013). These materials have been employed in SPE and also exhibited good extraction performance in SPME.

3.4.1 Nanostructured Materials

Nanostructured materials are defined as a set of substances with at least one external dimension in the size range of approximately 1–100 nm. Nanostructured particles that are naturally occurring (*e.g.*, volcanic ash, soot from forest fires) or are the incidental byproducts of combustion processes (*e.g.*, welding, diesel engines) are usually physically and chemically heterogeneous and often termed ultrafine particles. Engineered nanostructured particles are intentionally produced and designed with very specific properties related to shape, size, surface properties and chemistry. These properties are reflected in aerosols, colloids and powders. Often, the behavior of nanomaterials may depend more on surface area than particle composition itself. Relative surface area is one of the principal factors that enhance its reactivity, strength and electrical properties.

Due to the small size effect, the number of surface atoms on nanomaterials increases rapidly with the decrease in the particle size, and the binding sites of surface atoms are unsaturated, causing a tendency for nanomaterials to adsorb to external substances and become saturated. On the other hand, nanostructured materials can provide high adsorption capacity because of their much larger surface area over other adsorbent materials. From this point of view, nanostructured materials are good alternatives to certain adsorbents and their application potential in preconcentration of interest analytes has been demonstrated extensively (Tian et al., 2013).

Table 3.1 Commercial SPME fibers (Mills & Walker, 2000).

Fiber coating	Film thickness (μm)	Polarity	Coating stability	Maximum temperature (°C)	Analytical application	Recommended uses
PDMS	100	non-polar	non-bonded	280	GC/HPLC	volatiles
		non-polar	non-bonded		GC/HPLC	
	30	non-polar	bonded	280	GC/HPLC	non-polar semi-volatiles
	7			340		mid- to non-polar semi-volatiles
PDMS-DVB	65	bi-polar	cross-linked	270	GC	polar volatiles
(StableFlex fiber)	60	bi-polar	cross-linked	270	HPLC	general purpose
	65	bi-polar	cross-linked	270	GC	polar volatiles
PA	85	polar	cross-linked	320	GC/HPLC	polar semi-volatiles (phenols)
CAR-PDMS	75	bi-polar	cross-linked	320	GC	gases and volatiles
(StableFlex fiber)	85	bi-polar	cross-linked	320	GC	gases and volatiles
CW/DVB	65	polar	cross-linked	265	GC	polar analytes (alcohols)
(StableFlex fiber)	70	polar	cross-linked	265	GC	polar analytes (alcohols)
CW/TR	50	polar	cross-linked	240	HPLC	surfactants
DVB-PDMS-Carboxen	50/30	bi-polar	cross-linked	270	GC	odours and flavours

According to Richard W. Siegel, nanostructured materials can be classified as zero-dimensional (atomic clusters, filaments and cluster assemblies), one-dimensional (multilayers), two-dimensional (ultrafine-grained overlayers or buried layers), and three-dimensional nanostructures (nanophase materials consisting of equiaxed nanometer sized crystals). They can exist in single, fused, aggregated or agglomerated forms with spherical, tubular, and irregular shapes. Common types of nanomaterials include nanotubes, dendrimers, quantum dots and fullerenes. The morphologies (shapes) of various nanomaterials are depicted in Figure 3.1. Differing in their compositions, carbon nanomaterials and metal/metal oxide nanomaterials have caused great interest for SPME coatings preparation.

3.4.1.1 Carbon Nanomaterials

Novel carbon nanomaterials include zero-dimensional fullerenes, one-dimensional carbon nanotubes (CNTs) and two-dimensional graphenes, with graphene considered to be the basic unit of carbon materials (Geim & Novoselov, 2007). The structure of these carbon nanomaterials are depicted in **Figure 3.2**.

Fullerenes can be considered as a carbon cage formed by wrapping graphenes together. A series of isomers for fullerenes including C_{60}, C_{70}, C_{240}, C_{540} and C_{720} have been recognized. Fullerenes have high electron affinity, strong hydrophobicity, large surface area (Scida et al., 2011) and have been widely used for the separation and enrichment of analytes. Vallant et al. (2007) prepared C_{60}-bonded silica material and demonstrated its extraction capability for flavonoids, proteins, polypeptides and hydrophilic phosphorylated polypeptide. Gallego and co-workers (Jurado-Sánchez et al., 2009) separated aromatic amines from non-aromatic amines by using C_{60} fullerenes as the adsorbents, based on the π-π conjugation between C_{60} fullerenes and aromatic amines. The preparation of a fullerene-coated SPME fiber is hampered by the poor solubility of fullerene in solvents. Polymeric fullerene has been coated on a SPME fiber with polysiloxane using an epoxy resin glue and employed for the extraction of BTEX, naphthalene congeners, and phthalic acid diesters from the headspace of water samples (Xiao et al., 2000). A fullerene-polysiloxane surface-bonded porous coating on a fused-silica fiber surface was obtained by sol-gel technology (Yu et al., 2002). The coating had stable performance at high temperature (even to 360 °C) and good resistance to organic and inorganic solvents because of the properties of fullerene and the chemical binding between the coating and the fiber surface. Compared with commercial SPME stationary phases, it showed higher sensitivity, faster velocities of mass transfer for aromatic compounds and possessed planar molecular recognition for polychlorinated biphenyls (PCBs).

CNTs can be classified into SWCNTs and MWCNTs, which are circular tubes formed by curling single-layer graphenes and coaxial tubes formed by curling multilayer graphenes, respectively. The diameter of CNTs is generally between a few nanometers to dozens of nanometers, the length can be up to several centimeters and both ends generally have a fullerene-like structure. Similar to fullerene, CNTs have a strong adsorption affinity for organic substances, providing a good alternative for adsorbents (Sae-Khow & Mitra, 2009; See et al., 2010). Sae-Khow and Mitra (2009) packed CNTs into a micro-syringe needle, and preconcentrated target analytes by pulling the plunger back and forth. CNTs provided much higher EE over the same amount of C_{18}, and was much easier to achieve miniaturized devices. See et al. (2010) developed a membrane protected device in which CNTs were flame-sealed in a conical polypropylene film. The polypropylene film was fitted onto a micro-syringe tip and a dynamic extraction performed by a controlling device. CNTs exhibit weak adsorption for inorganic metal ions, which can be greatly improved by oxidizing CNTs and introducing some oxygen-containing functional groups (Li et al., 2002, 2003). MWCNT-coated fibers were prepared for determination of polybrominated diphenyl ethers (PBDEs) in water samples (Wang et al., 2006; Tian & Feng, 2008) by direct and headspace SPME. The MWCNT powder was dispersed in dimethylformamide and then it was coated onto the fused-silica fibers through a physical deposition method. To improve the prepara-

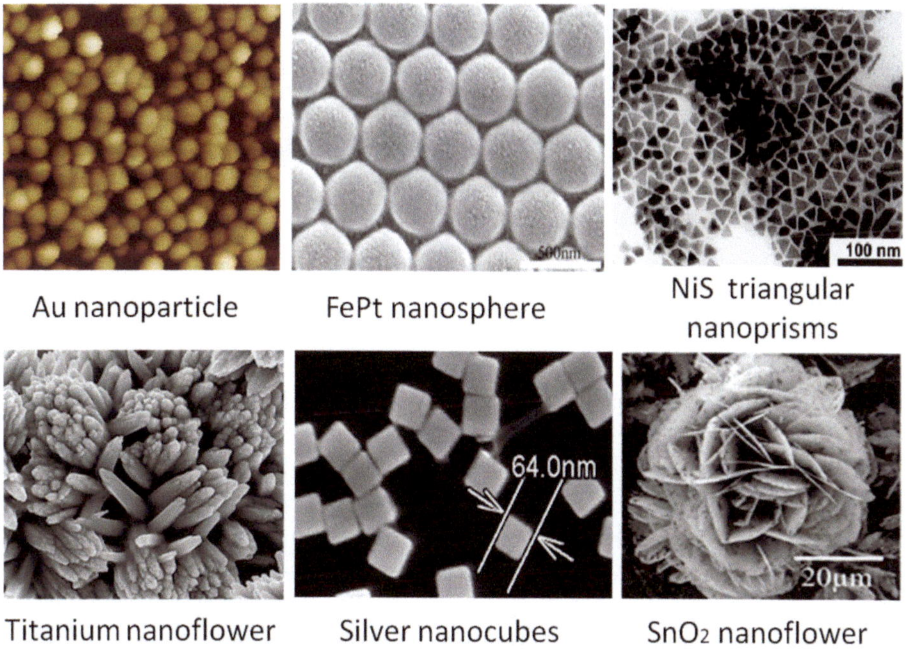

Figure 3.1 Morphologies (shapes) of various nanomaterials. (https://www.itrc.narl.org.tw/Research/Product/Vacuum/nanosphere-e.php, last accesed, August 2014; http://www.mse.umd.edu/research/nanotechnology last accesed, August 2014; http://www.ammrf.org.au/access/facilities/high-resolution-sem-microanalysis-facility-unsw/ last accesed, August 2014. Reprinted from Chen et al. (2004) with permission from The Royal Society of Chemistry.

tion reproducibility of the CNTs-based coatings, W.Y. Zhang et al. (2009) prepared a polymer-functionalized SWCNT-TSO-OH SPME coating by sol-gel technology, which strengthened the thermal stability of the coating. The device presented higher EE for target PBDEs than commercial fibers and MWCNT-based fibers. One possible reason is that SWCNTs are smaller than MWCNTs in diameter and have larger surface area. More reaction sites on SWCNTs over MWCNTs favor the derivatization reaction as well.

Graphene is a honeycomb lattice of carbon atoms (Geim & Novoselov, 2007). It is comprised of a single layer of sp^2-hybridized carbon atoms arranged in six-membered rings. Usually, the graphene sheet tends to curl, corrugate, and fold, resulting in a wrinkled structure. This two-dimensional form of carbon displays excellent mechanical, electrical, and optical properties which make it the most promising carbon-based nanomaterial after fullerene and CNTs. Graphene exhibits a high theoretical surface area close to 2700 m^2 g^{-1} (Stoller et al., 2008), more than twice that of SWCNTs. Graphene inherits most of the advantages of CNTs. Moreover, it is prepared from graphite without any residual heterogeneous materials, which usually include the residual metallic impurities and nanoparticles derived from catalysts. Combining these superior characteristics, graphene is expected to have better performance

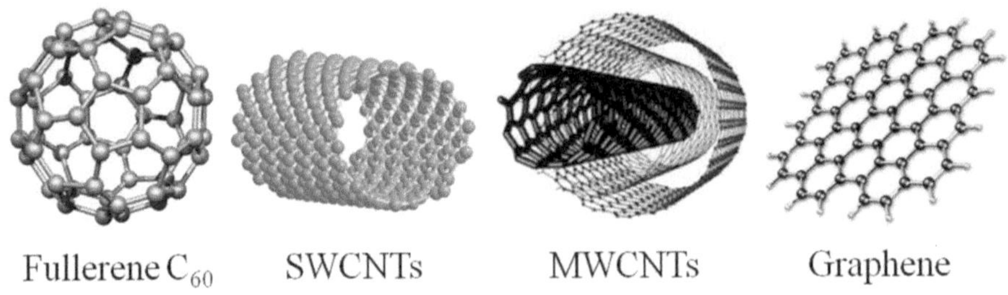

Figure 3.2 Some examples of nanoscale carbon-based materials.

for micro-organic substance analysis. A graphene-based sol-gel plunger-in-needle SPME coating was prepared for the preconcentration of PBDEs in canal water samples (Zhang & Lee, 2011). Graphene was coated on the metallic surface through a physical depositing method (J.M. Chen et al., 2010), and the new fiber possessed a homogeneous, porous and wrinkled surface that showed excellent thermal (over 330 °C), chemical and mechanical stability as well as a long lifespan (over 250 extractions).

In general, the lack of chemical bonding between the SPME coating and the substrate may lead to relative low stability and lifetime. On the other hand, graphene sheets have poor solubility in water and organic solvents, owing to the lack of proper functional groups on its surface. As a result, S.L. Zhang et al. (2011) used graphite oxides (GOs), which have rich oxygen-containing groups in the starting coating material, and chemically bonded them to the fused-silica substrate using 3-aminopropyltriethoxysilane (APTES) as cross-linking agent, then deoxidized the coating using hydrazine to give the graphene coating. The obtained graphene-coated fiber gave higher adsorption affinity to PAHs (EFs> 6300).

As adsorbents for certain analytes, these three carbon nanomaterials possess their own advantages and disadvantages. The common structure or properties among them include the sp^2-hybridized carbon atoms and π-electron conjugated system, allowing the adsorption of benzene ring-containing compounds through π-π conjugation and a large specific surface area which provides high adsorption capacity. They differ in terms of microstructure, purity, preparation cost and adsorption performance. For example, the single sheet structure of graphenes make their theoretical surface area 2 times of that of the SWCNTs (Pumera et al., 2010). Graphenes can sufficiently contact the target analyte during adsorption process through both sides of the sheet-like structure, while CNTs and fullerene contact with target analytes through their outer surface due to steric hindrance (X.Q. Liu et al., 2004), causing a difference in the adsorption performance. Metal impurities would be introduced during the preparation of carbon nanotubes, affecting the properties of the products during adsorption and sensing (Pumera et al., 2010), while graphenes do not have such problems. Graphenes can be synthesized by chemical oxidation-reduction from cheap graphite, and is suitable for large-scale preparation with low cost. Graphenes prepared by chemical methods have a number of oxygen-containing polar groups, favoring the adsorption of polar compounds.

3.4.1.2 Metal Oxide Nanomaterials

Generally, commercial SPME fibers consist of various polar coatings for the extraction of different substances for which the polymer coating is physically deposited on a fused silica fiber. However, these fibers are usually unable to exhibit a good thermal and chemical stability. The operating temperatures for these fibers are often limited to the range of 200–270 °C. On the other hand, the commonly used fused silica fiber features several drawbacks which promoted the development of metallic supporting substrates in novel SPME fiber preparations. Combined with electrochemical deposition methods, metallic wires have been employed as fiber supports featuring improved mechanical stability and reproducibility superior to the traditional fused silica fiber. Some inorganic coatings based on metallic compounds have exhibited good performance in SPME. In particular, metal oxides such as Al_2O_3 (Djozan et al., 2001), ZnO (Djozan & Abdollahi, 2003), ZrO_2 (Alhooshani et al., 2005), and nanostructured PbO_2 (Mehdinia et al., 2006) -based SPME coatings, prepared by electro-oxidizing or electrodeposition techniques, were found to be selective for polar and semi-polar organic compounds and showed improved features such as low cost, durability, sensitivity and a vast range of applications.

The metal oxides are stable over a wide pH range and at high temperature, providing good candidates for SPME coating fabrication. Djozan and co-workers prepared anodized aluminum (Djozan et al., 2001) and zinc (Djozan & Abdollahi, 2003) as metallic SPME fibers for the extraction of polar and semi-polar organic compounds. Aluminum/zinc wire was anodized by use of direct current in a solution of sulfuric acid/sodium hydroxide at room temperature, followed by conditioning at a prescribed temperature. The potential of anodized aluminum/zinc wire as a new fiber for the extraction and sampling of some organic compounds results from the porous layer of metallic oxide formed on the metal surface.

For those synthesized nanostructured metal oxides used as coatings in the fabrication of fibers (Mehdinia et al., 2006; Cao et al., 2008), the nanostructured coating involves a high surface area-to-volume ratio which improves the EE. Nanostructured titania-based SPME fibers were fabricated through *in situ* oxidation of titanium wires with H_2O_2 (30%, w/w) at 80 °C for 24 h (Cao et al., 2008), as shown in **Figure 3.3**. The obtained SPME fibers possess a ~1.2 μm thick nanostructured coating consisting of ~100 nm titania walls and 100–200 nm pores. As the nanostructured titania was formed *in situ* on the surface of a titanium wire, the coating was uniformly and strongly adhered on the wire. Because of the inherent chemical stability of the titania coating and the mechanical durability of the titanium wire substrate, this new SPME

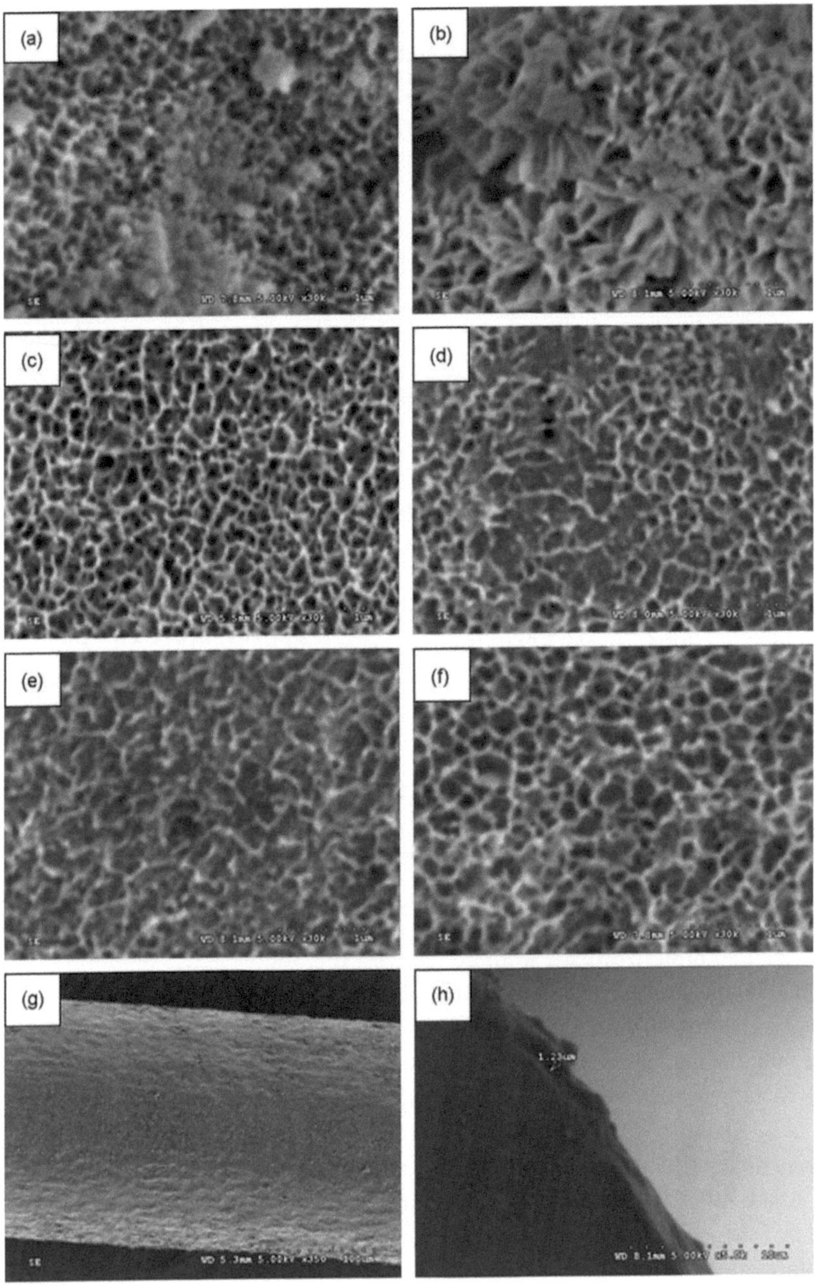

Figure 3.3 SEM images of the titania SPME fiber prepared by oxidization of titanium wire in H_2O_2 (30%, w/w) solution at various conditions. (A) 55 °C, 24 h; (B) 80 °C, 12 h; (C) 80 °C, 24 h; (D) 80 °C, 36 h; (E) 80 °C, 48 h; (F) 80 °C, 72 h; (G) longitudinal image (80 °C, 12 h); (H) image of part of the transect (80 °C, 12 h). Reprinted from Cao et al. (2008) with permission from Elsevier.

fiber exhibited a long life span (over 150 times). A novel Au nanoparticles/SPME fiber prepared by a layer-by-layer (LBL) self-assembly process (J. J. Feng et al., 2010) was also reported. The Au nanoparticles SPME fiber showed high stability to acid, alkali and high temperature, and special selectivity to some analytes based on the hydrophobic interactions and the electron transference effect between the π-donor system and the valency shell of Au. For coinage metals such as Au and Ag, their relativistic effect and relevance in chemistry brings about s-d hybridization, increases the mobilization of their electrons and facilitates the formation of chemical bonds. In addition, they are easily modified by organic molecules containing -SH, resulting in a tendency to adsorb -SH-containing substances.

For the fabrication of SPME coatings on the surface of fused silica (or recently metal alloy) fiber, some pre-treatments are needed prior to sol-gel deposition to create active sites on the substrate to facilitate interactions between the stationary phase and substrate. Briefly, the surface of the fiber is hydrolyzed by NaOH and then neutralized by HCl to create OH functional groups at the fiber surface. Next, it is inserted into a coating media that usually includes a mixture of alkoxysilane precursors, small amounts of water and an acid or base catalyst for a period of time to coat the modified fiber. Yan et al. prepared an etched stainless steel wire fiber for SPME without the need for any additional coatings (Xu et al., 2009). Comparison of the scanning electron microscopy (SEM) images of the stainless steel wire before and after etching (**Figure 3.4**, parts **A** and **B** *vs* parts **C** and **D**) shows that the surface of the stainless steel wire was smooth before etching but became rough and porous with a fine flower-like structure after etching. Such rough and porous flower-like structure of the etched stainless steel wire should significantly increase the surface area and sorption capacity of the fiber.

The electroless plating technique is one of the most frequently adopted industrial processes for metallization to pattern two- and three-dimensional structures and can fabricate uniform metallic thin films on either conductive or nonconductive substrates. Jiang et al. introduced an electroless plating process based on the classic silver mirror reaction into SPME coating fabrication (Feng et al., 2011) and prepared a silver-coated SPME fiber as shown in **Figure 3.5**. The coating had a porous structure, providing a high surface area which is favorable for efficient extraction. The relative thin coating with a thickness of about 12 μm also favors a fast mass transfer during extraction and desorption processes.

Compared with the commonly used silica fibers which are expensive, fragile and therefore must be handled with great care, these inorganic coatings based on metallic compounds are more robust, have lower cost and provide longer lifetimes.

3.4.1.3 Mesoporous Materials

Mesoporous materials have openings within their structure that are between 2 and 50 nm in diameter. In terms of porousness, they are in between* microporous mate-

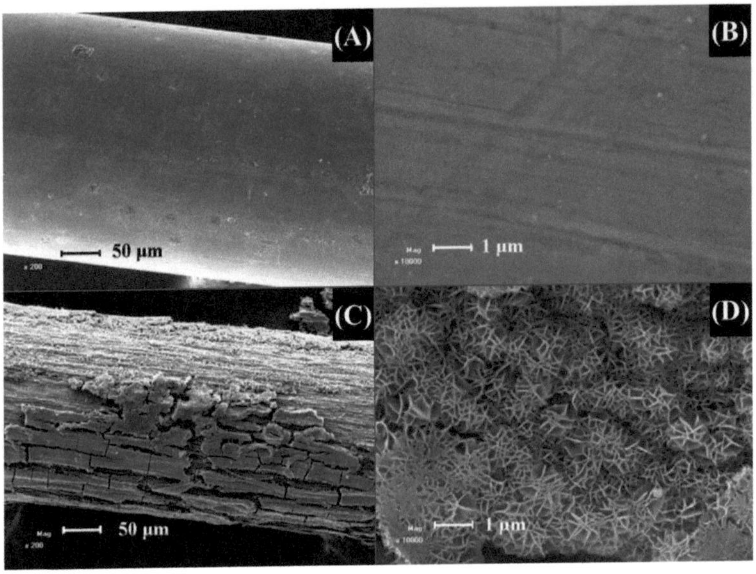

Figure 3.4 Scanning electron micrographs of the surface of the stainless steel wire before (A and B) and after (C and D) etching. The images A and C are at a magnification of 200x; images B and D are at a magnification of 10 000x. Reprinted from Xu et al. (2009), Copyright 2009, with permission from the American Chemical Society.

rial, which has openings less than 2 nm, and macroporous material, which has openings greater than 50 nm. Since their discovery in 1992, mesoporous materials have gained interest throughout the scientific community. These kinds of materials possess large surface area, mesoporous structure, very tight pore size distributions and hence are regarded as attractive candidates for a wide range of applications including shape-selective catalysis, sorption of large organic molecules and chromatographic separations. The potential use of a mesoporous material as an adsorbent can be viewed in three perspectives: (1) As a new nanometer material with unsaturated surface atoms that can bind with other atoms, it possesses high chemical activity, very high adsorption capacity and selective adsorption of metal ions. (2) Due to its large surface area, a mesoporous material provides more active sites, favoring the quantitative adsorption in short time. (3) The mesoporous structure of these materials ensures fast adsorption and desorption.

Ordered mesoporous film is one of the morphologies of ordered mesoporous materials that hold great promise for use as a separation media. Many researchers concentrated on the formation of the mesoporous metal oxide films with periodic pore structures using poly(ethylene oxide)-block-poly(propylene oxide)-block-poly(ethylene oxide) triblock co-polymer nonionic surfactants as the structure directing agents in conjunction with dip-coating (Yang et al., 1998). Those mesoporous materials also exhibit highly ordered and oriented mesostructures with variable pore size and porosity like that of bulk mesoporous material and can be used as separation media and

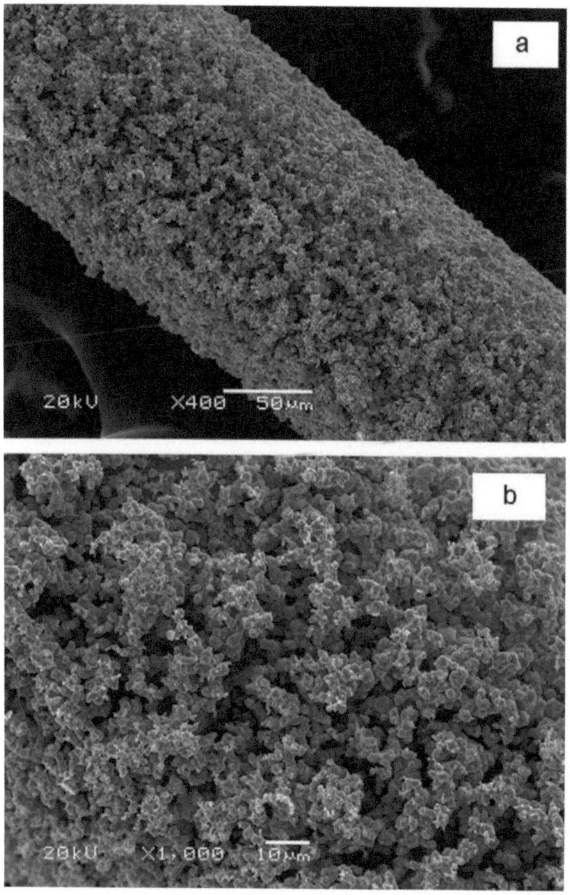

Figure 3.5 SEM images of the silver-coated SPME fiber at the magnification of (A) 400; (B) 1000. Reprinted from Feng et al. (2011) with permission from Elsevier.

chemical sensors. It is worth pursuing further exploitation of this unique material as a separation medium, especially for microextraction techniques.

Ordered mesostructure silica materials have a large surface area, highly ordered pore structure, very tight pore size distributions and thus have been considered as attractive candidates for a wide range of applications. Today, a variety of functionalized mesoporous silica materials have been introduced to SPME coatings. MCM-41 and phenyl functionalized MCM-41 mesoporous organosilica as a fiber coating in SPME exhibited better adsorption and selectivity than a bonded silica phase coating for the extraction of aromatic compounds (Hou et al., 2004; Du et al., 2005). Compared with MCM-41, SBA-15 materials are more desirable because they have better thermostability owing to their more regular structure, larger pore sizes and thicker pore wall. Hashemi et al. (2009) synthesized amino-ethyl-functionalized SBA-15 as a SPME fiber coating and good extraction ability for phenolic compounds was observed

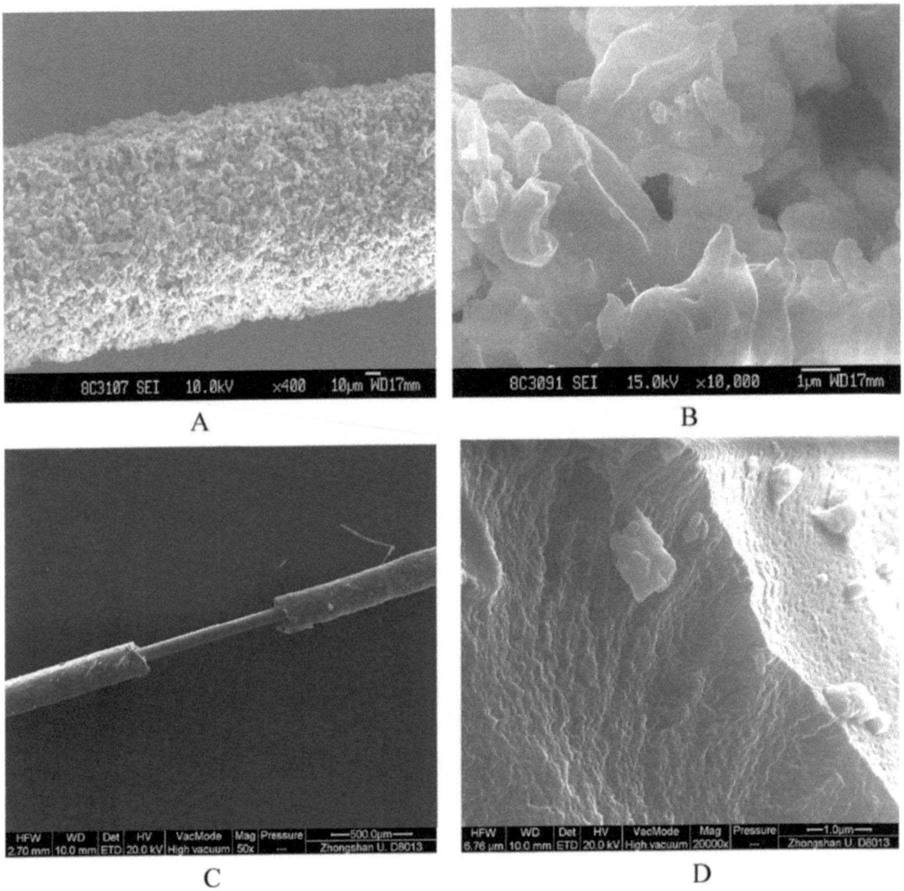

Figure 3.6 SEM images of the direct-coated vinyl-SBA-15 fiber (A, 400×; B, 10,000×) and SEM images of the sol-gel-coated vinyl-SBA-15 fiber (C, the image of the surface, 50×; D, the image of the coating crack of the fiber, 20,000×). Reprinted from Zhu et al. (2012) with permission from Elsevier.

due to the introduction of amino groups. Based on the hydrophobicity of vinyl groups, vinyl-functionalized SBA-15 was prepared by a one-step synthesis method and used as coating material of SPME based on sol-gel and direct coating techniques (Zhu et al., 2012) (**Figure 3.6**). The synthesized vinyl-SBA-15 organosilica had a highly ordered mesoporous structure, good thermal stability and a specific surface area of 688 $m^2\ g^{-1}$. The fibers prepared by either direct coating or sol-gel method exhibited high thermal stability (310 °C for direct-coated and 350 °C for sol-gel) and excellent solvent durability in methanol and acetonitrile.

These functionalized mesoporous silica materials can be prepared by post-grafting or direct synthesis. In the former method, organic functional groups are covalently attached to the pore surface by the reaction with the existing high concentration of surface silanol groups. Amino, thiol, cyclodextrin and alkyl groups have been attached to the mesoporous structure, and this method has been identified as a convenient method to obtain highly effective sorbents. Compared to the post-grafting method, the direct synthesis, which involves the one-step co-condensation of tetraalkoxysilanes and organosilanes, offers higher and more uniform surface coverage of functional groups and better control of the surface properties of the resultant materials.

3.4.1.4 Application of Nanomaterials in Solid-Phase Microextraction and Related Approaches

The extraction capacity of SPME fibers, which is limited by the tiny diameter of the supporting fiber, can be improved significantly by the large specific surface area and multiple active sites of nanostructured materials for adsorbing analytes. As such, there is a growing interest in recent years in developing SPME devices based on nanostructured (nanotubes, nanoporous thin film or layered nanospheres) sorbents, which possess strong extraction capacity. These nanostructured materials in SPME have important characteristics such as large modificative surfaces and multiple active sites for analyte recognition. Since specific surface area is inversely proportional to the particle size, fibers with a nanostructured coating can greatly increase the effective surface area. Consequently, the extraction capacity is enhanced while the extraction time to reach equilibrium decreases. The application of various nanomaterials in SPME is listed in **Table 3.2**.

3.4.2 Molecularly Imprinted Materials

MIPs are a type of synthetic materials that contain artificially generated binding sites to recognize a target molecule in preference to other compounds with similar structures and they have been extensively used in catalysis, sensors, drug carriers, artificial antibodies, and sample pre-treatment techniques including SPE (Sellergren, 1994), SPME (Prasad et al., 2008; Huang et al., 2009b) and SBSE (Zhu et al., 2006; Zhu & Zhu, 2008; Hu et al., 2010a). MIP synthesis involves three steps: (1) the template molecule integrates with the functional monomer by covalent or non-covalent binding forces; (2) the formed complexes react with a cross-linking agent to form the polymer material; (3) the template molecule/ion is removed (Fan et al., 2009). **Figure 3.7** illustrates the general scheme for the preparation of a molecularly imprinted, boronate-functionalized, monolithic column and its recognition mechanism toward a target glycoprotein.

In general, hydrogen bonds are frequently employed for the establishment of recognition sites in MIPs. However, the imprinting effect may be weakened or even damaged by strongly polar solvents, especially in aqueous media (Martín-Esteban, 2013). Thus, the development of new strategies for the use of MIPs in aqueous environments is desirable.

Table 3.2 Application of nanomaterials to SPME.

Coatings	Preparation method	Analytical Method	Analytes	Sample	Ref.
carbon NPs-stainless steel wire	flame-based preparation process	SPME-GC	PAEs PAHs	PAEs in food-wrap PAHs in cigarette ash and snow water	(Sun et al., 2013a)
epoxide polymer and carbon nanofiber	electrospun	DI-SPME-LC	nonvolatile analytes		(Newsome et al., 2012)
CMK-5 CMK-3 Pt/Cu/Fe CMK-1		(HS)-SPME-GC-FID/MS	chlorophenols; phenolic compounds; benzene, toluene; BTEX	tap and sea water samples; petrochemical samples	(Anbia & Khazaei, 2011; Rahimi et al., 2011; Anbia et al., 2012)
nano-fibrous structures of PPY	electrochemical method	HS-SPME-GC-FID	MNT isomers	real water samples	(Ebrahimzadeh et al., 2010)
carbon monolith	polymerization-carbonization method	SPME-GC-MS	phenols	real water samples	(Shi et al., 2009)
activated charcoal-PVC fiber-Ag		SPME-GC-FID; Capillary GC	organophosphorus pesticides (OPPs); n-alkanes	fruit juice; soil samples	(Farajzadeh & Matin, 2002; Farajzadeh & Hatami, 2004)
LTGC macrofiber		SPME-GC-MS	five lung cancer-related 2-methylheptane, styrene, propylbenzene, decane, undecane	simulating human breath	(Giardina & Olesik, 2003)
CNTs-HF-SPME		HF-SPME- HPLC-DAD	carbamate pesticides	apples	(Song et al., 2013)
PEG-g-MWCNTs	covalent functionalization	SPME-GC-FID	ibuprofen, naproxen and diclofenac; furan	real water samples; environmental food samples	(Sarafraz-Yazdi et al., 2012a, 2012b)

continued **Table 3.2** Application of nanomaterials to SPME.

Coatings	Preparation method	Analytical Method	Analytes	Sample	Ref.
MIPPy/MWCNTs composite Pt wire	electrochemical method	SPME-HPLC	fluoroquinolone antibiotics	urine and soil samples	(X. Liu et al., 2012)
SWCNTs	sol-gel	HS-SPME-GC-ECD/MS	chlorophenols and OCPs; monobutyltin, dibutyltin and tributyltin; PBDEs;VOCs	water samples; exhaled breath of healthy states and of chronic renal failure states	(Haick et al., 2009; Rastkari et al., 2010; Sun et al., 2011; W. Y. Zhang et al., 2009)
FGO with a controllable number of layers PTFE	LBL assembly of graphene	online SPME-HPLC- fluorescent detection	PAHs	environmental samples	(W. P. Zhang et al., 2013)
GO fused-silica	covalently bonding	HS-SPME-GC-MS	PAHs	water samples and soil samples	(S. L. Zhang et al., 2011)
MOF-199-GO		HS-SPME-GC-ECD	OCPs	river water, soil, water convolvulus and longan	(S. L. Zhang et al., 2013)
graphite fiber		SPME-GC-FID	alcohols and BTEX		(Luo et al., 2009)
fullerol fused-silica fiber	sol-gel	HS-SPME-GC-ECD/FID	PCBs, PAHs, polar aromatic amines (PAAs)		(Yu et al., 2002)
polymeric fullerene		HS-SPME-GC-FID	BTEX, naphthalene congeners, and phthalic acid diesters	water samples	(Xiao et al., 2000)
Au NPs fused silica fiber		HS-SPME	PAHs	seawater of Persian Gulf and Caspian Sea	(Karimi et al., 2013)
porous gold fiber	a rod of 585-gold etched by using nitric acid	SPME-GC-MS	dodecanethiol volatile organic sulfur-containing compounds	ethanolic solution in the headspace of the fresh onion sample	(Hafez & Wenclawiak, 2013)

continued **Table 3.2** Application of nanomaterials to SPME.

Coatings	Preparation method	Analytical Method	Analytes	Sample	Ref.
Au NPs-stainless steel wire	LBL self-assembly process	SPME-GC	PAHs, diphenyl and terphenyls	rainwater and soil extract	(J. J. Feng et al., 2010)
gold wire		HS-SPME-ETAAS	mercury	industrial wastewaters and tuna fish samples	(Hashemi & Rahimi, 2007)
Ag	electroless plating technique		PAEs and PAHs	aqueous extracts of disposable paper cup and instant noodle barrel	(J. J. Feng et al., 2011)
(3TMSPMA)/Cu nanocomposite	electrochemically co-deposited	Fiber SPME-GC-MS	aromatic pollutants as model compounds		(Tehrani et al., 2013)
Cu on a Cu wire	electrodeposition of sol-gel coatings using negative potentials on porous Cu wire	HS-SPME	aromatic hydrocarbons	aqueous samples	(Azar et al., 2012)
Fe	bare stainless steel wire etching with hydrofluoric acid	GC-FID	PAHs	local river water and wastewater samples	(Xu et al., 2009)
Al	anodization- chemical etching	alumina nanowire SPME-GC-MS	volatile esters and alcohols	banana and fermented glutinous rice	(Z. M. Zhang et al., 2013)
nanoporous array anodic alumina	two-step anodic oxidization method	Fiber GC-MS	mixed standards of VOCs	Bailan flower, stinkbug and orange peel samples	(Z. M. Zhang et al., 2012)
Al_2O_3/TiO_2 composite-Cu	sol-gel	fiber	chloroform, carbon tetrachloride, trichloroethene, and tetrachloroethene	Urine	(Farhadi et al., 2011)

continued **Table 3.2** Application of nanomaterials to SPME.

Coatings	Preparation method	Analytical Method	Analytes	Sample	Ref.
alumina-OH-TSO hybrid materials	hybrid organic-inorganic sol-gel	fiber	volatile alcohols and fatty acids	Beer	(M. M. Liu et al., 2006)
3MPTMOS-PEG-Cu	self assembled monolayers (SAM) and sol-gel(chemically bonded)	a copper tube online in-tube SPME-HPLC	PAHs	real water samples	(Bagheri et al., 2011a)
SU-8 2100-stainless steel wire carbon nanofibers by pyrolyzing SU-8	electrospinning polymers into nanofibrous mats pyrolyzing	HS-SPME-GC-FID	nonpolar compounds (BTEX and polar compounds (phenol, 4-chlorophenol and 4-nitrophenol)		(Zewe et al., 2010)
APTMS/PDMS-silica	a sol-gel organically modified silica	HS-SPME		synthetic samples and beer	(Biajoli & Augusto, 2008)
CN-PDMS coating -fused silica capillary	sol-gel 3-cyanopropyltriethoxysilane and hydroxy-terminated PDMS	CME-GC	polar and nonpolar analytes simultaneously	aqueous samples	(Kulkarni et al., 2006)
nanoporous silica-NH_2/ stainless steel wire		HS-SPME-GC-MS	volatile component	Citrus aurantium L. leaves. A	(Gholivand et al., 2013a)
porous flower-like silica / stainless steel wire	hydrothermal process	SPME LPME-GC-NPD	organophosphorus pesticides	tap, river and waste water samples	(Saraji & Farajmand, 2012)
aniline-silica nanocomposite/stainless steel wire	electrodeposited	HS-SPME-GC	PAHs	Kalan dam, rain and tap water samples	(Bagheri & Roostaie, 2012)

continued **Table 3.2** Application of nanomaterials to SPME.

Coatings	Preparation method	Analytical Method	Analytes	Sample	Ref.
phosphonate grafted silica NPs/capillaries	liquid phase deposition	IT-SPME (OT-IMAC)	phosphopeptides	a-casein	(Wu et al., 2010)
silica NPs/capillary	liquid-phase deposition	in-tube SPME-HPLC	endocrine disruptors and PAHs		(T. Li et al., 2009)
silica-bonded phase/ stainless steel fiber		SPME-GC	permethrin residues	Cucumber	(Noroozian et al., 2004)
SBA-15/PANI/Fe		Fiber MA-HS-SPME	volatile component of Teucrium polium L		(Gholivand et al., 2013a, 2013b)
PT/SBA-15 nanocomposite stainless-steel wire		SPME fiber-GC (laboratory-designed SPME device)	PAHs	water samples	(Abolghasemi & Yousefi, 2014)
vinyl-SBA-15 mesoporous organosilica	direct coating and sol-gel	SPME fiber	non-polar compounds (BTEX) polar compounds (phenols)	tap water, mineral water and lake water	(Zhu et al., 2012)
PPy/SBA15 stainless steel wire	*in situ* polymerization technique	SPME-GC-MS	PAHs	aqueous sample solutions	(Gholivand et al., 2011)
HPTES-SBA-15 copper wire	Functionalized with HPTES	SPME-GC-MS	BTEX and some phenolic compounds	spiked river water and sewage samples	(Hashemi et al., 2009)
methyl-, propyl- and octyl-MCM-41	one-step co-condensation	Fiber	DBP		(Rao et al., 2013)

continued **Table 3.2** Application of nanomaterials to SPME.

Coatings	Preparation method	Analytical Method	Analytes	Sample	Ref.
incorporation of silica NPs, CNTs and CNT-COOH to PDMS_Cu	SAM	HS-SPME-GC-MS	volatole organic compounds (VOCs)	eucalyptus leaf	(Azar et al., 2013)
CNT-TiO_2 composite	electroless plating and sol-gel	Fiber in direct-immersion mode GC	PAHs	real water samples	(Sun et al., 2013b)
nano-TiO_2/stainless steel wire	electrophoretic deposition	DI-SPME-GC-MS	PAEs	bottled mineral water sample	(Banitaba et al., 2013)
Al_2O_3/TiO_2 composite/Cu	sol-gel		chloroform, carbon tetrachloride, trichloroethene, and tetrachloroethene	urine	(Farhadi et al., 2011)
titania-chitin/silver wire	sol-gel	HS-SPME-GC	aliphatic alcohols	apple juice samples	(Farhadi et al., 2010)
nanostructured titania	*in situ* oxidation	HS-SPME-GC-ECD	DDT and its degradation products		(Cao et al., 2008)
ZrO_2- PEG/PDMS/ NiTi alloy	electrolytically deposited chemical bonding	SPME-GC-ECD	haloanisoles in red wine samples BTEX, PAEs	aqueous samples	(Budziak et al., 2007, 2008a, 2008b, 2008c, 2008d, 2009)
SnO_2 nanorods/fused silica fiber	a hydrothermal process	HS-SPME-GC-MS	1,4-dichloro-2-nitrobenzen, biphenyl, and acenaphthene	environmental water samples	(Alizadeh & Najafi, 2013)
oriented ZnO nanorods / porous PANI film	*in situ* hydrothermally grown	SPME-GC-FID	benzene homologues	three environmental water samples	(Zeng et al., 2013)

continued **Table 3.2** Application of nanomaterials to SPME.

Coatings	Preparation method	Analytical Method	Analytes	Sample	Ref.
ZnO/graphene coating/ silica fiber	deoxidized by hydrazine and dehydrated at high temperature sol-gel	SPME-GC-MS	several most abundant sulfur volatiles in Chinese chive and garlic sprout	chinese chive garlic sprout	(S. L. Zhang et al., 2012)
OTMS on the surface of ZnO nanorods		HS-SPME-GC	benzene homologues	limnetic water samples	(Zeng et al., 2012)
ZnO nano and micro rod on fused silica	hydrothermal process	SPME-GC-MS	1,4-dichloro-nitrobenzene, biphenyl and acenaphthene	environmental water samples	(Alizadeh et al., 2011)

3TMSPMA, 3-trimethoxysilyl propyl methacrylate; CMK, carbon mesoporous from Korea; CN-PDMS, cyano-polydimethylsiloxane; DAD, diode array detection; DBP, dibutyl phthalate; DI-SPME, direct solid-phase microextraction; ECD, electron capture detector; ETAAS, electrothermal atomic absorption spectrometry; FID, flame ionization detection; FGO, functional graphene oxide; GO, graphite oxide; HF-SPME, hollow fiber solid-phase microextraction; HPTES, 3-[Bis(2-hydroxyethyl)amino]propyl-triethoxysilane; HS-SPME, headspace solid-phase microextraction; NP, nanoparticles; LBL, layer by layer; LC, liquid chromatography; LPME, liquid-phase microextraction; LTGC, low-temperature glassy carbon; MA, microwave assisted; MIPPy, molecularly imprinted polypyrrole; MNT, mononitrotoluene; MS, mass spectrometry; NPD, nitrogen phosphorous detection;OCPs, organochlorine pesticides; OH-TSO, hydroxylterminated silicone oil; OPPs, organophosphorus pesticides; OT-IMAC, open tubular-immobilized metal-ion affinity chromatography; PAAs, polar aromatic amines; PAEs, phthalate esters; PVC, polyvinyl chloride.

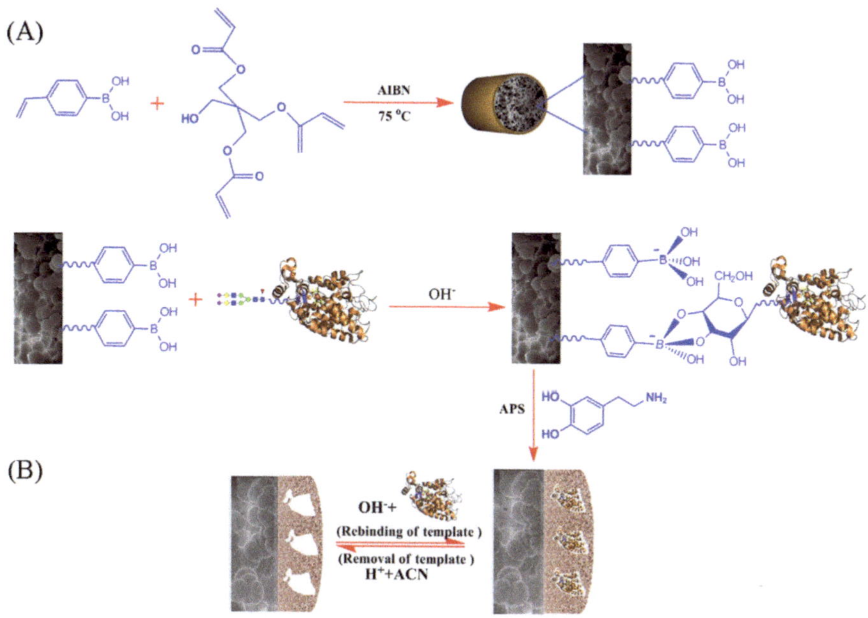

Figure 3.7 (A) Preparation of 4-vinylphenylboronic acid (VPBA)-based molecularly imprinted monolith with polydopamine coating and (B) its recognition mechanism toward glycoproteins. Reprinted from Lin et al. (2013b) with permission from Elsevier.

To address the water incompatibility of MIPs for SPME in real sample analysis, sol-gel technology has been employed to produce thermally stable imprinted stationary phases in SPME (Khorrami & Rashidpur, 2012; Farahani et al., 2009). M. K. Y. Li et al. (2009) imprinted an organically modified silicate (ORMOSIL) SPME stationary phase with decabromodiphenyl ether (BDE-209) by conventional sol-gel techniques from phenyltrimethoxysilane and tetraethoxysilane. The imprinted ORMOSIL sorbent was coated on fused silica fibers with a coating thickness of only 9.5 μm and volume of just 0.12 μL. Khorrami and Rashidpur (2012) prepared caffeine-imprinted sol-gel coated SPME fibers using a polymerization mixture composed of vinyl trimethoxysilane and methacrylic acid as the vinyl sol-gel precursor and functional monomer, respectively. The prepared coating showed good selectivity towards caffeine in the presence of some structurally-related compounds. Also, it offered high imprinting capability in comparison to a bare fiber and non-imprinted coating. Diazinon-imprinted sol-gel coated SPME fibers (Wang et al., 2013) exhibited a rough and porous surface and a larger extraction capability than the non-imprinted polymer and commercial fibers. Furthermore, the fiber exhibited excellent thermal (about 350 °C) and chemical stability.

A method involving the introduction of metal ions as a mediator to strengthen the interaction of the functional monomer and the template in the aqueous matrix can help

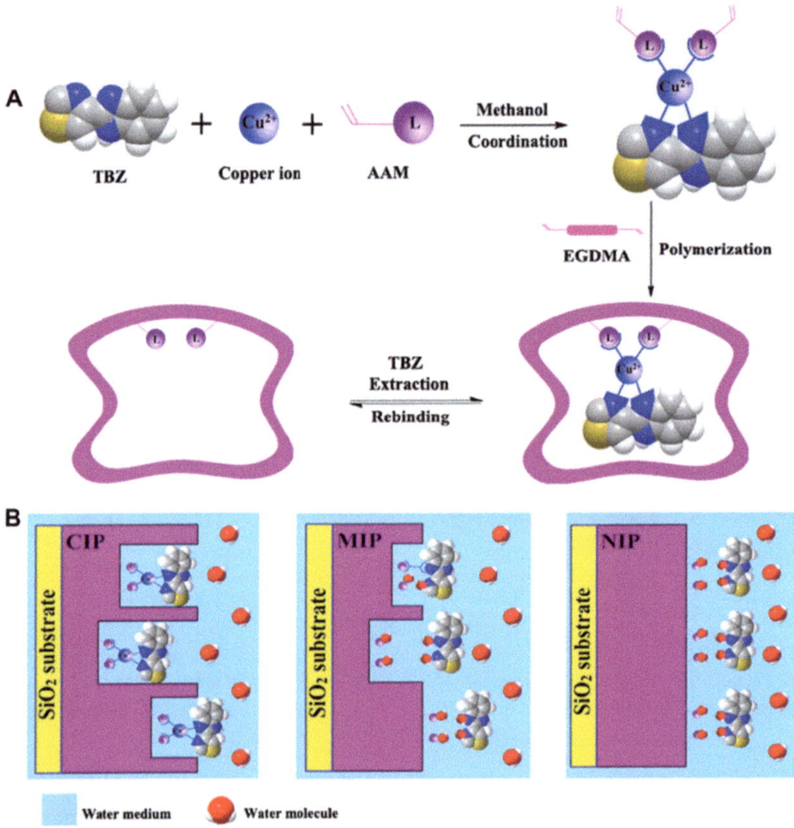

Figure 3.8 (A) Schematic illustration of metal ion mediated imprinting and rebinding of TBZ, (B) recognition mechanism of TBZ with CIP, MIP, and NIP in water medium. Reprinted from Lian et al. (2014) with permission from John Wiley & Sons.

MIP-SPME during aqueous analysis. In this approach, metal ions are employed as an assembly pivot that organizes the functional monomer and the template during prepolymerization. Thus, a complex based on strong ionic interactions to replace hydrogen bonding interactions between the template and functional monomer is created. After the removal of the template, a complex-imprinted polymer (CIP) is formed. In terms of directionality, specificity, and strength, the metal coordination interaction is stronger and more stable than hydrogen bonding or electrostatic interactions in polar systems, resulting in good water compatibility (Lian et al., 2013). J. Huang et al. (2011) synthesized a novel SPME fiber coated with a metal CIP that could recognize the complex template [Cu(Ac)$_2$(2,2'-dpy)] (where 2,2'-dpy is 2,2'-dipyridine) in an aqueous medium. A similar method has been employed to specifically recognize thiabendazole (TBZ) in citrus and soil samples (Lian et al., 2014). The extraction performance of the prepared CIP-SPME fiber in water was significantly improved based on the metal ion coordination interaction rather than relying on hydrogen bonding interactions

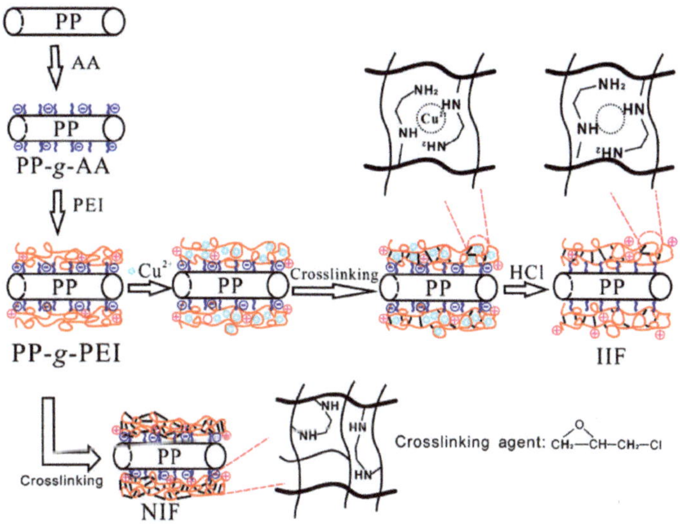

Figure 3.9 Schematic illustration of the preparation of IIF and NIF. Reprinted from Li et al. (2011) with permission from the American Chemical Society.

that are commonly applied for the MIP technique. The preparation scheme, including the Cu (II)-mediated CIP template interactions and the formation of CIP cavity is depicted in **Figure 3.8**, and the difference of the interaction between template and the SPME fiber of CIP, MIP, and a non-imprinted polymer (NIP) is also illustrated.

Ion-imprinted polymers (IIPs) are similar to MIPs, but they can recognize metal ions after imprinting and retain all the virtues of MIPs. IIPs have advantages such as pre-determined selectivity in addition to being simple and convenient to prepare. The main limitations of IIPs at present are the poor solubility of the analyte (template) in the imprinting mixture, inhomogeneity and leaching of the imprint ion.

There has been a rapid increase in publications related to synthesis of IIP materials and their use in separation or preconcentration of metals (Chang et al., 2007; Fan et al., 2011; Chen et al., 2012; Dakova et al., 2012; F. F. He et al., 2013), while the application of IIPs in SPME is still very scarce. Li et al. (2011) prepared a novel Cu (II)-imprinted fiber by grafting acrylic acid (AA) onto the surface of a polypropylene (PP) fiber which was subsequently modified with polyethylenimine (PEI) (**Figure 3.9**). The modification of PP fibers with AA was beneficial to the grafting of PEI onto the fibers. This ion-imprinted fiber (IIF) showed excellent tensile and chemical stability in acid solution which qualified the IIF for practical applications. Besides having a high adsorption capacity for Cu (II) (120 mg g^{-1}), the IIF adsorbent showed a high selectivity for Cu (II) as compared with that of the non-ion-imprinted fiber (NIF).

Zhu's group combined MIPs with SBSE for the first time and prepared a molecularly imprinted membrane (nylon-6) coated stir bar for the extraction of monocrotophos pesticide (Zhu et al., 2006) and amino acid enantiomers (Zhu & Zhu, 2008)

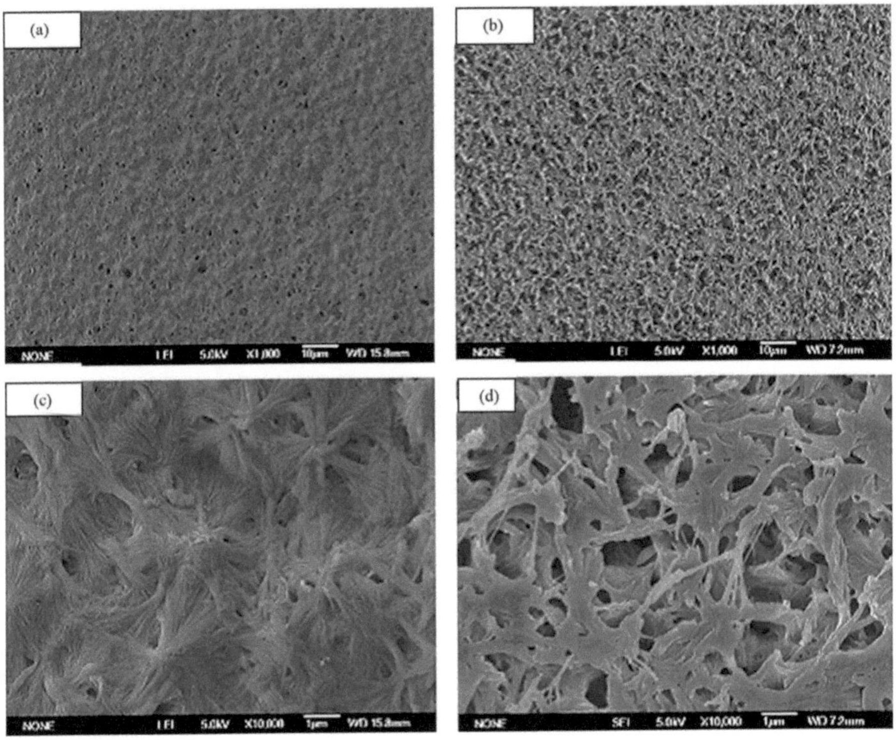

Figure 3.10 SEM image of the monocrotophos imprinted Nylon membrane. (A) Cross-section of the nonimprinted Nylon-6, (B) cross-section of the monocrotophos imprinted Nylon-6, (C) pores in the nonimprinted Nylon-6, (D) pores in the monocrotophos imprinted Nylon-6. Reprinted from Zhu et al. (2006) with permission from Elsevier.

in environmental and biological samples. The molecularly imprinted membrane was prepared from a formic acid solution of nylon-6 polymer imprinted with target analyte. The SEM characterization of the prepared MIP-nylon-6 membranes is shown in **Figure 3.10**. The MIP-film (Zhu et al., 2006) was prepared by the precipitation of the polymer in the presence of the template molecule, and coated onto the surface of a stir bar immersed in water by a phase inversion imprinting technique. The preparation of the stir bar coated with MIP-film was simple and rapid. The MIP-coated layer showed remarkably high affinity toward monocrotophos and the adsorption equilibrium was attained rapidly (within 60 min) in contrast to non-MIP coating in which adsorption equilibrium was attained after several hours. Compared with traditional MIP and SBSE, the MIP-coated film showed not only the high selectivity but also rapid adsorption dynamics. While a commercial PDMS coated stir bar is necessary as a substrate during the preparation of the MIP-coated stir bar.

The application of MIP/IIP-based coatings for SPME applications is listed in **Table 3.3**, along with the preparation methods and analytical performance of the developed methods. Related advantages or shortcomings of the prepared MIP/IIP-based coatings or the proposed method are also presented in the table.

Table 3.3 MIP/IIP SPME coatings and their applications.

Coating	Preparation method	Analytical method	Analytical performance	Analytes	Sample	Ref.
clenbuterol-MIP-coated silica fiber	*in situ* polymerization	MIP-SPME-LC	LODs: 10 ng mL^{-1}	brombuterol	human urine	(Koster et al., 2001)
prometryn MIP-coated fiber	co-polymerization synthesis of prometryn MIP on silylated silica fiber, multiple coating strategy	MI-SPME-HPLC	linear range: 0.1–5.0μg L^{-1}; LODs: 0.012 μg L^{-1}; RSD: 3.2%.	prometryn and its structural analogues	soybean and corn samples	(Hu et al., 2007b)
molecular sol-gel imprinted SPME coating	sol-gel technique	MI-SPME-GC/MS	linear range: 1–80 μg mL^{-1}; LODs: 0.01 μg mL^{-1}; RSD: 10% for intra-day 16% for inter-day	caffeine	human serum	(Khorrami & Rashidpur, 2012)
ascorbic acid MIP-coated fiber	sol-gel technique	MI-SPME-DPCSV	LODs: 0.0396 ng mL^{-1}; RSD: 2.3%.	ascorbic acid	blood serum	(Prasad et al., 2008)
linezolid-MIP coated SPME fibers	electrochemical polymerization	MIP-SPME-HPLC/MS	LODs: 0.029 μg mL^{-1}; RSD: 3.4–7.5% (within—batch), 3.8–11.2%(between-batch)	(antibiotic drugs) linezolid	simulated body fluid and human plasma samples	(Szultka et al., 2012)
MIPPy/MWCNT coated fiber	electrophoretic deposition	EE-SPME-HPLC	linear range: 18–1900 μg L^{-1}; LODs: 0.5–1.9 μgL^{-1}; RSD: 2.6–6.5%	Fluoroquinolone antibiotics	urine and soil samples	(X. Liu et al., 2012)
Cu(II)-mediated MIP coated SPME fiber	metal-ion-mediated polymerization	CIPF-SPME-HPLC	linear range: 1–80 μg mL^{-1}; LODs: 2.4 μg L^{-1}; RSD: 10%	TBZ	citrus and soil samples	(Lian et al., 2014)

continued **Table 3.3** MIP/IIP SPME coatings and their applications.

Coating	Preparation method	Analytical method	Analytical performance	Analytes	Sample	Ref.
prometryn-MIP-coated SPME fiber	co-polymerization	MIP-SPME-HPLC	linear range: 0.1–2μg L^{-1} LODs: 0.012–0.09 μg L^{-1}; RSD: 2.9% (n= 10).	triazines	soybean, corn, soil, and lettuce samples	(Hu et al., 2007a)
17β-estradiol-MIP-coated SPME fiber	multiple co-polymerization technique	MIP-SPME-HPLC	linear range: 5.0–40.0μg L^{-1}; LODs: 0.98–2.39 μg L^{-1}; RSD: below 10%.	17β-estradiol, estriol, estrone and 17α-ethynylestradiol	fishery samples	(Hu et al., 2010b)
tetracycline-imprinted coating fibers	multiple co-polymerization method	MIP-SPME-HPLC	linear range: 5–200 μg L^{-1} LODs 1.02–2.31 μg L^{-1}; RSD: 3.4–5.8%.	tetracyclines (TCs)	Chicken feed, chicken muscle, and milk	(Hu et al., 2008)
bisphenol A (BPA) MIP-coating	photoirradiation polymerization	MIP-SPME-HPLC	linear range: 5–1500 μg L^{-1}; LODs: 2.4–38.9 μg L^{-1}; RSD: 4.7–9.1%.	BPA	tap water, human urine and liquid milk samples	(Tan et al., 2009)
MIP-RAFT-agent-functionalized fiber	surface reversible addition-fragmentation chain transfer (RAFT) polymerization	SPME–LC-MS/MS	linear range: 0.05–100 μg L^{-1}; LODs: 20–55 μg L^{-1}; RSD: 5.1–7.8%.	Sudan I–IV dyes	chili tomato sauce and chili pepper samples	(X. G. Hu et al., 2012)
membrane-protected MWCNTs-MIP	thermal polymerization	MI-SPME-LPME-HPLC	linear range: 0.2–200 μg L^{-1}; LODs: 0.08–0.38 μg L^{-1}; RSD: 6.4–11.2%.	triazine herbicides	river water, wastewater, and liquid milk	(Tan et al., 2011)
MIP polypropylene membrane	precipitation polymerization	MI-MSPE-HPLC	linear range: 0.005–2.0 mg L^{-1}; LODs: 0.56–4.5 μg L^{-1}.	phenolic compounds	environmental water samples	(Feng et al., 2009)

continued **Table 3.3** MIP/IIP SPME coatings and their applications.

Coating	Preparation method	Analytical method	Analytical performance	Analytes	Sample	Ref.
carbon nanotubes reinforced molecularly imprinted sol-gel materials (MISGMs)	*in situ* polymerization combined with sol-gel technique	MIP-HF-SPME-HPLC	linear range: 0.02–1000 ng mL^{-1}; LODs: 0.08 ng mL^{-1}; RSD: 0.38%.	chlorogenic acid (CGA)	medicinal samples	(Golsefidi et al., 2012)
MIP hollow fiber membrane tube	photo polymerization	MIP-HFT-HPLC	linear range: 7.5–200 μg L^{-1}; LODs: 2.5–3.3 μg L^{-1}; RSD: 6.4–8.9%.	diethylstilbestrol (DES)	milk samples	(M. H. Liu et al., 2010)
monoliths MIP-fiber	*in situ* polymerization	HF-LPME-MI-SPME -HPLC-UV	linear range: 0.01–5.00 mg L^{-1}; LODs: 4 μg L^{-1}; RSD: below 10%.	TBZ	orange juice samples	(Barahona et al., 2011)
monolithic MIP fiber	*in situ* polymerization	MIP-SPME-CE	linear range: 5–500 μg L^{-1}; LODs: 0.96 μg L^{-1} (E), 1.1μg L^{-1} (PE); RSD: 3.8–9.1%.	ephedrine (E) and pseudo-ephedrine (PE)	urine and serum samples	(Deng et al., 2012)
monolithic MIP-fiber	thermal radical copolymerization	MI-SPME-GC/MS	linear range: 7–8000 ng mL^{-1}; LODs: 1–300 ng mL^{-1}; RSD: 5.1–15.6%.	Diacetylmorphine and its structural analogues	street heroin sample	(Djozan & Baheri, 2007)
monolithic ametryn-MIP fiber	thermal radical copolymerization	MI-SPME-GC/MS	linear range: 50–10000 ng mL^{-1}; LODs: 14–95 μg L^{-1}; RSD: 5.2–11.8%	triazines	tap water, rice, maize, and onion	(Djozan et al., 2009)
monolithic MIP-fiber	*in situ* polymerization	MI-SPME-LC-UV	--	propazine	soil and vegetable samples (potato and pea)	(Turiel et al., 2007)

continued **Table 3.3** MIP/IIP SPME coatings and their applications.

Coating	Preparation method	Analytical method	Analytical performance	Analytes	Sample	Ref.
MIP capillary column	Bulk polymerization	automated and on-line in-tube MIP-SPME-HPLC	linear range: 0.5–100μg mL^{-1}; LODs: 0.32 μg L^{-1}; RSD: <5%	propranolol	biological fluids	(Mullett et al., 2001)
MIP monolithic column	dummy template	in-tube MIP-SPME-HPLC/UV	linear range: 0.010–5.30 μmol L^{-1}; LODs: 3.2 nmol L^{-1}; RSD: 1.1–6.8% EF: 76	8-hydroxy-2′-deoxyguanosine (8-OHdG)	urinary samples	(S. W. Zhang et al., 2009)
HRP-imprinted poly(VPBA-co-PETA) monolithic column	reversible covalent complexation, surface imprinting	in-tube SPME-SDS-PAGE	--	horseradish peroxidase (HRP)	human serum	(Lin et al., 2013b)
Lyz-MIP hybrid monolith column	one-pot process	in-tube SPME-SDS-PAGE	--	lysozyme (Lyz)	egg white and human serum	(Lin et al., 2013a)
Mn(II) imprinted 3-MPTS-silica	double imprinting *in situ* polymerization	IIP-CME-ICP-MS	LODs: 10.3 ng L^{-1}; RSD: 4.3%; EF: 16.7	Mn(II)	human serum	(Zheng et al., 2011)
Cu(II)-imprinted fiber (IIF)	combining graft polymerization with chemical modification	FAAS	--	Cu(II)	--	(Li et al., 2011)
chiral imprinted polymer -coated stir bar	interpenetrating polymerization	SBSE-LC-ITMS	linear range: 25–103 μg L^{-1}; LODs: 6.26 μg L^{-1};	(+)-(S)-citalopram, (+)-(S)-desmethylcitalopram and (+)-(S)-didesmethylcitalopram	urine samples	(Gomez-Caballero et al., 2011)

continued **Table 3.3** MIP/IIP SPME coatings and their applications.

Coating	Preparation method	Analytical method	Analytical performance	Analytes	Sample	Ref.
sulfamethazine MIP coated stir bar	copolymerization	MIP-SBSE-HPLC	linear range: 1–100 μg L^{-1}; LODs: 0.20–0.72μg L^{-1}; RSD: 3.7–6.7%	sulfa drugs	pork, liver and chicken samples	(Xu et al., 2011)
ractopamine MIP coated stir bar	copolymerization	MIP-SBSE-ECL	linear range: 1.0×10^{-10}–5.0×10^{-8}mol L^{-1}; LODs: 3.5×10^{-12} mol L^{-1}; RSD: <5%.	ractopamine	pork samples	(S. Wang et al., 2012)
MIP nylon-6 films coated stir bar	phase-inversion imprinting technique	MIP-SBSE-GC-NPD	LODs: 12–24 μg kg^{-1}	monocrotophos and its structural close analogy	soil samples	(Zhu et al., 2006)
Cd(II) imprinted 3-MPTS-silica coated stir bar	Sol-gel technique and double-imprinting	IIP-SBSE-ICP-MS	LODs: 4.40 ng L^{-1}; RSD: 3.38%; EF: 21.	Cd(II)	water samples	(Zhang & Hu, 2012)

BPA, bisphenol A; CE, capillary electrophoresis; CGA, chlorogenic acid; CIPF, complex-imprinted polymer -coated fiber; DPCSV, differential pulse cathodic stripping voltammetry; ECL, electrochemiluminescence; EF, enrichment factor; FAAS, flame atomic absorption spectrometry; HFT, hollow fiber tube; HRP, horse-radish peroxidase; ICP-MS, inductively coupled plasma-mass spectrometry; ITMS, ion trap mass spectrometry; Liz, lysozyme; LOD, limit of detection; MISGMs, molecularly imprinted sol-gel materials; MSPE, micro-solid phase extraction; PETA, pentaerythritol triacrylate; RAFT, reversible addition-fragmentation chain transfer; RSD, relative standard deviation; SDS-PAGE, sodium dodecyl sulfate polyacrylamide gel electrophoresis; TCs, tetracyclines

3.4.3 Ionic Liquid Coatings

Ionic liquids (ILs) are organic salts which consist largely of organic cations paired with organic or inorganic anions. These compounds often possess melting points less than or equal to 100 °C. The structure of ILs can be designed to produce desired properties including high thermal and chemical stability, wide electrochemical window, tunable viscosity and miscibility with various solvents, as well as relatively high and adjustable polarities. The unique physicochemical properties of ILs have attracted increasing interests in electrochemistry, synthetic, catalytic and analytical chemistry (Blanchard et al., 1999; Maiti & Rogers, 2011). **Table 3.4** lists some commonly used ILs in analytical sciences, along with their characteristics. Good thermal stability, negligible vapor pressure, tunable viscosity and natural liquid nature at room temperature make ILs good candidates for SPME coatings. In addition, chemical functionality can be imparted to the IL to promote unique interactions between the analyte and solvent and help the extraction in SPME. ILs can be applied in SPME as extraction phase, desorption solvent and mediators (Martin-Calero et al., 2009; Shearrow et al., 2009; Zhou et al., 2011).

According to the preparation method, IL-based SPME coatings can be divided into physically coated-IL SPME coatings (J. F. Liu et al., 2005; He et al., 2009) and chemically-bonded IL-SPME coatings (Lopez-Darias et al., 2010b; Meng & Anderson, 2010; Amini et al., 2011b). Physically-coated ILs can flow off easily during extraction and desorption steps, which limits their lifetime and EE. Chemically-bonded IL coatings can overcome this problem and they present exceptional thermal and chemical stability as well as highly reproducible EE.

Physically coated-IL SPME sorbent coatings can be eliminated in direct immersion (DI)-SPME and liquid desorption since ILs can be easily washed away by organic solvents. Jiang and co-workers prepared a disposable HS-SPME coating with [C_8MIM][PF_6] by physically coating both a stainless steel wire and a fused-silica fiber, in which the dipping and evaporating process was repeated several times to obtain relatively thick coating and better repeatability of the coating (J. F. Liu et al., 2005).

In order to improve the reusability of IL-coated fibers and extend their application to HPLC detection, Amini et al. (2011b) proposed an approach to prepare chemically-bonded IL-fibers. The ILs with functional groups were bonded to a fused silica fiber to form a chemically-bonded IL-SPME sorbent coating. The chemically-bonded IL-fibers showed good thermal (up to 220°C) and chemical stability (can be reused up to dozens of times). However, the amount of IL bonded to the silica fiber was relatively small, resulting in low adsorption capacity. To engineer a thermally stable as well as a pH- and solvent-resistant SPME coating, two ILs, namely 1-allyl-3-methylimidazolium hexafluorophosphate ([AMIM][PF_6]) and 1-allyl-3-methylimidazolium bis(trifluoromethanesulfonyl)imide ([AMIM][NTf_2]), were chemically bonded to a derivatized fused silica fiber *via* a sol-gel free-radical cross-linking method (M. M. Liu et al., 2010). By reacting the [AMIM][PF_6] with the sol-gel precursors, hydroxyl-ter-

Table 3.4 Characteristics of commonly used ILs.

Chemical formula		Abbreviation	Density ($g\ mL^{-1}$) (25 °C)	Viscosity (cP) (25 °C)	Melting point (°C)
Cation	Anion				
$[C_2MIM]^+$	$[BF_4]^-$	$[C_2MIM][BF_4]$	1.248	66	6
	$[PF_6]^-$	$[C_2MIM][PF_6]$	1.373	450	-
	$[NTf_2]^-$	$[C_2MIM]\ [NTf_2]$	1.425	323	4
$[C_4MIM]^+$ ($[BMIM]^+$)	$[BF_4]^-$	$[C_4MIM][BF_4]$	1.208	233	-75–81
	$[PF_6]^-$	$[C_4MIM][PF_6]$	1.373	400	4–10
	$[Br]^-$	$[C_4MIM][Br]$	1.134	solid	55–79.2
	[Cl]	$[C_4MIM][Cl]$	1.120	solid	41–65
	$[NTf_2]^-$ $[TfO]^-$ $[CF_3CO_2]^-$	$[C_4MIM]\ [NTf_2]$ $[C_4MIM][TfO]$ $[C_4MIM][CF_3CO_2]$	1.420 1.29 1.2	52 90 73(20 °C)	-25–2 16 40
$[C_6MIM]^+$ ($[HMIM]^+$)	$[BF_4]^-$	$[C_6MIM][BF_4]$	1.075	211	-82
	$[PF_6]^-$	$[C_6MIM][PF_6]$	1.304	800	-61
	$[NTf_2]^-$	$[C_6MIM]\ [NTf_2]$	1.423	674	-7
$[C_8MIM]^+$ ($[OMIM]^+$)	$[BF_4]^-$	$[C_8MIM][BF_4]$	1.11	400	-79
	$[PF_6]^-$	$[C_8MIM][PF_6]$	1.000	16,000	-82–40
	$[Cl]^-$	$[C_8MIM][Cl]$	1.212	4232	-82–12.3
	$[NTf_2]^-$ $[TfO]^-$	$[C_8MIM][NTf_2]$ $[C_8MIM][TfO]$	1.242	5234	-
$[EEMIM]^+$	$[NTf_2]^-$	$[EEMIM][NTf_2]$	1.432	36	-22
$[BPy]^+$	$[BF_4]^-$	$[BPy][BF_4]$	1.21	-	-1–15
$[Phpromim]^+$	$[TfO]^-$	[Phpromim][TfO]	1.368		

minated silicone oil (OH-TSO), as well as a bridging precursor, $[AMIM][PF_6]$-OH-TSO was synthesized and applied as a sorbent coating. ($[AMIM][PF_6]$-OH-TSO and [AMIM] $[NTf_2]$-OH-TSO) demonstrated higher EE for target aromatic amines compared to the OH-TSO fiber due to superior electrostatic, hydrogen-bonding, and π-π interactions with the polar analytes. A kind of reusable IL-based SPME fiber was prepared by fixing IL *via* cross-linking of IL on the surface of a fused-silica fiber which was then applied to the forensic determination of methamphetamine (MAP) and amphetamine (AP) in human urine samples (He et al., 2009). By reacting the ILs with the sol-gel precursors and bridging precursors, the IL-sol-gel coating was synthesized and applied as the extraction sorbent. Here the ILs usually had functional groups and were bonded to the silica fiber by reaction with the bridging precursors. The IL-sol-gel coating presented high EE and good thermal and chemical stability. The extraction mechanism was thought to involve enhanced electrostatic, hydrogen-bonding and π-π interactions with target analytes.

The presence of ILs in sol-gel reaction could improve the morphology of the coating and ILs can be utilized as co-solvents for the synthesis of sol-gel SPME sorbent coatings. Generally, ILs act as porogens which alter the morphology of the coating, enabling a larger surface area for the partitioning of analytes. Shearrow et al. utilized ILs as co-solvents for the synthesis of sol-gel hybrid organic-inorganic SPME (Shearrow et al., 2009b). The hybrid coatings, namely hydroxylterminated PDMS, poly(tetrahydrofuran) (poly(THF)), bis[(3-methyldimethoxysilyl)propyl] (BMPO) were synthesized *via* a sol-gel method with the addition of trihexyltetradecylphosphonium tetrafluoroborate (TTPT) as an IL co-solvent. The IL-mediated BMPO coating exhibited superior EE and lower detection limits due to its reactive Si-OH terminal group in the presence of the IL co-solvent. An IL-mediated BMPO microextraction capillary was used for the extraction of various analytes including phenols, alcohols, amines, acids, ketones, aldehydes and PAHs (Shearrow et al., 2009a). The EE of the [OMIM][Cl]-mediated BMPO coating was superior to that of the TTPT-mediated coating and the BMPO coating with no IL. The [OMIM][Cl]-mediated BMPO capillary exhibited a more porous morphology in comparison with the other coatings. Thus, it must be stated that the choice of IL co-solvent will impact the porosity of the capillary sorbent coating.

Compared with ILs, PILs typically exhibit higher viscosity which prevents them from flowing into the GC injector at higher operating temperatures. Previous attempts to use IL sorbent coatings were restricted exclusively to headspace extraction mode. With PIL-based coatings, sampling *via* direct-immersion mode is suitable as the hydrophobic coatings are capable of maintaining structural integrity in aqueous matrices. Furthermore, utilizing a highly viscous material as an extractive phase can enable an even coating on the bare fiber in addition to a larger loading capacity. In comparison to commercial coatings, PILs possess the major advantage of structural tunability for the selective extraction of target analytes. Three PILs, namely poly([ViHIM][NTf_2]), poly([ViDDIM][NTf_2]), and poly([ViHDIM][NTf_2]) were successfully employed as SPME sorbent coatings (Zhao et al., 2008). The reactant 1-vinyl-imidazole was reacted with an alkyl halide to produce the imidazolium-based IL. The IL was subsequently subjected to free-radical polymerization using the initiator 2,2'-azo-bis(isobutyronitrile) (AIBN). The consistency of the PIL was much more viscous compared to the IL monomer. When used as a SPME coating, the PIL coating was not observed to flow during the desorption process.

The determination of high-molecular weight aliphatic hydrocarbons and fatty acid methyl esters (FAMEs) which possess high boiling points and low vapor pressures was obtained by HS-SPME-GC (Meng et al., 2009). The method utilizes three independently structurally engineered ILs in which the imparted physical and chemical properties make them compatible with the requirements of each component of the method. Component one is composed of a thermally stable IL solvent that is used to increase the equilibrium concentration of analytes in the HS. Component two is a SPME sorbent coating based on a PIL for the selective HS extraction of analytes. Component three is an IL-based low-bleed GC stationary phase that performs the selective separation of the analytes. The method demonstrates the versatility of ILs within separation science in addition to determining these analytes. The [HMIM][FAP] IL has been shown to be an excellent solvent in that the hydrophobic and refractory nature of the IL promotes dissolution of the apolar analytes while avoiding pressure build-up within the sample vial under extreme temperatures. As a selective sorbent coating for SPME, the PIL component exhibits acceptable EE of the studied analytes under the extreme experimental conditions. Finally, the structural design of the IL-based GC stationary phase produces a thermally stable material that exhibits high separation selectivity of the analytes while producing minimal column bleed. The overall method nicely demonstrates the versatility of ILs within separation science for the determination of low volatility analytes using the headspace extraction mode.

The application of ILs and PILs to SPME and related approaches are listed in **Table 3.5**.

3.4.4 Immunosorbents

To improve the selectivity of the extraction phase, an assortment of coatings or adsorbents involving antigen-antibody interactions have been prepared and selective extraction can be realized accordingly based on molecular recognition. Antibodies are covalently bonded onto an appropriate sorbent to form an immunosorbent which allows highly selective preconcentration of the antigen even in very complicated samples due to the high affinity and high selectivity of the antigen-antibody interactions.

Antibodies showed a high affinity and specific recognition ability towards their complementary antigens in biological systems. The immunoaffinity SPME technique uses analyte-specific antibodies covalently immobilized on the surface of a support. The immobilization of antibodies on the surface of fused silica fibers (Yuan et al., 2001) or into fused silica capillaries (Queiroz et al., 2007) has been obtained *via* modification of the silica surface. In one example study. silanized fused silica fibers with APTES were reacted with glutaraldehyde. Next, the glutaraldehyde surfaces were reacted with antibody solution in 0.1 mol L^{-1} carbonate buffer (pH = 9.2) for 15 h. Immunoaffinity SPME was used for drug determination in complex biomatrices (*e.g.*, serum, urine and plasma) with higher specificity than other SPME methods. Furthermore, the immunoaffinity in-tube SPME technique offered high sensitivity, selectivity and reproducibility for the quantification of fluoxetine in human serum at therapeutic levels (Queiroz et al., 2007). By using a room temperature ionic liquid

Table 3.5 Applications of ILs and PILs in SPME and related approaches.

Adsorbents	Preparation method	Analytes	Method	Sample	Ref.
[OMIM][PF_6]	physical coating	BTEX	HS-SPME-GC-FID	paints	(J. F. Liu et al., 2005)
[OMIM][TfO], [BMIM][TfO], [Phpromim][TfO]	physical coating	PAHs	HS-SPME-GC-MS	rain and ground water	(Hsieh et al., 2006)
[EeMim][NTf_2]	physical coating	MAP and AP	HS-SPME-GC-MS	human urine	(He et al., 2009)
[MTPIM][NTf_2]	chemical bonding	methyl tert-butyl ether (MTBE)	HS-SPME-GC-FID	methanol, gasoline	(Amini et al., 2011b)
dicationic ILs, styrene-based PILs	chemical bonding, polymerization	short chained alcohols and amines	HS-SPME-GC-FID	water	(Wanigasekara et al., 2010)
TTPT, N-butyl-4-methylpyridinium tetrafluoroborate (BMPT)	IL-mediated, sol-gel	aliphatic alcohol and aldehydes, aromatic ketones and PAHs	CME-GC-FID		(Shearrow et al., 2009b)
[OMIM][Cl]	IL-mediated	phenols, alcohols, amines, acids, ketones, aldehydes and PAHs	CME-GC-FID		(Shearrow et al., 2009a)
[AMIM][PF_6], [AMIM][NTf_2]	chemical bonding, sol-gel	phenolic environmental estrogen, aromatic amines	SPME	water	(M. M. Liu et al., 2010)
1-(2-hydroxylethyl)-3-methylimidazolium tetrafluoroborate, 1-(2-hydroxylethyl)-3-methylimidazolium bis-(trifluoromethanesulfonyl) imide	sol-gel	aromatic amines	HS-SPME-GC		(Zhou et al., 2010)

continued **Table 3.5** Applications of ILs and PILs in SPME and related approaches.

Adsorbents	Preparation method	Analytes	Method	Sample	Ref.
poly([ViHIM][NTf_2]), poly([ViDDIM][NTf_2]), poly([ViHDIM][NTf_2])	polymerization	FAMEs and esters	HS-SPME-GC-FID	synthetic wine and two real wine samples	(Zhao et al., 2008)
poly([ViHIM][NTf_2]), poly([ViHIM][Cl]), poly([ViHDIM][NTf_2]), poly([ViHDIM][Cl])	polymerization	volatile alcohols, xylenes, octylaldehyde, napthalene, ethyl phenyl ether, benzonitrile, 1,2-dichlorobenzene	HS-SPME-GC-FID	water	(Graham et al., 2011)
poly([ViHDIM][NTf_2])	polymerization	long-chained aliphatic hydrocarbons and FAMEs possessing high boiling points	HS-SPME-GC-FID	water	(Meng et al., 2009)
poly([ViHDIM][NTf_2])	polymerization	water pollutants including PAHs and substituted phenol	DI-SPME-GC-MS	water	(Lopez-Darias et al., 2010a)
poly([ViBHDIM][NTf_2])	polymerization	PAHs	DI-SPME-GC-MS	water	(Meng & Anderson, 2010)
poly([ViBHDIM][NTf_2])	polymerization	endocrine disruptors (PAHs, parabens, phenols)	DI-SPME-GC-FID	water	(Lopez-Darias et al., 2010b)
poly([ViHIM][NTf_2]), poly([ViHIM][taurate])	polymerization	CO_2	HS-SPME-GC-FID/TCD		(Zhao et al., 2010)
poly([ViHIM][taurate]),poly([ViHIM][NTf_2])	polymerization	CO_2	HS-SPME-GC-FID/TCD	gas mixture (N_2, CH_4, CO_2)	(X. L. Feng et al., 2010)
poly([ViHIM][NTf_2]), poly([ViHIM][Cl])	polymerization	volatile fatty acids (VFAs), alcohols	HS-SPME-GC-FID/TCD	water	(Meng et al., 2011)

continued **Table 3.5** Applications of ILs and PILs in SPME and related approaches.

Adsorbents	Preparation method	Analytes	Method	Sample	Ref.
$[C_8C_1IM][BF_4]$, $[C_2C_1IM][ETSO_4]$, $[C_8C_1IM][PF_6]$, $[C_4C_1IM][BF_4]$	chemical bonding	MTBE	HS-SPME-GC-FID	gasoline	(Amini et al., 2011b)
2-amino-2-hydroxymethyl-propane-1,3-dial (IRIS)	IL-mediated, sol-gel	pesticide residues	HF-SPME-HPLC-DAD	water and hair	(Ebrahimi et al., 2011)
1-butyl-3-methylimidazolium hexafluorophosphate ($BMIPF_6$)	electro-chemical polymerization	OCPs	HS-SPME-GC-ECD	water	(Gao et al., 2011)
[poly($VBHDIm^+NTf_2^-$)], [poly($ViHIm^+Cl^-$)]	polymerization	coffee aromas	HS-SPME- GC-MS	coffee beans	(Lopez-Darias et al., 2011)
$[C_4MIM][BF_4]$	electrochemical polymerization	benzene derivatives	HS-SPME- GC-FID	waste water and tap water	(Zhao et al., 2011)
1-allyl-3-(6'-oxo-benzo-15-crown-5 hexyl) imidazolium hexafluorophosphate	sol-gel	alcohols, PAEs, phenolic environmental estrogens, fatty acids and aromatic amines	SPME-GC-FID	Methanol and acetonitrile	(Zhou et al., 2011)
1-octyl-3-methylimidazolium hexafluorophosphate	electrochemical fabrication	phenolic compounds	HS-SPME-GC-FID	waste water from a chemical factory	(Ai et al., 2012)
1,1'-(1,6-hexanediyl)bis(1-vinylimidazolium) bibromide and 1-vinyl-3-octylimidazolium bromide	polymerization	polar alcohols	HS-SPME-GC-FID	liquors	(Feng et al., 2012)
poly([DDMGlu][MTFSI]), poly([VPPIM][Cl])	polymerization	alkyl halides and aromatics	HS-SPME-GC-FID	aqueous matrices	(T. D. Ho et al., 2012a)

continued **Table 3.5** Applications of ILs and PILs in SPME and related approaches.

Adsorbents	Preparation method	Analytes	Method	Sample	Ref.
$[(VIM)_2C_8]$ 2[Br], $[(VIM)_2C_{12}]$ 2[Br], [VHIM][Cl]	polymerization	alcohols, aldehydes, and esters	HS-SPME-GC-MS, DI-SPME-GC-MS	deionized, well, and river water	(T. D. Ho et al., 2012b)
$[C_4MIM][PF_6]$	microwave irradiation synthesis	chlorophenols	HS-SPME-GC-MS	landfill leachate	(T. T. Ho et al., 2012)
1-vinyl-3-(3-triethoxysilylpropyl)-4,5-dihydroimidazolium chloride	chemical binding	PAHs	DI-SPME-GC-MS	water	(Pang & Liu, 2012)
1-vinyl-3-hexadecylimidazolium hexafluorophosphate ($ViHDIm^+PF_6^-$)	chemical binding	pyrethroids	SPME-GC-ECD	vegetables	(Y. Zhang et al., 2012)
$[AMIM][N(SO_2CF_3)_2]$-OH-TSO	sol-gel	PAEs	UE-SPME-GC-FID	agricultural plastic films	(Zhou et al., 2012)
alkylimidazolium ionic liquid (PMO-IL)	physical coating	PAHs	HS-SPME-GC-MS	aqueous sample solutions	(Abolghasemi et al., 2013)
1-allyl-3-methylimidazolium tetrafluoroborate($[AMIM][BF_4]$)	sol-gel	organophosphate esters	HS-SPME-GC-FID	lake water, wastewater, sewage treatment plant effluent, and tap water	(Gao et al., 2013)

continued **Table 3.5** Applications of ILs and PILs in SPME and related approaches.

Adsorbents	Preparation method	Analytes	Method	Sample	Ref.
3-(but-3''-en-1''-yl)-1[2'-hydroxycyclohexyl]-1H-imidazol-3-ium bis(trifluoromethanesulfonyl) imide (IL-1),1-(2'-hydroxycyclohexyl)-3-(4''-vinylbenzyl)-1H-imidazol-3-ium bis(trifluoromethylsulfonyl)imide (IL-2)	polymerization	beer volatiles	HS-SPME-GC-FID	lemon beer	(Gonzalez-Alvarez et al., 2013)
$[C_9OHVIm]Br$, $[(VIM)_2C_4]2[Br]$,	*in situ* polymerization	hydrogen bonding compounds: polar alcohols and VFAs	GS-SPME-GC-FID, HS-SPME-GC-FID	aqueous matrix	(Jia et al., 2013)
$[C_{18}VIm]Br$	surface radical chain-transfer addition	amino acids	SPME-LC-MS/MS	grape juice and pulp	(Mousavi & Pawliszyn, 2013)
$[C_6MIM][PF_6]$	sol-gel	BTEX	HS-SPME-GC-FID	water	(Sarafraz-Yazdi & Vatani, 2013)
$[TPMIM][NTf_2]$	sol-gel	PAHs	HS-SPME-GC-FID	lake water	(Zhou & Xing, 2013)
[VBHDIM] $[NTf_2]$(monomer),$[(DVIM)_2C1_2]_2[NTf_2]$ (crosslinker) ,$[(DVBIM)_2C1_2]_2[NTf_2]$(crosslinker)	crosslinking polymerization	PCBs	HS-SPME-GC-ECD/MS	ocean water, bovine milk	(Joshi et al., 2014)
1-vinyl-3-butylimidazolium ditrifluoro methyl sulfimide	sol-gel	Triazines	DI-SPME-GC-FID	fruit and vegetable samples	(Tian et al., 2014)
poly(1-vinyl-3-hexylimidazolium chloride)	polymerization	VFAs	HS-SPME-GC-MS	dairy wastewater	(Wasielewska et al., 2014)

BMPT, N-butyl-4-methylpyridinium tetrafluoroborate; GS-SPME, gaseous sampling solid-phase microextraction; IRIS, 2-amino-2-hydroxymethyl-propane-1,3-dial; MTBE, methyl tert-butyl ether; TCD, thermal conductivity detection; UE-SPME, ultrasonic extraction solid-phase microextraction

(RTIL)-mediated sol-gel method, a non-isotopic immunoassay based on APTES-silica hybrid immunoaffinity monolithic in-tube SPME-ICP-MS was developed for the quantitative analysis of human IgG in real human serum sample with quantum dot labels (B. B. Chen et al., 2010).

In contrast to antibodies, aptamers, a new class of single stranded DNA/RNA molecules, have the advantage of high affinity, specificity, stability with targets and are easily prepared and modified at low cost. Nevertheless, digestion of the aptamer by nucleases and contamination of the fiber by nonspecific proteins restricted the application of aptamer coatings in SPME. Mu et al. (2013) introduced aptamer sol-gel SPME to overcome this barrier. In this method, the aptamer is protected from digestion by the mesoporous sol-gel matrix. In addition, the porous sol-gel coating avoids nonspecific bindings. This aptamer sol-gel fiber was applied over 20 times for microextraction of adenosine from human plasma without significant loss of efficiency.

Selected applications of immunoadsorbents to SPME analysis are listed in **Table 3.6.**

3.4.5 Metal-organic Frameworks

Metal-organic frameworks (MOFs) are a new class of hybrid inorganic-organic microporous crystalline materials that self-assemble straightforwardly from metal ions with organic linkers *via* coordination bonds (Gu et al., 2012). **Figure 3.11** presents the three-dimensional structure of representative MOFs, and **Table 3.7** lists the characteristics of several MOFs, along with selected applications.

Due to their unique physical and chemical properties such as permanent nanoscale pore structure, uniform pore size, large specific surface area and good thermal stability, MOFs demonstrate great application prospect in hydrogen storage, gas separation, catalysis, sensors and biological imaging. The inherent high specific surface area (appr. 5000 $m^2 g^{-1}$) (Koh et al., 2009) indicates a large adsorption capacity whereas the uniform pore structure with specific pore size allows a selective adsorption of molecular with less pore size. Moreover, functionalization in the nanopores or on the outer surface can be achieved easily. These features make MOFs particularly favored alternatives for the adsorption or separation of analytes of interest, and MOFs have found great application in chromatographic stationary phases (Maes, et al., 2010a, 2010b; Zhang et al., 2010), adsorbents for on-site sampling (Zheng et al., 2007; Gu et al., 2010a), as well as the preconcentration of organic/inorganic substances in SPE (Zhou et al., 2006; Gu et al., 2011a; Huo & Yan, 2012) and SPME (Cui et al., 2009; Chang et al., 2011).

The feasibility of MOF-5 (Zn as the metals and N,N'-dimethylformamide or N,N'-diethylformamide as solvents) in the field for adsorbing formaldehyde from the atmosphere has been demonstrated (Gu et al., 2010a) and the material can be reused more than 200 times. Compared with the organic polymer and graphite carbon, MOF-5

Table 3.6 Application of immunoadsorbents in SPME and related approaches.

Immunoadsorbents	Pre-treatment technique	Analyte	Sample	Detection	Ref.
monolith conjugated with synthetic peptides	monolith	monospecific anti-peptide immunoglobulins	blood serum	UV, SDS-PAGE	(Platonova et al., 1999)
silica fiber immobilized with theophylline antiserum	fiber SPME	theophylline	human serum	scintillation counter	(Yuan et al., 2001)
porous hollow-fiber membrane anti-estrogen antibody	immunoaffinity porous hollow-fiber membrane	17β-estradiol		ELISA	(Nishiyama et al., 2002)
monolithic materials that contain the monoclonal anti-fluorescein antibody	capillary monolith	fluorescein		laser induced fluorescence (LIF)	(Hodgson et al., 2005)
GMA/EDMA Monolith immobilized with anti-FITC antibodies	monolith	fluorescein		FLD	(Jiang et al., 2005)
convective interaction media monoliths immobilized with mAbs against rat liver plasma membrane proteins	monolith	rat liver plasma membrane proteins	rat liver	UV, SDS-PAGE , LC-MS/MS	(Rucevic et al., 2006)
monolith immobilized with anti-NT-proBNP- or anti-myoglobin	monolith, HPLC	myoglobin, NT-proBNP	blood serum	electrochemiluminescence, HPLC-ESI-MS	(Delmotte et al., 2007)
chip based immunoadsorbent with 12 antibodies	chip extraction, electrophoresis	inflammatory biomarkers	human skin biopsies	LIF	(Phillips & Wellner, 2007)
silica capillary immobilized with anti-fluoxetine antibody	in-tube SPME, HPLC	fluoxetine	human serum	LC-MS	(Queiroz et al., 2007)

continued **Table 3.6** Application of immunoadsorbents in SPME and related approaches.

Immunoadsorbents	Pre-treatment technique	Analyte	Sample	Detection	Ref.
agarose gel beads with immobilized mouse monoclonal antibody	chip extraction	murine recombinant Fab' antibody fragment		LIF	(Shimura et al., 2007)
polystyrene-based stationary phase immobilized with monoclonal anti-D-hydroxy acid antibody	HPLC	α-hydroxy acids		UV	(Franco et al., 2008)
monolith immobilized with antibody against bisphenol A	monolith on-line extraction	bisphenol A	tap water, lake water and effluent from municipal sewage treatment plant	LC-ESI/MS/MS	(Wang et al., 2008)
monolith immobilized with anti-FITC on chip	monolith on a chip, CE	FITC-tagged HSA		FLD	(Sun et al., 2008)
fiber glass immobilized 4 antibodies against analyte on a chip	chip based SPE, CE	follicle-stimulating hormone-FSH, luteinizing hormone-LH, testosterone, thyroid-stimulating hormone- TSH	whole human blood, saliva, and urine	LIF	(Wellner & Kalish, 2008)
capillary monolith immobilized with anti-testosterone polyclonal antibody	capillary monolith extraction on line	testosterone		LIF	(H.X. Chen et al., 2009)
monolith immobilized with antibodies against pyrethroids	monolith extraction on line, HPLC	deltamethrin, flumethrin, flucythrinate and cis/trans permethrin		UV	(Liang et al., 2010)
monolith immobilized with anti-haptoglobin antibody	monolith HPLC	haptoglobin	human serum	LC-MS/MS	(Gunasena & El Rassi, 2011)

continued **Table 3.6** Application of immunoadsorbents in SPME and related approaches.

Immunoadsorbents	Pre-treatment technique	Analyte	Sample	Detection	Ref.
anti-EPO monolith membrane	monolith, HPLC	recombinant human erythropoietin	horse plasma and urine	LC-FAIMS-MS/MS	(Bailly-Chouriberry et al., 2012)
capillary monolith immobilized with anti-ochratoxin A	capillary monolith, CZE	ochratoxin A		UV, LIF	(Chamieh et al., 2012)
magnetic beads functionalized with antibodies against for α-lactoglobulin and β-lactalbumin	magnetic beads trapping, IACE	α-lactoglobulin, β-lactalbumin	milk sample	UV, MALDI-MS	(Gasilova et al., 2012)
capillary monolith immobilized with mAb against microcystin-LR	on-line CME, nanoHPLC	microcystin-LR	cyanobacteria	UV	(Brothier & Pichon, 2013)
capillary silica tube immobilized with mAb against interferon-α	in-tube SPE, HPLC	interferon-α	plasma	FLD	(Chaves & Queiroz, 2013)

CZE, capillary zone electrophoresis; ESI-MS, electrospray ionization mass spectrometry; FAIMS, Field asymmetric ion mobility spectrometry; FSH, follicle-stimulating hormone; LH, luteinizing hormone; LIF, Laser induced fluorescence; NT-proBNP, N-terminal pro-natriuretic peptide; TSH, thyroid-stimulating hormone; IACE, immunoaffinity capillary electrophoresis

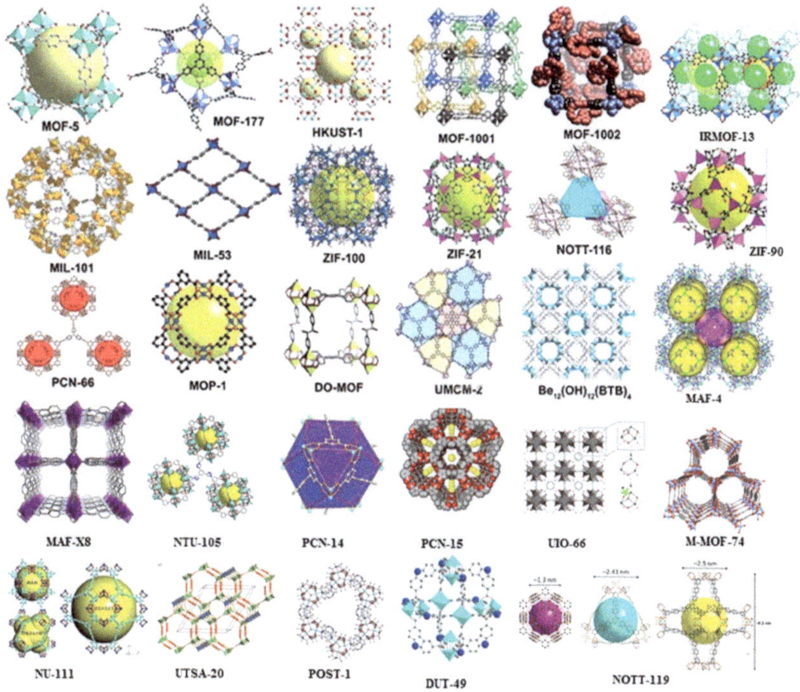

Figure 3.11 The three-dimensional structures of representative MOFs. Reprinted from Li et al. (1999), Chae et al. (2004), and X. J. Wang et al. (2013) with permission from Macmillan Publishers Ltd.; reprinted from Eddaoudi et al. (2001), Serre et al. (2002), Rowsell & Yaghi (2006), Morris et al. (2008), Koh et al. (2009), Ma et al. (2009), Mulfort et al. (2009), Sumida et al. (2009), Yan et al. (2010), and Farha et al. (2012) with permission from the American Chemical Society; reprinted from D. Zhao et al. (2008), Stoeck et al. (2012), and Peng et al. (2013) with permission from the Royal Society of Chemistry; reprinted from A.-X. Zhu et al. (2012), and Glover et al. (2013) with permission from Elsevier; reprinted from Guo et al., (2011), and Ameloot et al. (2013) with permission from John Wiley & Sons; reprinted from Chui et al (1999), Férey et al. (2005), Banerjee et al. (2008), and Q. Li et al. (2009) with permission from AAAS.

provided better EFs (53 and 73 times higher) for formaldehyde, corresponding to a molecular sieve 13X, due to the large specific surface area and porous structure. The sorption mechanism likely involves interactions between the Zn metal in MOF-5 and formaldehyde.

The first MOF- coated SPME device was fabricated by Cui et al. (2009) using microporous Cu(II) trimesate (HKUST-1), which showed high enrichment and a low limit of detection for benzene homologues. However, owing to the poor water stability of HKUST-1, the EE of this device was very low when working in an environment with a relative humidity over 30%. Metal azolate frameworks (MAFs) constructed by imidazolate or pyrazolate derivatives feature very high thermal and chemical stability, thus Zn(II) 2-methylimidazolate (MAF-4) (Huang et al., 2006) and benzimidazolate

Table 3.7 Characteristics of several MOFs.

		Characteristic data						Analytical application
MOFs	**Formula**	**BET surface area ($m^2 g^{-1}$)**	**Pore/channel diameter (Å)**	**Window Diameter (Å)**	**Open Metal sites**	**Thermo-stability (°C)**	**Moisture stability**	
MOF-5,IRMOF-1	$Zn_4O(BDC)_3$, BDC =terephthalic acid	630–2900	11,15	7.5, 11.2	no	400–480	no	sorbent for sampling and trapping (Ni et al., 2007; Gu et al., 2010a); stationary phase for GC (Gu et al., 2010b; Chang et al., 2011; Gu et al., 2011b; Munch et al., 2011), chromatography on a single MOF-5 crystal (Han et al., 2010)
copper(II) isonicotinate	$Cu(4\text{-}C_5H_4N\text{-}COO)_2(H_2O)_4$	146			no		yes	sorbent for SPE of PAHs in water (Zhou et al., 2006)
	Gd(DPA)(HDPA), DPA = pyridine-2,6-dicarboxylicate						yes	sorbent for SPE of pesticides (de Carvalho et al., 2009)
	$(La_{0.9}Eu_{0.1})_2(DPA)_3(H_2O)_3$						yes	sorbent for SPE of pesticides (Barreto et al., 2010)
IRMOF-3	$Zn_4O(NH_2\text{-}BDC)_3$	1957	9.6		no	320	no	stationary phase for GC (Gu et al., 2011b)
HKUST-1, MOF-199	$Cu_3(BTC)_2$,BTC= 1,3,5-benzenetricarboxylate	1000–1458	12	8, 9	yes	280	no	engineered films for SPME (Cui et al., 2009; Chang et al., 2011), stationary phase for LC (Ahmad et al., 2009; Ameloot et al., 2010)
MIL-47(V)	$V^{IV}O(BDC)$	800	8.5	8.5	no	350	yes	stationary phase for GC (Finsy et al., 2008) and LC (Alaerts et al., 2007)

continued **Table 3.7** Characteristics of several MOFs

		Characteristic data						Analytical application
MOFs	**Formula**	**BET surface area ($m^2 g^{-1}$)**	**Pore/channel diameter (Å)**	**Window Diameter (Å)**	**Open Metal sites**	**Thermo-stability (°C)**	**Moisture stability**	
MIL-53(Al)	$Al^{III}(OH)(BDC)$	940–1038	8.5	8.5	no	330	yes	sorbent for enriching peptides and removing proteins (Gu et al., 2011a), stationary phase for LC (Alaerts et al., 2008; S. S. Liu et al., 2012; Yang et al., 2012) Sensor detection of Fe^{3+} (C. X. Yang et al., 2013)
MIL-100(Cr)	$Cr_3O(H_2O)_2F(BTC)_2$	1595	25, 29	5.6, 8.6	yes	350	yes	sorbent for enriching peptides and removing proteins (Gu et al., 2011a)
MIL-101(Cr)	$Cr_3O(H_2O)_2F(BDC)_3$	2736–2907	29, 34	12, 16	yes	300–330	yes	engineered films for QCM (C. Y. Huang et al., 2011), sorbents for enriching peptides and removing proteins (Gu et al., 2011a), stationary phase for GC (Gu & Yan, 2010; Chang et al., 2011) and LC (Yang et al., 2011; Yang & Yan, 2011)
ZIF-7	$Zn(benzimidazolate)_2$		4.3	2.9	no	480	yes	engineered films for SPME (Chang et al., 2011), stationary phase for GC (Chang et al., 2010, 2011)

continued **Table 3.7** Characteristics of several MOFs

		Characteristic data						Analytical application
MOFs	**Formula**	**BET surface area ($m^2 g^{-1}$)**	**Pore/channel diameter (Å)**	**Window Diameter (Å)**	**Open Metal sites**	**Thermo-stability (°C)**	**Moisture stability**	
ZIF-8	Zn(2-methylimidazole)$_2$	1504	11.4	3.4	no	380–550	yes	engineered films for SPME (Chang et al., 2011), sorbent for μ-SPE (Ge & Lee, 2011), stationary phase for GC (Chang et al., 2010; Luebbers et al., 2010; Chang et al., 2011)
UIO-66	$Zr_6O_4(OH)_4(CO_2)_{12}$	1187–1580	8–11	5–7	yes	375–500	yes	stationary phase for GC and LC (Chang & Yan, 2012; Fu et al., 2013)
ZIF-90	$C_8H_6N_4O_2Zn$	1270	3.5	11.2	no	300–500	yes	SPME of phenolic endocrine disruptors in environmental samples (Yu & Yan, 2013)

(MAF-3) (Huang et al., 2003) were prepared and demonstrated to be highly water-stable. Due to the very small aperture size, MAF-4 (3.2 Å) shows superior selectivity for *n*-alkanes over branched alkanes, but requires a relatively long extraction time (ca. 20 min). With even smaller aperture size, MAF-3 (2.9 Å) can adsorb neither *n*-alkanes nor branched alkanes. At the same extraction conditions, benzene homologues can be only adsorbed on the outer surfaces of both MAF-4 and MAF-3 (Chang et al., 2011). Considering that carboxylate ligands can generally facilitate pore size modulation while azolate ligands can enhance framework stability, the rational combination of these coordination groups may be a simple strategy for constructing new MOFs suitable for SPME applications. A porous metal azolate framework [Zn(mpba)] (MAF-X8, H_2mpba=4-(3,5-dimethylpyrazol-4-yl)benzoic acid) with large, hydrophobic, one-dimensional channels and good thermal/chemical stability was synthesized, and high-quality MAF-X8 thin films were grown on stainless-steel fibers for SPME, which showed high sensitivity and selectivity towards non-polar VOCs (C. T. He et al., 2013). These examples highlight the importance of both ligand functionality and pore size of MOFs for SPME applications. **Table 3.8** represents the application of MOFs for SPME analysis.

MOFs that are stable in water and other solvents are needed not only for applications in atmospheric science but also in environmental and biological research for the analysis of pollutants and biomolecules in aqueous solution. Precise control for engineering MOFs to films, two-dimensional patterns, and three-dimensional structures is necessary for facilitating the fabrication of smart multifunctional devices to explore the practical applications and to meet analytical challenges for biological and environmental research. The combination of MOF and other materials to build multifunctional composites such as MOF/nanoparticles, MOF/graphene, MOF/silica, and MOF/organic polymers is also a solution to improving analytical performance. In the near future, effort should be made to explore more practical analytical uses of MOFs, to make full use of currently available MOFs to solve analytical problems and to design specific MOFs for special analytical challenges.

3.5 Other Novel Materials

3.5.1 Monolithic Materials

A monolithic stationary phase is the continuous unitary porous structure prepared by *in situ* polymerization or consolidation inside the column tubing where, if necessary, the surface is functionalized to convert it into a sorbent with the desired chromatographic binding properties (Gusev et al., 1999). Monolithic materials can be easily prepared by initiating the polymerization mixture in a mold by heat or radiation and the preparation is simple, inexpensive and reproducible. Monolithic material possesses a bimodal pore size distribution consisting of both large micrometer-sized pores and

Table 3.8 Applications of MOFs for SPME and related approaches.

Coatings	Method	Synthesis approach	Analytes	Sample	Analytical method	LODs	Ref.
MOF-199	fiber-SPME	*in situ* hydrothermal growth	gaseous benzene homologues	indoor air	GC-FID	8.3–23.3 ng L^{-1}	(Cui et al., 2009)
ZIF-8	fiber-SPME	LBL deposition	n-alkanes	petroleum-based fuel and human serum	GC-MS	0.46–1.06 ng L^{-1}	(Chang et al., 2011)
MIL-53(Al)	fiber-SPME	adhesive	16 PAHs	water samples	GC-MS/MS	0.10–0.73 ng L^{-1}	(Chen et al., 2012)
ZIF-90	fiber-SPME	covalent bonding	phenolic endocrine disruptors	water and soil samples	GC-FID	28.9–196 ng L^{-1}	(Yu & Yan, 2013)
MAF-X8	fiber-SPME	*in situ* hydrothermal growth	BTEX	-	GC-MS	6–60 ng L^{-1}	(C. T. He et al., 2013)
MOF-199/GO	fiber-SPME	covalent bonding	OCPs	water, soil, water convolvulus and longan	GC-ECD	2.3–6.9 ng L^{-1}	(S. L. Zhang et al., 2013)
Fe^{3+}-BTC	*in situ* SPME	*in situ* hybrid gel	PAHs amphetamines drugs	water samples, urine	HPLC-UV	PAHs:0.24–0.40 μg L^{-1} amphetamines drugs: 2.0–3.0 μg L^{-1}	(Y. Hu et al., 2012)
IRMOF-3	SBSE	sol-gel	estrogens	water samples	HPLC-UV	0.15–0.35 μg L^{-1}	(Hu et al., 2013)

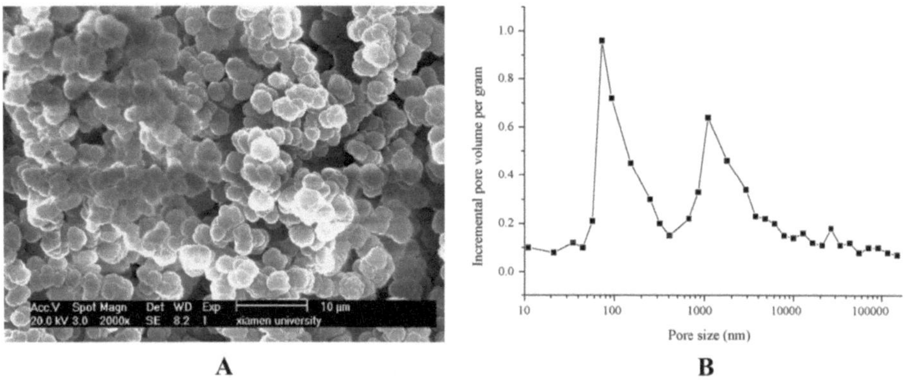

Figure 3.12 Scanning electron microphotograph (A) and pore size distribution profile (B) of the poly (MASE-EDMA) monolithic material (Huang et al., 2008b, 2009d). Reprinted from Huang et al. (2008b) wit permission from Elsevier.

much smaller pores in the 10-nm size range (see **Figure 3.12**). The large pores allow liquid to flow through these materials under low pressures even at high flow rates, while the small pores contribute substantially to the overall surface area. With the advantages of good permeability and fast mass transfer rates, monolithic materials have been widely applied in various chromatographic separation techniques such as reverse phase chromatography, ion exchange chromatography, hydrophobic interaction chromatography and affinity chromatography (Bai et al., 2011; Kortz et al., 2011; Sadilek et al., 2009)) as well as in-tube SPME (Domingues et al., 2009; F. Zheng & Hu, 2009).

Essentially, monolithic columns are divided into two groups: rigid organic polymer-based monoliths and silica-based monoliths. Both formats have been adopted as the sorbents of in-tube SPME. The drawback of the silica-based monoliths is that they are apt to hydrolysis of the Si-O-C linkage, especially under moderately acidic or slightly alkaline conditions (Shintani et al., 2003). Organic polymer monoliths, which show stability over the entire pH range and exhibit excellent biocompatibility, are therefore very suitable to serve as in-tube SPME media. Feng and co-workers (Fan, et al., 2004; Zheng et al., 2007, 2009) have prepared a series of organic polymer monoliths, including poly(methacrylic acid-co-ethylene glycol dimethacrylate), poly(MAA-co-EGDMA), poly(glycidyl methacrylate-co-ethylene glycol dimethacrylate), poly (GMA-co-EGDMA), poly(acrylamide-co-vinylpyridine-co-N,N'-methylene bisacrylamide) and poly(AA-VP-Bis) monoliths and have applied them for the analysis of basic drugs, angiotensin II receptor antagonists, and sulfonamides in several kinds of edible animal-based products. The organic-inorganic hybrid silica monoliths, which combine the advantages of silica with organic polymer monoliths, have also been reported as an extraction sorbent for in-tube SPME in dealing with water, milk and urine samples (M. L. Chen et al., 2010; Bai et al., 2011; Wu et al., 2011).

Monolithic capillary microextraction is one type of in-tube SPME that has displayed unique characteristics such as fast mass transportation, high adsorption capacity and easy preparation, making monolith a superior support in SPME for the analysis of biological macromolecules, small organic molecules and inorganic ions in biological samples. In comparison, published studies focusing on the analysis of trace elements by monolithic capillary microextraction are still scarce. Zhang et al. (L. Zhang et al., 2011) used aminopropyl-triethoxysilane (APTES)-silica hybrid monolithic capillary to preconcentrate trace heavy elements from human serum and urine samples. Zheng and Hu (2008) developed a method for on-line speciation of Al in rainwater and fruit juice based on a silica monolithic capillary, and in a subsequent work, they proposed a model of dual silica monolithic capillary microextraction for sequential speciation of inorganic arsenic and selenium in natural waters (Zheng & Hu, 2009).

In 2007, Huang and Yuan (2007) introduced monolithic materials into SBSE coatings and prepared an octyl methacrylate-ethylene dimethacrylate monolithic stir bar for the extraction of eight non-polar PAHs and four polar steroids. The prepared monolithic stir bar exhibited fast adsorption/desorption kinetics, good inter-assay preparation reproducibility (3.6–5.2%, n=4), good mechanical strength, chemical stability, and could be reused more than 10 times. The same group has prepared a series of monolithic stir bar coatings such as polymethacrylate stearyl acrylate-vinyl dimethacrylate poly(MASE-EDMA) (Huang et al., 2008b), polyvinyl pyridine-vinyl dimethacrylate (poly (VP-EDMA)) (Huang et al., 2008a, 2009a), vinyl pyrrolidone-divinylbenzene (poly (VPL-DVB)) (Huang et al., 2009c, 2009d), polyvinyl imidazole-divinylbenzene (poly (VIDB)) (Huang et al., 2009b, 2009d), poly [2-(methacryloyloxy) ethyl trimethylammonium chloride]-divinylbenzene (poly (META-*co*-DB)) (Huang et al., 2010b), methyl acrylate, 3 sulfonate-divinylbenzene (poly (MASE-*co*-DB)) (Huang et al., 2010a) for SBSE of polar substances (hormones, phenols, aromatic amines, sulfonamides, quinolones, and inorganic ions) and non-polar substances (*e.g.*, PAHs). Feng's group prepared a poly 2-acrylamido-2-methylpropane sulfonic acid-stearyl methacrylate-ethylene glycol dimethacrylate (poly (AMPS-*co*-OCMA-*co*-EDMA)) monolithic stir rod and achieved the extraction of four fluoroquinolones in honey samples based on hydrophobic and ion exchange interactions (Luo et al., 2010).

3.5.2 Restricted Access Materials

Restricted access materials (RAM) are a class of biocompatible adsorbent particles which enable the direct extraction of analytes from biological fluids (*e.g.*, plasma and urine) (Souverain et al., 2004). In particular, alkyl-diol-silica (ADS) RAM particles with controlled pore size can fractionate a sample into the protein matrix and the analytes, excluding macromolecules (>15,000 molecular weight) and extract and enrich low molecular weight compounds (Mullett & Pawliszyn, 2002). The exterior of the sil-

Figure 3.13 Scanning electron micrographs of (A) glass stir bar and (B) RAM-SBSE coating on a stir bar. Reprinted from Lambert et al. (2005) with permission from Elsevier.

ica-based particles can be modified with diol moieties to prevent irreversible adsorption of proteins and hence acts as a biocompatible surface, enabling direct exposure to biological fluids. Mullett's group prepared a biocompatible RAM-coated stir bar by adhering 25 μm ADS to the surface of a glass stir rod with Epo-Tek 353ND epoxy glue (Lambert et al., 2005). The scanning electron micrograph of the RAM stir bar coating is shown in **Figure 3.13**. The stir bar can directly extract caffeine and its metabolites (1,7-dimethyl xanthine, 1-methyl urea and 1-methyl xanthine) from plasma samples. There was no requirement to precipitate proteins from the sample prior to extraction, therefore minimizing sample preparation time and eliminating potential sample preparation artifacts. The utilization of the RAM material for many classes of drugs ensures the potential versatility and usefulness of this approach. More fundamentally, the extraction phase located inside the pores of the coating can be designed according to the properties of target analytes. For example, phases with C_4, C_8 or ion exchange functional groups will enable the extraction of analytes over a wide range of polarities (Rbeida et al., 2004).

3.6 Application of Various Materials in Solid-phase Microextraction-related Approaches

The applications of different novel materials in SPME-related approaches, namely in-tube SPME, SBSE, TFME, NTD and MEPS are represented in the following Tables.

For in-tube SPME, the various coating/monolithic materials that are usually applied can be classified into four categories: inorganic, organic, organic-inorganic hybrid and polymer (**Table 3.9**). The inherent shortcoming of inorganic materials lies in their poor selectivity, which could be improved by coupling with other chemical reagents or biological molecules. The adsorption mechanism of analytes on most organic material is mainly attributed to complexation, ion-exchange, hydrogen bonding, electrostatic effects, hydrophobic effects and van der Waals forces. However, unlike inorganic substances, most organic materials have weak mechanical strength and limited pH, thermal and solvent stability. Organic-inorganic hybrid coating/monolithic materials exhibit good mechanical strength, thermal/solvent stability and good selectivity. The most appealing advantage of polymer adsorbents is their excellent stability in strong acidic or basic conditions, however, they tend to easily shrink or swell when exposed to some organic mobile phases, leading to poor stability.

Only a few types of stir bars are commercially available at present, including PDMS, ethylene glycol (EG)-silicone, and PA twister (Twister; Gerstel GmbH, Germany). The primary homemade SBSE coatings are organic polymers and modified polymers. The extraction mechanisms of these coatings for the target analytes are mainly based on hydrophobic interactions, hence only non-polar and weak polar compounds can be extracted, though polar compounds could be extracted with the aid of derivatization. The further development of the SBSE technique is highly dependent on the exploration of new extraction materials (coatings), especially those novel SBSE coatings with good affinities for polar compounds and good mechanical strength. Therefore, great efforts have been directed towards the development of new coating materials for SBSE in recent years. Homemade stir bars are mainly based on sol-gel technology, monolithic technology, molecularly imprinting technology, and the physical adhesion method. **Table 3.10** lists the reported new SBSE coatings and applied preparation technologies.

In TFME, a thin and wide membrane is employed as a sorbent for a microextraction procedure. The high surface area-to-volume ratio, together with the increase of extraction phase volume, can enhance the sensitivity of the technique without sacrificing the sampling time as compared to other SPME approaches. The equilibrium extraction time is shortened when the thickness of the extraction phase is decreased. Thin sheets of PDMS membrane, PDMS tape, polyaniline-nylon-6 nanofiber sheet, thin film of C_{18} and ethylene/vinyl acetate film have been used as the extracting phase in TFME, as shown in **Table 3.11**.

For NTD and MEPS, the adsorbents employed for diverse analysis are summarized in **Tables 3.12** and **3.13**, respectively.

Table 3.9 Coating/monolithic materials employed for in-tube SPME.

Coating/monolithic materials	Target analytes	Ref.
Organic		
2-acrylamido-2-methyl-1-propanesulfonic acid	Pb(II)	(Garcia-Sanchez et al., 2001)
β-cyclodextrin	non-steroidal anti-inflammatory drugs	(Fan et al., 2005a)
Antibody	fluoxetine	(Queiroz et al., 2007)
methyl coating	PAHs, ketones, phenols, alcohols, amines	(Ma et al., 2008)
Inorganic		
Al_2O_3	metal ions (Co, Ni, Cd, As(III)/As(V), Cr(III)/Cr(VI))	(Malik et al., 2002; Hu et al., 2006)
ZrO_2	Cr, Cu, Cd, Pb	(Chen et al., 2009)
TiO_2	V, Cr, Cu phosphopeptide	(Wu et al., 2007) (Lin et al., 2008)
GeO_2	phenanthrene, trans-chalcone, m-toluidine, 9-anthracene-methanol, 2,4-dichlorophenol	(Segro et al., 2010)
SWCNTs	alkanes, alcohols, aromatic compounds, ketones La, Eu, Dy, Y	(Yuan et al., 2006; Wu et al., 2013)
Organic-inorganic hybrid		
PDMS/PDMDPS/APTMS-GeO_2	PAHs, aldehydes, ketones, alcohols, phenols	(Fang et al., 2007)
AAPTS-silica	Cu, Zn, Ni, Hg, Cd	(Zheng & Hu, 2007)
poly THF-silica	PAHs, aldehydes, ketones, chlorophenols, alcohols	(Fan et al., 2005)
poly THF-TiO_2	o-chlorobenzoic acid, *o*-toluidine, 3,5-dimethylphenol, benzhydrol, trans-chalcone, acenaphthene	(Segro et al., 2009)
PDMS-TiO_2	PAHs, ketones and alkylbenzenes	(Kim et al., 2004)

continued **Table 3.9** Coating/monolithic materials employed for in-tube SPME.

Coating/monolithic materials	Target analytes	Ref.
AAPTS-silica/3-mercaptopropyltrimethoxysilane (MPTS)-silica	As(III), As(V), MMA(V) and DMA(V) species	(Zheng & Hu, 2010)
phosphonium-/pyridinium-based IL-mediated materials	dodecanal, heptanophenone, pyrene	(Shearrow et al., 2009b)
diazonium salts	atrazine, ametryn and terbutryn	(Bagheri et al., 2013)
triazines molecularly imprinted xerogel	atrazine, ametryn, terbutryn	(Bagheri & Piri-Moghadam, 2012)
Mn(II) imprinted MPTS-silica	Mn(II)	(Zheng et al., 2011)
RTIL-APTES-silica	human IgG	(B. B. Chen et al., 2010)
C_{18}-SiO_2	bisphenol A	(Minakuchi et al., 1996)
Polymer		
PDMS	nonpolar and moderately polar compounds (PAHs, aldehydes, ketones),	(Bigham et al., 2002)
PA	polar thermally labile analytes, phenylurea pesticides	(Eisert & Pawliszyn, 1997)
PPY	organoarsenic compounds	(Jinno et al., 2001)
poly(AA-VP-Bis)	Cd, Tl, Pb	(Yin et al., 2009)
poly(methacrylic acid-ethylene glycol dimethacrylate) (MAA-EGDMA)	fluoroquinolones, angiotensin II receptor antagonists	(Louch et al., 1992; Zhang et al., 2006)
poly(acrylamide-vinylpyridine) (AA-VP)	endocrine disruptors	(Zhang et al., 2007)
poly(lauryl methacrylate-co-ethylenedime thacrylate) and poly(styrene-co-divinyl benzene)	proteins and peptides	(Levkin et al., 2008)

AAPTS, N-(2-Aminoethyl)-3-aminopropyltrimethoxysilane; AA-VP, acrylamide-vinylpyridine; MAA-EGDMA, methacrylic acid-ethylene glycol dimethacrylate; MPTS, 3-mercaptopropyltrimethoxysilane; PDMDPS, poly(dimethyldiphenylsiloxane)

Table 3.10 Novel SBSE coatings and their applications.

Home-made coatings	Preparation techniques	Analytes	Samples	Method LODs	Ref.
PDMS	sol-gel	*n*-alkanes, PAHs and OPPs	aqueous samples	*n*-alkanes:0.74–20.0 ng L^{-1} PAHs:0.18–2.76 ng L^{-1} OPPs:0.3–8.0 ng L^{-1}	(W. M. Liu et al., 2004)
PDMS	sol-gel	OPPs	cucumber and potato	0.06–1.22 ng L^{-1}	(W. Liu et al., 2005)
CW-PDMS-PVA	sol-gel	VOSs	water	0.04–4.8 µg L^{-1}	(Yu et al., 2008)
PDMS-PVA	sol-gel	OPPs	honey	0.013–0.081 µg L^{-1}	(Yu & Hu, 2009)
PDMS/β-CD	sol-gel	BFRs	soil and dust	2.9–4.2 µg L^{-1}	(Yu & Hu, 2007)
PDMS/β-CD/DVB	sol-gel	PAHs and PASHs	lake water and soil	0.007–0.103 µg L^{-1}	(Yu et al., 2009)
PDMS/β-CD	sol-gel	estrogens BPA	environmental water drinking water leachate of one-off dishware	0.04–0.11 µg L^{-1}	(Y. L. Hu et al., 2007)
PDMS-PVA	sol-gel	DMSe and DMDSe	garlic, onion and their juice	DMSe: 33 ng L^{-1} DMDSe: 7.1 ng L^{-1}	(Duan et al., 2009)
titania-OH-TSO	sol-gel	drugs of abuse	urine	2.3–9.1 µg L^{-1}	(Lan et al., 2010)
poly(MOAE-*co*-EDMA)	monolithic	PAHs and anabolic steroids	seawater urine	PAHs: 1.86–6.61 ng L^{-1} anabolic steroids: 62.3–212 ng L^{-1}	(Huang & Yuan, 2007)
poly(MASE-EDMA)	monolithic	steroid sex hormones	urine	0.062–0.38 µg L^{-1}	(Huang et al., 2008b)

continued **Table 3.10** Novel SBSE coatings and their applications.

Home-made coatings	Preparation techniques	Analytes	Samples	Method LODs	Ref.
poly(VP-EDMA)	monolithic	phenols	lake and sea water	0.98–2.57 μg L^{-1}	(Huang et al., 2008a)
poly(VP-EDMA)	monolithic	steroid sex hormones	wastewater	0.10–0.41 μg L^{-1}	(Huang et al., 2009a)
poly(VPL-DVB)	monolithic	PAHs, hormones, aromatic amines, phenols, and heavy metal ions (Cu^{2+}, Pb^{2+}, Cr^{3+} and Cd^{2+})	-	PAHs: 0.045–98 μg L^{-1} hormones: 0.036–0.048 μg L^{-1} aromatic amines: 0.038–0.095 μg L^{-1}	(Huang et al., 2009c)
poly(VPL-DVB)	monolithic	polar phenols	tap water, sea water and waste water	0.72–1.37 μg L^{-1}	(Huang et al., 2009d)
poly(VIDB)	monolithic	PAAs	lake and sea waters	0.09–0.28 μg L^{-1}	(Huang et al., 2009d)
poly(VIDB)	monolithic	sulfonamide antibacterial residues	milk	1.30–7.90 μg L^{-1}	(Huang et al., 2009b)
poly(META-*co*-DB)	monolithic	inorganic anions (Br^-, NO_3^-, PO_4^{3-} and SO_4^{2-})	commercial purified water	0.92–2.62 μg L^{-1}	(Huang et al., 2010b)
poly(MASE-*co*-DB)	monolithic	quinolones	wastewater	0.37–0.56 μg L^{-1}	(Huang et al., 2010a)
poly(AMPS-*co*-OCMA-*co*-EDMA)	monolithic	fluoroquinolones	honey	to 0.14 ng g^{-1}	(Luo et al., 2010)
monocrotophos imprinted nylon-6	MIP	OPPs	soil	12–24 μg kg^{-1}	(Zhu et al., 2006)
L-glutamine imprinted nylon-6	MIP	D,L-glutamine enantioseparation	-	-	(Zhu & Zhu, 2008)

continued **Table 3.10** Novel SBSE coatings and their applications.

Home-made coatings	Preparation techniques	Analytes	Samples	Method LODs	Ref.
terbuthylazine imprinted polymer	MIP	nine triazines	rice, apple, lettuce and soil	0.04–0.12 μg L^{-1}	(Hu et al., 2010a)
ractopamine imprinted polymer	MIP	β_2-agonists	pork, liver and feed	0.10–0.21 μg L^{-1}	(Xu et al., 2010)
nicosulfuron imprinted polymer	MIP	nicosulfuron	tap water and soil	tap water: 0.75 nM soil: 12.0 nmol kg^{-1}	(Yang et al., 2010)
RAM-ADS	adhension	caffeine and metabolites	plasma	caffeine: 25 μg L^{-1}	(Lambert et al., 2005)
PPESK	immersion precipitation	OCPs and OPPs	sea water grape and peach juice	organochlorine:0.05–2.53 ng L^{-1} OPPs: 0.17–2.25 ng L^{-1} 2.47–10.3 ng L^{-1}	(Guan et al., 2008)
PDMS-PPY	-	antidepressants	plasma	5–20 μg L^{-1}	(Melo et al., 2009)
PU	self-synthesis	atrazine, 2,3,4,5-tetrachlorophenol and fluorine	-	-	(Neng et al., 2007)
PU	self-synthesis	triazinic herbicides	ground and superficial water	0.1–0.5 μg L^{-1}	(Portugal et al., 2008)
PU	self-synthesis	acidic pharmaceuticals	environmental water	0.4–1.7 μg L^{-1}	(Silva et al., 2008)
PU	self-synthesis	triazinic metabolites	water	0.4–1.7 μg L^{-1}	(Portugal et al., 2010)

BFRs, brominated flame-retardant compounds; DMSe, dimethylselenide; DMDSe; dimethyldiselenide; PPESK, poly(phthalazine ether sulfone ketone); PU, polyurethane; PVA, poly (vinyl alcohol); β-CD, β-cyclodextrin

Table 3.11 Adsorbents and their applications for TFME.

Adsorbents	Preparation methods	Analytes	Sample	Ref.
ZNRs/PANI	dipping	six benzene homologues	water	(Zeng et al., 2013)
cellulose paper modified with phenyl isocyanate	chemically modified	synthetic and natural estrogenic hormones	wastewater, human urine, pool and river water	(Saraji & Farajmand, 2013)
C18-PAN	brush painting	cocaine, methadone	urine	(Rodriguez-Lafuente et al., 2013)
C18-PAN	sol-gel technology	polar and non-polar compounds	samples	(Mirnaghi et al., 2013)
Zn/Al LDH and TiO_2 composite	sol-gel technology	valproic acid	human serum , tablet and syrup	(Matin et al., 2013)
MWCNTs-COOH/PDMS	dipping	phenolic compounds	river water samples, an upstream and a downstream sample	(Kueseng & Pawliszyn, 2013)
PDMS	dipping	EEDCs	biosolids	(Engler & Lemley, 2013)
C18-PAN	spraying	pharmaceuticals	fish tissue	(Togunde et al., 2012b)
CNT	physical coating	pharmaceuticals	environmental water	(Togunde et al., 2012a)
C18/SCX	dipping	carbamazepine and triclosan	wastewater	(Strittmatter et al., 2012)
CAR/PDMS PDMS/DVB	spinning	N-nitrosamines	water	(Riazi Kermani & Pawliszyn, 2012)
C18-silica glass	sol-gel technology	benzodiazepines	human plasma	(Mirnaghi et al., 2012)

continued **Table 3.11** Adsorbents and their applications for TFME.

Adsorbents	Preparation methods	Analytes	Sample	Ref.
PEG-DA	dipping	parabens	cosmetic products	(Fei et al., 2011)
EVA	dipping	organic chemicals	sediments	(Meloche et al., 2009)
MWCNTs–PANI	electrochemical deposition	phenolic compounds	water	(Du et al., 2009)
PDMS	spinning	PAHs	river water	(Qin et al., 2008)
PDMS	spin casting	standard analytes	classes of samples	(Lange et al., 2008)
nanostructured PPy/CNT	N.D.	ClO_4^-	wastewater	(Lin et al., 2006)
PDMS	dipping	PAHs	field	(Bragg et al., 2006)
1-octanol (alkyl-phenols)	dipping	humic acids, bovine serum albumin	water	(J. F. Liu et al., 2006)
PDMS	N.D.	PAHs	lake water	(Bruheim et al., 2003)
EVA	rotating stage	poorly volatile hydrophobic organic chemicals	biological tissues	(Wilcockson & Gobas, 2001)

EECDs, estrogen-like endocrine disrupting compounds; EVA, ethylene vinyl acetate; PAN, polyacrylonitrile; PANI, polyaniline; PEG-DA, poly(ethylene glycol) diacrylate ZNRs, ZnO nanorods; LDH, layered double hydroxide

Table 3.12 Adsorbents for NTD.

Adsorbents	Analytes	Sample	Detection method	LODs	Ref.
HF-etched stainless steel plunger wire (extraction solvent toluene)	PAHs	water	GC-MS	0.003–0.136 μg L^{-1}	(Zhang et al., 2014)
graphene-based sol-gel coating	PBDEs	water	GC-MS	0.2–5.3 ng L^{-1}	(Zhang & Lee, 2011)
carbon adsorbent film	VOCs	---	GC-FID	---	(Wong et al., 2012)
PDMS-DVB-Carboxen, Carboxen 1000	VOCs	gases	GC-FID	benzene: 0.23; ethylbenzene: 2.10; o-xylene: 1.12 ng L^{-1}	(Wang et al., 2005)
Tenax TA and Carboxen 1000	VOCs	natural and waste waters	GC-MS	0.1–0.3 μg L^{-1}	(Alonso et al., 2011)
Au-wire	Hg	ambient air	GC-MS	appr. 0.2 pg m^{-3}	(Cai et al., 2008)
divinylbenzene	BTEX	aqueous samples	GC-FID	1 ng mL^{-1}	(Eom et al., 2008)
SWCNTs/silica	HVOCs	air	GC-MS	0.001–0.01 ng mL^{-1}	(Heidari et al., 2012)]
MWCNTs/silica	HVOCs	air	GC-MS	0.01–0.05 ng mL^{-1}	(Heidari et al., 2013)
divinylbenzene and activated carbon	VOCs	tap water	GC-MS	0.01–0.3 μg L^{-1}	(Ueta et al., 2013)
SiO_2-OA particles	PAHs	water	GC-MS	2–5 ng L^{-1}	(Bagheri et al., 2011b)
PPY	PAHs	water	GC-MS	0.002–0.01 ng mL^{-1}	(Bagheri et al., 2009)
polymer-coated fiber	EO	air	GC-MS	ng L^{-1}	(Ueta et al., 2009a)
copolymer of methacrylic acid and ethylene glycol dimethacrylate	acetone	human breath and urine sample	GC-MS	gaseous sample: 5.0×10^{-4}; aqueous sample: 1.0×10^{-2} ppmv	(Ueta et al., 2009b)

continued **Table 3.12** Adsorbents for NTD.

Adsorbents	Analytes	Sample	Detection method	LODs	Ref.
DVB particle	formic and acetic acids	aqueous solution	GC-FID	acetic acid: 87.2; formic acid: 234.8 μg L^{-1}	(Lou et al., 2008)
DVB-CAR mixed phase	thiol compounds	garlic	GC-MS	derivatized thiols: 11–15 nonderivatized thiol compounds:0.85–2.90 μg L^{-1}	(Warren et al., 2013)
graphene	three sulfonamide antibiotics	environmental water samples	LC-FD	0.5–1.7 pg mL^{-1}	(Sun et al., 2014)
IDA-bonded monolith	22 trace elements	natural waters	ICP-MS	0.000003–0.18 μg L^{-1}	(Rahmi et al., 2010)
poly(MAA-EGDMA) monolith	ARA-IIs (telmisartan (T), irbesartan (I) and losartan (L))	human urine	CZE	15–20 ng mL^{-1}	(Zhang et al., 2006)
molecularly imprinted MIPPy/CNTs	ochratoxin A	red wine	HPLC-FD	ng mL^{-1}	(Wei et al., 2007)
PDMS/AC	BTEX and n-aldehydes	melted snow water	GC-MS	BTEX: 19–30; n-aldehydes: 21–63 ng L^{-1}	(Sieg et al., 2008)
macroporous poly(styrene-divinylbenzene) (PS-DVB) monoliths	phenolic compounds	water samples	GC	---	(Pietrzynska et al., 2013)
Porapak Q and alumina	BTEX	aqueous sample	GC-FID	0.059–0.125 μg L^{-1}	(Prikryl et al., 2006)
Tenax TA	volatile organic hydrocarbons	aqueous sample	GC-MS	28–799 ng L^{-1}	(Jochmann et al., 2008)

AC, activated carbon; ARA-IIs, angiotensin II receptor antagonists; EO, ethylene oxide; HVOCs, volatile organohalogen compounds; IDA, iminodiacetic acid; OA, cis-9-octadecenoic acid; PS-DVB, poly(styrene-divinylbenzene)`

Table 3.13 Adsorbents for the MEPS.

Adsorbents	Analytes	Sample	Detection method	LODs	Ref.
CMK-3 nanoporous carbon	olive biophenols CA,TY and OE	rat plasma	HPLC-UV	CA: 0.7; TY: 4.7; OE:0.25 μM	(Khoshdel et al., 2013)
C18	isoxanthohumol (IXN), xanthohumol (XN)	beers	UHPLC	IXN:0.4; XN:0.9 ng mL^{-1}	(Goncalves et al., 2013)
divinylbenzene–methacrylate copolymeric resin containing PAPC groups	La, Ce, Pr, Nd, Sm, Eu, Gd, Tb, Dy, Ho, Er, Tm, Yb, Lu	seawater samples	ICP-MS	0.005–0.09 pg mL^{-1}	(Zhu et al., 2009)
C18	biogenic amines serotonin (5-HT), dopamine (DA) and norepinephrine (NE)	urine	HPLC-ED	5-HT:20; DA and NE:2 ng mL^{-1}	(Oppolzer et al., 2013)
C18	Salvinorin A	urine	GC MS-MS	5 ng mL^{-1}	(Moreno et al., 2013)
C8	8-hydroxy-29-deoxyguanosine (8-oxodG), 5-hydroxymethyluracil (5-HMUra)	urine	UHPLC	8-oxodG:0.04; 5-HMUra:0.00005 μg mL^{-1}	(Mendes et al., 2013)
silica-C8	dopamine(DA) 5-hydroxytryptamine (5-HT)	urine	LC-MS-MS	DA and 5-HTz:1.0 μg L^{-1}	(El-Beqqali et al., 2007)
C8	ethyl carbamate (EC)	fortified wines	GC-MS	μg L^{-1}	(Leca et al., 2014)
silica-C2	lidocaine, prilocaine, ropivacaine and mepivacaine	human plasma	GC-MS	---	(Abdel-Rehim, 2004)

continued **Table 3.13** Adsorbents for the MEPS.

Adsorbents	Analytes	Sample	Detection method	LODs	Ref.
silica based benzenesulphonic acid cation exchanger	local anaesthetics	human plasma	HPLC-MS-MS	---	(Altun et al., 2004)
Oasis HLB	phenoxyacid herbicides (2,4-D, silvex, and haloxyfop)	water samples	negative ESI-IMS	2,4-D:60; Silvex:70; Haloxyfop:90 ng L^{-1}	(Jafari et al., 2012)
silica gel sorbents modified with C18	PAHs, PCBs, PEs, NPs, BPA, steroid hormones	snow and waste-water	LVI-GC-MS	0.2–266 ng L^{-1}	(Prieto et al., 2010)
PPY nanowires network	triazine herbicides (Atrazine, Ametryn, Terbutryn)	water samples	GC-MS	Atrazine: 0.04; Ametryn: 0.05; Terbutryn: 0.02 ng mL^{-1}	(Bagheri & Ayazi, 2011)
silica-C8	polycyclic hydrocarbons (PAHs)	water samples	GC-MS	1–5 ng L^{-1}	(El-Beqqali et al., 2006)
PDPA/CNT nanocomposite	triazine, OPPs, OCPs and aryloxy-phenoxy propionic acid pesticides	water samples	GC-MS	distilled water: 0.01–0.1; river water: 0.02–0.1 ng mL^{-1}	(Bagheri et al., 2012c)
MIP	fluoroquinolone antibiotics (CIP, norfloxacin (NOR) and ofloxacin (OFLO))	wastewater samples	LC-MS-MS	CIP: 0.8; NOR: 3.8; OFLO: 0.5 ng L^{-1}	(Prieto et al., 2011a)
MIPs	ropivacaine	plasma	LC-MS-MS	---	(Abdel-Rehim et al., 2006)
polystyrene polymer	roscovitine	plasma and urine	LC-MS-MS	ng mL^{-1}	(Vita et al., 2005)
polystyrene polymer	AZD3409	plasma samples	LC-MS-MS	---	(Abdel-Rehim et al., 2008)
C8,C18	linezolid (LIN) and amoxicillin (AMOX)	human plasma	HPLC-MS-MS	LIN: 0.1407; AMOX: 0.1341 ng mL^{-1}	(Szultka et al., 2013)

continued **Table 3.13** Adsorbents for the MEPS.

Adsorbents	Analytes	Sample	Detection method	LODs	Ref.
M1:C8-SCX	22 pesticides	honey samples	GC-MS	---	(Salami & Queiroz, 2013)
C18	12 chlorobenzene congeners	water samples	PTV-GC-MS	0.0003–0.07 μg L^{-1}	(Noche et al., 2013)
tantalum-based sol-gel material	phosphopeptides	non-fat milk	MALDI-MS	---	(Celikbicak et al., 2013)
carbon nanotubes modified filters	ILs	river water samples	NACE	mg L^{-1}	(Polo-Luque et al., 2012)
polyaniline nanowires network	multiclass pesticides	river water samples	GC-MS	0.07–0.3 ng mL^{-1}	(Bagheri et al., 2012a)
PPY/polyamide electrospun-based nanofiber	OPPs	river water samples	GC-MS	0.04–0.1 ng mL^{-1}	(Bagheri et al., 2012b)
17 beta-Estradiol- MIP, silica gel	estrogenic compounds	water samples	LVI-derivatiza-tion-GC-MS	MIP: 1.3–22; C18: 0.02–87 ng L^{-1}	(Prieto et al., 2011b)
C18 sorbent and an ionic liquid stationary phase(SLB-IL59)	haloanisoles	wines	GC-μECD	1.2–4.8 ng L^{-1}	(Pinto et al., 2012)

3.7 Conclusions and Prospects

Diverse novel materials including nanostructured materials, molecularly-imprinted materials, ILs/PILs, immunosorbents and MOFs have been applied in the preparation of SPME coatings which significantly widen the application field of SPME.

In most cases, application of nanomaterials enhances analytical performance compared to commercial SPME fibers due to the high surface area of nanomaterials. This aspect of nanomaterials along with their thermal and mechanical stabilities may lead to their wide application in SPME. In addition, some nanomaterial-containing composites are under development for an improved extraction performance and coating stability. Although IL-based SPME is still relatively new and in its early development stage, analytical results demonstrated that IL-based SPME fiber is a very promising technique. IL-based SPME will be more specific than conventional SPME because the fiber properties can be fine-tuned according to the characteristics of target analytes by choosing or designing the molecular structure of the IL. This feature provides more flexibility and versatility in terms of choosing the fiber coating material. MOF-based SPME coatings are also a new development. Interest is focused on the preparation of water stable MOFs for real sample analysis by SPME. The combination of MOF and other materials to build multifunctional composites such as MOF/nanoparticles, MOF/graphene, MOF/silica, and MOF/organic polymers is also a solution to improving analytical performance. To improve the extraction selectivity of SPME, MIPs are a good choice, however, great efforts are still needed to improve their performance in aqueous solution.

The analytical chemist has a variety of materials to choose from in SPME coatings fabrication and related method development. However, there are aspects that still remain to be improved, especially the selectivity for target analytes and the applicability of certain materials in samples with complex matrices.

Abbreviations

2,2'-dpy	2,2'-dipyridine
3TMSPMA	3-trimethoxysilyl propyl methacrylate
5-HT	serotonin
AA	acrylic acid
AAPTS	N-(2-Aminoethyl)-3-aminopropyltrimethoxysilane
AA-VP	acrylamide-vinylpyridine
AA-VP-bis	acrylamide-vinylpyridine-N,N'-methylene bisacrylamide
AC	activated carbon
ADS	alkyl-diol-silica
AIBN	2,2'-azo-bis(isobutyronitrile)

AMPS-*co*-OCMA-*co*-EDMA	poly 2-acrylamido-2-methylpropane sulfonic acid-stearyl methacrylate-ethylene glycol dimethacrylate
anti-FITC	antifluorescein isothiocyanate
AP	amphetamine
APTES	3-aminopropyltriethoxysilane
ARA-IIs	angiotensin II receptor antagonists
BDC	terephthalic acid
BFRs	brominated flame-retardant compounds
BMPO	bis[(3-methyldimethoxysilyl)propyl]
BMPT	N-butyl-4-methylpyridinium tetrafluoroborate
BPA	bisphenol A
BTEX	benzene, toluene, ethylbenzene, and xylenes
CA	caffeic acid
CAR	Carboxen
CE	capillary electrophoresis
CGA	chlorogenic acid
CIP	complex-imprinted polymer
CIPF	complex-imprinted polymer -coated fiber
CME	capillary microextraction
CMK	carbon mesoporous from Korea
CN-PDMS	cyano-polydimethylsiloxane
CNTs	carbon nanotubes
CP	conductive polymer
CVD	chemical vapor deposition
CW	Carbowax
CW/TR	Carbowax/Templated resin
CZE	capillary zone electrophoresis
DA	dopamine
DAD	diode array detection
DBP	dibutyl phthalate
DMDSe	dimethyldiselenide
DMSe	dimethylselenide
DPCSV	differential pulse cathodic stripping voltammetry
DVB	divinylbenzene
ECD	electron capture detector
ECL	electrochemiluminescence
EDMA	ethylene dimethacrylate
EE	extraction efficiency
EECDs	estrogen-like endocrine disrupting compounds
EE-SPME	electrosorption-enhanced-solid-phase microextraction
EF	enrichment factor
EG	ethylene glycol

EO	ethylene oxide
EPD	electrophoretic deposition
ESI-MS	electrospray ionization mass spectrometry
ETAAS	electrothermal atomic absorption spectrometry
EVA	ethylene vinyl acetate
FAAS	flame atomic absorption spectrometry
FAIMS	Field asymmetric ion mobility spectrometry
FAMEs	fatty acid methyl esters
FGO	functional graphene oxide
FID	flame ionization detection
FLD	fluorescence detection
FSH	follicle-stimulating hormone
GC	gas chromatography
GMA	glycidyl methacrylate
GMA-co-EGDMA	glycidyl methacrylate-co-ethylene glycol dimethacrylate
GOs	graphite oxides
GS-SPME	gaseous sampling solid-phase microextraction
HF-SPME	hollow fiber solid-phase microextraction
HFT	hollow fiber tube
HPLC	high performance liquid chromatography
HPTES	3-[Bis(2-hydroxyethyl)amino]propyl-triethoxysilane
HRP	horseradish peroxidase
HS-SPME	headspace solid-phase microextraction
HVOCs	volatile organohalogen compounds
IACE	immunoaffinity capillary electrophoresis
ICP-MS	inductively coupled plasma-mass spectrometry
IDA	iminodiacetic acid
IIF	ion imprinted fiber
IIPs	ion-imprinted polymers
IL	ionic liquid
IRIS	2-amino-2-hydroxymethyl-propane-1,3-dial
ITMS	ion trap mass spectrometry
IXN	isoxanthohumol, LBL, layer by layer
LC	liquid chromatography
LDH	layered double hydroxide
LH	luteinizing hormone
LIF	Laser induced fluorescence
Liz	lysozyme
LOD	limit of detection
LPD	liquid phase deposition
LPME	liquid-phase microextraction
LTGC	low-temperature glassy carbon

MA	microwave assisted
MAA-co-EGDMA	methacrylic acid-co-ethylene glycol dimethacrylate
MAFs	metal azolate frameworks
MAP	methamphetamine
MASE-*co*-DB	methyl acrylate, 3 sulfonate-divinylbenzene
MASE-EDMA	methacrylate stearyl acrylate-vinyl dimethacrylate
MEPS	microextraction in a packed syringe
META-*co*-DB	poly [2-(methacryloyloxy) ethyl trimethylammonium chloride]-divinylbenzene
MIPPy	molecularly imprinted polypyrrole
MIPs	molecularly imprinted polymers
MISGMs	molecularly imprinted sol-gel materials
MNT	mono-nitrotoluene
MOFs	metal-organic frameworks
MPTS	3-mercaptopropyltrimethoxysilane
MS	mass spectrometry
MSPE	micro-solid phase extraction
MTBE	methyl tert-butyl ether
MWCNTs	multi-walled carbon nanotubes
NE	norepinephrine
NIF	non-ion-imprinted fiber
NIP	nonimprinted polymer
NPD	nitrogen phosphorous detection
NPs	nanoparticles
NTD	needle trap device
NT-proBNP	N-terminal pro-natriuretic peptide
OA	cis-9-octadecenoic acid
OCPs	organochlorine pesticides
OE	oleuropein
OH-TSO	hydroxylterminated silicone oil
OPPs	organophosphorus pesticides
OT-IMAC	open tubular-immobilized metal-ion affinity chromatography
PA	polyacrylate
PAAs	polar aromatic amines
PAEs	phthalate esters
PAHs	polycyclic aromatic hydrocarbons
PAN	polyacrylonitrile
PANi	polyaniline
PBDEs	polybrominated diphenyl ethers
PCBs	polychlorinated biphenyls
PDMDPS	poly(dimethyldiphenylsiloxane)

PDMS	polydimethylsiloxane
PDMS	polydimethylsiloxane
PDMS/DVB	polydimethylsiloxane/divinylbenzene
PDVB	poly(divinylbenzene)
PEG	polyethylene glycol
PEG-DA	poly(ethylene glycol) diacrylate
PEI	polyethylenimine
PETA	pentaerythritol triacrylate
PIL	polymeric ionic liquid
poly(THF)	poly(tetrahydrofuran)
PP	polypropylene
PPESK	poly(phthalazine ether sulfone ketone)
PPY	polypyrrole
PS-DVB	poly(styrene-divinylbenzene)
PTH	polythiophene
PU	polyurethane
PVA	poly (vinyl alcohol)
PVC	polyvinyl chloride
RAFT	reversible addition-fragmentation chain transfer
RAM	restricted access materials
RSD	relative standard deviation
RTIL	room temperature ionic liquid
SBSE	stir-bar sorptive extraction
SDS-PAGE	sodium dodecyl sulfate polyacrylamide gel electrophoresis
SEM	scanning electron microscopy
SPE	solid phase extraction
SPME	solid-phase microextraction
SWCNTs	single-walled carbon nanotubes
TBZ	thiabendazole
TCD	thermal conductivity detection
TCs	tetracyclines
TFME	thin-film microextraction
TSH	thyroid-stimulating hormone
TTPT	trihexyltetradecylphosphonium tetrafluoroborate
TY	tyrsol
UE-SPME	ultrasonic extraction solid-phase microextraction
VFAs	volatile fatty acids
VIDB	vinyl imidazole-divinylbenzene
VOCs	volatile organic compounds
VPBA	4-vinylphenylboronic acid
VP-EDMA	vinyl pyridine-vinyl dimethacrylate
VPL-DVB	vinyl pyrrolidone-divinylbenzene

XN	xanthohumol
ZNRs	ZnO nanorods
β-CD	β-cyclodextrin

Acknowledgements

The National Nature Science Foundation of China (No.21375097).

References

Abdel-Rehim, M. (2004). New trend in sample preparation: on-line microextraction in packed syringe for liquid and gas chromatography applications - I. Determination of local anaesthetics in human plasma samples using gas chromatography-mass spectrometry. *Journal of Chromatography B-Analytical Technologies in the Biomedical and Life Sciences, 801*, 317–321.

Abdel-Rehim, M., Andersson, A., Breitholtz-Emanuelsson, A., Sandberg-Stall, M., Brunfelter, K., Pettersson, K. J., & Norsten-Hoog, C. (2008). MEPS as a rapid sample preparation method to handle unstable compounds in a complex matrix: Determination of AZD3409 in plasma samples utilizing MEPS-LC-MS-MS. *Journal of Chromatographic Science, 46*, 518–523.

Abdel-Rehim, M., Andersson, L. I., Altun, Z., & Blomberg, L. G. (2006). Microextraction in packed syringe online with liquid chromatography-tandem mass spectrometry: Molecularly imprinted polymer as packing material for MEPS in selective extraction of ropivacaine from plasma. *Journal of Liquid Chromatography & Related Technologies, 29*, 1725–1736.

Abolghasemi, M. M., Karimi, B., & Yousefi, V. (2013). Periodic mesoporous organosilica with ionic liquid framework as a novel fiber coating for headspace solid-phase microextraction of polycyclic aromatic hydrocarbons. *Analytica Chimica Acta, 804*, 280–286.

Abolghasemi, M. M., & Yousefi, V. (2014). Polythiophene/hexagonally ordered silica nanocomposite coating as a solid-phase microextraction fiber for the determination of polycyclic aromatic hydrocarbons in water. *Journal of Separation Science, 37*, 120–126.

Agrawal, Y. K. (2007). Poly(beta-styryl)-(1,2-methanofullerene-C60)-61-formo, hydroxamic acid for the solid phase extraction, separation and preconcentration of rare earth elements. *Fullerenes Nanotubes and Carbon Nanostructures, 15*, 353–365.

Ahmad, R., Wong-Foy, A. G., & Matzger, A. J. (2009). Microporous coordination polymers as selective sorbents for liquid chromatography. *Langmuir*, 25), 11977–11979.

Ai, Y., Zhao, F., & Zeng, B. (2012). Electrochemical fabrication of hydrophobic ionic liquid-polyaniline composite coating for solid-phase microextraction and its application in the GC determination of phenolic compounds. *Acta Chimica Sinica, 70*, 2440–2444.

Alaerts, L., Kirschhock, C. E. A., Maes, M., van der Veen, M. A., Finsy, V., Depla, A., Martens, J. A., Baron, G. V., Jacobs, P. A., Denayer, J. E. M., De Vos, D. E. (2007). Selective adsorption and separation of xylene isomers and ethylbenzene with the microporous vanadium(IV) terephthalate MIL-47. *Angewandte Chemie International Edition, 46*, 4293–4297.

Alaerts, L., Maes, M., Giebeler, L., Jacobs, P. A., Martens, J. A., Denayer, J. F. M., Kirschhock, C. E. A., De Vos, D. E. (2008). Selective adsorption and separation of ortho-substituted alkylaromatics with the microporous aluminum terephthalate MIL-53. *Journal of the American Chemical Society, 130*, 14170–14178.

Alhooshani, K., Kim, T. Y., Kabir, A., & Malik, A. (2005). Sol-gel approach to *in situ* creation of high pH-resistant surface-bonded organic-inorganic hybrid zirconia coating for capillary microextraction (in-tube SPME). *Journal of Chromatography A, 1062*, 1–14.

Alizadeh, R., & Najafi, N. M. (2013). Quantification of PAHs and chlorinated compounds by novel solid-phase microextraction based on the arrays of tin oxide nanorods. *Environmental Monitoring and Assessment, 185*, 7353–7363.

Alizadeh, R., Najafi, N. M., & Kharrazi, S. (2011). A new solid phase micro extraction for simultaneous head space extraction of ultra traces of polar and non-polar compounds. *Analytica Chimica Acta, 689*, 117–121.

Alonso, M., Cerdan, L., Godayol, A., Antico, E., & Sanchez, J. M. (2011). Headspace needle-trap analysis of priority volatile organic compounds from aqueous samples: Application to the analysis of natural and waste waters. *Journal of Chromatography A, 1218*, 8131–8139.

Altun, Z., Abdel-Rehim, M., & Blomberg, L. G. (2004). New trends in sample preparation: on-line microextraction in packed syringe (MEPS) for LC and GC applications Part III: Determination and validation of local anaesthetics in human plasma samples using a cation-exchange sorbent, and MEPS-LC-MS-MS. *Journal of Chromatography B-Analytical Technologies in the Biomedical and Life Sciences, 813*, 129–135.

Ameloot, R., Aubrey, M., Wiers, B. M., Gómora-Figueroa, A. P., Patel, S. N., Balsara, N. P., & Long, J. R. (2013). Ionic conductivity in the metal–organic framework UiO-66 by dehydration and insertion of lithium *tert*-butoxide. *Chemistry - A European Journal, 19*, 5533–5536.

Ameloot, R., Liekens, A., Alaerts, L., Maes, M., Galarneau, A., Coq, B., Desmet, G., Sels, B. F., Denayer, J. F. M., De Vos, D. E. (2010). Silica-MOF composites as a stationary phase in liquid chromatography. *European Journal of Inorganic Chemistry*, 3735–3738.

Amini, R., Rouhollahi, A., Adibi, M., & Mehdinia, A. (2011a). A new disposable ionic liquid based coating for headspace solid-phase microextraction of methyl tert-butyl ether in a gasoline sample followed by gas chromatography-flame ionization detection. *Talanta, 84*, 1–6.

Amini, R., Rouhollahi, A., Adibi, M., & Mehdinia, A. (2011b). A novel reusable ionic liquid chemically bonded fused-silica fiber for headspace solid-phase microextraction/gas chromatography-flame ionization detection of methyl tert-butyl ether in a gasoline sample. *Journal of Chromatography A, 1218*, 130–136.

Anbia, M., Haghi, A., & Shariati, S. (2012). Novel fiber coated with nanoporous carbons for headspace solid-phase microextraction of chlorophenols from aqueous media. *Analytical Methods, 4*, 2555–2561.

Anbia, M., & Khazaei, M. (2011). Ordered nanoporous carbon-based SPME and determination by GC. *Chromatographia, 73*, 379–384.

Aranda, R., Kruus, P., & Burk, R. C. (2000). Assessment of polycrystalline graphites as sorbents for solid-phase microextraction of nonionic surfactants. *Journal of Chromatography A, 888*, 35–41.

Arthur, C. L., & Pawliszyn, J. (1990). Solid phase microextraction with thermal desorption using fused silica optical fibers. *Analytical Chemistry, 62*, 2145–2148.

Azar, P. A., Mohammadiazar, S., Tehrani, M. S., & Haghi, J. N. (2013). Nanoparticle-incorporated PDMS films as an improved performance SPME fibre for analysis of volatile components of eucalyptus leaf. *Journal of Chemistry*., 347927.

Azar, P. A., Tehrani, M. S., Mohammadiazar, S., & Husain, S. W. (2012). A novel strategy to increase performance of solid-phase microextraction fibers: Electrodeposition of sol-gel films on highly porous substrate. *Journal of Separation Science, 35*, 3354–3360.

Bagheri, H., Alipour, N., & Ayazi, Z. (2012a). Multiresidue determination of pesticides from aquatic media using polyaniline nanowires network as highly efficient sorbent for microextraction in packed syringe. *Analytica Chimica Acta, 740*, 43–49.

Bagheri, H., & Ayazi, Z. (2011). Polypyrrole nanowires network for convenient and highly efficient microextraction in packed syringe. *Analytical Methods, 3*, 2630–2636.

Bagheri, H., Ayazi, Z., Aghakhani, A., & Alipour, N. (2012b). Polypyrrole/polyamide electrospun-based sorbent for microextraction in packed syringe of organophosphorous pesticides from aquatic samples. *Journal of Separation Science, 35*, 114–120.

Bagheri, H., Ayazi, Z., Es'haghi, A., & Aghakhani, A. (2012c). Reinforced polydiphenylamine nanocomposite for microextraction in packed syringe of various pesticides. *Journal of Chromatography A, 1222*, 13–21.

Bagheri, H., Babanezhad, E., & Khalilian, F. (2009). An interior needle electropolymerized pyrrole-based coating for headspace solid-phase dynamic extraction. *Analytica Chimica Acta, 634*, 209–214.

Bagheri, H., Bayat, P., & Piri-Moghadam, H. (2013). Grafting the sol-gel based sorbents by diazonium salts: A novel approach toward unbreakable capillary microextraction. *Journal of Chromatography A, 1318*, 58–64.

Bagheri, H., & Piri-Moghadam, H. (2012). Sol-gel-based molecularly imprinted xerogel for capillary microextraction. *Analytical and Bioanalytical Chemistry, 404*, 1597–1602.

Bagheri, H., Piri-Moghadam, H., & Es'haghi, A. (2011a). An unbreakable on-line approach towards sol-gel capillary microextraction. *Journal of Chromatography A, 1218*, 3952–3957.

Bagheri, H., & Roostaie, A. (2012). Aniline-silica nanocomposite as a novel solid phase microextraction fiber coating. *Journal of Chromatography A, 1238*, 22–29.

Bagheri, H., Roostaie, A., & Babanezhad, E. (2011b). New grafted nanosilica-based sorbent for needle trap extraction of polycyclic aromatic hydrocarbons from water samples followed by GC/MS. *Chromatographia, 74*, 429–436.

Bai, L., Liu, H., Liu, Y., Zhang, X., Yang, G., & Ma, Z. (2011). Preparation of a novel hybrid organic–inorganic monolith for the separation of lysozyme by high performance liquid chromatography. *Journal of Chromatography A, 1218*, 100–106.

Bailly-Chouriberry, L., Cormant, F., Garcia, P., Lonnberg, M., Szwandt, S., Bondesson, U., Popot, M. A., Bonnaire, Y. (2012). A new analytical method based on anti-EPO monolith column and LC-FAIMS-MS/MS for the detection of rHuEPOs in horse plasma and urine samples. *The Analyst, 137*, 2445–2453.

Banerjee, R., Phan, A., Wang, B., Knobler, C., Furukawa, H., O'Keeffe, M., & Yaghi, O. M. (2008). High-throughput synthesis of zeolitic imidazolate frameworks and application to CO_2 capture. *Science, 319*, 939–943.

Banitaba, M. H., Davarani, S. S. H., & Pourahadi, A. (2013). Solid-phase microextraction of phthalate esters from aqueous media by electrophoretically deposited TiO_2 nanoparticles on a stainless steel fiber. *Journal of Chromatography A, 1283*, 1–8.

Barahona, F., Turiel, E., & Martin-Esteban, A. (2011). Supported liquid membrane-protected molecularly imprinted fibre for solid-phase microextraction of thiabendazole. *Analytica Chimica Acta, 694*, 83–89.

Barreto, A. S., da Silva, R. L., Silva, S. C. G. D., Rodrigues, M. O., de Simone, C. A., de Sa, G. F., Junior, S. A., Navickiene, S., de Mesquita, M. E. (2010). Potential of a metal-organic framework as a new material for solid-phase extraction of pesticides from lettuce (Lactuca sativa), with analysis by gas chromatography-mass spectrometry. *Journal of Separation Science, 33*, 3811–3816.

Belardi, R. P., & Pawliszyn, J. B. (1989). The application of chemically modified fused silica fibers in the extraction of organics from water matrix samples and their rapid transfer to capillary columns. *Water Pollution Research Journal of Canada, 24*, 179–191.

Biajoli, A. F. P., & Augusto, F. (2008). Solid phase microextraction fibers coated with sol-gel aminopropylsilica/polydimethylsiloxane: Development and its application to screening of beer headspace. *Analytical Sciences, 24*, 1141–1146.

Bigham, S., Medlar, J., Kabir, A., Shende, C., Alli, A., & Malik, A. (2002). Sol-gel capillary microextraction. *Analytical chemistry, 74*, 752–761.

Blanchard, L. A., Hancu, D., Beckman, E. J., & Brennecke, J. F. (1999). Green processing using ionic liquids and CO_2. *Nature, 399*, 28–29.

Bragg, L., Qin, Z., Alaee, M., & Pawliszyn, J. (2006). Field sampling with a polydimethylsiloxane thin-film. *Journal of Chromatographic Science, 44*.

Brothier, F., & Pichon, V. (2013). Immobilized antibody on a hybrid organic-inorganic monolith: Capillary immunoextraction coupled on-line to nanoLC-UV for the analysis of microcystin-LR. *Analytica Chimica Acta, 792*, 52–58.

Budziak, D., Martendal, E., & Carasek, E. (2007). Application of NiTi alloy coated with ZrO_2 as a new fiber for solid-phase microextraction for determination of halophenols in water samples. *Analytica Chimica Acta, 598*, 254–260.

Budziak, D., Martendal, E., & Carasek, E. (2008a). Application of poly(dimethylsiloxane) fiber sol-gel coated onto NiTi alloy electrodeposited with zirconium oxide for the determination of organochlorine pesticides in herbal infusions. *Journal of Separation Science, 31*, 2875–2881.

Budziak, D., Martendal, E., & Carasek, E. (2008b). Application of robust NiTi-ZrO_2-PEG SPME fiber in the determination of haloanisoles in cork stopper samples. *Analytica Chimica Acta, 629*, 92–97.

Budziak, D., Martendal, E., & Carasek, E. (2008c). New poly(ethylene glycol) solid-phase microextraction fiber employing zirconium oxide electrolytically deposited onto a NiTi alloy as substrate for sol-gel reactions. *Journal of Chromatography A, 1198*, 54–58.

Budziak, D., Martendal, E., & Carasek, E. (2008d). Preparation and characterization of new solid-phase microextraction fibers obtained by sol-gel technology and zirconium oxide electrodeposited on NiTi alloy. *Journal of Chromatography A, 1187*, 34–39.

Budziak, D., Martendal, E., & Carasek, E. (2009). Application of an NiTi alloy coated with ZrO_2 solid-phase microextraction fiber for determination of haloanisoles in red wine samples. *Microchimica Acta, 164*, 197–202.

Cai, J. B., Ouyang, G., Gong, Y., & Pawliszyn, J. (2008). Simultaneous sampling and analysis for vapor mercury in ambient air using needle trap coupled with gas chromatography-mass spectrometry. *Journal of Chromatography A, 1213*, 19–24.

Cao, D. D., Lü, J. X., Liu, J. F., & Jiang, G. B. (2008). *In situ* fabrication of nanostructured titania coating on the surface of titanium wire: A new approach for preparation of solid-phase microextraction fiber. *Analytica Chimica Acta, 611*, 56–61.

Celikbicak, O., Atakay, M., Guler, U., & Salih, B. (2013). A novel tantalum-based sol-gel packed microextraction syringe for highly specific enrichment of phosphopeptides in MALDI-MS applications. *The Analyst, 138*, 4403–4410.

Chae, H. K., Siberio-Pérez, D. Y., Kim, J., Go, Y. B., Eddaoudi, M., Matzger, A. J., O'Keeffe, M., & Yaghi, O. M. (2004). A route to high surface area, porosity and inclusion of large molecules in crystals. *Nature, 427*, 523–527.

Chamieh, J., Faye, C., Dugas, V., Moreau, T., Vandenabeele-Trambouze, O., & Demesmay, C. (2012). Preparation and full characterization of a micro-immunoaffinity monolithic column and its in-line coupling with capillary zone electrophoresis with Ochratoxin A as model solute. *Journal of Chromatography A, 1232*, 93–100.

Chang, N., Gu, Z. Y., Wang, H. F., & Yan, X. P. (2011). Metal-organic-framework-based tandem molecular sieves as a dual platform for selective microextraction and high-resolution gas chromatographic separation of n-alkanes in complex matrixes. *Analytical Chemistry, 83*, 7094–7101.

Chang, N., Gu, Z. Y., & Yan, X. P. (2010). Zeolitic imidazolate framework-8 nanocrystal coated capillary for molecular sieving of branched alkanes from linear alkanes along with high-resolution chromatographic separation of linear alkanes. *Journal of the American Chemical Society, 132*, 13645–13647.

Chang, N., & Yan, X. P. (2012). Exploring reverse shape selectivity and molecular sieving effect of metal-organic framework UIO-66 coated capillary column for gas chromatographic separation. *Journal of Chromatography A, 1257*, 116–124.

Chang, X. J., Jiang, N., Zheng, H., He, Q., Hu, Z., Zhai, Y. H., & Cui, Y. M. (2007). Solid-phase extraction of iron(III) with an ion-imprinted functionalized silica gel sorbent prepared by a surface imprinting technique. *Talanta, 71*, 38–43.

Chaves, A. R., & Queiroz, M. E. C. (2013). Immunoaffinity in-tube solid phase microextraction coupled with liquid chromatography with fluorescence detection for determination of interferon alpha in plasma samples. *Journal of Chromatography B-Analytical Technologies in the Biomedical and Life Sciences, 928*, 37–43.

Chen, A., Peng, X., Koczkur, K., Miller, B. (2004). Super-hydrophobic tin oxide nanoflowers. *Chemical Communications*, 1964–1965.

Chen, A. W., Zeng, G. M., Chen, G. Q., Hu, X. J., Yan, M., Guan, S., Shang, C., Lu, L. H., Zou, Z. J., Xie, G. X. (2012). Novel thiourea-modified magnetic ion-imprinted chitosan/TiO_2 composite for simultaneous removal of cadmium and 2,4-dichlorophenol. *Chemical Engineering Journal, 191*, 85–94.

Chen, B. B., Peng, H. Y., Zheng, F., Hu, B., He, M., Zhao, W., & Pang, D. W. (2010). Immunoaffinity monolithic capillary microextraction coupled with ICP-MS for immunoassay with quantum dot labels. *Journal of Analytical Atomic Spectrometry, 25*, 1674–1681.

Chen, H. X., Huang, T., & Zhang, X. X. (2009). Immunoaffinity extraction of testosterone by antibody immobilized monolithic capillary with on-line laser-induced fluorescence detection. *Talanta, 78*, 259–264.

Chen, J. M., Zou, J., Zeng, J. B., Song, X. H., Ji, J. J., Wang, Y. R., Ha, J., Chen, X. (2010). Preparation and evaluation of graphene-coated solid-phase microextraction fiber. *Analytica Chimica Acta, 678*, 44–49.

Chen, M. L., Zheng, M. M., & Feng, Y. Q. (2010). Preparation of organic–inorganic hybrid silica monolith with octyl and sulfonic acid groups for capillary electrochromatograhpy and application in determination of theophylline and caffeine in beverage. *Journal of Chromatography A, 1217*, 3547–3556.

Chen, S., Liu, C., Yang, M., Lu, D., Zhu, L., & Wang, Z. (2009). Solid-phase extraction of Cu, Co and Pb on oxidized single-walled carbon nanotubes and their determination by inductively coupled plasma mass spectrometry. *Journal of Hazardous Materials, 170*, 247–251.

Chen, X. F., Zang, H., Wang, X., Cheng, J. G., Zhao, R. S., Cheng, C. G., & Lu, X. Q. (2012). Metal-organic framework MIL-53(Al) as a solid-phase microextraction adsorbent for the determination of 16 polycyclic aromatic hydrocarbons in water samples by gas chromatography-tandem mass spectrometry. *The Analyst, 137*, 5411–5419.

Choi, K., Ng, A. H. C., Fobel, R., Chang-Yen, D. A., Yarnell, L. E., Pearson, E. L., Oleksak, C. M., Fischer, A. T., Luoma, R. P., Robinson, J. M., Audet, J., Wheeler, A. R. (2013). Automated digital microfluidic platform for magnetic-particle-based immunoassays with optimization by design of experiments. *Analytical Chemistry, 85*, 9638–9646.

Chong, S. L., Wang, D. X., Hayes, D., Wilhite, B. W., & Malik, A. (1997). Sol-gel coating technology for the preparation of solid-phase microextraction fibers of enhanced thermal stability. *Analytical Chemistry, 69*, 3889–3898.

Chui, S. S.-Y., Lo, S. M.-F., Charmant, J. P. H., Guy Orpen, A., & Williams, I. D. (1999). A chemically functionalizable nanoporous material $[Cu_3(TMA)_2(H_2O)_3]_n$. *Science, 283*, 1148–1150.

Cui, X. Y., Gu, Z. Y., Jiang, D. Q., Li, Y., Wang, H. F., & Yan, X. P. (2009). *In situ* hydrothermal growth of metal–organic framework 199 films on stainless steel fibers for solid-phase microextraction of gaseous benzene homologues. *Analytical Chemistry, 81*, 9771–9777.

Dakova, I., Karadjova, I., Georgieva, V., & Georgiev, G. (2012). Synthesis and application of vinylpyridine containing ion-imprinted copolymer gel microbeads for Cu(II) solid-phase extraction. *Journal of Separation Science, 35*, 2805–2812.

de Carvalho, P. H. V., Barreto, A. S., Rodrigues, M. O., Prata, V. D., Alves, P. B., de Mesquita, M. E., Júnior, S. A., & Navickiene, S. (2009). Two-dimensional coordination polymer matrix for solid-phase extraction of pesticide residues from plant Cordia salicifolia. *Journal of Separation Science, 32*, 2132–2138.

de Oliveira, A. F., da Silveira, C. B., de Campos, S. D., de Campos, E. A., & Carasek, E. (2005). Niobium(V) oxide coated on thin glass-ceramic rod as a solid phase microextraction fiber. *Talanta, 66*, 74–79.

Delmotte, N., Kobold, U., Meier, T., Gallusser, A., Strancar, A., & Huber, C. G. (2007). Miniaturized monolithic disks for immunoadsorption of cardiac biomarkers from serum. *Analytical and Bioanalytical Chemistry, 389*, 1065–1074.

Deng, D. L., Zhang, J. Y., Chen, C., Hou, X. L., Su, Y. Y., & Wu, L. (2012). Monolithic molecular imprinted polymer fiber for recognition and solid phase microextraction of ephedrine and pseudoephedrine in biological samples prior to capillary electrophoresis analysis. *Journal of Chromatography A, 1219*, 195–200.

Dietz, C., Sanz, J., & Camara, C. (2006). Recent developments in solid-phase microextraction coatings and related techniques. *Journal of Chromatography A, 1103*, 183–192.

Ding, T. H., Lin, H. H., & Whang, C. W. (2005). Determination of chromium(III) in water by solid-phase microextraction with a polyimide-coated fiber and gas chromatography-flame photometric detection. *Journal of Chromatography A, 1062*, 49–55.

Djozan, D., & Abdollahi, L. (2003). Anodized zinc wire as a solid-phase microextraction fiber. *Chromatographia, 57*, 799–804.

Djozan, D., Assadi, Y., & Haddadi, S. H. (2001). Anodized aluminum wire as a solid-phase microextraction fiber. *Analytical Chemistry, 73*, 4054–4058.

Djozan, D., & Baheri, T. (2007). Preparation and evaluation of solid-phase microextraction fibers based on monolithic molecularly imprinted polymers for selective extraction of diacetylmorphine and analogous compounds. *Journal of Chromatography A, 1166*, 16–23.

Djozan, D., Mahkam, M., & Ebrahimi, B. (2009). Preparation and binding study of solid-phase microextraction fiber on the basis of ametryn-imprinted polymer Application to the selective extraction of persistent triazine herbicides in tap water, rice, maize and onion. *Journal of Chromatography A, 1216*, 2211–2219.

Domingues, V., Cabral, M., Alves, A., & Delerue-Matos, C. (2009). Use and reuse of SPE disks for the determination of pyrethroids in water by GC-ECD. *Analytical Letters, 42*, 706–726.

Du, W., Zhao, F., & Zeng, B. (2009). Novel multiwalled carbon nanotubes-polyaniline composite film coated platinum wire for headspace solid-phase microextraction and gas chromatographic determination of phenolic compounds. *Journal of Chromatography A, 1216*, 3751–3757.

Du, X. Z., Wang, Y. R., Tao, X. J., & Deng, H. L. (2005). An approach to application of mesoporous hybrid as a fiber coating of solid-phase microextraction. *Analytica Chimica Acta, 543*, 9–16.

Duan, J. K., Li, X., Yu, C. H., & Hu, B. (2009). Headspace stir bar sorptive extraction combined with GC-ICP-MS for the speciation of dimethylselenide and dimethyldiselenide in biological samples. *Journal of Analytical Atomic Spectrometry, 24*, 297–303.

Ebrahimi, M., Es'haghi, Z., Samadi, F., & Hosseini, M. S. (2011). Ionic liquid mediated sol-gel sorbents for hollow fiber solid-phase microextraction of pesticide residues in water and hair samples. *Journal of Chromatography A, 1218*, 8313–8321.

Ebrahimzadeh, H., Mehdinia, A., Yamini, Y., Kasraee, S., & Gholizade, A. (2010). Analysis of mono-nitrotoluenes in water samples by using nano-structured polypyrrol as a sorbent of solid-phase microextraction. *International Journal of Environmental Analytical Chemistry, 90*, 963–975.

Eddaoudi, M., Kim, J., Wachter, J. B., Chae, H. K., O'Keeffe, M., & Yaghi, O. M. (2001). Porous metal-organic polyhedra: 25 Å cuboctahedron constructed from 12 $Cu_2(CO_2)_4$ paddle-wheel building blocks. *Journal of the American Chemical Society, 123*, 4368–4369.

Eisert, R., & Pawliszyn, J. (1997). Automated in-tube solid-phase microextraction coupled to high-performance liquid chromatography. *Analytical Chemistry, 69*, 3140–3147.

El-Beqqali, A., Kussak, A., & Abdel-Rehim, M. (2006). Fast and sensitive environmental analysis utilizing microextraction in packed syringe online with gas chromatography-mass spectrometry - Determination of polycyclic aromatic hydrocarbons in water. *Journal of Chromatography A, 1114*, 234–238.

El-Beqqali, A., Kussak, A., & Abdel-Rehim, M. (2007). Determination of dopamine and serotonin in human urine samples utilizing microextraction online with liquid chromatography/electrospray tandem mass spectrometry. *Journal of Separation Science, 30*, 421–424.

Engler, K. N., & Lemley, A. T. (2013). Development of an in vitro thin-film solid-phase microextraction method to determine the bioavailability of xenoestrogens in soil. *Environmental Toxicology and Chemistry, 32*, 1962–1968.

Eom, I. Y., Niri, V. H., & Pawliszyn, J. (2008). Development of a syringe pump assisted dynamic headspace sampling technique for needle trap device. *Journal of Chromatography A, 1196*, 10–14.

Fan, J., Wei, Y., Wang, J., Wu, C., & Shi, H. (2009). Study of molecularly imprinted solid-phase extraction of diphenylguanidine and its structural analogs. *Analytica Chimica Acta, 639*, 42–50.

Fan, L. L., Luo, C. N., Lv, Z., Lu, F. G., & Qiu, H. M. (2011). Removal of Ag^+ from water environment using a novel magnetic thiourea-chitosan imprinted Ag^+. *Journal of Hazardous Materials, 194*, 193–201.

Fan, Y., Feng, Y. Q., Da, S. L., & Shi, Z. G. (2004). Poly (methacrylic acid–ethylene glycol dimethacrylate) monolithic capillary for in-tube solid phase microextraction coupled to high performance liquid chromatography and its application to determination of basic drugs in human serum. *Analytica Chimica Acta, 523*, 251–258.

Fan, Y., Feng, Y. Q., Da, S. L., & Wang, Z. H. (2005a). In-tube solid phase microextraction using a β-cyclodextrin coated capillary coupled to high performance liquid chromatography for determination of non-steroidal anti-inflammatory drugs in urine samples. *Talanta, 65*, 111–117.

Fan, Y., Zhang, M., & Feng, Y. Q. (2005b). Poly(acrylamide-vinylpyridine-N,N'-methylene bisacrylamide) monolithic capillary for in-tube solid-phase microextraction coupled to high performance liquid chromatography. *Journal of Chromatography A, 1099*, 84–91.

Fang, L., Kulkarni, S., Alhooshani, K., & Malik, A. (2007). Germania-based, sol-gel hybrid organic-inorganic coatings for capillary microextraction and gas chromatography. *Analytical chemistry, 79*, 9441–9451.

Farahani, H., Norouzi, P., Dinarvand, R., & Ganjali, M. R. (2009). Liquid-phase microextraction by solidification of floating organic microdrop and GC-MS detection of trihalomethanes in drinking water. *Journal of Separation Science, 32*, 314–320.

Farajzadeh, M., & Matin, A. A. (2002). A new PVC-activated charcoal fiber coated on silver wire; Application in determination of n-alkanes in the headspace of soil samples by SPME-GC. *Analytical Sciences, 18*, 77–81.

Farajzadeh, M. A., & Hatami, M. (2004). Solid-phase microextraction gas chromatography for determination of some organophosphorus pesticides. *Chromatographia, 59*, 259–262.

Farha, O. K., Eryazici, I., Jeong, N. C., Hauser, B. G., Wilmer, C. E., Sarjeant, A. A., Snurr, R. Q., Nguyen, S. T., Özgür Yazaydın, A., & Hupp, J. T. (2012). Metal-organic framework materials with ultrahigh surface areas: Is the sky the limit? *Journal of the American Chemical Society, 134*, 15016–15021.

Farhadi, K., Maleki, R., & Tahmasebi, R. (2010). A sol-gel based solid phase microextraction fiber for the analysis of aliphatic alcohols in apple juices. *Journal of Separation Science, 33*, 88–92.

Farhadi, K., Maleki, R., & Tahmasebi, R. (2011). Preparation of Al_2O_3/TiO_2 composite sol-gel fiber for headspace solid-phase microextraction of chlorinated organic solvents from urine. *Journal of Separation Science, 34*, 1669–1674.

Fei, T., Li, H., Ding, M., Ito, M., & Lin, J. M. (2011). Determination of parabens in cosmetic products by solid-phase microextraction of poly(ethylene glycol) diacrylate thin film on fibers and ultra high-speed liquid chromatography with diode array detector. *Journal of Separation Science, 34*, 1599–1606.

Feng, J., Sun, M., Wang, X., Liu, X., & Jiang, S. (2012). Ionic liquids-based crosslinked copolymer sorbents for headspace solid-phase microextraction of polar alcohols. *Journal of Chromatography A, 1245*, 32–38.

Feng, J. J., Sun, M., Li, J. B., Liu, X., & Jiang, S. X. (2011). A novel silver-coated solid-phase microextraction metal fiber based on electroless plating technique. *Analytica Chimica Acta, 701*, 174–180.

Feng, J. J., Sun, M., Liu, H. M., Li, J. B., Liu, X., & Jiang, S. X. (2010). Au nanoparticles as a novel coating for solid-phase microextraction. *Journal of Chromatography A, 1217*, 8079–8086.

Feng, Q., Zhao, L., & Lin, J. M. (2009). Molecularly imprinted polymer as micro-solid phase extraction combined with high performance liquid chromatography to determine phenolic compounds in environmental water samples. *Analytica Chimica Acta, 650*, 70–76.

Feng, X. L., Tian, M. C., Li, A. M., Zhao, X. N., & Zhang, Y. P. (2010). Multiwalled carbon nanotube coated on stainless steel wire for solid-phase microextraction of organochlorine pesticides in water. *Analytical Letters, 43*, 2477–2486.

Férey, G., Mellot-Draznieks, C., Serre, C., Millange, F., Dutour, J., Surblé, S., & Margiolaki, I. (2005). A chromium terephthalate-based solid with unusually large pore volumes and surface area. *Science, 309*, 2040–2042.

Finsy, V., Verelst, H., Alaerts, L., De Vos, D., Jacobs, P. A., Baron, G. V., & Denayer, J. F. M. (2008). Pore-filling-dependent selectivity effects in the vapor-phase separation of xylene isomers on the metal-organic framework MIL-47. *Journal of the American Chemical Society, 130*, 7110–7118.

Fontanals, N., Marce, R. M., & Borrull, F. (2005). New hydrophilic materials for solid-phase extraction. *TrAC Trends in Analytical Chemistry, 24*, 394–406.

Franco, E. J., Hofstetter, H., & Hofstetter, O. (2008). Enantiomer separation of alpha-hydroxy acids in high-performance immunoaffinity chromatography. *Journal of Pharmaceutical and Biomedical Analysis, 46*, 907–913.

Fu, Y. Y., Yang, C. X., & Yan, X. P. (2013). Incorporation of metal-organic framework UiO-66 into porous polymer monoliths to enhance the liquid chromatographic separation of small molecules. *Chemical Communications, 49*, 7162–7164.

Gao, Z., Deng, Y., Hu, X., Yang, S., Sun, C., & He, H. (2013). Determination of organophosphate esters in water samples using an ionic liquid-based sol-gel fiber for headspace solid-phase microextraction coupled to gas chromatography-flame photometric detector. *Journal of Chromatography A, 1300*, 141–150.

Gao, Z., Li, W., Liu, B., Liang, F., He, H., Yang, S., & Sun, C. (2011). Nano-structured polyaniline-ionic liquid composite film coated steel wire for headspace solid-phase microextraction of organochlorine pesticides in water. *Journal of Chromatography A, 1218*, 6285–6291.

Garcia-Sanchez, R., Feldhaus, R., Bettmer, J., & Ebdon, L. (2001). Lead speciation in rainwater samples by modified fused silica capillaries coupled to a direct injection nebulizer (DIN) for sample introduction in ICP-MS. *Journal of Analytical Atomic Spectrometry, 16*, 1028–1034.

Gasilova, N., Gassner, A. L., & Girault, H. H. (2012). Analysis of major milk whey proteins by immunoaffinity capillary electrophoresis coupled with MALDI-MS. *Electrophoresis, 33*, 2390–2398.

Ge, D., & Lee, H. K. (2011). Water stability of zeolite imidazolate framework 8 and application to porous membrane-protected micro-solid-phase extraction of polycyclic aromatic hydrocarbons from environmental water samples. *Journal of Chromatography A, 1218*, 8490–8495.

Geim, A. K., & Novoselov, K. S. (2007). The rise of graphene. *Nature Materials, 6*, 183–191.
Gholivand, M. B., Abolghasemi, M. M., & Fattahpour, P. (2011). Polypyrrole/hexagonally ordered silica nanocomposite as a novel fiber coating for solid-phase microextraction. *Analytica Chimica Acta, 704*, 174–179.
Gholivand, M. B., Piryaei, M., & Abolghasemi, M. M. (2013a). Analysis of volatile oil composition of Citrus aurantiumL. by microwave-assisted extraction coupled to headspace solid-phase microextraction with nanoporous based fibers. *Journal of Separation Science, 36*, 872–877.
Gholivand, M. B., Piryaei, M., Abolghasemi, M. M., & Maassoumi, S. M. (2013b). Rapid analysis of volatile components from Teucrium polium L. by nanoporous silica-polyaniline solid phase microextraction fibre. *Phytochemical Analysis, 24*, 69–74.
Giardina, M., & Olesik, S. V. (2003). Application of low-temperature glassy carbon-coated macrofibers for solid-phase microextraction analysis of simulated breath volatiles. *Analytical Chemistry, 75*, 1604–1614.
Glover, T. G., Peterson, G. W., Schindler, B. J., Britt, D., & Yaghi, O. (2011). MOF-74 building unit has a direct impact on toxic gas adsorption. *Chemical Engineering Science, 66*, 163–170.
Golsefidi, M. A., Es'haghi, Z., & Sarafraz-Yazdi, A. (2012). Design, synthesis and evaluation of a molecularly imprinted polymer for hollow fiber-solid phase microextraction of chlorogenic acid in medicinal plants. *Journal of Chromatography A, 1229*, 24–29.
Gomez-Caballero, A., Guerreiro, A., Karim, K., Piletsky, S., Goicolea, M. A., & Barrio, R. J. (2011). Chiral imprinted polymers as enantiospecific coatings of stir bar sorptive extraction devices. *Biosensors & Bioelectronics, 28*, 25–32.
Goncalves, J. L., Alves, V. L., Rodrigues, F. P., Figueira, J. A., & Camara, J. S. (2013). A semi-automatic microextraction in packed sorbent, using a digitally controlled syringe, combined with ultra-high pressure liquid chromatography as a new and ultra-fast approach for the determination of prenylflavonbids in beers. *Journal of Chromatography A, 1304*, 42–51.
Gonzalez-Alvarez, J., Blanco-Gomis, D., Arias-Abrodo, P., Pello-Palma, J., Rios-Lombardia, N., Busto, E., Gotor-Fernandez, V., Dolores Gutierrez-Alvarez, M. (2013). Analysis of beer volatiles by polymeric imidazolium-solid phase microextraction coatings: Synthesis and characterization of polymeric imidazolium ionic liquids. *Journal of Chromatography A, 1305*, 35–40.
Graham, C. M., Meng, Y. J., Ho, T., & Anderson, J. L. (2011). Sorbent coatings for solid-phase microextraction based on mixtures of polymeric ionic liquids. *Journal of Separation Science, 34*, 340–346.
Gu, Z. Y., Chen, Y. J., Jiang, J. Q., & Yan, X. P. (2011a). Metal-organic frameworks for efficient enrichment of peptides with simultaneous exclusion of proteins from complex biological samples. *Chemical Communications, 47*, 4787–4789.
Gu, Z. Y., Wang, G., & Yan, X. P. (2010a). MOF-5 Metal-organic framework as sorbent for in-field sampling and preconcentration in combination with thermal desorption GC/MS for determination of atmospheric formaldehyde. *Analytical Chemistry, 82*, 1365–1370.
Gu, Z. Y., Yang, C. X., Chang, N., & Yan, X. P. (2012). Metal-organic frameworks for analytical chemistry: From sample collection to chromatographic separation. *Accounts of Chemical Research, 45*, 734–745.
Gu, Z. Y., Jiang, D. Q., Wang, H. F., Cui, X. Y., & Yan, X. P. (2010b). Adsorption and separation of xylene isomers and ethylbenzene on two Zn-terephthalate metal-organic frameworks. *Journal of Physical Chemistry C, 114*, 311–316.
Gu, Z. Y., Jiang, J. Q., & Yan, X. P. (2011b). Fabrication of isoreticular metal-organic framework coated capillary columns for high-resolution gas chromatographic separation of persistent organic pollutants. *Analytical Chemistry, 83*, 5093–5100.
Gu, Z. Y., & Yan, X. P. (2010). Metal-organic framework MIL-101 for high-resolution gas-chromatographic separation of xylene isomers and ethylbenzene. *Angewandte Chemie International Edition, 49*, 1477–1480.

Guan, W., Wang, Y. J., Xu, F., & Guan, Y. F. (2008). Poly(phthalazine ether sulfone ketone) as novel stationary phase for stir bar sorptive extraction of organochlorine compounds and organophosphorus pesticides.. *Journal of Chromatography A, 1177*, 28–35.

Gunasena, D. N., & El Rassi, Z. (2011). Hydrophilic diol monolith for the preparation of immunosorbents at reduced nonspecific interactions. *Journal of Separation Science, 34*, 2097–2105.

Guo, Z., Wu, H., Srinivas, G., Zhou, Y., Xiang, S., Chen, Z., Yang, Y., Zhou, W., O'Keeffe, M., & Chen, B. (2011). A metal-organic framework with optimized open metal sites and pore spaces for high methane storage at room temperature. *Angewandte Chemie International Edition, 50*, 3178–3181.

Gusev, I., Huang, X., & Horváth, C. (1999). Capillary columns with *in situ* formed porous monolithic packing for micro high-performance liquid chromatography and capillary electrochromatography. *Journal of Chromatography A, 855*, 273–290.

Hafez, A. M., & Wenclawiak, B. W. (2013). Development of a new porous gold SPME fiber for selective and efficient extraction of dodecanethiol followed by GC-MS analysis. *Analytical and Bioanalytical Chemistry, 405*, 1753–1758.

Haick, H., Hakim, M., Patrascu, M., Levenberg, C., Shehada, N., Nakhoul, F., & Abassi, Z. (2009). Sniffing chronic renal failure in rat model by an array of random networks of single-walled carbon nanotubes. *ACS Nano, 3*, 1258–1266.

Hall, B. J., & Brodbelt, J. S. (1997). Determination of barbiturates by solid-phase microextraction (SPME) and ion trap gas chromatography-mass spectrometry. *Journal of Chromatography A, 777*, 275–282.

Han, S. B., Wei, Y. H., Valente, C., Lagzi, I., Gassensmith, J. J., Coskun, A., Stoddart, J. F., Grzybowski, B. A. (2010). CChromatography in a single metal-organic framework (MOF) crystal. *Journal of the American Chemical Society, 132*, 16358–16361.

Hashemi, P., & Rahimi, A. (2007). A highly sensitive method for the determination of mercury using vapor generation gold wire microextraction and electrothermal atomic absorption spectrometry. *Spectrochimica Acta Part B: Atomic Spectroscopy, 62*, 423–428.

Hashemi, P., Shamizadeh, M., Badiei, A., Poor, P. Z., Ghiasvand, A. R., & Yarahmadi, A. (2009). Amino ethyl-functionalized nanoporous silica as a novel fiber coating for solid-phase microextraction. *Analytica Chimica Acta, 646*, 1–5.

He, C. T., Tian, J. Y., Liu, S. Y., Ouyang, G., Zhang, J. P., & Chen, X. M. (2013). A porous coordination framework for highly sensitive and selective solid-phase microextraction of non-polar volatile organic compounds. *Chemical Science, 4*, 351.

He, F. F., Wang, H. Q., Wang, Y. Y., Wang, X. F., Zhang, H. S., Li, H. L., & Tang, J. H. (2013). Magnetic Th(IV)-ion imprinted polymers with salophen schiff base for separation and recognition of Th(IV). *Journal of Radioanalytical and Nuclear Chemistry, 295*, 167–177.

He, Y., Pohl, J., Engel, R., Rothman, L., & Thomas, M. (2009). Preparation of ionic liquid based solid-phase microextraction fiber and its application to forensic determination of methamphetamine and amphetamine in human urine. *Journal of Chromatography A, 1216*, 4824–4830.

Heidari, M., Bahrami, A., Ghiasvand, A. R., Shahna, F. G., & Soltanian, A. R. (2012). A novel needle trap device with single wall carbon nanotubes sol-gel sorbent packed for sampling and analysis of volatile organohalogen compounds in air. *Talanta, 101*, 314–321.

Heidari, M., Bahrami, A., Ghiasvand, A. R., Shahna, F. G., & Soltanian, A. R. (2013). A needle trap device packed with a sol-gel derived, multi-walled carbon nanotubes/silica composite for sampling and analysis of volatile organohalogen compounds in air. *Analytica Chimica Acta, 785*, 67–74.

Ho, T. T., Chen, C. Y., Li, Z. G., Yang, T. C. C., & Lee, M. R. (2012). Determination of chlorophenols in landfill leachate using headspace sampling with ionic liquid-coated solid-phase microextraction fibers combined with gas chromatography-mass spectrometry. *Analytica Chimica Acta, 712*, 72–77.

Ho, T. D., Joshi, M. D., Silver, M. A., & Anderson, J. L. (2012a). Selective extraction of genotoxic impurities and structurally alerting compounds using polymeric ionic liquid sorbent coatings in solid-phase microextraction: Alkyl halides and aromatics. *Journal of Chromatography A, 1240*, 29–44.

Ho, T. D., Yu, H., Cole, W. T. S., & Anderson, J. L. (2012b). Ultraviolet photoinitiated on-fiber copolymerization of ionic liquid sorbent coatings for headspace and direct immersion solid-phase microextraction. *Analytical Chemistry, 84*, 9520–9528.

Hodgson, R. J., Brook, M. A., & Brennan, J. D. (2005). Capillary-scale monolithic immunoaffinity columns for immunoextraction with in-line laser-induced fluorescence detection. *Analytical Chemistry, 77*, 4404–4412.

Hou, J. G., Ma, Q., Du, X. Z., Deng, H. L., & Gao, J. Z. (2004). Inorganic/organic mesoporous silica as a novel fiber coating of solid-phase microextraction. *Talanta, 62*, 241–246.

Hsieh, Y. N., Huang, P. C., Sun, I. W., Whang, T. J., Hsu, C. Y., Huang, H. H., & Kuei, C. H. (2006). Nafion membrane-supported ionic liquid-solid phase microextraction for analyzing ultra trace PAHs in water samples. *Analytica Chimica Acta, 557*, 321–328.

Hu, C., He, M., Chen, B., Zhong, C., & Hu, B. (2013). Polydimethylsiloxane/metal-organic frameworks coated stir bar sorptive extraction coupled to high performance liquid chromatography-ultraviolet detector for the determination of estrogens in environmental water samples. *Journal of Chromatography A, 1310*, 21–30.

Hu, L., Boos, K.-S., Ye, M., Wu, R. a., & Zou, H. (2009). Selective on-line serum peptide extraction and multidimensional separation by coupling a restricted-access material-based capillary trap column with nanoliquid chromatography–tandem mass spectrometry. *Journal of Chromatography A, 1216*, 5377–5384.

Hu, W., Hu, B., & Jiang, Z. (2006). On-line preconcentration and separation of Co, Ni and Cd *via* capillary microextraction on ordered mesoporous alumina coating and determination by inductively plasma mass spectrometry (ICP-MS). *Analytica Chimica Acta, 572*, 55–62.

Hu, X., Hu, Y., & Li, G. (2007a). Development of novel molecularly imprinted solid-phase microextraction fiber and its application for the determination of triazines in complicated samples coupled with high-performance liquid chromatography. *Journal of Chromatography A, 1147*, 1–9.

Hu, X., Pan, J., Hu, Y., Huo, Y., & Li, G. (2008). Preparation and evaluation of solid-phase microextraction fiber based on molecularly imprinted polymers for trace analysis of tetracyclines in complicated samples. *Journal of Chromatography A, 1188*, 97–107.

Hu, X. G., Fan, Y. N., Zhang, Y., Dai, G. M., Cai, Q. L., Cao, Y. J., & Guo, C. J. (2012). Molecularly imprinted polymer coated solid-phase microextraction fiber prepared by surface reversible addition-fragmentation chain transfer polymerization for monitoring of Sudan dyes in chilli tomato sauce and chilli pepper samples. *Analytica Chimica Acta, 731*, 40–48.

Hu, X. G., Hu, Y. L., & Li, G. K. (2007b). Preparation and characterization of prometryn molecularly imprinted solid-phase microextraction fibers. *Analytical Letters, 40*, 645–660.

Hu, Y., Fan, Y., Huang, Z., Song, C., & Li, G. (2012). *In situ* fabrication of metal-organic hybrid gels in a capillary for online enrichment of trace analytes in aqueous samples. *Chemical Communications, 48*, 3966–3968.

Hu, Y., Li, J., Hu, Y., & Li, G. (2010a). Development of selective and chemically stable coating for stir bar sorptive extraction by molecularly imprinted technique. *Talanta, 82*, 464–470.

Hu, Y., Wang, Y., Chen, X., Hu, Y., & Li, G. (2010b). A novel molecularly imprinted solid-phase microextraction fiber coupled with high performance liquid chromatography for analysis of trace estrogens in fishery samples. *Talanta, 80*, 2099–2105.

Hu, Y. L., Zheng, Y. J., Zhu, F., & Li, G. K. (2007). Sol-gel coated polydimethylsiloxane/beta-cyclodextrin as novel stationary phase for stir bar sorptive extraction and its application to analysis of estrogens and bisphenol A. *Journal of Chromatography A, 1148*, 16–22.

Hu, Y. L., Zheng, Y. X., & Li, G. K. (2004). Solid-phase microextraction of phenol compounds using a fused-silica fiber coated with beta-cyclodextrin-bonded silica particles. *Analytical Sciences, 20*, 667–671.

Huang, C. Y., Song, M., Gu, Z. Y., Wang, H. F., & Yan, X. P. (2011). Probing the adsorption characteristic of metal-organic framework MIL-101 for volatile organic compounds by quartz crystal microbalance. *Environmental Science & Technology, 45*, 4490–4496.

Huang, C. Z., Hu, B., Jiang, Z., & Zhang, N. (2007). Simultaneous on-line preconcentration and determination of trace metals in environmental samples using a modified nanometer-sized alumina packed micro-column by flow injection combined with ICP-OES. *Talanta, 71*, 1239–1245.

Huang, J., Hu, Y., Hu, Y., & Li, G. (2011). Development of metal complex imprinted solid-phase microextraction fiber for 2,2′-dipyridine recognition in aqueous medium. *Talanta, 83*, 1721–1729.

Huang, X., Lin, J., Yuan, D., & Hu, R. (2009a). Determination of steroid sex hormones in wastewater by stir bar sorptive extraction based on poly(vinylpyridine-ethylene dimethacrylate) monolithic material and liquid chromatographic analysis. *Journal of Chromatography A, 1216*, 3508–3511.

Huang, X., Qiu, N., Yuan, D., & Lin, Q. (2010a). Preparation of a mixed stir bar for sorptive extraction based on monolithic material for the extraction of quinolones from wastewater. *Journal of Chromatography A, 1217*, 2667–2673.

Huang, X. C., Lin, Y. Y., Zhang, J. P., & Chen, X. M. (2006). Ligand-directed strategy for zeolite-type metal-organic frameworks: Zinc(II) imidazolates with unusual zeolitic topologies. *Angewandte Chemie International Edition, 45*, 1557–1559.

Huang, X. C., Zhang, J. P., & Chen, X. M. (2003). [$Zn(bim)_2$] center dot (H_2O)(1.67): A metalorganic open-framework with sodalite topology. *Chinese Science Bulletin, 48*, 1531–1534.

Huang, X. J., Lin, J. B., & Yuan, D. X. (2010b). A new anionic exchange stir bar sorptive extraction coating based on monolithic material for the extraction of inorganic anion. *Journal of Chromatography A, 1217*, 4898–4903.

Huang, X. J., Qiu, N. N., & Yuan, D. X. (2008a). Direct enrichment of phenols in lake and sea water by stir bar sorptive extraction based on poly (vinylpyridine-ethylene dimethacrylate) monolithic material and liquid chromatographic analysis. *Journal of Chromatography A, 1194*, 134–138.

Huang, X. J., Qiu, N. N., & Yuan, D. X. (2009b). Simple and sensitive monitoring of sulfonamide veterinary residues in milk by stir bar sorptive extraction based on monolithic material and high performance liquid chromatography analysis. *Journal of Chromatography A, 1216*, 8240–8245.

Huang, X. J., Qiu, N. N., Yuan, D. X., & Huang, B. L. (2009c). A novel stir bar sorptive extraction coating based on monolithic material for apolar, polar organic compounds and heavy metal ions. *Talanta, 78*, 101–106.

Huang, X. J., Qiu, N. N., Yuan, D. X., & Lin, Q. M. (2009d). Sensitive determination of strongly polar aromatic amines in water samples by stir bar sorptive extraction based on poly(vinylimidazole-divinylbenzene) monolithic material and liquid chromatographic analysis. *Journal of Chromatography A, 1216*, 4354–4360.

Huang, X. J., & Yuan, D. X. (2007). Preparation of stir bars for sorptive extraction based on monolithic material. *Journal of Chromatography A, 1154*, 152–157.

Huang, X. J., Yuan, D. X., & Huang, B. L. (2008b). Determination of steroid sex hormones in urine matrix by stir bar sorptive extraction based on monolithic material and liquid chromatography with diode array detection. *Talanta, 75*, 172–177.

Huo, S. H., & Yan, X. P. (2012). Facile magnetization of metal-organic framework MIL-101 for magnetic solid-phase extraction of polycyclic aromatic hydrocarbons in environmental water samples. *The Analyst, 137*, 3445–3451.

Bruheim, I., Liu, X. & Pawliszyn, J. (2003). Thin-film microextraction. *Analytical Chemistry* 75, 1002–1010.

Jafari, M. T., Saraji, M., & Yousefi, S. (2012). Negative electrospray ionization ion mobility spectrometry combined with microextraction in packed syringe for direct analysis of phenoxyacid herbicides in environmental waters. *Journal of Chromatography A, 1249*, 41–47.

Jia, J., Liang, X., Wang, L., Guo, Y., Liu, X., & Jiang, S. (2013). Nanoporous array anodic titanium-supported co-polymeric ionic liquids as high performance solid-phase microextraction sorbents for hydrogen bonding compounds. *Journal of Chromatography A, 1320*, 1–9.

Jiang, T., Mallik, R., & Hage, D. S. (2005). Affinity monoliths for ultrafast immunoextraction. *Analytical Chemistry, 77*, 2362–2372.

Jinno, K., Kawazoe, M., Saito, Y., Takeichi, T., & Hayashida, M. (2001). Sample preparation with fiber-in-tube solid-phase microextraction for capillary electrophoretic separation of tricyclic antidepressant drugs in human urine. *Electrophoresis, 22*, 3785–3790.

Jochmann, M. A., Yuan, X., Schilling, B., & Schmidt, T. C. (2008). In-tube extraction for enrichment of volatile organic hydrocarbons from aqueous samples. *Journal of Chromatography A, 1179*, 96–105.

Joshi, M. D., Ho, T. D., Cole, W. T. S., & Anderson, J. L. (2014). Determination of PCBs in ocean water and bovine milk using crosslinked polymeric ionic liquid sorbent coatings by solid-phase microextraction. *Talanta, 118*, 172–179.

Jurado-Sánchez, B., Ballesteros, E., & Gallego, M. (2009). Fullerenes for aromatic and non-aromatic N-nitrosamines discrimination. *Journal of Chromatography A, 1216*, 1200–1205.

Karimi, M., Aboufazeli, F., Zhad, H., Sadeghi, O., & Najafi, E. (2013). Determination of polycyclic aromatic hydrocarbons in Persian Gulf and Caspian Sea: Gold nanoparticles fiber for a head space solid phase micro extraction. *Bulletin of Environmental Contamination and Toxicology, 90*(3), 291–295.

Khorrami, A. R., & Rashidpur, A. (2012). Development of a fiber coating based on molecular sol-gel imprinting technology for selective solid-phase micro extraction of caffeine from human serum and determination by gas chromatography/mass spectrometry. *Analytica Chimica Acta, 727*, 20–25.

Khoshdel, Z., Hashemi, P., Safdaryan, M., Delfan, B., Rashidipour, M., & Badiei, A. (2013). Microextraction in a packed syringe for the analysis of olive biophenols in rat plasma using CMK-3 nanoporous sorbent. *Analytical Sciences, 29*, 527–532.

Kim, T. Y., Alhooshani, K., Kabir, A., Fries, D. P., & Malik, A. (2004). High pH-resistant, surface-bonded sol-gel titania hybrid organic-inorganic coating for effective on-line hyphenation of capillary microextraction (in-tube solid-phase microextraction) with high-performance liquid chromatography. *Journal of Chromatography A, 1047*, 165–174.

Koh, K., Wong-Foy, A. G., & Matzger, A. J. (2009). A porous coordination copolymer with over 5000 m^2/g BET surface area. *Journal of the American Chemical Society, 131*, 4184–4185.

Kortz, L., Helmschrodt, C., & Ceglarek, U. (2011). Fast liquid chromatography combined with mass spectrometry for the analysis of metabolites and proteins in human body fluids. *Analytical and Bioanalytical Chemistry, 399*, 2635–2644.

Koster, E. H. M., Crescenzi, C., den Hoedt, W., Ensing, K., & de Jong, G. J. (2001). Fibers coated with molecularly imprinted polymers for solid-phase microextraction. *Analytical Chemistry, 73*, 3140–3145.

Kueseng, P., & Pawliszyn, J. (2013). Carboxylated multiwalled carbon nanotubes/polydimethyl-siloxane, a new coating for 96-blade solid-phase microextraction for determination of phenolic compounds in water. *Journal of Chromatography A, 1317*, 199–202.

Kulkarni, S., Fang, L., Alhooshani, K., & Malik, A. (2006). Sol-gel immobilized cyano-polydimethyl-siloxane coating for capillary microextraction of aqueous trace analytes ranging from polycyclic aromatic hydrocarbons to free fatty acids. *Journal of Chromatography A, 1124*, 205–216.

Lambert, J. P., Mullett, W. M., Kwong, E., & Lubda, D. (2005). Stir bar sorptive extraction based on restricted access material for the direct extraction of caffeine and metabolites in biological fluids. *Journal of Chromatography A, 1075*, 43–49.

Lan, L., Hu, B., & Yu, C. (2010). pH-resistant titania hybrid organic-inorganic coating for stir bar sorptive extraction of drugs of abuse in urine samples followed by high performance liquid chromatography-ultraviolet visible detection. *Journal of Chromatography A, 1217*, 7003–7009.

Lange, J. J., Culbertson, C. T., & Higgins, D. A. (2008). Single molecule studies of solvent-dependent diffusion and entrapment in poly(dimethylsiloxane) thin films. *Analytical Chemistry*, 9726–9734.

Leca, J. M., Pereira, V., Pereira, A. C., & Marques, J. C. (2014). Rapid and sensitive methodology for determination of ethyl carbamate in fortified wines using microextraction by packed sorbent and gas chromatography with mass spectrometric detection. *Analytica Chimica Acta, 811*, 29–35.

Levkin, P. A., Eeltink, S., Stratton, T. R., Brennen, R., Robotti, K., Yin, H., & Fréchet, J. M. (2008). Monolithic porous polymer stationary phases in polyimide chips for the fast high-performance liquid chromatography separation of proteins and peptides. *Journal of Chromatography A, 1200*, 55–61.

Li, H., Eddaoudi, M., O'Keeffe, M., & Yaghi, O. M. (1999). Design and synthesis of an exceptionally stable and highly porous metal-organic framework. *Nature, 402*, 276–279.

Li, L., Wang, J. B., Zhou, S. A., & Zhao, M. P. (2008). Development and characterization of an immunoaffinity monolith for selective on-line extraction of bisphenol A from environmental water samples. *Analytica Chimica Acta, 620*, 1–7.

Li, M. K. Y., Lei, N. Y., Gong, C. B.,Yu, Y. J., Lam, K. H., Lam, M. H. W.,Yu, H. X., Lam, P. K. S. (2009). An organically modified silicate molecularly imprinted solid-phase microextraction device for the determination of polybrominated diphenyl ethers. *Analytica Chimica Acta,633*, 197–203.

Li, Q., Zhang, W., Miljanić, O. Š., Sue, C.-H., Zhao, Y.-L., Liu, L., Knobler, C. B., Fraser Stoddart, J., & Yaghi, O. M. (2009). Docking in metal-organic frameworks, *Science, 325*, 855–859.

Li, T., Chen, S., Li, H., Li, Q., & Wu, L. (2011). Preparation of an ion-imprinted fiber for the selective removal of Cu^{2+}. *Langmuir, 27*, 6753–6758.

Li, T., Xu, J., Wu, J. H., & Feng, Y. Q. (2009). Liquid-phase deposition of silica nanoparticles into a capillary for in-tube solid-phase microextraction coupled with high-performance liquid chromatography. *Journal of Chromatography A, 1216*, 2989–2995.

Li, X., Li, C. M., Chen, J. M., Li, C. L., & Sun, C. (2008). Polythiophene as a novel fiber coating for solid-phase microextraction. *Journal of Chromatography A, 1198*, 7–13.

Li, X., Zhong, M., & Chen, J. M. (2008). Electrodeposited polyaniline as a fiber coating for solid-phase microextraction of organochlorine pesticides from water. *Journal of Separation Science, 31*, 2839–2845.

Li, Y. H., Wang, S., Luan, Z., Ding, J., Xu, C., & Wu, D. (2003). Adsorption of cadmium(II) from aqueous solution by surface oxidized carbon nanotubes. *Carbon, 41*, 1057–1062.

Li, Y. H., Wang, S., Wei, J., Zhang, X., Xu, C., Luan, Z., & Wei, B. (2002). Lead adsorption on carbon nanotubes. *Chemical Physics Letters, 357*, 263–266.

Lian, H., Hu, Y., & Li, G. (2013). Novel metal ion-mediated complex imprinted membrane for selective recognition and direct determination of naproxen in pharmaceuticals by solid surface fluorescence. *Talanta, 116*, 460–467.

Lian, H. X., Hu, Y. L., & Li, G. K. (2014). Novel metal-ion-mediated, complex-imprinted solid-phase microextraction fiber for the selective recognition of thiabendazole in citrus and soil samples. *Journal of Separation Science, 37*, 106–113.

Liang, P., Qin, Y., Hu, B., Peng, T., & Jiang, Z. (2001). Nanometer-size titanium dioxide microcolumn on-line preconcentration of trace metals and their determination by inductively coupled plasma atomic emission spectrometry in water. *Analytica Chimica Acta, 440*, 207–213.

Liang, Y., Zhou, S., Hu, L. M., Li, L., Zhao, M. P., & Liu, H. W. (2010). Class-specific immunoaffinity monolith for efficient on-line clean-up of pyrethroids followed by high-performance liquid chromatography analysis. *Journal of Chromatography B-Analytical Technologies in the Biomedical and Life Sciences, 878*, 278–282.

Ligor, M., & Buszewski, B. (1999). Determination of menthol and menthone in food and pharmaceutical products by solid-phase microextraction-gas chromatography. *Journal of Chromatography A, 847*, 161–169.

Lin, B., Li, T., Zhao, Y., Huang, F. K., Guo, L., & Feng, Y. Q. (2008). Preparation of a TiO_2 nanoparticle-deposited capillary column by liquid phase deposition and its application in phosphopeptide analysis. *Journal of Chromatography A, 1192*, 95–102.

Lin, Y., Cui, X., & Bontha, J. (2006). Electrically controlled anion exchange based on polypyrrole and carbon nanotubes nanocomposite for perchlorate removal. *Environmental Science & Technology,*40, 4004–4009.

Lin, Z. A., Lin, Y., Sun, X. B., Yang, H. H., Zhang, L., & Chen, G. N. (2013a). One-pot preparation of a molecularly imprinted hybrid monolithic capillary column for selective recognition and capture of lysozyme. *Journal of Chromatography A, 1284*, 8–16.

Lin, Z. A., Wang, J., Tan, X. Q., Sun, L. X., Yu, R. F., Yang, H. H., & Chen, G. N. (2013b). Preparation of boronate-functionalized molecularly imprinted monolithic column with polydopamine coating for glycoprotein recognition and enrichment. *Journal of Chromatography A, 1319*, 141–147.

Liu, J. F., Hu, X. L., Peng, J. F., Jonsson, J. Å., Mayer, P., & Jiang, G. B. (2006). Equilibrium sampling of freely dissolved alkylphenols into a thin film of 1-octanol supported on a hollow fiber membrane. *Analytical Chemistry,* 78, 8526–8534.

Liu, J. F., Li, N., Jiang, G. B., Li, J. M., Jonsson, J. A., & Wen, M. J. (2005). Disposable ionic liquid coating for headspace solid-phase microextraction of benzene, toluene, ethylbenzene, and xylenes in paints followed by gas chromatography-flame ionization detection. *Journal of Chromatography A, 1066*, 27–32.

Liu, M. H., Li, M. J., Qiu, B., Chen, X., & Chen, G. N. (2010). Synthesis and applications of diethylstilbestrol-based molecularly imprinted polymer-coated hollow fiber tube. *Analytica Chimica Acta, 663*, 33–38.

Liu, M. M., Liu, Y., Zeng, Z. R., & Peng, T. Y. (2006). Preparation and characteristics of high pH-resistant sol-gel alumina-based hybrid organic-inorganic coating for solid-phase microextraction of polar compounds. *Journal of Chromatography A, 1108*, 149–157.

Liu, M. M., Zhou, X., Chen, Y. Y., Liu, H. L., Feng, X. H., Qiu, G. H., Liu, F., Zeng, Z. R. (2010). Innovative chemically bonded ionic liquids-based sol-gel coatings as highly porous, stable and selective stationary phases for solid phase microextraction. *Analytica Chimica Acta, 683*, 96–106.

Liu, S. S., Yang, C. X., Wang, S. W., & Yan, X. P. (2012). Metal-organic frameworks for reverse-phase high-performance liquid chromatography. *The Analyst, 137*, 816–818.

Liu, W., Hu, Y., Zhao, J., Xu, Y., & Guan, Y. (2005). Determination of organophosphorus pesticides in cucumber and potato by stir bar sorptive extraction. *Journal of Chromatography A, 1095*, 1–7.

Liu, W. M., Wang, H. M., & Guan, Y. F. (2004). Preparation of stir bars for sorptive extraction using sol-gel technology. *Journal of Chromatography A, 1045*, 15–22.

Liu, X., Wang, X. C., Tan, F., Zhao, H. X., Quan, X., Chen, J. W., & Li, L. J. (2012). An electrochemically enhanced solid-phase microextraction approach based on molecularly imprinted polypyrrole/multi-walled carbon nanotubes composite coating for selective extraction of fluoroquinolones in aqueous samples. *Analytica Chimica Acta, 727*, 26–33.

Liu, X. Q., Ma, Z. Y., Xing, J. M., & Liu, H. Z. (2004). Preparation and characterization of amino-silane modified superparamagnetic silica nanospheres. *Journal of Magnetism and Magnetic Materials, 270*, 1–6.

Lopez-Darias, J., Anderson, J. L., Pino, V., & Afonso, A. M. (2011). Developing qualitative extraction profiles of coffee aromas utilizing polymeric ionic liquid sorbent coatings in headspace solid-phase microextraction gas chromatography-mass spectrometry. *Analytical and Bioanalytical Chemistry, 401*, 2965–2976.

Lopez-Darias, J., Pino, V., Anderson, J. L., Graham, C. M., & Afonso, A. M. (2010a). Determination of water pollutants by direct-immersion solid-phase microextraction using polymeric ionic liquid coatings. *Journal of Chromatography A, 1217*, 1236–1243.

Lopez-Darias, J., Pino, V., Meng, Y. J., Anderson, J. L., & Afonso, A. M. (2010b). Utilization of a benzyl functionalized polymeric ionic liquid for the sensitive determination of polycyclic aromatic hydrocarbons, parabens and alkylphenols in waters using solid-phase microextraction coupled to gas chromatography-flame ionization detection. *Journal of Chromatography A, 1217*, 7189–7197.

Lou, D. W., Lee, X. Q., & Pawliszyn, J. (2008). Extraction of formic and acetic acids from aqueous solution by dynamic headspace-needle trap extraction - Temperature and pH optimization. *Journal of Chromatography A, 1201*, 228–234.

Louch, D., Motlagh, S., & Pawliszyn, J. (1992). Dynamics of organic compound extraction from water using liquid-coated fused silica fibers. *Analytical Chemistry, 64*, 1187–1199.

Luebbers, M. T., Wu, T. J., Shen, L. J., & Masel, R. I. (2010). Effects of molecular sieving and electrostatic enhancement in the adsorption of organic compounds on the zeolitic imidazolate framework ZIF-8. *Langmuir, 26*, 15625–15633.

Luo, F., Wu, Z. C., Tao, P., & Cong, Y. Q. (2009). Preparation by low-temperature nonthermal plasma of graphite fiber and its characteristics for solid-phase microextraction. *Analytica Chimica Acta, 631*, 62–68.

Luo, Y. B., Ma, Q. A., & Feng, Y. Q. (2010). Stir rod sorptive extraction with monolithic polymer as coating and its application to the analysis of fluoroquinolones in honey sample. *Journal of Chromatography A, 1217*, 3583–3589.

Lv, Y. K., Wang, L. M., Yang, L., Zhao, C. X., & Sun, H. W. (2012). Synthesis and application of molecularly imprinted poly(methacrylic acid)–silica hybrid composite material for selective solid-phase extraction and high-performance liquid chromatography determination of oxytetracycline residues in milk. *Journal of Chromatography A, 1227*, 48–53.

Ma, J., Liang, Z., Qiao, X., Deng, Q., Tao, D., Zhang, L., & Zhang, Y. (2008). Organic-inorganic hybrid silica monolith based immobilized trypsin reactor with high enzymatic activity. *Analytical Chemistry, 80*, 2949–2956.

Ma, S., Simmons, J. M., Sun, D., Yuan, D., & Zhou, H.-C. (2009). Porous metal-organic frameworks based on an anthracene derivative: Syntheses, structure analysis, and hydrogen sorption studies. *Inorganic Chemistry, 48*, 5263–5268.

Maes, M., Alaerts, L., Vermoortele, F., Ameloot, R., Couck, S., Finsy, V., Denayer, J. F. M., De Vos, D. E. (2010a). Separation of C5-hydrocarbons on microporous materials: Complementary performance of MOFs and zeolites. *Journal of the American Chemical Society, 132*, 2284–2292.

Maes, M., Vermoortele, F., Alaerts, L., Couck, S., Kirschhock, C. E. A., Denayer, J. F. M., & De Vos, D. E. (2010b). Separation of styrene and ethylbenzene on metal–organic frameworks: Analogous structures with different adsorption mechanisms. *Journal of the American Chemical Society, 132*, 15277–15285.

Maiti, A., & Rogers, R. D. (2011). A correlation-based predictor for pair-association in ionic liquids. *Physical Chemistry Chemical Physics, 13*, 12138–12145.

Malik, A.-K., Gómez, M., Cámara, C., Riepe, H.-G., Bettmer, J. (2002). On-line chloride interference removal for arsenic determination in waste water and urine by ICP-MS using a modified capillary. *International Journal of Environmental & Analytical Chemistry, 82*, 795–804.

Mani, V. (1999). *Applications of solid phase microextraction*. Cambridge, UK: Royal Society of Chemistry.

Martín-Esteban, A. (2013). Molecularly-imprinted polymers as a versatile, highly selective tool in sample preparation. *TrAC Trends in Analytical Chemistry, 45*, 169–181.

Martin-Calero, A., Ayala, J. H., Gonzalez, V., & Afonso, A. M. (2009). Ionic liquids as desorption solvents and memory effect suppressors in heterocyclic aromatic amines determination by SPME-HPLC fluorescence. *Analytical and Bioanalytical Chemistry, 394*, 937–946.

Matin, A. A., Biparva, P., Amanzadeh, H., & Farhadi, K. (2013). Zinc/Aluminum layered double hydroxide-titanium dioxide composite nanosheet film as novel solid phase microextraction fiber for the gas chromatographic determination of valproic acid. *Talanta, 103*, 207–213.

Mehdinia, A., Mousavi, M. F., & Shamsipur, M. (2006). Nano-structured lead dioxide as a novel stationary phase for solid-phase microextraction. *Journal of Chromatography A, 1134*, 24–31.

Melo, L. R., Nogueira, A. M., Lancas, F. M., & Queiroz, M. E. C. (2009). Polydimethylsiloxane/polypyrrole stir bar sorptive extraction and liquid chromatography (SBSE/LC-UV) analysis of antidepressants in plasma samples. *Analytica Chimica Acta, 633*, 57–64.

Meloche, L. M., Debruyn, A. M. H., Otton, S. V., Ikonomou, M. G., & Gobas, F. A. P. C. (2009). Assessing exposure of sediment biota to organic contaminants by thin-film solid phase extraction. *Environmental Toxicology and Chemistry, 28*, 247–253.

Mendes, B., Silva, P., Aveiro, F., Pereira, J., & Camara, J. S. (2013). A micro-extraction technique using a new digitally controlled syringe combined with UHPLC for assessment of urinary biomarkers of oxidatively damaged DNA. *PLoS One, 8, 0058366.*

Meng, Y., Pino, V., & Anderson, J. L. (2011). Role of counteranions in polymeric ionic liquid-based solid-phase microextraction coatings for the selective extraction of polar compounds. *Analytica Chimica Acta, 687*, 141–149.

Meng, Y. J., & Anderson, J. L. (2010). Tuning the selectivity of polymeric ionic liquid sorbent coatings for the extraction of polycyclic aromatic hydrocarbons using solid-phase microextraction. *Journal of Chromatography A, 1217*, 6143–6152.

Meng, Y. J., Pino, V., & Anderson, J. L. (2009). Exploiting the versatility of ionic liquids in separation science: Determination of low-volatility aliphatic hydrocarbons and fatty acid methyl esters using headspace solid-phase microextraction coupled to gas chromatography. *Analytical Chemistry, 81*, 7107–7112.

Mills, G. A., & Walker, V. (2000). Headspace solid-phase microextraction procedures for gas chromatographic analysis of biological fluids and materials. *Journal of Chromatography A, 902*, 267–287.

Minakuchi, H., Nakanishi, K., Soga, N., Ishizuka, N., & Tanaka, N. (1996). Octadecylsilylated porous silica rods as separation media for reversed-phase liquid chromatography. *Analytical Chemistry, 68*, 3498–3501.

Mirnaghi, F. S., Gorynski, K., Rodriguez-Lafuente, A., Boyaci, E., Bojko, B., & Pawliszyn, J. (2013). Microextraction versus exhaustive extraction approaches for simultaneous analysis of compounds in wide range of polarity. *Journal of Chromatography A, 1316*, 37–43.

Mirnaghi, F. S., Monton, M. R., & Pawliszyn, J. (2012). Thin-film octadecyl-silica glass coating for automated 96-blade solid-phase microextraction coupled with liquid chromatography-tandem mass spectrometry for analysis of benzodiazepines. *Journal of Chromatography A, 1246*, 2–8.

Moreno, I., da Fonseca, B., Oppolzer, D., Martinho, A., Barroso, M., Cruz, A., Queiroz, J. A., Gallardo, E. (2013). Analysis of Salvinorin A in urine using microextraction in packed syringe and GC-MS/MS. *Bioanalysis, 5*, 661–668.

Morris, W., Doonan, C. J., Furukawa, H., Banerjee, R., & Yaghi, O. M. (2008). Crystals as molecules: Postsynthesis covalent functionalization of zeolitic imidazolate frameworks. *Journal of the American Chemical Society, 130*, 12626–12627.

Mousavi, F., & Pawliszyn, J. (2013). Silica-based ionic liquid coating for 96-blade system for extraction of aminoacids from complex matrixes. *Analytica Chimica Acta, 803*, 66–74.

Mu, L., Hu, X., Wen, J., & Zhou, Q. (2013). Robust aptamer sol-gel solid phase microextraction of very polar adenosine from human plasma. *Journal of Chromatography A, 1279*, 7–12.

Mulfort, K. L., Farha, O. K., Stern, C. L., Sarjeant, A. A., & Hupp, J. T. (2009). Post-synthesis alkoxide formation within metal–organic framework materials: A strategy for incorporating highly coordinatively unsaturated metal ions. *Journal of the American Chemical Society, 131*, 3866–3868.

Mullett, W. M., Martin, P., & Pawliszyn, J. (2001). In-tube molecularly imprinted polymer solid-phase microextraction for the selective determination of propranolol. *Analytical Chemistry, 73*, 2383–2389.

Mullett, W. M., & Pawliszyn, J. (2002). Direct determination of benzodiazepines in biological fluids by restricted-access solid-phase microextraction. *Analytical Chemistry, 74*, 1081–1087.

Munch, A. S., Seidel, J., Obst, A., Weber, E., & Mertens, F. O. R. L. (2011). High-separation performance of chromatographic capillaries coated with MOF-5 by the controlled SBU approach. *Chemistry - A European Journal, 17*, 10958–10964.

Neng, N. R., Pinto, M. L., Pires, J., Marcos, P. M., & Nogueira, J. M. F. (2007). Development, optimisation and application of polyurethane foams as new polymeric phases for stir bar sorptive extraction. *Journal of Chromatography A, 1171*, 8–14.

Newsome, T. E., Zewe, J. W., & Olesik, S. V. (2012). Electrospun nanofibrous solid-phase microextraction coatings for preconcentration of pharmaceuticals prior to liquid chromatographic separations. *Journal of Chromatography A, 1262*, 1–7.

Ni, Z., Jerrell, J. P., Cadwallader, K. R., & Masel, R. I. (2007). Metal-organic frameworks as adsorbents for trapping and preconcentration of organic phosphonates. *Analytical Chemistry, 79*, 1290–1293.

Nishiyama, S., Goto, A., Saito, K., Sugita, K., Tamada, M., Sugo, T., Funami, T., Goda, Y., Fujimoto, S. (2002). Concentration of 17 beta-estradiol using an immunoaffinity porous hollow-fiber membrane. *Analytical Chemistry, 74*, 4933–4936.

Noche, G. G., Laespada, M. E. F., Pavon, J. L. P., Cordero, B. M., & Lorenzo, S. M. (2013). Determination of chlorobenzenes in water samples based on fully automated microextraction by packed sorbent coupled with programmed temperature vaporization-gas chromatography-mass spectrometry. *Analytical and Bioanalytical Chemistry, 405*, 6739–6748.

Noroozian, E., Kazemipour, M., Tehrani, M. S., & Mahmoudian, M. (2004). Solid-phase microextraction (SPME) of permethrin residues from cucumber using a silica-bonded phase-coated stainless steel fibre. *Food Additives and Contaminants Part A-Chemistry Analysis Control Exposure & Risk Assessment, 21*, 222–231.

Oppolzer, D., Moreno, I., da Fonseca, B., Passarinha, L., Barroso, M., Costa, S., Queiroz, J.A., & Gallardo, E. (2013). Analytical approach to determine biogenic amines in urine using microextraction in packed syringe and liquid chromatography coupled to electrochemical detection. *Biomedical Chromatography, 27*, 608–614.

Pang, L., & Liu, J.-F. (2012). Development of a solid-phase microextraction fiber by chemical binding of polymeric ionic liquid on a silica coated stainless steel wire. *Journal of Chromatography A, 1230*, 8–14.

Pawliszyn, J. (1997). *Solid Phase Microextraction: Theory and Practice*. New York, USA: Wiley-VCH, Inc.

Pawliszyn, J. (2009). *Handbook of Solid Phase Microextraction*. Beijing, PR China: Chemical Industry Press.

Peng, Y., Srinivas, G., Wilmer, C. E., Eryazici, I., Snurr, R. Q., Hupp, J. T., Yildirim, T., & Farha, O. K. (2013). Simultaneously high gravimetric and volumetric methane uptake characteristics of the metal-organic framework NU-111. *Chemical Communications, 49*, 2992–2994.

Phillips, T. M., & Wellner, E. F. (2007). Analysis of inflammatory biomarkers from tissue biopsies by chip-based immunoaffinity CE. *Electrophoresis, 28*, 3041–3048.

Pietrzynska, M., Voelkel, A., & Bielicka-Daszkiewicz, K. (2013). Preparation and examination of monolithic in-needle extraction (MINE) device for the direct analysis of liquid samples. *Analytica Chimica Acta, 776*, 50–56.

Pinho, O., Ferreira, I., & Santos, L. (2006). Method optimization by solid-phase microextraction in combination with gas chromatography with mass spectrometry for analysis of beer volatile fraction. *Journal of Chromatography A, 1121*, 145–153.

Pinto, C. G., Anton, A. P., Pavon, J. L. P., & Cordero, B. M. (2012). Coupling of microextraction by packed sorbents with gas chromatography with ionic liquid stationary phases for the determination of haloanisoles in wines. *Journal of Chromatography A, 1260*, 200–205.

Platonova, G. A., Pankova, G. A., Il'ina, I. Y., Vlasov, G. P., & Tennikova, T. B. (1999). Quantitative fast fractionation of a pool of polyclonal antibodies by immunoaffinity membrane chromatography. *Journal of Chromatography A, 852*, 129–140.

Polo-Luque, M. L., Simonet, B. M., & Valcarcel, M. (2012). Combination of carbon nanotubes modified filters with microextraction by packed sorbent for the NACE analysis of trace levels of ionic liquids in river water samples. *Talanta, 89*, 124–128.

Portugal, F. C. M., Pinto, M. L., & Nogueira, J. M. F. (2008). Optimization of polyurethane foams for enhanced stir bar sorptive extraction of triazinic herbicides in water matrices. *Talanta, 77*, 765–773.

Portugal, F. C. M., Pinto, M. L., Pires, J., & Nogueira, J. M. F. (2010). Potentialities of polyurethane foams for trace level analysis of triazinic metabolites in water matrices by stir bar sorptive extraction. *Journal of Chromatography A, 1217*, 3707–3710.

Prasad, B. B., Tiwari, K., Singh, M., Sharma, P. S., Patel, A. K., & Srivastava, S. (2008). Molecularly imprinted polymer-based solid-phase microextraction fiber coupled with molecularly imprinted polymer-based sensor for ultratrace analysis of ascorbic acid. *Journal of Chromatography A, 1198–1199*, 59–66.

Prieto, A., Schrader, S., Bauer, C., & Moder, M. (2011a). Synthesis of a molecularly imprinted polymer and its application for microextraction by packed sorbent for the determination of fluoroquinolone related compounds in water. *Analytica Chimica Acta, 685*, 146–152.

Prieto, A., Schrader, S., & Moeder, M. (2010). Determination of organic priority pollutants and emerging compounds in wastewater and snow samples using multiresidue protocols on the basis of microextraction by packed sorbents coupled to large volume injection gas chromatography-mass spectrometry analysis. *Journal of Chromatography A, 1217*, 6002–6011.

Prieto, A., Vallejo, A., Zuloaga, O., Paschke, A., Sellergen, B., Schillinger, E., Schrader, S., Moder, M. (2011b). Selective determination of estrogenic compounds in water by microextraction by packed sorbents and a molecularly imprinted polymer coupled with large volume injection-in-port-derivatization gas chromatography-mass spectrometry. *Analytica Chimica Acta, 703*, 41–51.

Prikryl, P., Kubinec, R., Jurdakova, H., Sevcik, J., Ostrovsky, I., Sojak, L., & Berezkin, V. (2006). Comparison of needle concentrator with SPME for GC determination of benzene, toluene, ethylbenzene, and xylenes in aqueous samples. *Chromatographia, 64*, 65–70.

Pu, X., Hu, B., Jiang, Z., & Huang, C. (2005). Speciation of dissolved iron(II) and iron(III) in environmental water samples by gallic acid-modified nanometer-sized alumina micro-column separation and ICP-MS determination. *The Analyst, 130*, 1175–1181.

Pumera, M., Ambrosi, A., Bonanni, A., Chng, E. L. K., & Poh, H. L. (2010). Graphene for electrochemical sensing and biosensing. *TrAC Trends in Analytical Chemistry, 29*, 954–965.

Qin, Z., Bragg, L., Ouyang, G., & Pawliszyn, J. (2008). Comparison of thin-film microextraction and stir bar sorptive extraction for the analysis of polycyclic aromatic hydrocarbons in aqueous samples with controlled agitation conditions. *Journal of Chromatography A, 1196–1197*, 89–95.

Queiroz, M. E. C., Oliveira, E. B., Breton, F., & Pawliszyn, J. (2007). Immunoaffinity in-tube solid phase microextraction coupled with liquid chromatography-mass spectrometry for analysis of fluoxetine in serum samples. *Journal of Chromatography A, 1174*, 72–77.

Rahimi, A., Hashemi, P., Badiei, A., Arab, P., & Ghiasvand, A. R. (2011). CMK-3 nanoporous carbon as a new fiber coating for solid-phase microextraction coupled to gas chromatography-mass spectrometry. *Analytica Chimica Acta, 695*, 58–62.

Rahmi, D., Takasaki, Y., Zhu, Y. B., Kobayashi, H., Konagaya, S., Haraguchi, H., & Umemura, T. (2010). Preparation of monolithic chelating adsorbent inside a syringe filter tip for solid phase microextraction of trace elements in natural water prior to their determination by ICP-MS. *Talanta, 81*, 1438–1445.

Rao, H. H., Wang, X. M., Du, X. Z., & Xue, Z. H. (2013). The relationship between the hydrophobic capacity and extraction efficiency of organic-inorganic hybrid mesoporous silicas for the determination of dibutyl phthalate by SPME-HPLC. *Journal of Porous Materials, 20*, 1231–1238.

Rastkari, N., Ahmadkhaniha, R., Samadi, N., Shafiee, A., & Yunesian, M. (2010). Single-walled carbon nanotubes as solid-phase microextraction adsorbent for the determination of low-level concentrations of butyltin compounds in seawater. *Analytica Chimica Acta, 662*, 90–96.

Rbeida, O., Christiaens, B., Hubert, P., Lubda, D., Boos, K. S., Crommen, J., & Chiap, P. (2004). Evaluation of a novel anion-exchange restricted-access sorbent for on-line sample clean-up prior to the determination of acidic compounds in plasma by liquid chromatography. *Journal of Chromatography A, 1030*, 95–102.

Riazi Kermani, F., & Pawliszyn, J. (2012). Sorbent coated glass wool fabric as a thin film microextraction device. *Analytical Chemistry, 84*, 8990–8995.

Rodriguez-Lafuente, A., Mirnaghi, F. S., & Pawliszyn, J. (2013). Determination of cocaine and methadone in urine samples by thin-film solid-phase microextraction and direct analysis in real time (DART) coupled with tandem mass spectrometry. *Analytical and Bioanalytical Chemistry, 405*, 9723–9727.

Rowsell, J. L. C., & Yaghi, O. M. (2006). Effects of functionalization, catenation, and variation of the metal oxide and organic linking units on the low-pressure hydrogen adsorption properties of metal-organic frameworks. *Journal of the American Chemical Society, 128*, 1304–1315.

Rucevic, M., Clifton, J. G., Huang, F. L., Li, X. S., Callanan, H., Hixson, D. C., & Josic, D. (2006). Use of short monolithic columns for isolation of low abundance membrane proteins. *Journal of Chromatography A, 1123*, 199–204.

Sadilek, P., Satinsky, D., Ot'apka, M., Sladkovsky, R., & Solich, P. (2009). Rapid and simple determination of Vitamin A and Vitamin E in human plasma by column-switching high-performance liquid chromatography. *Current Analytical Chemistry, 5*, 311–315.

Sae-Khow, O., & Mitra, S. (2009). Carbon nanotubes as the sorbent for integrating μ-solid phase extraction within the needle of a syringe. *Journal of Chromatography A, 1216*, 2270–2274

Salami, F. H., & Queiroz, M. E. C. (2013). Microextraction in packed sorbent for the determination of pesticides in honey samples by gas chromatography coupled to mass spectrometry. *Journal of Chromatographic Science, 51*, 899–904.

Sarafraz-Yazdi, A., Abbasian, M., & Amiri, A. (2012a). Determination of furan in food samples using two solid phase microextraction fibers based on sot-gel technique with gas chromatography-flame ionisation detector. *Food Chemistry, 131*, 698–704.

Sarafraz-Yazdi, A., Amiri, A., Rounaghi, G., & Eshtiagh-Hosseini, H. (2012b). Determination of non-steroidal anti-inflammatory drugs in water samples by solid-phase microextraction based sol-gel technique using poly(ethylene glycol) grafted multi-walled carbon nanotubes coated fiber. *Analytica Chimica Acta, 720*, 134–141.

Sarafraz-Yazdi, A., & Vatani, H. (2013). A solid phase microextraction coating based on ionic liquid sol-gel technique for determination of benzene, toluene, ethylbenzene and o-xylene in water samples using gas chromatography flame ionization detector. *Journal of Chromatography A, 1300*, 104–111.

Saraji, M., & Farajmand, B. (2012). Microporous silica with nanolayer structure coated with renewable organic solvent film as a novel extracting phase: A combination of solid- and liquid-phase microextraction. *Analytica Chimica Acta, 721*, 61–67.

Saraji, M., & Farajmand, B. (2013). Chemically modified cellulose paper as a thin film microextraction phase. *Journal of Chromatography A, 1314*, 24–30.

Scida, K., Stege, P. W., Haby, G., Messina, G. A., & García, C. D. (2011). Recent applications of carbon-based nanomaterials in analytical chemistry: Critical review. *Analytica Chimica Acta, 691*, 6–17.

See, H. H., Marsin Sanagi, M., Ibrahim, W. A. W., & Naim, A. A. (2010). Determination of triazine herbicides using membrane-protected carbon nanotubes solid phase membrane tip extraction prior to micro-liquid chromatography. *Journal of Chromatography A, 1217*, 1767–1772.

Segro, S. S., Cabezas, Y., & Malik, A. (2009). Ultra-high-stability, pH-resistant sol-gel titania poly(tetrahydrofuran) coating for capillary microextraction on-line coupled to high-performance liquid chromatography. *Journal of Chromatography A, 1216*, 4329–4338.

Segro, S. S., Triplett, J., & Malik, A. (2010). Sol-gel germania triblock polymer coatings of exceptional pH stability in capillary microextraction online-coupled to high-performance liquid chromatography. *Analytical Chemistry, 82*, 4107–4113.

Sellergren, B. (1994). Direct drug determination by selective sample enrichment on an imprinted polymer. *Analytical Chemistry, 66*, 1578–1582.

Serre, C., Millange, F., Thouvenot, C., Noguès, M., Marsolier, G., Louër, D., & Férey, G. (2002). Very large breathing effect in the first nanoporous chromium(III)-based solids: MIL-53 or $Cr^{III}(OH)\cdot\{O_2C-C_6H_4-CO_2\}\cdot\{HO_2C-C_6H_4-CO_2H\}_x\cdot H_2O_y$. *Journal of the American Chemical Society, 124*, 13519–13526.

Shearrow, A. M., Bhansali, S., & Malik, A. (2009a). Ionic liquid-mediated bis (3-methyldimethoxysilyl)propyl polypropylene oxide-based polar sol-gel coatings for capillary microextraction. *Journal of Chromatography A, 1216*, 6349–6355.

Shearrow, A. M., Harris, G. A., Fang, L., Sekhar, P. K., Nguyen, L. T., Turner, E. B., Bhansali, S., Malik, A. (2009b). Ionic liquid-mediated sol-gel coatings for capillary microextraction. *Journal of Chromatography A, 1216*, 5449–5458.

Shi, Z. G., Chen, F., Xing, J., & Feng, Y. Q. (2009). Carbon monolith: Preparation, characterization and application as microextraction fiber. *Journal of Chromatography A, 1216*, 5333–5339.

Shimura, K., Koyama, Y., Sato, K., & Kitamori, T. (2007). On-chip connector valve for immunoaffinity chromatography in a microfluidic chip. *Journal of Separation Science, 30*, 1477–1481.

Shintani, Y., Zhou, X., Furuno, M., Minakuchi, H., & Nakanishi, K. (2003). Monolithic silica column for in-tube solid-phase microextraction coupled to high-performance liquid chromatography. *Journal of Chromatography A, 985*, 351–357.

Sieg, K., Fries, E., & Puttmann, W. (2008). Analysis of benzene, toluene, ethylbenzene, xylenes and n-aldehydes in melted snow water *via* solid-phase dynamic extraction combined with gas chromatography/mass spectrometry. *Journal of Chromatography A, 1178*, 178–186.

Silva, A. R. M., Portugal, F. C. M., & Nogueira, J. M. F. (2008). Advances in stir bar sorptive extraction for the determination of acidic pharmaceuticals in environmental water matrices Comparison between polyurethane and polydimethylsiloxane polymeric phases. *Journal of Chromatography A, 1209*, 10–16.

Song, X. Y., Shi, Y. P., & Chen, J. (2013). Carbon nanotubes-reinforced hollow fibre solid-phase microextraction coupled with high performance liquid chromatography for the determination of carbamate pesticides in apples. *Food Chemistry, 139*, 246–252.

Souverain, S., Rudaz, S., & Veuthey, J. L. (2004). Restricted access materials and large particle supports for on-line sample preparation: an attractive approach for biological fluids analysis.

Journal of Chromatography B-Analytical Technologies in the Biomedical and Life Sciences, 801, 141–156.

Souza Silva, E. A., Risticevic, S., & Pawliszyn, J. (2013). Recent trends in SPME concerning sorbent materials, configurations and *in vivo* applications. *TrAC Trends in Analytical Chemistry, 43*, 24–36.

Stoeck, U., Krause, S., Bon, V., Senkovska, I., & Kaskel, S. (2012). A highly porous metal-organic framework, constructed from a cuboctahedral super-molecular building block, with exceptionally high methane uptake. *Chemical Communications, 48*, 10841–10843.

Stoller, M. D., Park, S. J., Zhu, Y. W., An, J. H., & Ruoff, R. S. (2008). Graphene-based ultracapacitors. *Nano Letters, 8*, 3498–3502.

Strittmatter, N., During, R. A., & Takats, Z. (2012). Analysis of wastewater samples by direct combination of thin-film microextraction and desorption electrospray ionization mass spectrometry. *The Analyst, 137*, 4037–4044.

Su, S. W., Chen, B. B., He, M., Hu, B. (2014). Graphene oxide-silica composite coating hollow fiber solid phase microextraction online coupled with inductively coupled plasma mass spectrometry for the determination of trace heavy metals in environmental water samples. *Talanta, 123*, 1–9.

Suleiman, J. S., Hu, B., Pu, X. L., Huang, C. Z., & Jiang, Z. C. (2007). Nanometer-sized zirconium dioxide microcolumn separation/preconcentration of trace metals and their determination by ICP-OES in environmental and biological samples. *Microchimica Acta, 159*, 379–385.

Sumida, K., Hill, M. R., Horike, S., Dailly, A., & Long, J. R. (2009). Synthesis and hydrogen storage properties of $Be_{12}(OH)_{12}$(1,3,5-benzenetribenzoate)$_4$. *Journal of the American Chemical Society, 131*, 15120–15121.

Sun, M., Feng, J. J., Qiu, H. M., Fan, L. L., Li, L. L., & Luo, C. N. (2013a). A solid-phase microextraction fiber with carbon nanoparticles as sorbent material prepared by a simple flame-based preparation process. *Journal of Chromatography A, 1300*, 173–179.

Sun, M., Feng, J. J., Qiu, H. M., Fan, L. L., Li, X. J., & Luo, C. N. (2013b). CNT-TiO_2 coating bonded onto stainless steel wire as a novel solid-phase microextraction fiber. *Talanta, 114*, 60–65.

Sun, N., Han, Y. H., Yan, H. Y., & Song, Y. X. (2014). A self-assembly pipette tip graphene solid-phase extraction coupled with liquid chromatography for the determination of three sulfonamides in environmental water. *Analytica Chimica Acta, 810*, 25–31.

Sun, X. H., Yang, W. C., Pan, T., & Woolley, A. T. (2008). Affinity monolith integrated poly(methyl methacrylate) microchips for on-line protein extraction and capillary electrophoresis. *Analytical Chemistry, 80*, 5126–5130.

Sun, Y., Zhang, W. Y., Xing, J., & Wang, C. M. (2011). Solid-phase microfibers based on modified single-walled carbon nanotubes for extraction of chlorophenols and organochlorine pesticides. *Microchimica Acta, 173*, 223–229.

Szultka, M., Krzeminski, R., Szeliga, J., Jackowski, M., & Buszewski, B. (2013). A new approach for antibiotic drugs determination in human plasma by liquid chromatography-mass spectrometry. *Journal of Chromatography A, 1272*, 41–49.

Szultka, M., Szeliga, J., Jackowski, M., & Buszewski, B. (2012). Development of novel molecularly imprinted solid-phase microextraction fibers and their application for the determination of antibiotic drugs in biological samples by SPME-LC/MSn. *Analytical and Bioanalytical Chemistry, 403*, 785–796.

Tan, F., Deng, M. J., Liu, X., Zhao, H. X., Li, X. N., Quan, X., & Chen, J. W. (2011). Evaluation of a novel microextraction technique for aqueous samples: Porous membrane envelope filled with multiwalled carbon nanotubes coated with molecularly imprinted polymer. *Journal of Separation Science, 34*, 707–715.

Tan, F., Zhao, H. X., Li, X. N., Quan, X., Chen, J. W., Xiang, X. M., & Zhang, X. (2009). Preparation and evaluation of molecularly imprinted solid-phase microextraction fibers for selective extraction of bisphenol A in complex samples. *Journal of Chromatography A, 1216*, 5647–5654.

Tehrani, M. S., Azar, P. A., & Mohammadiazar, S. (2013). A single step technique for preparation of porous solid phase microextraction fibers by electrochemically co-deposited silica based sol-gel/Cu nanocomposite. *Journal of Chromatography A, 1278*, 1–7.

Tian, J., Xu, J.Q., Zhu F., Lu, T.B., Su, C.Y., Ouyang G.F. (2013). Application of nanomaterials in sample preparation. *Journal of Chromatography A, 1300*, 2–16.

Tian, M., Cheng, R., Ye, J., Liu, X., & Jia, Q. (2014). Preparation and evaluation of ionic liquid-calixarene solid-phase microextraction fibres for the determination of triazines in fruit and vegetable samples. *Food Chemistry, 145*, 28–33.

Tian, M. K., & Feng, X. L. (2008). Determination of polybrominated diphenyl ethers in water samples by headspace solid-phase microextraction technique with carbon nanotube coated fibers. *Chinese Journal of Chemistry, 26*, 1251–1256.

Togunde, O. P., Cudjoe, E., Oakes, K. D., Mirnaghi, F. S., Servos, M. R., & Pawliszyn, J. (2012a). Determination of selected pharmaceutical residues in wastewater using an automated open bed solid phase microextraction system. *Journal of Chromatography A, 1262*, 34–42.

Togunde, O. P., Oakes, K. D., Servos, M. R., & Pawliszyn, J. (2012b). Optimization of solid phase microextraction for non-lethal *in vivo* determination of selected pharmaceuticals in fish muscle using liquid chromatography-mass spectrometry. *Journal of Chromatography A, 1261*, 99–106.

Tumbiolo, S., Gal, J. F., Maria, P. C., & Zerbinati, O. (2004). Determination of benzene, toluene, ethylbenzene and xylenes in air by solid phase micro-extraction/gas chromatography/mass spectrometry. *Analytical and Bioanalytical Chemistry, 380*, 824–830.

Turiel, E., Tadeo, J. L., & Martin-Esteban, A. (2007). Molecularly imprinted polymeric fibers for solid-phase microextraction. *Analytical Chemistry, 79*, 3099–3104.

Ueta, I., Abd Razak, N., Mizuguchi, A., Kawakubo, S., Saito, Y., & Jinno, K. (2013). Needle-type extraction device for the purge and trap analysis of 23 volatile organic compounds in tap water. *Journal of Chromatography A, 1317*, 211–216.

Ueta, I., Saito, Y., Ghani, N. B. A., Ogawa, M., Yogo, K., Abe, A., Shirai, S., Jinno, K. (2009a). Rapid determination of ethylene oxide with fiber-packed sample preparation needle. *Journal of Chromatography A, 1216*, 2848–2853.

Ueta, I., Saito, Y., Hosoe, M., Okamoto, M., Ohkita, H., Shirai, S., Tamura, H., Jinno, K. (2009b). Breath acetone analysis with miniaturized sample preparation device: In-needle preconcentration and subsequent determination by gas chromatography-mass spectroscopy. *Journal of Chromatography B-Analytical Technologies in the Biomedical and Life Sciences, 877*, 2551–2556.

Vallant, R. M., Szabo, Z., Bachmann, S., Bakry, R., Najam-ul-Haq, M., Rainer, M., Heigl, N., Peeter, C., Huck, C., Bonn, G. K. (2007). Development and application of C60-Fullerene bound silica for solid-phase extraction of biomolecules. *Analytical Chemistry, 79*, 8144–8153.

Van der Perre, S., Duerinck, T., Valvekens, P., De Vos, D. E., Baron, G. V., & Denayer, J. F. M. (2013). Chromatographic separation through confinement in nanocages. *Microporous and Mesoporous Materials, 189*, 216–221.

Vita, M., Skansen, P., Hassan, M., & Abdel-Rehim, M. (2005). Development and validation of a liquid chromatography and tandem mass spectrometry method for determination of roscovitine in plasma and urine samples utilizing on-line sample preparation. *Journal of Chromatography B-Analytical Technologies in the Biomedical and Life Sciences, 817*, 303–307.

Wang, A. P., Fang, F., & Pawliszyn, J. (2005). Sampling and determination of volatile organic compounds with needle trap devices. *Journal of Chromatography A, 1072*, 127–135.

Wang, F., Guan, Y., Zhang, S., & Xia, Y. (2012). Hydrophilic modification of silica–titania mesoporous materials as restricted-access matrix adsorbents for enrichment of phosphopeptides. *Journal of Chromatography A, 1246*, 76–83.

Wang, J. X., Jiang, D. Q., Gu, Z. Y., & Yan, X. P. (2006). Multiwalled carbon nanotubes coated fibers for solid-phase microextraction of polybrominated diphenyl ethers in water and milk samples

before gas chromatography with electron-capture detection. *Journal of Chromatography A, 1137*, 8–14.
Wang, S., Wei, J., Hao, T. T., & Guo, Z. T. (2012). Determination of ractopamine in pork by using electrochemiluminescence inhibition method combined with molecularly imprinted stir bar sorptive extraction. *Journal of Electroanalytical Chemistry, 664*, 146–151.
Wang, X. J., Li, P. Z., Chen, Y., Zhang, Q., Zhang, H., Ganguly, R., Li, Y., Jing, J., & Zhao, Y. (2013). A rationally designed nitrogen-rich metal-organic framework and its exceptionally high CO_2 and H_2 uptake capability, *Scientific Reports, 3*, 1149.
Wang, Y. L., Gao, Y. L., Wang, P. P., Shang, H., Pan, S. Y., & Li, X. J. (2013). Sol-gel molecularly imprinted polymer for selective solid phase microextraction of organophosphorous pesticides. *Talanta, 115*, 920–927.
Wanigasekara, E., Perera, S., Crank, J. A., Sidisky, L., Shirey, R., Berthod, A., & Armstrong, D. W. (2010). Bonded ionic liquid polymeric material for solid-phase microextraction GC analysis. *Analytical and Bioanalytical Chemistry, 396*, 511–524.
Warren, J. M., Parkinson, D. R., & Pawliszyn, J. (2013). Assessment of thiol compounds from garlic by automated headspace derivatized in-needle-NTD-GC-MS and derivatized in-fiber-SPME-GC-MS. *Journal of Agricultural And Food Chemistry, 61*, 492–500.
Wasielewska, M., Zygmunt, B., & Anderson, J. L. (2014). Application of HS-SPME with poly(1-vinyl-3-hexylimidazolium chloride) polymeric ionic liquid coating using GC-MS to determine volatile fatty acids in dairy wastewater. *Chromatographia, 77*, 151–158.
Wei, F., Fan, Y., Zhang, M., & Feng, Y. Q. (2005). Poly(methacrylic acid-ethylene glycol dimethacrylate) monolith in-tube solid-phase microextraction applied to simultaneous analysis of some amphetamine derivatives in urine by capillary zone electrophoresis. *Electrophoresis, 26*, 3141–3150.
Wei, Y., Qiu, L. H., Yu, J. C. C., & Lai, E. P. C. (2007). Molecularly imprinted solid phase extraction in a syringe needle packed with polypyrrole-encapsulated carbon nanotubes for determination of ochratoxin a in red wine. *Food Science and Technology International, 13*, 375–380.
Wellner, E. F., & Kalish, H. (2008). A chip-based immunoaffinity capillary electrophoresis assay for assessing hormones in human biological fluids. *Electrophoresis, 29*, 3477–3483.
Werciski, S. A. S. (1999). *Solid Phase Microextraction*. New York: Mercel Dekker Inc.
Wilcockson, J. B., & Gobas, F. A. P. C. (2001). Thin-Film solid-phase extraction to measure fugacities of organic chemicals with low volatility in biological samples. *Environmental Science & Technology*. 35, 1425–1431.
Wong, M. Y., Cheng, W. R., Liu, M. H., Tian, W. C., & Lu, C. J. (2012). A preconcentrator chip employing mu-SPME array coated with *in-situ*-synthesized carbon adsorbent film for VOCs analysis. *Talanta, 101*, 307–313.
Wu, J. C., Mullett, W. M., & Pawliszyn, J. (2002). Electrochemically controlled solid-phase microextraction based on conductive polypyrrole films. *Analytical Chemistry, 74*, 4855–4859.
Wu, J. H., Zhao, Y., Li, T., Xu, C., Xiao, K. A., Feng, Y. Q., & Guo, L. (2010). The use of liquid phase deposition prepared phosphonate grafted silica nanoparticle-deposited capillaries in the enrichment of phosphopeptides. *Journal of Separation Science, 33*, 1806–1815.
Wu, M. H., Wu, R. A., Zhang, Z. B., & Zou, H. F. (2011). Preparation and application of organic-silica hybrid monolithic capillary columns. *Electrophoresis, 32*, 105–115.
Wu, S. W., Hu, C. G., He, M., Chen, B. B., & Hu, B. (2013). Capillary microextraction combined with fluorinating assisted electrothermal vaporization inductively coupled plasma optical emission spectrometry for the determination of trace lanthanum, europium, dysprosium and yttrium in human hair. *Talanta, 115*, 342–348.
Wu, Y., Hu, B., Hu, W., Jiang, Z., & Li, B. (2007). A novel capillary microextraction on ordered mesoporous titania coating combined with electrothermal vaporization inductively coupled

plasma mass spectrometry for the determination of V, Cr and Cu in environmental and biological samples. *Journal of Mass Spectrometry, 42*, 467–475.

http://www.ammrf.org.au/access/facilities/high-resolution-sem-microanalysis-facility-unsw/ last accesed, August 2014.

http://www.mse.umd.edu/research/nanotechnology last accesed, August 2014.

https://www.itrc.narl.org.tw/Research/Product/Vacuum/nanosphere-e.php, last accesed, August 2014.

Xia, L. B., Hu, B., Jiang, Z. C., Wu, Y. L., & Liang, Y. (2004). Single-drop microextraction combined with low-temperature electrothermal vaporization ICPMS for the determination of trace Be, Co, Pd, and Cd in biological samples. *Analytical Chemistry, 76*, 2910–2915.

Xiao, C. H., Liu, Z. L., Wang, Z. Y., Wu, C. Y., & Han, H. M. (2000). Use of polymeric fullerene as a new coating for solid-phase microextraction. *Chromatographia, 52*, 803–809.

Xu, H. L., Li, Y., Jiang, D. Q., & Yan, X. P. (2009). Hydrofluoric acid etched stainless steel wire for solid-phase microextraction. *Analytical Chemistry, 81*, 4971–4977.

Xu, Z., Hu, Y., Hu, Y., & Li, G. (2010). Investigation of ractopamine molecularly imprinted stir bar sorptive extraction and its application for trace analysis of β_2-agonists in complex samples. *Journal of Chromatography A, 1217*, 3612–3618.

Xu, Z., Song, C., Hu, Y., & Li, G. (2011). Molecularly imprinted stir bar sorptive extraction coupled with high performance liquid chromatography for trace analysis of sulfa drugs in complex samples. *Talanta, 85*, 97–103.

Yan, Y., Telepeni, I., Yang, S., Lin, X., Kockelmann, W., Dailly, A., Blake, A. J., Lewis, W., Walker, G. S., Allan, D. R., Barnett, S. A., Champness, & N. R., Schröder, M. (2010). Metal-organic polyhedral frameworks: High H_2 adsorption capacities and neutron powder diffraction studies. *Journal of the American Chemical Society, 132*, 4092–4094.

Yang, C. X., Chen, Y. J., Wang, H. F., & Yan, X. P. (2011). High-performance separation of fullerenes on metal-organic framework MIL-101(Cr). *Chemistry - A European Journal, 17*, 11734–11737.

Yang, C. X., Liu, S. S., Wang, H. F., Wang, S. W., & Yan, X. P. (2012). High-performance liquid chromatographic separation of position isomers using metal-organic framework MIL-53(Al) as the stationary phase. *The Analyst, 137*, 133–139.

Yang, C. X., Ren, H. B., & Yan, X. P. (2013). Fluorescent metal organic framework MIL-53(Al) for highly selective and sensitive detection of Fe^{3+} in aqueous solution. *Analytical Chemistry, 85*, 7441–7446.

Yang, C. X., & Yan, X. P. (2011). Metal-organic framework MIL-101(Cr) for high-performance liquid chromatographic separation of substituted aromatics. *Analytical Chemistry, 83*, 7144–7150.

Yang, L., Zhao, X., & Zhou, J. (2010). Selective enrichment and determination of nicosulfuron in water and soil by a stir bar based on molecularly imprinted polymer coatings. *Analytica Chimica Acta, 670*, 72–77.

Yang, P. D., Zhao, D. Y., Margolese, D. I., Chmelka, B. F. & Stucky, G. D. (1998). Generalized syntheses of large-poremesoporous metal oxides with semicrystalline frameworks. *Nature, 396*, 152–155.

Yang, X. Q., Yang, C. X., & Yan, X. P. (2013). Zeolite imidazolate framework-8 as sorbent for on-line solid-phase extraction coupled with high-performance liquid chromatography for the determination of tetracyclines in water and milk samples. *Journal of Chromatography A, 1304*, 28–33.

Yin, J., Hu, B., He, M., Zheng, M., & Feng, Y. Q. (2009). Polymer monolith microextraction combined with electrothermal vaporization inductively coupled plasma mass spectrometry for the determination of trace Cd, Tl, and Pb in human serum and urine. *Journal of Analytical Atomic Spectrometry, 24*, 76–82.

Yin, J., Jiang, Z., Chang, G., & Hu, B. (2005). Simultaneous on-line preconcentration and determination of trace metals in environmental samples by flow injection combined with

inductively coupled plasma mass spectrometry using a nanometer-sized alumina packed micro-column. *Analytica Chimica Acta, 540*, 333–339.
Yu, C., Li, X., & Hu, B. (2008). Preparation of sol-gel polyethylene glycol-polydimethylsiloxane-poly(vinyl alcohol)-coated sorptive bar for the determination of organic sulfur compounds in water. *Journal of Chromatography A, 1202*, 102–106.
Yu, C. H., & Hu, B. (2007). Novel combined stir bar sorptive extraction coupled with ultrasonic assisted extraction for the determination of brominated flame retardants in environmental samples using high performance liquid chromatography. *Journal of Chromatography A, 1160*, 71–80.
Yu, C. H., & Hu, B. (2009). Sol-gel polydimethylsiloxane/poly(vinylalcohol)-coated stir bar sorptive extraction of organophosphorus pesticides in honey and their determination by large volume injection GC. *Journal of Separation Science, 32*, 147–153.
Yu, C. H., & Hu, B. (2012). C-18-coated stir bar sorptive extraction combined with high performance liquid chromatography-electrospray tandem mass spectrometry for the analysis of sulfonamides in milk and milk powder. *Talanta, 2012*, 77–84.
Yu, C. H., Yao, Z. M., & Hu, B. (2009). Preparation of polydimethylsiloxane/beta-cyclodextrin/divinylbenzene coated "dumbbell-shaped" stir bar and its application to the analysis of polycyclic aromatic hydrocarbons and polycyclic aromatic sulfur heterocycles compounds in lake water and soil by high performance liquid chromatography. *Analytica Chimica Acta, 641*, 75–82.
Yu, H. X., Dong, L., Wu, C. Y., Wu, L., & Xing, J. (2002). Hydroxyfullerene as a novel coating for solid-phase microextraction fiber with sol-gel technology. *Journal of Chromatography A, 978*, 37–48.
Yu, L. Q., & Yan, X. P. (2013). Covalent bonding of zeolitic imidazolate framework-90 to functionalized silica fibers for solid-phase microextraction. *Chemical Communications, 49*, 2142–2144.
Yuan, H. D., Mullett, W. M., & Pawliszyn, J. (2001). Biological sample analysis with immunoaffinity solid-phase microextraction. *The Analyst, 126*, 1456–1461.
Yuan, L. M., Ren, C. X., Li, L., Li, L., Ai, P., Yan, Z. H., Zi, M., Li, Z. Y. (2006). Single-walled carbon nanotubes used as stationary phase in GC. *Analytical Chemistry, 78*, 6384–6390.
Zambrzycka, E., Roszko, D., Leśniewska, B., Wilczewska, A. Z., & Godlewska-Żyłkiewicz, B. (2011). Studies of ion-imprinted polymers for solid-phase extraction of ruthenium from environmental samples before its determination by electrothermal atomic absorption spectrometry. *Spectrochimica Acta Part B: Atomic Spectroscopy, 66*, 508–516.
Zeng, J. B., Liu, H. H., Chen, J. M., Huang, J. L., Yu, J. F., Wang, Y. R., & Chen, X. (2012). Octadecyltrimethoxysilane functionalized ZnO nanorods as a novel coating for solid-phase microextraction with strong hydrophobic surface. *The Analyst, 137*, 4295–4301.
Zeng, J. B., Zhao, C. Y., Chong, F. Y., Cao, Y. Y., Subhan, F., Wang, Q. R., Yu, J., Zhang, M., Luo, L., Ren, W., Chen, X., Yan, Z. F. (2013). Oriented ZnO nanorods grown on a porous polyaniline film as a novel coating for solid-phase microextraction. *Journal of Chromatography A, 1319*, 21–26.
Zewe, J. W., Steach, J. K., & Olesik, S. V. (2010). Electrospun fibers for solid-phase microextraction. *Analytical Chemistry, 82*, 5341–5348.
Zhang, H., & Lee, H. K. (2011). Plunger-in-needle solid-phase microextraction with graphene-based sol-gel coating as sorbent for determination of polybrominated diphenyl ethers. *Journal of Chromatography A, 1218*, 4509–4516.
Zhang, H., Lv, X. J., Li, Y. M., Wang, Y., & Li, J. H. (2010). P25-graphene composite as a high performance photocatalyst. *ACS Nano, 4*, 380–386.
Zhang, H., Ng, B. W. L., & Lee, H. K. (2014). Development and evaluation of plunger-in-needle liquid-phase microextraction. *Journal of Chromatography A, 1326*, 20–28.
Zhang, L., Chen, B. B., Peng, H. Y., He, M., & Hu, B. (2011). Aminopropyltriethoxysilane-silica hybrid monolithic capillary microextraction combined with inductively coupled plasma

mass spectrometry for the determination of trace elements in biological samples. *Journal of Separation Science, 34*, 2247–2254.

Zhang, L., Chen, B. B., He, M., & Hu, B. (2013). Polymer monolithic capillary microextraction combined on-line with inductively coupled plasma MS for the determination of trace rare earth elements in biological samples. *Journal of Separation Science, 36*, 2158–2167.

Zhang, M., Wei, F., Zhang, Y. F., Nie, J., & Feng, Y. Q. (2006). Novel polymer monolith microextraction using a poly (methacrylic acid-ethylene glycol dimethacrylate) monolith and its application to simultaneous analysis of several angiotensin II receptor antagonists in human urine by capillary zone electrophoresis. *Journal of Chromatography A, 1102*, 294–301.

Zhang, N., & Hu, B. (2012). Cadmium (II) imprinted 3-mercaptopropyltrimethoxysilane coated stir bar for selective extraction of trace cadmium from environmental water samples followed by inductively coupled plasma mass spectrometry detection. *Analytica Chimica Acta, 723*, 54–60.

Zhang, S. L., Du, Z., & Li, G. K. (2013). Metal-organic framework-199/graphite oxide hybrid composites coated solid-phase microextraction fibers coupled with gas chromatography for determination of organochlorine pesticides from complicated samples. *Talanta, 115*, 32–39.

Zhang, S. L., Du, Z., & Li, G. K. (2011). Layer-by-layer fabrication of chemical-bonded graphene coating for solid-phase microextraction. *Analytical Chemistry, 83*, 7531–7541.

Zhang, S. L., Du, Z., & Li, G. K. (2012). Graphene-supported zinc oxide solid-phase microextraction coating with enhanced selectivity and sensitivity for the determination of sulfur volatiles in Allium species. *Journal of Chromatography A, 1260*, 1–8.

Zhang, S. W., Xing, J., Cai, L. S., & Wu, C. Y. (2009). Molecularly imprinted monolith in-tube solid-phase microextraction coupled with HPLC/UV detection for determination of 8-hydroxy-2'-deoxyguanosine in urine. *Analytical and Bioanalytical Chemistry, 395*, 479–487.

Zhang, W. P., Zhang, J., Bao, T., Zhou, W., Meng, J. W., & Chen, Z. L. (2013). Universal multilayer assemblies of graphene in chemically resistant microtubes for microextraction. *Analytical Chemistry, 85*, 6846–6854.

Zhang, W. Y., Sun, Y., Wu, C. Y., Xing, J., & Li, J. Y. (2009). Polymer-functionalized single-walled carbon nanotubes as a novel sol-gel solid-phase micro-extraction coated fiber for determination of polybrominated diphenyl ethers in water samples with gas chromatography-electron capture detection. *Analytical Chemistry, 81*, 2912–2920.

Zhang, Y. P., Fan, L. Q., Lee, K. P., Zhang, Y. J., Choi, S. H., & Gong, W. J. (2007). Preparation of monolithic capillary columns for capillary electrochromatography by γ-ray irradiation. *Microchimica Acta, 158*, 353–360.

Zhang, Y., Liu, H., Zhang, X., Lei, H., Bai, L., & Yang, G. (2013). On-line solid phase extraction using organic–inorganic hybrid monolithic columns for the determination of trace β-lactam antibiotics in milk and water samples. *Talanta, 104*, 17–21.

Zhang, Y., Wang, X., Lin, C., Fang, G., & Wang, S. (2012). A novel SPME fiber chemically linked with 1-vinyl-3-hexadecylimidazolium hexafluorophosphate ionic liquid coupled with GC for the simultaneous determination of pyrethroids in vegetables. *Chromatographia, 75*, 789–797.

Zhang, Z. M., Ma, Y. J., Wang, Q. T., Chen, A., Pan, Z. Y., & Li, G. K. (2013). Preparation of novel alumina nanowire solid-phase microextraction fiber coating for ultra-selective determination of volatile esters and alcohols from complicated food samples. *Journal of Chromatography A, 1290*, 27–35.

Zhang, Z. M., Wang, Q. T., & Li, G. K. (2012). Fabrication of novel nanoporous array anodic alumina solid-phase microextraction fiber coating and its potential application for headspace sampling of biological volatile organic compounds. *Analytica Chimica Acta, 727*, 13–19.

Zhao, D., Yuan, D., & Zhou, H.-C. (2008). The current status of hydrogen storage in metal-organic frameworks. *Energy & Environmental Science, 1*, 222–235.

Zhao, F., Meng, Y. J., & Anderson, J. L. (2008). Polymeric ionic liquids as selective coatings for the extraction of esters using solid-phase microextraction. *Journal of Chromatography A, 1208*, 1–9.

Zhao, F., Wang, M., Ma, Y., & Zeng, B. (2011). Electrochemical preparation of polyaniline-ionic liquid based solid phase microextraction fiber and its application in the determination of benzene derivatives. *Journal of Chromatography A, 1218*, 387–391.
Zhao, Q. C., Wajert, J. C., & Anderson, J. L. (2010). Polymeric ionic liquids as CO_2 selective sorbent coatings for solid-phase microextraction. *Analytical Chemistry, 82*, 707–713.
Zheng, F., & Hu, B. (2007). Preparation of a high pH-resistant AAPTS-silica coating and its application to capillary microextraction (CME) of Cu, Zn, Ni, Hg and Cd from biological samples followed by on-line ICP-MS detection. *Analytica Chimica Acta, 605*, 1–10.
Zheng, F., & Hu, B. (2008). Novel bimodal porous N-(2-aminoethyl)-3-aminopropyltrimethoxysilane-silica monolithic capillary microextraction and its application to the fractionation of aluminum in rainwater and fruit juice by electrothermal vaporization inductively coupled plasma mass spectrometry. *Spectrochimica Acta Part B: Atomic Spectroscopy, 63*, 9–18.
Zheng, F., & Hu, B. (2009). Dual silica monolithic capillary microextraction (CME) on-line coupled with ICP-MS for sequential determination of inorganic arsenic and selenium species in natural waters. *Journal of Analytical Atomic Spectrometry, 24*, 1051–1061.
Zheng, F., & Hu, B. (2010). Dual-column capillary microextraction (CME) combined with electrothermal vaporization inductively coupled plasma mass spectrometry (ETV-ICP-MS) for the speciation of arsenic in human hair extracts. *Journal of Mass Spectrometry, 45*, 205–214.
Zheng, F., Zhang, N., & Hu, B. (2011). Mn(II) imprinted 3-mercaptopropyltrimethoxysilane (MPTS)-silica coated capillary microextraction on-line hyphenated with inductively coupled plasma mass spectrometry for the determination of trace Mn(II) in biological samples. *Journal of Analytical Atomic Spectrometry, 26*, 1521–1525.
Zheng, M. M., Lin, B., & Feng, Y. Q. (2007). Hybrid organic–inorganic octyl monolithic column for in-tube solid-phase microextraction coupled to capillary high-performance liquid chromatography. *Journal of Chromatography A, 1164*, 48–55.
Zheng, M. M., Ruan, G. D., & Feng, Y. Q. (2009). Evaluating polymer monolith in-tube solid-phase microextraction coupled to liquid chromatography/quadrupole time-of-flight mass spectrometry for reliable quantification and confirmation of quinolone antibacterials in edible animal food. *Journal of Chromatography A, 1216*, 7510–7519.
Zhou, R. J., & Xing, J. (2013). Ionic liquid and benzyl double functional moiety bonded coating for determination of polycyclic aromatic hydrocarbons in water Samples by solid-phase microextraction coupled with gas chromatography. *Chinese Journal of Analytical Chemistry, 41*, 1115–1119.
Zhou, X., Shang, J., Liu, M. M., Liu, H. L., Hao, R., Feng, X. H., & Liu, F. (2010). Preparation of novel ionic liquids-based sol-gel coatings for solid-phase microextraction. *Acta Chimica Sinica, 68*, 1749–1757.
Zhou, X., Shao, X., Shu, J. J., Liu, M. M., Liu, H. L., Feng, X. H., & Liu, F. (2012). Thermally stable ionic liquid-based sol-gel coating for ultrasonic extraction solid-phase microextraction-gas chromatography determination of phthalate esters in agricultural plastic films. *Talanta, 89*, 129–135.
Zhou, X., Xie, P. F., Wang, J., Zhang, B. B., Liu, M. M., Liu, H. L., & Feng, X. H. (2011). Preparation and characterization of novel crown ether functionalized ionic liquid-based solid-phase microextraction coatings by sol-gel technology. *Journal of Chromatography A, 1218*, 3571–3580.
Zhou, Y. Y., Yan, X. P., Kim, K. N., Wang, S. W., & Liu, M. G. (2006). Exploration of coordination polymer as sorbent for flow injection solid-phase extraction on-line coupled with high-performance liquid chromatography for determination of polycyclic aromatic hydrocarbons in environmental materials. *Journal of Chromatography A, 1116*, 172–178.
Zhu, A.-X., Lin, R.-B, Qi, X.-L., Liu, Y., Lin, Y.-Y., Zhang, J.-P., & Chen, X.-M. (2012). Zeolitic metal azolate frameworks (MAFs) from $ZnO/Zn(OH)_2$ and monoalkyl-substituted imidazoles and 1,2,4-triazoles: Efficient syntheses and properties. *Microporous and Mesoporous Materials, 157*, 42–49.

Zhu, F., Liang, Y. J., Xia, L. Y., Rong, M. Z., Su, C. Y., Lai, R., Li, R. Y., Ouyang, G. F. (2012). Preparation and characterization of vinyl-functionalized mesoporous organosilica-coated solid-phase microextraction fiber. *Journal of Chromatography A, 1247*, 42–48.

Zhu, X. L., Cai, J. B., Yang, J., Su, Q. D., & Gao, Y. (2006). Films coated with molecular imprinted polymers for the selective stir bar sorption extraction of monocrotophos. *Journal of Chromatography A, 1131*, 37–44.

Zhu, X. L., & Zhu, Q. S. (2008). Molecular imprinted Nylon-6 stir bar as a novel extraction technique for enantioseparation of amino acids. *Journal of Applied Polymer Science, 109*, 2665–2670.

Zhu, Y. B., Umemura, T., Haraguchi, H., Inagaki, K., & Chiba, K. (2009). Determination of REEs in seawater by ICP-MS after on-line preconcentration using a syringe-driven chelating column. *Talanta, 78*, 891–895.

Zuba, D., Parczewski, A., & Reichenbächer, M. (2002). Optimization of solid-phase microextraction conditions for gas chromatographic determination of ethanol and other volatile compounds in blood. *Journal of Chromatography B-Analytical Technologies in the Biomedical and Life Sciences, 773*, 75–82.

4 Liquid-phase Microextraction Techniques

Elena Fernández and Lorena Vidal*
Departamento de Química Analítica, Nutrición y Bromatología e Instituto Universitario de Materiales, Universidad de Alicante, Apdo. 99, 03080 Alicante, Spain
*e-mail address: lorena.vidal@ua.es

4.1 Introduction

4.1.1 History

Every analytical chemist knows that "*the best sample preparation is the one that does not exist*", however, it is considered a utopia because samples usually need to be adapted to the measurement instrument. Sample preparation has always been considered the Achilles heel of the analytical procedure due to its drawbacks such as tediousness, high degree of manipulation, risk of losses and contamination, the employment of large amounts of sample, solvents and sorbents, and therefore, generation of large amounts of wastes. For this reason, many efforts in recent decades have been focused on the reduction of this negative impact over the analytical procedure.

Sample preparation includes extraction and/or preconcentration of the target compound, interferences separation from the target compound and other operations such as derivatization or dilution. Nowadays, there are many sample preparation strategies available for these purposes, with liquid-liquid extraction (LLE) and solid-phase extraction (SPE) being the most commonly employed techniques for many years. However, these classical techniques present numerous disadvantages. For example, LLE requires large volumes of toxic organic solvents and samples, involves high degrees of sample manipulation such as glassware transfers, provides limited enrichment factors (EFs) and utilizes tedious procedures. On the other hand, SPE is also a tedious procedure, sorbents are limited and although the volume of toxic organic solvents is reduced, those employed in the procedure are still essential. For the reasons described above, these techniques have been replaced in the last two decades by their miniaturized techniques, maintaining their advantages and reducing or eliminating the drawbacks. Solid-phase microextraction (SPME) is the miniaturized technique of SPE developed by Pawliszyn in 1990 (Arthur & Pawliszyn, 1990). Otherwise, liquid-phase microextraction (LPME) or solvent microextraction (SME) are the given names to the miniaturized LLE technique introduced in the mid-1990s (S. Liu & Dasgupta, 1995; H. Liu & Dasgupta 1996; Jeannot & Cantwell, 1996). The great success of both techniques in the last two decades may be attributed to their fundamental properties, which have helped to overcome the shortcoming of sample preparation. In addition, microextraction techniques cover most of the twelve principles of green analytical chemistry (GAC) (Gałuszka et al., 2013) and have helped to recognize the importance of extraction technology on the generation of quality analytical information (Raynie, 2006).

LPME can be defined as a miniaturization of LLE technique where the volume of the extractant phase is equal or below 100 μL (Kokosa, 2013). The main advantages of LPME techniques are the rapidity and easiness, low cost, low sample volume, extremely low or even no solvent consumption, reduced generation of wastes, high EFs and its affordability to any laboratory. Numerous developments and applications about LPME techniques have appeared in the last two decades as reflected in the excellent book (Kokosa, Przyjazny, & Jeannot, 2009) and numerous reviews published on this topic (Psillakis & Kalogerakis, 2002, 2003; Raynie, 2004, 2006; Pawliszyn & Pedersen-Bjergaard, 2006; L. Xu et al., 2007; Nerín et al., 2009; Jeannot et al., 2010; Pena-Pereira et al., 2010b, 2011; Sarafraz-Yazdi & Amiri, 2010; Jain & Verma, 2011; Ramos, 2012; Andruch et al., 2013b; Kocúrová et al., 2013; Kokosa, 2013; Szreniawa-Sztajnert et al., 2013; Spietelun et al., 2014).

Most of the scientific literature gives the year 1996 as the starting point of LPME, however, the history of LPME really starts in 1995 when S. Liu and Dasgupta (1995) introduced the first droplet-based analysis systems. This system proposed the exposition of a microdrop at the end of a capillary tube to collect NH_3 and SO_2 contained in gaseous samples. The same year, Cardoso and Dasgupta (1995) substituted the spherical drop with a liquid film (14–57 μL) to determine NO_2. One year later, Liu and Dasgupta develop the well known drop-in-drop system, where a droplet of chloroform (1.3 μL) is suspended inside a bigger aqueous drop (H. Liu & Dasgupta, 1996). At the same time, Jeannot and Cantwell (1996) introduced another miniaturized system where a small drop (8 μL) of a water-immiscible organic solvent is located at the end of a Polytetrafluoroethylene (PTFE) rod which is immersed in a stirred aqueous sample solution. After the extraction, the organic phase is sampled with a microsyringe and injected into the gas chromatograph for quantification. Since the use of the PTFE rod presented some inconvenience, the same authors described in 1997 the use of a microsyringe not only as a holder of a microdrop, but also for injection into the analytical instrument (Jeannot & Cantwell, 1997). At this point, this configuration is known as single-drop microextraction (SDME) technique. The same year, He and Lee introduce the terms 'static' and 'dynamic' for LPME techniques (He & Lee, 1997). Both systems employ a conventional microsyringe. The term static LPME is given to the technique developed by Jeannot and Cantwell (Jeannot & Cantwell, 1997). Dynamic LPME changes the exposition of the droplet to the sample by introducing the sample into the syringe, which contains the extractant phase. The microsyringe is used as a microseparation funnel for extraction as well as for injection into the analytical system. The sample is introduced in the microsyringe for a controlled time, and then discharged to the sample. This procedure is repeated many times and provides better EFs, but relatively poorer precision (12.8%) due to manual operation. For this reason, Myung et al. (2003) proposed an automated procedure.

To date, two-phase systems (i.e., aqueous sample and organic extractant phase) had been considered; however, Ma and Cantwell proposed a three-phase system where the final extractant phase is an aqueous droplet (Ma & Cantwell, 1998, 1999). The configuration commonly termed liquid-liquid-liquid microextraction (LLLME) employed a 30 μL *n*-octane liquid membrane confined inside a small PTFE ring, layered over 1.60 mL of aqueous sample buffered at pH = 13. The acceptor phase was a 1 μL aqueous drop buffered at pH = 2.1 which was suspended in the 30 μL membrane directly from the tip of a microsyringe needle (Ma & Cantwell, 1999).

Since the use of a droplet caused some inconvenience (*e.g.*, instability), Pedersen-Bjergaard and Rasmussen integrated the basic principle of a supported liquid membrane (SLM) into a simple extraction unit for LLLME using a polypropylene porous hollow fiber (HF) to protect the extractant phase (Pedersen-Bjergaard & Rasmussen, 1999). They presented the three-phase system where the pores of the fiber were impregnated with an organic solvent and the lumen contained an acceptor aqueous solution. The two-phase system was developed one year later (Rasmussen et al., 2000) and an electric field has been used to assist the extraction (Pedersen-Bjergaard & Rasmussen, 2006), leading to a technique known as electromembrane extraction (EME).

In 2000, Liu and Lee (2000) introduced the continuous-flow microextraction (CFME) mode where an organic drop (1–5 μL) is held at the outlet tip of a polyetheretherketone (PEEK) connecting tubing which is immersed in a continuously flowing sample solution and acts as the fluid delivery duct and as a solvent holder. Once the extraction is finished, the organic drop is withdrawn by a microsyringe and injected into the analytical instrument.

Until now, the developed LPME systems had maintained the droplet or film of extractant phase (i.e., organic or aqueous) in direct contact with the sample or through a SLM. The direct contact presents more instability problems due to the stirring, solvent solubility or dirty matrices. For this reason, Theis et al. (2001) and Tankeviciute et al. (2001) proposed in 2001 the headspace SDME (HS-SDME) mode where the droplet is exposed in the headspace above the sample. This mode allows the extraction of volatile and semi-volatile analytes contained in complex and dirty matrices while avoiding instability problems.

In 2004, Jiang and Lee (2004) proposed a modification of HF-LPME termed solvent-bar microextraction (SBME). This extraction mode fills the lumen of a short piece of a porous HF with organic solvent and seals the HF at both ends. The solvent bar is introduced into the sample solution together with a stirring bar and after extraction for a prescribed period of time the solvent bar is taken out and the extractant phase analyzed. A three-phase system was proposed by Melwanki and Huang (2006).

Since SDME and HF-LPME presented some drawbacks such as drop instability, slow kinetics and the use of holders (i.e., microsyringes, hollow fibers, capillaries), Rezaee et al. (2006) and Yangcheng et al. (2006) proposed the elimination of any holder, disposing the extractant phase (i.e., organic solvent) directly into the sample. These modalities of extraction were named dispersive liquid-liquid microextrac-

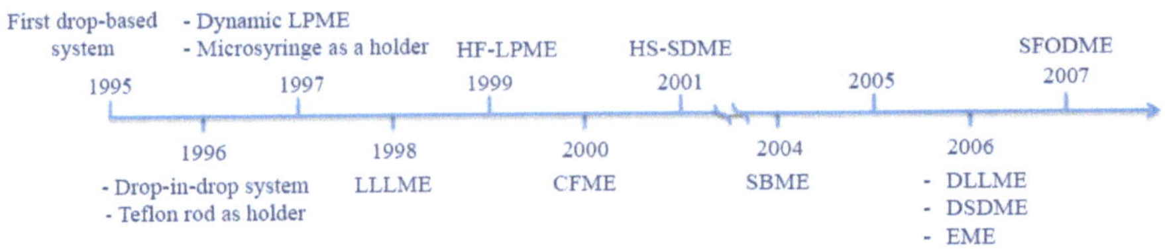

Figure 4.1 Timeline of LPME techniques.

tion (DLLME) (Rezaee et al., 2006) and directly suspended droplet microextraction (DSDME) (Yangcheng et al., 2006). The former technique is based on a ternary component solvent system where a dispersing agent is used to disperse the extractant phase, infinitely increasing the surface area of the interface. A cloudy solution is formed and the phase separation is carried out by centrifugation. This technique provides more simplicity of operation, decreased equilibration time, higher recovery and therefore, higher EFs. All these advantages make DLLME the most and widely used technique today. On the other hand, the DSDME technique avoids the use of the disperser agent and the drop maintains its integrity along the extraction procedure. After the extraction, the droplet, floating on the water surface, is removed with a syringe for analysis. The difficulty on the manipulation of the floating droplet promotes the development of another technique named solidification of floating organic drop microextraction (SFODME) (Khalili Zanjani et al., 2007). The difference with DSDME is that this technique employs an ice bath for 5 min to solidify the floating drop once the extraction is finished; meanwhile the sample maintains its liquid state. The solidified solvent is transferred into a conical vial where it melts immediately before analysis.

The timeline of the LPME techniques is shown in **Figure 4.1**. Since its appearance, LPME techniques have undergone important modifications where different solvents (i.e., ionic liquids, surfactants, etc.), dispersion modes (i.e., ultrasound (US) energy, vortex, etc.), energy and radiation (i.e., US, microwave, vortex, etc.), analytical detection systems or automated procedures have been employed. These modifications carry out the appearance of new names with new acronyms when each scientist tends to vary a general method (Kokosa, 2013). This turns into a negative fact because the LPME literature contains a confusing array of more than 100 acronyms (Kokosa, 2013). Hence, classifying all the developed methods in a few groups turns up a difficult task. Therefore, in this chapter LPME techniques have been classified in three main groups (**Figure 4.2**): (i) Single-drop microextraction (SDME); (ii) hollow fiber liquid-phase microextraction (HF-LPME); and (iii) dispersive liquid-liquid microextraction (DLLME). It is important to point out that some techniques that employ a

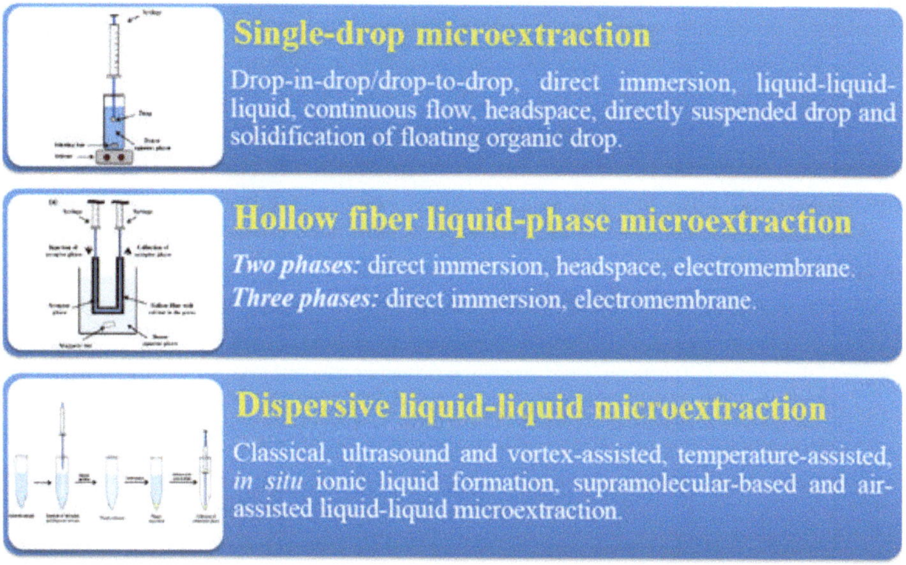

Figure 4.2 Classification of the LPME techniques.

film of extractant phase (i.e., LLLME or dynamic-LPME) have been included in the SDME group since these are not appropriate to be included in the HF-LPME or DLLME groups.

4.1.2 Solvents

Organic solvents and aqueous solutions had been the main extractant phases until 2003, when ionic liquids (ILs) were proposed as extractant phases in SDME technique (Liu et al., 2003). ILs are a group of organic salts with melting points below 100 °C. They possess unique properties such as their immeasurable low vapor pressure, good chemical and thermal stability, nonflammability, high ionic conductivity, and a wide electrochemical potential window. These properties make them widely used in every chemistry field. However, the negligible volatility and the adjustable hydrophobicity, polarity and selectivity propose them as excellent "green" solvents to replace harmful organic solvents in extraction and separation techniques. It is important to remark that some studies about the "green" aspect of ILs have proven their toxicity in aqueous media (Pham et al., 2007). For this reason, more criticism should be shown when the green aspect of a new developed method/technique that uses ILs is highlighted. However, the use of ILs in LPME techniques allows bigger and more stable droplets and higher stirring speeds, what results in higher extraction efficiency (EE) and therefore, better analytical parameters. In addition, ILs are also called "designer" solvents due to the 10^{18} possible combinations (Carmichael & Seddon, 2000), what provides ILs with different properties designed for many applications. These different properties enable the extraction of organic compounds, both polar and non-polar, as well as metals. From an analytical point of view, the use of ILs in extraction techniques provides more pros than cons but additional larger studies related to toxicity are needed before firm conclusions can be drawn. Since 2003, the number of publications related to ILs and LPME has increased exponentially. This fact is reflected in the numerous and excellent reviews recently published (Koel, 2005; Liu et al., 2005; Zhao et al., 2005; Pandey, 2006; Han & Armstrong, 2007; R. Liu et al., 2009; Soukup-Hein et al., 2009; Aguilera-Herrador et al., 2010; Poole & Poole, 2010; Sun & Armstrong, 2010; Zhao & Anderson, 2011; Joshi & Anderson, 2012; Patel & Lee, 2012; Tan et al., 2012; Vičkačkaitė & Padarauskas, 2012; Escudero et al., 2013; Kokosa, 2013). In addition, the use of ILs as extractant phases in LPME has helped not only to overcome problems associated with LPME techniques using organic solvents (i.e., instability, volatility and irreproducibility) and has allowed researchers to extract a wider range of analytes (Aguilera-Herrador et al., 2010), but also has enabled the development of new methods, including temperature-controlled IL dispersive liquid-liquid microextraction (TC-IL-DLLME) (Zhou et al., 2008), and *in situ* IL-DLLME (Baghdadi & Shemirani, 2009; Yao & Anderson, 2009).

Surfactants, often presented as safer and more economical alternatives to hazardous and expensive organic solvents, have also been used as extractant phases in LPME techniques (Ballesteros-Gómez et al., 2010). The first use of surfactants was introduced by Kanyanee et al. who proposed soap bubbles of Triton-X 100 to extract SO_2 (Kanyanee et al., 2006). However, the terms surfactant and LPME technique were not jointed until 2008 when López-Jiménez et al. (2008) employed coacervates as the extractant phase in SDME. Coacervates made of surfactant aggregates were first used as solvents and proposed for the extraction and concentration of chlorophenols prior to liquid chromatography (LC). Since then, many other applications have appeared and three reviews can be found in the literature (Ballesteros-Gómez et al., 2010; Yazdi, 2011; Moradi & Yamini, 2012b). As the authors claim, these solvents are a valuable strategy to extend the applicability of LPME to areas where the use of conventional water immiscible organic solvents is less effective or not adequate, *e.g.*, when LC is used for the separation of analytes or when the extraction of polar or ionic organic compounds is of interest (López-Jiménez et al., 2008). These solvents present the ability to extract a wider range of compounds than organic solvents and with different configurations as aqueous micelles, reverse micelles and vesicles have been tested (Ballesteros-Gómez et al., 2010; Yazdi, 2011; Moradi & Yamini, 2012b).

Water solutions as extractant phases have been employed since the development of the three-phase system (Ma & Cantwell, 1998, 1999). These solutions can be considered as the greenest solvents compared to the others (i.e., ILs, organic solvents), however, their extraction is limited to ionizable compounds. The different pH values between the sample and extractant phase are responsible for the extraction. Analytes should be undissociated in the sample to diffuse towards the organic solvent layer and then to the droplet or film, where the pH is adjusted to ionize the analytes, being back-extracted. Although aqueous solutions as extractant films (Ma & Cantwell, 1998) or droplets (Ma & Cantwell, 1999) had been published previously, Zhang et al. later introduced the term headspace water-based LPME (Zhang et al., 2005). In comparison with previous works, a drop of 5 μL of NaOH solution (1 M) was disposed in the headspace to extract five phenols contained in water samples. The analytical instrumentation was capillary electrophoresis (CE) and the authors presented the systems as an entire analytical process that was completely organic solvent-free. However, one year before Fragueiro et al. (2004) had already introduced the use of a Pd(II)-containing aqueous drop to determine arsine. The same research group has extracted methylmercury (Gil et al., 2005) and selenium (Figueroa et al., 2005, 2006) with noble metal-containing aqueous drops.

ILs combined with other reagents or materials, such as surfactants or quantum dots (QDs), have been proposed as extractant phases in SDME. Carrillo-Carrión et al. (2012) proposed the use of $[C_6mim][PF_6]$ IL-modified with CdSe/ZnS QD for the combination of IL-HS-SDME and QD-based fluorimetric detection. Trimethylamine from fish samples was extracted and preconcentrated directly onto a (QD)IL microdrop. Then, the microdrop was transferred to a microcuvette with 300 μL of acetonitrile and the fluorescence was recorded. In another study, Yao et al. (2010) studied micellar ILs as extraction solvents in HS-SDME to perform the extraction of 17 aromatic compounds. Two different micellar solutions were formed by dissolving $[C_{10}mim][Br]$ IL and the traditional surfactant sodium dodecyl sulphate in $[C_4mim][Cl]$ IL. EE increased with the micellar systems compared to the net $[C_4mim][Cl]$.

4.1.3 Separation and Detection Systems

The use of organic solvents in LPME techniques established gas chromatography (GC) coupled to different detectors (*e.g.*, flame ionization detector (FID), electron capture detector (ECD) or mass spectrometry (MS)) as the most suitable analysis system. Afterwards, the use of aqueous solutions as extractant phases extended the use of its homologous LC technique coupled to UV-Vis spectrophotometry or spectrofluorimetry. In addition, the compatibility of ILs and surfactants with LC has increased its number of applications. On the other hand, Fang et al. introduced in 2006 the first use of CE as analytical system for LPME (Fang et al., 2006), and three recent reviews summarize the developments reached with LPME and CE (Arce et al., 2009; Xie & He, 2010; ALOthman et al., 2012).

Apart from the wider range of compounds that are extractable by ILs and surfactants, their compatibility with spectroscopic techniques (*e.g.*, UV-Vis spectrophotometry, spectrofluorimetry, electrothermal atomic absorption spectrometry (ETAAS), flame atomic absorption spectrometry (FAAS), inductively coupled plasma-optical emission spectroscopy (ICP-OES) or ICP-mass spectrometry (ICP-MS)) proposes these as the optimum systems for analysis (Pena-Pereira et al., 2010a; Andruch et al., 2012b; Dehghani et al., 2012; Miró & Hansen, 2013). UV-Vis spectrophotometry had been used mainly coupled to a separation technique (*e.g.*, LC), however, the direct measurement of the extractant phase by a UV-Vis spectrophotometer, avoiding a dilution process, was a difficult task until Pena-Pereira et al. (2009a) introduced the cuvetteless microspectrophotometer. This instrument enable the analysis of a 1–2 μL drop held during measurements between a pair of sample pedestals made of stainless steel and quartz fiber by surface tension only. The same research group has also reported the use of a cuvetteless microvolume fluorospectrometer for formaldehyde determination (Sáenz et al., 2011). An excellent review summarized all the developed works that combine SDME and DLLME with UV-Vis and other related detection techniques until 2012 (Andruch et al., 2012b). Microvolume measurements have also been carried out with FAAS after LPME, where a hand-made system was employed for the determination of lead in water samples (Naseri et al., 2008).

The low vapor pressure of ILs limits their direct analysis by GC, and different systems were developed for that purpose (Aguilera-Herrador et al., 2008; Zhao et al., 2008a, 2008b; Chisvert et al., 2009). The coupling of ILs with GC not only improves analytical parameters (i.e., limits of detection), but also avoids the interference of broad organic solvent peaks, allowing the separation and identification of a wide range of analytes with different polarities and boiling points (Aguilera-Herrador et al., 2010). The first system designed by Aguilera-Herrador et al. (2008) in 2008 proposed a removable interface that enabled the introduction of the extracted analytes, contained in IL, into the GC while preventing the IL from entering the column. The system developed by Zhao et al. (2008a) proposed the exposition of the IL drop in the injection port and then it was retracted into the syringe. For this system an improvement in the injection system was required and the upper diameter of the split inlet liner of the GC instrument had to be enlarged (Zhao et al., 2008). Another modification was carried out by the same research group by placing a small glassy tube in the sample-injection part of the GC to avoid the IL entering the column (Zhao et al., 2008). An external option for direct coupling was proposed by Chisvert et al. (2009). The approach was based on thermal desorption of the analytes from the IL droplet to the GC system, by using a robust and commercially-available thermodesorption system. For this purpose, a two-glass-tube concentrically disposed system was designed.

The vast majority of the conventional detection systems (i.e., LC-UV/MS, FAAS, ETAAS, GC-MS or ICP-OES/MS) are slow, expensive and bulky, so analytical instrumentation used for detection has not achieved miniaturization to the same extent as sample preparation methods. For this reason, alternative detection systems have been proposed. Fernández et al. (2014) have presented the use of electrochemical sensors as an attractive detection option for IL-LPME techniques. Screen-printed electrodes (SPELs) (Metters et al., 2011) are inexpensive, mass-produced, disposable devices, which are ideal for low volume sample analysis. The combination of IL-LPME with SPEL for 2,4,6-trinitrotoluene extraction is shown in **Figure 4.3**. Ahmar et al. also pro-

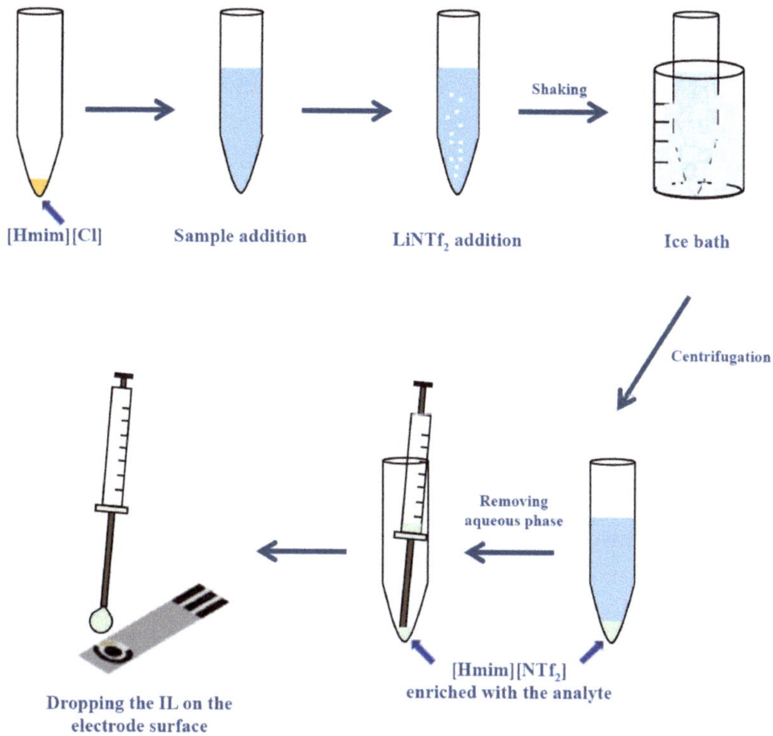

Figure 4.3 *In situ* IL-DLLME coupled with screen-printed graphite electrodes. Reprinted from Fernández et al. (2014) with kind permission from Springer Science and Business Media. Copyright (2014) Springer.

posed the use of SPELs after EME (Ahmar et al., 2013, 2014; Fakhari et al., 2014). The research group of Valcárcel (Aguilera-Herrador et al., 2009; Márquez-Sillero et al., 2011) proposed the combination of IL-LPME with ion mobility spectrometry (IMS). IMS is an analytical technique that characterizes molecules by the gas phase mobility of their ions formed at ambient pressure in a weak electric field (Eiceman, 2005). This system suffers from the limitation that liquids cannot be directly injected in IMS detector. Therefore, an injection unit was designed where the IL was trapped and the analytes were volatilized and transported to IMS. On the other hand, a tungsten coil electrothermal atomic absorption spectrometer (W-coil ETAAS) has also been coupled to LPME (Wen et al., 2013a, 2013b). The system enabled the determination of different metals by employing ILs as extractant phases. Due to the viscosity of the ILs, a dilution process with tetrahydrofuran was needed before the deposition on the tungsten coil for analysis.

These alternative detection systems together with the cuvetteless micro-spectrophotometer provide simplicity, portability, sensitivity, fast response and relatively low cost to any LPME methodology.

4.1.4 Energy and Radiation

Over the past two decades, different energy and radiation sources such as US energy, microwave radiation and vortex agitation have been employed to enhance the efficiency of LPME techniques. Microwave radiation has mainly been applied for heating of the sample in the HS-SDME technique. It is well known that extraction is an exothermic process and heating of the droplet should be prevented to avoid EE decreases. Conventional heating systems are slow and heat the entire system, including the sample, headspace and droplet, meanwhile microwave radiation focuses the heating just on the sample. The developed microwave heating system introduces the sample inside the microwave cavity disposing the droplet, suspended on the tip of the syringe, outside the cavity at room temperature. This system was firstly used by Deng et al. for the distillation and extraction of volatile compounds from Chinese herbs employing a droplet of dodecane (Deng et al., 2007), and by Vidal et al. who used an IL droplet to extract chlorobenzene compounds from water samples (Vidal et al., 2007a). US energy produces a cavitation phenomenon that accelerates mass transfer enhancing EE, therefore, US energy has also been used for treating the sample and maintaining the droplet outside of the ultrasonic system (H. Xu et al., 2007).

In addition, cavitation can be used to produce very fine emulsions from immiscible liquids, which result in very large interfacial contact areas between the liquids. For this reason, US energy has been employed to assess the dispersion of the extractant phase in DLLME techniques. In addition, vortex agitation has been used for the same purpose, and two recent publications review the developments about SME techniques that employ ultrasonic irradiation and vortex agitation (Andruch et al., 2013b; Szreniawa-Sztajnert et al., 2013). The first research about using US energy to assist the dispersion of the extractant phase was published by Huang et al. (Huang et al., 2006). Since then, different modalities of extraction such as ultrasound-assisted emulsification microextraction (USAEME), ultrasound-assisted dispersive liquid-liquid microextraction (UA-DLLME) or vortex-assisted liquid-liquid microextraction (VALLME), among others, have been developed. US energy was also used in HF-LPME technique for the extraction of benzene and toluene in beverage samples (Yang et al., 2010).

4.1.5 Optimization Strategies

The parameters that mainly affect the extraction are stirring speed, extraction time, sample and headspace volumes, pH, ionic strength of the sample, temperature, disperser solvent, centrifugation time and speed, drop volume and extractant solvent (Pena-Pereira et al., 2010a). Two different strategies can be followed to optimize these parameters (Stalikas et al., 2009): (i) step-by-step and (ii) experimental design. The former approach is most commonly used and studies the variation of one parameter while all the others kept constant. This approach is inefficient, since the interaction between the parameters and the statistical significance of the parameters are excluded. Furthermore, it requires a large number of experiments, which leads to higher sample and reagent volumes and longer experiment times. The second strategy has increased in use during recent years due to its numerous advantages over the step-by-step approach. This strategy involves the variation of all the parameters in the different experiments to study their effect upon the extraction and provides information such as interactions between parameters and the statistical significance. In addition, this approach reduces the number of experiments and, therefore, reduces experiment time, effort, the amount of reagents and samples required, energy and costs. For the reasons described above, experimental design should be the unique strategy chosen for optimization following the guidelines of environmental friendly of microextraction techniques according to the 12 principles of GAC (Gałuszka et al., 2013).

The authors would like to highlight that due to the numerous articles in LPME field, applications have generally been excluded, except those that report novel and unique developments.

4.2 Single-drop Microextraction

Single-drop microextraction (SDME), as described above, was the first LPME technique developed by S. Liu and Dasgupta (1995). Since then, SDME has undergone different modifications and the system introduced by Jeannot and Cantwell (1997) that employs a conventional GC syringe as a support of the drop and for the injection in the analytical system has been the most widely used. More recent techniques avoid any holder for the drop, disposing it directly into the sample (Khalili Zanjani et al., 2007; Yangcheng et al., 2006). In general terms, SDME employs just a few microliters of solvent as a single drop disposed either to the headspace or directly immersed in the sample. The analytes within the sample partition to the drop and after the extraction, the drop is transferred to the analytical instrument for analysis. Initially, depending on the position of the drop, the SDME technique was defined as: (i) headspace SDME (HS-SDME) (Tankeviciute et al., 2001; Theis et al., 2001) or (ii) direct immersion SDME (DI-SDME) (Jeannot & Cantwell, 1997). However, evolution of the technique has provided different direct immersion modalities which are termed drop-to-drop/drop-in-drop, CFME, LLLME, DSDME and SFODME. Therefore, DI-SDME is considered as an individual technique where the droplet is held on the tip of the microsyringe in a two-phase system. Since the boundary between the different direct immersion techniques is not clearly defined, drop-to-drop/drop-in-drop, CFME, LLLME, DSDME and SFODME are included in the direct immersion section. On the other hand, the authors of this chapter consider both DSDME and SFODME as SDME techniques bearing in mind that the extractant phase maintains its integrity as a single drop throughout the extraction procedure.

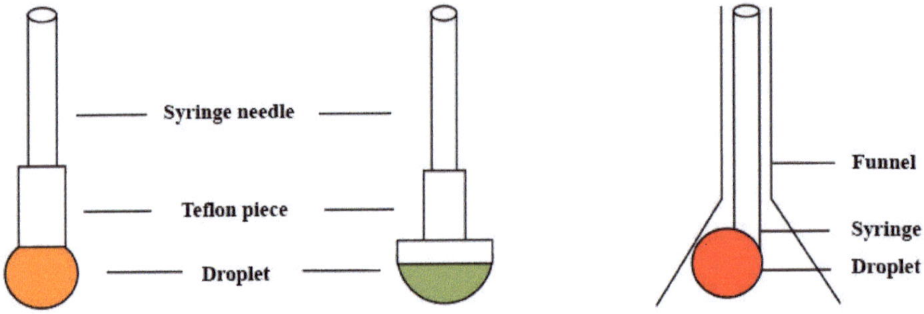

Figure 4.4 Devices used as holders for SDME.

The volume of the drop typically used with organic solvents is around 1–3 μL, however, viscosity, surface tension and density of ILs and surfactants can enable bigger drops (*e.g.*, 3–10 μL). The syringe usually employed with ILs or surfactants is a LC syringe with flat tip, in contrast to the bevel tip of the GC syringes mainly used with organic solvents. Different systems (**Figure 4.4**) employing a short 3 mm tube (Liu et al., 2003), a PTFE flange (Batlle et al., 2008; López-Jiménez et al., 2008) a bell-mouthed (Ye et al., 2007) or a brass funnel (Qian & He, 2006) have been developed since the flat tip prevents the formation of the drop or to maximize the contact area between the needle and the drop allowing bigger volumes (*e.g.*, 20 μL). These arrangements augment the adhesion force between the drop and the improvised device, increasing drop stability over a longer duration and allowing use of higher stirring rates.

Organic and inorganic compounds, small and large molecules, polar and non-polar analytes and charged and non-charged compounds have successfully been extracted by the SDME technique due to the different properties of solvents used. As a representative example of the wide range of extraction ability, organic compounds (Tobiszewski et al., 2009; Poole & Poole, 2010; Martín-Calero et al., 2011; Tankiewicz et al., 2011; Ruiz-Aceituno et al., 2013), inorganic compounds (Pena-Pereira et al., 2009b; Dadfarnia & Shabani, 2010; Martinis et al., 2010; Martín-Calero et al., 2011; Hu et al., 2013) and emerging pollutants (Mahugo-Santana et al., 2011) have successfully been extracted. Regarding sample matrices, numerous simple and complex matrices have been studied along these almost twenty years, including environmental (Tankiewicz et al., 2011; Han & Row, 2012; Ruiz-Aceituno et al., 2013), food (Asensio-Ramos et al., 2011b; Martín-Calero et al., 2011; Ruiz-Aceituno et al., 2013) and biological samples (Han & Row, 2012; Escudero et al., 2013).

Simplicity in operation, time savings, low cost, virtually organic solventless nature, freedom from analyte carryover, and compatibility of microextraction for direct analysis by a large variety of instrumental techniques make SDME one of the preferred techniques for sample preparation.

4.2.1 Headspace Single-drop Microextraction

The headspace single-drop microextraction (HS-SDME) technique disposes the drop to the headspace of the sample thus avoiding direct contact with the sample. This technique limits the extraction to volatile and semi-volatile compounds, but enables the extraction from complex and dirty samples. For instance, this method allows rapid stirring of the sample solution with no adverse impact on the stability of the droplet. Additionally, non-volatile matrix interferences are reduced, if not eliminated. The system is made up of a conventional syringe, a vial containing the sample, a stirring bar and a cap to avoid losses of the analytes and solvent (**Figure 4.5A**). This technique can be considered a three-phase system because analytes are distributed among the liquid sample, headspace and drop of solvent. The analytes should be transferred from the sample to the headspace and then to the drop. The rate-determining step of the extraction is the transfer from the sample to the headspace since the diffusion in a gas phase is faster than in a liquid medium. Hence, a high stirring speed of the sample solution is required to facilitate mass transfer among the three phases.

Organic solvents were the first solvents of choice for HS-SDME. However, these solvents presented a limitation related to the evaporation problems. This limitation was solved by ILs and surfactants (section 4.1.2.) and by dynamic methods. The first dynamic method was proposed by Shen and Lee (Shen & Lee, 2003). Extraction took place within the microsyringe barrel as in dynamic LPME and the spherical drop form is transformed in a film. The organic solvent film formed in a microsyringe barrel was used as an extraction interface in headspace liquid-phase microextraction (HS-LPME) of chlorobenzenes. The extraction consisted of drawing 2 µL of organic solvent into the microsyringe, then the microsyringe needle was passed through the vial septum and kept the needle suspended over the liquid sample. Next, 5 µL of gaseous sample was withdrawn and maintained for a short period of time, and then the plunger was depressed back to the original mark. The same process was repeated 25 times. Finally, the syringe needle removed from the vial was injected into the GC for analysis. In general, the reasons for successful extraction were the very small space within the microsyringe barrel and the fast equilibrium between gaseous analytes and the organic solvent film. Both of these factors significantly reduced the risk of solvent loss during extraction due to evaporation or dislodgement of the drop. Thus, dynamic HS-LPME was shown to be an inexpensive, fast, and simple sample preparation method for volatile compounds, improving efficiency and avoiding evaporation problems.

Since the manual operation of dynamic HS-LPME presents high imprecision, Saraji proposed a technique called semiautomatic dynamic HS-LPME (Saraji, 2005). The author proposed the system to improve ease of operation and to achieve greater reproducibility in the sample extraction. A variable speed stirring motor was used for automation of sample extraction step to ensure uniform movement of syringe plunger through the barrel. In comparison to manually operated extraction, semiautomation

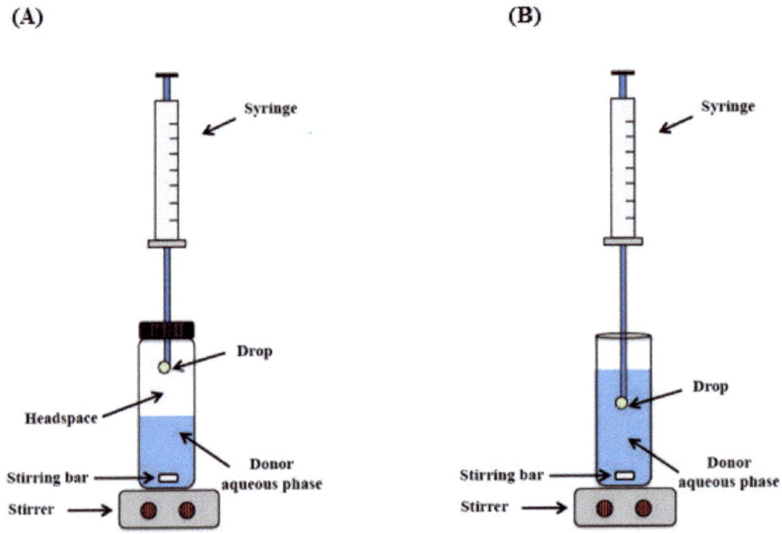

Figure 4.5 SDME modes. (A) HS-SDME and (B) DI-SDME.

of the method led to a better precision. Ouyang et al. (2005, 2007) went a step further and developed fully-automated SDME methods using the Combipal autosampler (CTC Analytics, Zwingen, Switzerland) (Kokosa, 2007). In the first work (Ouyang et al., 2005), the authors automated HS-LPME, named static and dynamic HS-LPME. The latter was operated with exposition of the droplet to the headspace instead of withdrawing the headspace gaseous sample into the syringe barrel. The 1-octanol droplet of 1 μL volume was withdrawn in and out of the barrel for a repeated number of times. Agitation of the sample was carried out when the drop was inside the barrel to avoid instability problems. All the operational parameters involved in this process were precisely and conveniently controlled by the autosampler. The same research group automated all the SDME modes and the HF-LPME technique. SDME included static HS-LPME, dynamic HS-LPME (exposed and unexposed drop) and DI-LPME (static and dynamic) (Ouyang et al., 2007).

Related to automation, Vallecillos et al. have developed a fully automated IL-HS-SDME procedure to preconcentrate trace amounts of ten musk fragrances from wastewater (Vallecillos et al., 2012b) and sewage sludge (Vallecillos et al., 2012a) samples prior to analysis by GC and ion trap tandem mass spectrometry. A CTC Combipal autosampler was also used for the fully automated IL-HS-SDME. The [C_8mim][PF_6] IL was exposed in the headspace of a 10 mL sample. The main problem associated with the use of ILs in automated processes is its high viscosity, therefore, to easily take 1 μL of IL, the fill and ejection speed of the syringe during all the HS-SDME procedure was 1 μL s^{-1}. A large internal diameter (3.4 mm) GC inlet liner with a piece of glass wool was used to avoid the introduction of the ILs to the GC column and a guard column was used to prevent analytical column damage. The authors also highlighted that the non-modification of the GC injector permitted the development of a completely automated, simple, and environmentally friendly method. On the whole, automated systems are preferred because they offer important advantages, such as minimizing the errors associated with manual handling, reducing time and improving sensitivity and precision.

Microwave radiation and US energy, as described in section 4.1.4., have been employed for the enhancement of the extraction, speeding up the sample heating and maintaining the drop at room temperature. Vidal et al. proposed a one step, *in situ* sample preparation method coupling microwave radiation and HS-SDME (Vidal et al., 2007a). A homemade experimental setup with a domestic microwave oven was used for the extraction of chlorobenzenes from water samples. The chlorobenzenes were extracted directly onto a [C_6mim][PF_6] IL single-drop in the vial headspace under the aid of microwave radiation. Limits of detection one order of magnitude below those obtained without microwave heating (Vidal et al., 2007b) demonstrated the favorable effect of the microwave radiation upon SDME. On the other hand, H. Xu et al. (2007) proposed the extraction of chlorophenols from water samples assisted by US energy. The vial containing the sample was introduced in the ultrasonic cavity while the extractant phase was suspended on the bottom of a cone-shaped polychloroprene rubber (PCR) tube outside of the cavity. Ice was placed around the bulge section of PCR tube to maintain the relatively low temperature of extractant. EE was found to be 21 times higher in US-assisted HS-LPME than with the conventional HS-LPME. Part of this improvement was related to the larger volume (20 μL) of solvent suspended on the bottom of the cone-shaped PCR tube.

4.2.2 Direct Immersion

4.2.2.1 Direct Immersion Single-drop Microextraction

Direct immersion single-drop microextraction (DI-SDME) was the term given to the direct immersion of the drop into the sample to differentiate from HS-SDME. The DI-SDME technique is used when the solvent drop is held from the tip of a syringe or capillary in a two-phase system (**Figure 4.5B**). The main drawback of this technique is that the drop at the needle or capillary tip limits the use of extended extraction times, high stirring rates and sample temperatures, and samples should be filtered before use to avoid drop instability by suspended particulate matter.

For the reasons above, dynamic mode is preferred because the inconvenience of accidently losing the drop is eliminated and also results in higher enrichment of analytes and much faster extractions are achieved (He & Lee, 1997; Wang et al., 1998). In dynamic mode, the solvent is moved inside the needle and the barrel of a syringe, and the sample solution is withdrawn and ejected. Pulling up the syringe plunger permits the formation of a renewable microfilm along the inside walls of the microsyringe into which the analytes are transferred quickly from the aqueous sample. After a dwell time of several seconds to achieve equilibrium, the aqueous sample is discharged and during this step the microfilm is merged with the organic solvent. The same process is repeated several times on fresh portions of aqueous sample, always retaining the organic solvent in the syringe. Finally, the solvent is transferred to the instrument for analysis (Jain & Verma, 2011). The parameters that mostly affect the extraction are the plunger withdrawal speed, sample volume and the number of cycles. As in dynamic HS-SDME, the manual operation reduces the precision, hence, the semiautomation of the technique was also developed (Hou & Lee, 2002). A syringe pump was used to automate the repetitive procedure of filling a microsyringe barrel with fresh aliquots of sample and expelling them after extraction. The number of cycles carried out were 40 during 20 min, obtaining an EF higher than 280 for the extraction of polycyclic aromatic hydrocarbons (PAHs). Fully static (Ouyang et al., 2007) and dynamic DI-LPME (Myung et al., 2003; Ouyang et al., 2007) have been automated. The developed homemade dynamic LPME device was a programmable automated syringe dispenser to overcome deteriorating precision and difficulties in manually manipulating the plunger repeatedly (Myung et al., 2003). The authors claim that accuracies had a high percentage bias, but the precision, which is one of major problems in manual LPME, showed better results than manual dynamic LPME. Thus, the automation greatly simplified the operations and produced better sensitivity and precision than manual DI-SDME.

On the other hand, Pena-Pereira et al. (2008) and Anthemidis and Adam (2009) have proposed an on-line semi-automated system based on sequential injection analysis (SIA) coupled to atomic techniques. The former hyphenates the DI-SDME technique and an SIA system for ultratrace quantification Cr(VI) by ETAAS (Pena-Pereira et al., 2008). The extraction is carried out in a homemade flow-through extraction cell connected to the SIA valve. The sampling capillary with the inserted needle is fitted with an autosampler arm and connected to a high-precision microsyringe. The capillary filled with 10 μL is immersed at a 5 mm depth into the microextraction cell containing the sample and a 3 μL drop volume is suspended at the tip. After a microextraction time in the range of 5–20 min, depending on the sought preconcentration factor, the drop is retracted back into the sampling capillary, the furnace autosampler arm is directed to the graphite tube and the whole volume present inside the PTFE capillary is injected into the instrument. Finally, the extraction cell is emptied and washed. Anthemidis and Adam also proposed the on-line extraction of metals in a homemade flow-cell connected directly to an SIA manifold (Anthemidis & Adam, 2009). The system employed a flow-through cell made of a polyethylene and glass capillary tube to suspend the microdrop of extraction solvent. After the extraction is completed, the drop was retracted into the holding coil and subsequently delivered into a graphite tube for analyte atomization and quantification. The efficiency of this system was demonstrated for cadmium determination using ammonium diethyldithiophosphate as a chelating agent.

Even when ILs are employed, the volume is restricted to the stability of the drop on the needle or capillary tip. Therefore, Cruz-Vera et al. eliminated the syringe and proposed the use of ILs in dynamic liquid-phase microextraction (dLPME) for the efficient extraction of non-steroidal anti-inflammatory drugs (Cruz-Vera et al., 2008) and phenothiazine derivatives (Cruz-Vera et al., 2009a) in urine. The extraction was carried out in a Pasteur pipette where the volumes and flow rates of different solutions involved in the extraction process were controlled by an automatic flow configuration. First, 100 µL of the mixture IL:acetonitrile (50:50, v/v) were picked up in the Pasteur pipette at a flow rate of 0.5 mL min^{-1}. In a second step, the pipette was immersed in the sample vial to draw in a fixed volume of sample at a flow rate of 0.5 mL min^{-1}. The sample was continuously introduced in the system and passed through the IL plug, allowing a dynamic extraction of the analytes since the extractant remained in the lower part of the extraction unit due to its higher density. In this cycle of sample loading, the flow rate of the syringe emptying step was fixed to 5 mL min^{-1}. Finally, the extract was delivered to an Eppendorf vial, diluted with acetonitrile and analyzed by LC. The overall extraction process takes place in less than 35 min. Since a new Pasteur pipette was used for each extraction, no analyte carry-over was observed. This technique also avoids repeated filling and emptying of a syringe with sample.

The most recent approach in DI-SDME has been the use of a microfluidic system where a new chip-based liquid–liquid extraction technique is coupled to GC (Peroni et al., 2012). Extraction is performed in a segmented flow system with additional mixing provided by an etched channel structure (**Figure 4.6**). The dimensions of the device were optimized to benefit from the advantages of chip technology without suffering from the limitations of over-miniaturization. Depending on the sample volume, two different instrumental modes were used, continuous introduction LLE and plug-injection LLE. The results obtained for selected test analytes show that the extraction is quantitative (recoveries = 92–110%, RSD < 6%) for a wide range of hydrophobicities (Log K_{ow} = 0.86–4.79). the authors highlighted that the instrumental set-up is simple and mechanically strong. This simplicity and robustness, combined with the stability typical of segmented flow, enables automated operation and/or use in the field.

4.2.2.2 Drop-in-drop and Drop-to-drop

Drop-in-drop system was the pioneering LPME technique developed by H. Liu and Dasgupta (1996) where an organic microdrop was suspended inside a flowing aqueous sample drop from which analyte was extracted. The analytical system was validated for sodium dodecyl sulphate by its ion-pair formation with methylene blue. After the sampling/extraction period, a wash solution replaced the sample/reagent in the aqueous layer, resulting in a clear outer aqueous drop housing a colored organic drop containing the extracted analyte. This also resulted in an automatic backwash. The color intensity of the organic drop, related to the analyte concentration, was

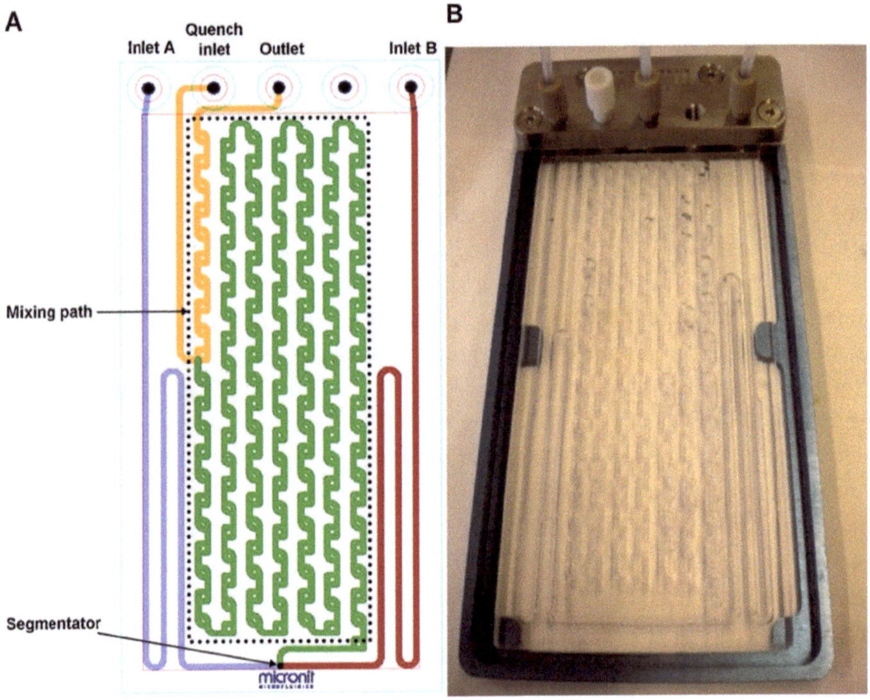

Figure 4.6 Schematic representation (A) and photograph (B) of the microextractor. Reprinted from Peroni et al. (2012), Copyright (2012), with permission from Elsevier.

monitored by a light-emitting diode-based absorbance detector. This system had the advantage of performing both processes of extraction and analysis in the same chamber.

Drop-to-drop is an analogous technique where the sample is a static microdroplet of just a few μL in volume (Wu et al., 2006). The proposed system reduces the size of the sample to typically 7–8 μL, into which was placed less than 1 μL of extraction solvent drop hanging at the needle tip of microsyringe (**Figure 4.7**). Once the extraction is finished, the drop is withdrawn back into the syringe and injected into the analytical instrument. Equilibrium between the sample and solvent is established quickly due to the small sample and solvent volume. Stirring is not necessary, and this also contributes to drop stability and simplifies the experimental setup. However, the larger ratio of volumes of organic microdrop to aqueous sample reduces the EF. Another application of drop-to-drop has been the determination of trimeprazine involving its extraction in 0.6 μL of toluene from 8 μL of urine and blood samples of rats (Agrawal & Wu, 2007). Caffeine in beverages (Shrivas & Wu, 2007) and nicotine in nightshade vegetables (Shrivas & Patel, 2010) have also been extracted by this technique.

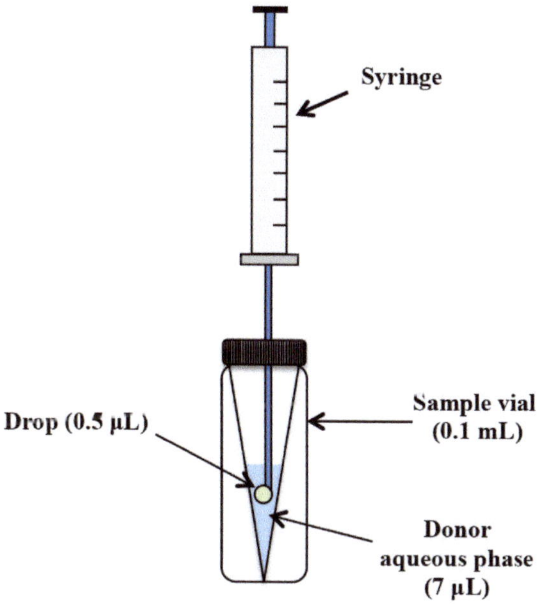

Figure 4.7 Drop-to-drop system.

US energy has also been used in the drop-to-drop technique to determine benzene, toluene and xylene in water samples (Zhang et al., 2010). This method maintains the low volumes of extractant solvent (3 µL) and sample (20 µL), however, the extractant solvent is completely dispersed by means of the US energy and the spherical drop loses its integrity, being more DLLME than drop-to-drop approach.

4.2.2.3 Continuous Flow Microextraction

Continuous flow microextraction (CFME) is a technique developed by Liu and Lee where the microdrop of organic solvent is in contact with a continuously flowing aqueous sample solution (Liu & Lee, 2000). The organic drop (1–5 µL) is held at the outlet tip of a PEEK connecting tubing that is immersed in a continuously flowing sample solution and acts as the fluid delivery duct and as a solvent holder. A pump was used to introduce the aqueous sample into the extraction chamber at a constant flow rate *via* a line made of PEEK tubing and terminating at the center of the chamber. The solvent drop emerges from the PEEK tubing and stays at the end whilst the sample solution keeps flowing around the drop. Once the extraction is finished, the organic drop is picked up with a GC microsyringe and injected into GC. EFs from 260- to 1600-fold were achieved with 10 min of extraction time. Two different modifications of this system have been proposed. One modification proposed the use of a unique syringe to act as a holder of the drop and for injection in the analytical instrument. The syringe was placed just above the tube outlet in the extraction chamber

(Xiao et al., 2006). The other proposed modification, named cycle-flow microextraction, returns the sample from the extraction chamber back to the sample reservoir and it is used repeatedly for extraction (Xia et al., 2004). The re-circulation of sample solution permitted the analysis of a smaller sample size (1–2 mL) and avoided inadvertently running the sample reservoir dry.

Another dynamic system named IL-based cycle-flow SDME was developed by Xia et al. for preconcentration of Co, Hg and Pb from biological and environmental samples (Xia et al., 2008). This system extracted analytes by exposing an IL droplet (2.5 μL) to a flowing stream of sample. $[C_4mim][PF_6]$ IL was the extraction solvent and 1-(2-pyridylazo)-2-naphthol was used as complexing agent and chemical modifier. The EE and EF were higher than in static conditions since continuous contact between the IL phase and the fresh flowing sample solution was achieved.

4.2.2.4 Liquid-liquid-liquid Microextraction

Liquid-liquid-liquid microextraction (LLLME) is a three-phase mode best suited for the extraction of hydrophilic organic compounds, such as phenols, fatty acids or amines. Analytes are extracted from an aqueous sample to an organic solvent and simultaneously back-extracted from the organic solvent to the acceptor solution, usually a few microliters of an aqueous solution at the appropriate pH (Jeannot et al., 2010). The organic solvent, with lower density than water, is therefore an interface between the two aqueous solutions. In order to achieve analyte isolation and enrichment, the acid-base properties of the analytes are used. For acidic analytes, the pH of the donor solution (sample) is adjusted to a low value so that ionization of the analytes is suppressed and they can be extracted as neutral species into the organic solvent. At the same time, the pH of the acceptor solution is maintained at a high value to promote ionization of the analytes. This way, the analytes are converted into ionic species, which are excluded from the liquid organic membrane and therefore accumulated in the aqueous acceptor solution (Jeannot et al., 2010). The first system employed a PTFE ring (Ma & Cantwell, 1998, 1999), however, other systems avoiding the PTFE ring (He & Kang, 2006) or employing a volumetric flask have been used (Sarafraz-Yazdi et al., 2005; Fan & Liu, 2008). The latter system placed a layer of an organic solvent on top of the aqueous sample and a microdrop of the acceptor solution was immersed into the organic solvent layer with a syringe (**Figure 4.8**).

Choi et al. have proposed an automated three-phase SDME coupled to CE using a commercial instrument (Choi et al., 2009). The CE instrument provided adjustable forward and backward pressures and a single drop of an aqueous acceptor phase covered with a thin layer of an organic phase was formed at the capillary tip. Acidic analytes from an acidic donor phase were concentrated into a basic acceptor phase yielding 2000-fold enrichment in 10 min with agitation of the donor phase using a microstirrer. The end surface of the fused silica capillary was hydrophobically treated by silanization to increase drop adhesion, and therefore, the failure rate of drop

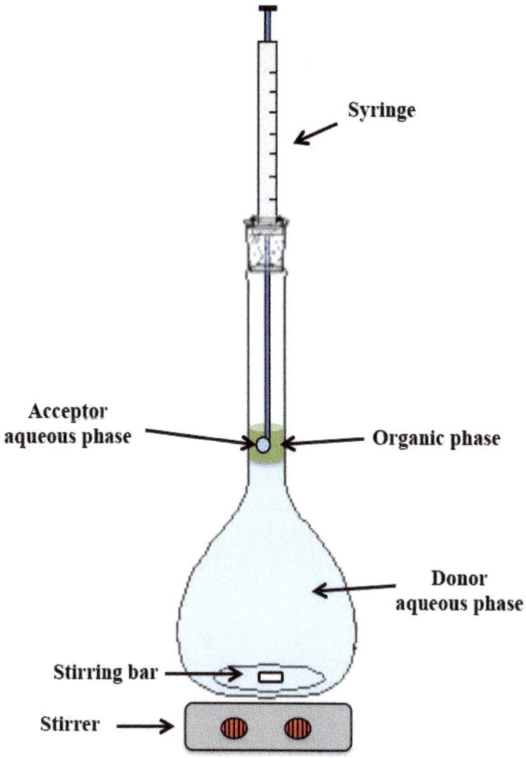

Figure 4.8 LLLME system employing a volumetric flask.

formation was negligible (Choi et al., 2009). Comparison studies with a two-phase system and without stirring were carried out, obtaining lower EF.

The most recent approach combines LLLME with DSDME. One proposal avoids the use of the microsyringe as a support of the aqueous drop and a large aqueous droplet is freely suspended at the top-center position of a layer of immiscible organic solvent, which is laid over the aqueous sample solution while being agitated (Sarafraz-Yazdi et al., 2009). The procedure involved the stirring of the sample with the organic solvent for a period of time before the acceptor phase was delivered at the top-center position of the immiscible organic solvent layer. Then, the three phases were stirred for 6 min, and finally, the aqueous droplet was withdrawn into the LC microsyringe and then injected into the LC system. Another combination uses a syringe as a holder of the aqueous droplet but the organic phase is a spherical suspended drop instead of a layer (Gao et al., 2011). This system also employs a two times stirring procedure, being the organic drop and the sample stirred before the aqueous drop is disposed inside the organic drop. After a second stirring and once the extraction is finished, the aqueous back-extractive phase was retracted into the syringe and transferred into a microvial (**Figure 4.9**).

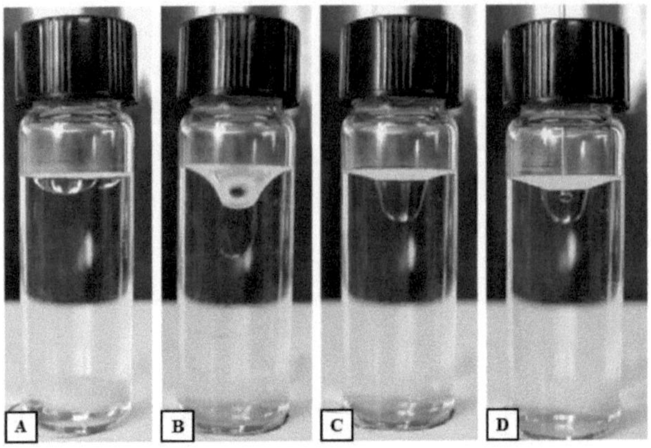

Figure 4.9 Picture of the different steps for DSDME and single drop back-extraction: (A) addition of the organic extractive phase to the aqueous sample phase, (B) DSDME procedure at 1150 rpm, (C) stirring rate at 800 rpm and the larger droplet of organic phase kept steady, and (D) back-extraction procedure at 800 rpm. Reprinted from Gao et al. (2011), Copyright (2011), with permission from Elsevier.

4.2.2.5 Directly Suspended Droplet Microextraction

The use of a holder (i.e., microsyringes, capillaries) presents drawbacks such as drop instability and slow kinetics. For this reason, different approaches proposed the elimination of the holder and the droplet was directly suspended on the sample. One of these techniques, named directly suspended droplet microextraction (DSDME), was introduced by Yangcheng et al. (2006). In this technique a free microdroplet (5–100 µL) of solvent is delivered to the surface of an immiscible aqueous sample while being agitated (typically at 1000 rpm) by a stir bar placed on the bottom of the sample cell (**Figure 4.10**). After extraction, the microdroplet of solvent is withdrawn by a syringe and analyzed. Under the proper stirring conditions, the suspended droplet can remain in a top-center position of the aqueous sample. The droplet can become partly engulfed within the sample while maintaining a stable shape with mechanical equilibrium and the mass transfer could be effectively intensified.

Dislodgement of the drop, small organic solvent volumes and limited stirring speed are not drawbacks in DSDME. Relatively high stirring rates, limited by the disintegration of the droplet, can be used since the drop freely rotates on the top centre of the sample. This technique usually employs volumes higher than 4 µL due to the difficulty of withdrawing the drop after the extraction. However, the technique presents a limitation related to the solvent properties, which should be poor water miscibility, density lower than water and low vapor pressure. Other limitations that the authors noted were the size and shape of the stir bar used, which had a distinct effect on the shape of the drop, which in turn can lead to difficulty in sampling. In addition, they considered that analyte adsorption on the surface of the stir bar was inevitable.

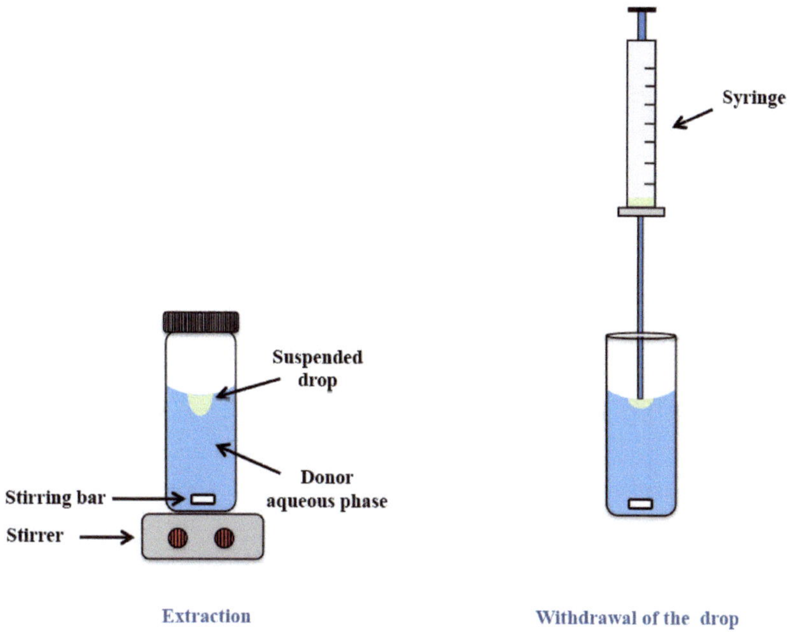

Figure 4.10 DSDME procedure.

Therefore, a modified method of DSDME was proposed in which the extraction vial was rotated to produce a vortex in the aqueous sample instead of using a stirring bar (Mingyuan et al., 2009). During extraction, the rotating vial provided a very stable flow field, and solvent spreading along the parabolic surface of the aqueous phase reduced mass transfer resistance. A modified two-way magnetic stirrer device was used for rotating the vial. A hole was drilled through the tray of the magnetic stirrer, and a rotating shaft was inserted. A circular deck made of PTFE with a circular groove was fixed onto the rotating shaft with a screw. A cylindrical extraction vial was snugly placed inside the groove such that the vial rotated along with the deck. For sampling convenience, a hole was made at the centre of the extraction vial lid through which a needle could be inserted. A piece of sealing film was placed inside of the lid to prevent the release of volatile compounds. Potential emulsification was avoided using the centrifugation effect of the rotating vial. During sampling, the shape of the organic solvent droplet changed upon insertion of a needle, causing the droplet height to increase approximately 3–4-fold. This increase in droplet height made solvent sampling much more convenient and allowed for the use of smaller volumes, which enhanced mass transfer and enrichment. In addition, the absence of a stir bar enables more stable and reproducible drop shapes providing better precision values and lower drop disintegration.

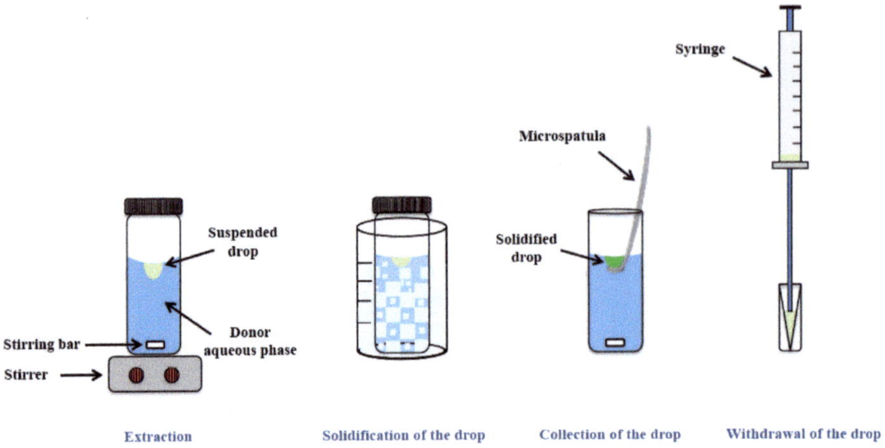

Figure 4.11 SFODME procedure.

4.2.2.6 Solidification of Floating Organic Drop Microextraction

Khalili Zanjani et al. (2007) proposed the technique named solidification of floating organic drop microextraction (SFODME) to eliminate or reduce the transfer of some aqueous sample when the organic solvent is being withdrawn by a syringe in DSDME. The difference with DSDME technique is that once the extraction is finished, the directly suspended solvent drop is solidified by introducing the vial into an ice bath for approximately 5 min. Then, the solidified organic solvent is transferred into a small conical vial using a microspatula (**Figure 4.11**). The solid organic solvent known as the 'solid drop' melts quickly at room temperature. Finally, it is retracted by a micro-syringe and injected into an analytical instrument for final analysis. This method has initially been tested with PAHs using 1-undecanol, and the extract was analyzed by GC-FID (Khalili Zanjani et al., 2007). Preconcentration factors in the range of 560–1940 were obtained which was better than some other extraction methods. Different applications published since then are discussed in two recent reviews (Ganjali et al., 2010; Ghambarian et al., 2013).

The main limitation of SFODME is that not only requires an extractant phase less dense than water, but also the melting point should be near room temperature (in the range 10–30 °C). The different solvents available for this technique are: 1-undecanol, 1-dodecanol, 2-dodecanol, 1-bromohexadecane, 1-hexadecane, 1,10-dichlorodecane and 1-chlorooctanedecane, with 1-undecanol and 1-dodecanol being the most commonly used. Moreover, the complete removal of trace water is hard to accomplish and an extra centrifugation step of the melted drop is sometimes needed (Zheng et al., 2011). Therefore, the several steps needed to perform the microextraction process make SFODME a tedious and time-consuming technique. In addition, automation of DSDME and SFODME is hardly feasible.

4.3 Membrane-based Liquid-phase Microextraction

4.3.1 Hollow Fiber Liquid-phase Microextraction

SDME gained a widespread interest since its appearance and obtained an undoubted relevance as a starting point toward miniaturized liquid-phase extraction techniques. However, some problems related to the instability of the hanging drop, which can be easily dislodged from the holder during extraction, needed to be solved. In 1999, Pedersen–Bjergaard and Rasmussen (1999) introduced the HF-LPME technique which easily overcame this shortcoming through the stabilization of the extractant phase using a porous membrane. In HF-LPME, an organic solvent is immobilized in the pores of a polymeric HF, normally made of polypropylene, and forming a SLM. The final acceptor phase is introduced within the lumen of the fiber, being protected and physically separated from the sample by the SLM. HF-LPME possesses some remarkable advantages (Pedersen-Bjergaard & Rasmussen, 2008; Ghambarian et al., 2012), namely: (i) high stirring rates can be employed because the problem of the dislodgment of the drop is avoided, although undesirable air bubbles can be generated on the surface of the fiber at very high stirring rates; (ii) the contact area between the extractant phase and the sample is higher, favoring mass transfer; (iii) the HF acts as a protection barrier for the extractant phase, therefore, dirty samples can be analyzed while obtaining very clean extracts; (iv) the fiber can be disposable after an only one use due to its low cost, avoiding carry-over effects between extractions; and (v) the technique presents automation options. Fundamental theory and basic equations about HF-LPME have been presented in detail in the literature (Ghambarian et al., 2012; Gjelstad et al., 2012). As major shortcomings, the manual preparation of the fiber can lead to deviations between extractions and the relatively long extraction times. Ultrasound energy has been proposed to reduce extraction times by accelerating mass transfer rate of analytes in a methodology called US-enhanced HF-LPME (Yang et al., 2010).

Two extraction modes, two-phase and three-phase, are described for HF-LPME. In two-phase HF-LPME (Rasmussen et al., 2000), the same organic solvent is immobilized in the pores of the fiber and introduced in its lumen (**Figure 4.12A**). Thereby, analytes are extracted from the donor phase through the SLM into the organic solvent in the lumen of the fiber (acceptor phase). Two-phase mode is suitable for hydrophobic analytes with a substantial solubility in organic solvents, since high partition coefficients are essential for an efficient extraction. GC followed by different detectors (*e.g.*, MS, FID, ECD) is normally selected as the quantification technique due to its compatibility with the organic acceptor phase. Two-phase HF-LPME can be performed in two different ways: direct immersion (Rasmussen et al., 2000) or headspace (Jiang et al., 2005). Strictly from the point of view of phases, headspace two-phase HF-LPME is not a two-phase mode due to the donor aqueous phase, the gaseous phase and the acceptor organic phase which are involved in the extraction process. However, if an

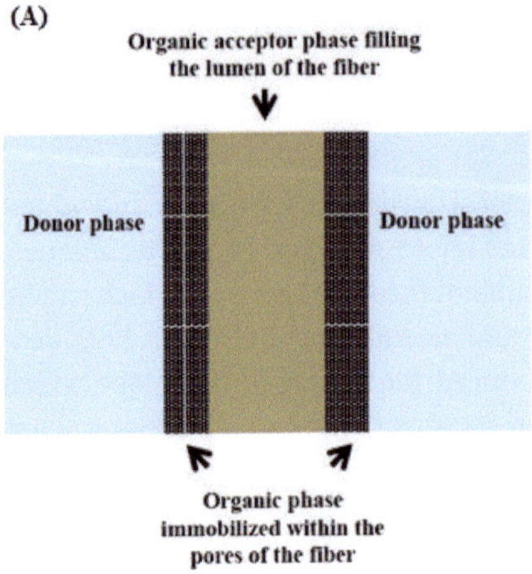

Figure 4.12 Cross-section view of the HF inside the sample. (a) Two-phase HF-LPME; and (b) three-phase HF-LPME.

organic solvent fills the pores and lumen of the fiber, the methodology is included in the two-phase classification. Both conventional sample heating using hot plates (Jiang et al., 2005) and microwave radiation (Shi et al., 2008) have been employed to assist headspace extractions.

In three-phase HF-LPME (Pedersen-Bjergaard & Rasmussen, 1999) analytes are extracted from the aqueous sample (donor phase) through the immobilized organic solvent in the pores of the fiber, into another aqueous solution (acceptor phase) within the lumen of the fiber (**Figure 4.12B**). Three-phase HF-LPME is normally coupled with LC and CE, usually followed by UV or MS detectors, due to the aqueous nature of the acceptor phase. Three-phase HF-LPME is intended for the extraction of hydrophilic or ionizable compounds by three different mechanisms, namely: (i) pH gradient (ii) carrier mediated and (iii) EME. The solubility of basic and acid analytes in donor and acceptor aqueous solutions can be adjusted by pH changes. For example, the extraction of basic compounds implies the basification of the sample solution to decrease the solubility, and the acidification of the acceptor phase to promote it. Conversely, for the effective extraction of acid compounds, the pH of the donor phase is adjusted into the acid range whereas the acceptor phase is basified. Thereby, basic and acid analytes can be easily extracted into the immobilized organic phase, diffuse through the SLM and, finally, be back-extracted into the acceptor phase without returning into the sample. Highly hydrophilic compounds are poorly extracted by partitioning and diffusion mechanisms and require an active transport to cross the SLM and reach the acceptor phase. The use of carriers can enhance EEs in these cases (Ho et al., 2003). Carriers form hydrophobic ion-pairs with analytes, which are extracted into the immobilized organic solvent and diffuse through the SLM. In the contact region with the acceptor phase, analytes are released from the ion-pairs into the acceptor phase, and carriers form new ion-pairs with other counter ions present in the acceptor phase. The new ion-pair complexes are back extracted into the sample, thus completing the cycle (Rasmussen & Pedersen-Bjergaard, 2004). Three-phase HF-LPME assisted by an electric field (i.e., EME) will be discussed separately in another section of this chapter.

Although classical modes of two-phase and three-phase HF-LPME are described above, some authors have proposed alternative methodologies. For example, Bedendo et al. (2012) modified classical two-phase HF-LPME by disposing organic solvent in the sample and an empty fiber. Pesticides from orange juice were firstly extracted into the organic solvent, which were then absorbed into the fiber due to their high affinity, forming a renewable SLM and providing a good sample clean-up. A final solvent desorption step is included in the procedure to extract analytes into methanol-acetone mixture for its injection in a LC system. Three-phase HF-LPME generally involves an aqueous acceptor phase, but the use of two immiscible organic solvents in the pores and lumen of the fiber has also been presented. Ghambarian et al. (2010) devised a new approach compatible with GC in which *n*-dodecane was immobilized in the pores of the fiber whereas methanol or acetonitrile was introduced into the lumen for the extraction of chlorophenols from honey and water samples. Zhang et al. (2006) proposed a new organic solvent-free three-phase mode (named liquid-gas-liquid microextraction) in which the pores of the hydrophobic HF, not filled with a solvent, separated donor and acceptor aqueous phases. Therefore, analytes crossed the fiber by gas diffusion from a heated sample. This work included a comparison with classical three-phase HF-LPME, obtaining better EF for the most volatile compounds studied (phenols). Finally, Carletto et al. (2009) employed a method called HF renewal liquid membrane based on three-phase HF-LPME for the extraction of cadmium from water samples. The authors added organic solvent to the donor phase as well as in the pores of the fiber for the continuous renewal of SLM and replenishment of lost solvent, obtaining a high stability. Complexes of cadmium were extracted into the organic solvent in the donor phase and carried to the fiber. The organic solvent was then solubilized into the fiber forming a homogeneous phase. Cadmium finally reached the acceptor solution where the complex was released at a high pH.

Three general configurations of fiber exposure to the sample have been proposed in the literature. The first work about HF-LPME (Pedersen-Bjergaard & Rasmussen, 1999) employed a configuration based on an U-shaped fiber with each end connected to a needle, one for the injection of the acceptor phase, and the other for its collection after extraction (**Figure 4.13A**). A rod-like configuration appeared later, becoming the most interesting and widely employed option, especially when the fiber is held by a syringe (Zhu & Lee, 2001) (**Figure 4.13B**). In this configuration, the acceptor phase is introduced and removed through the same side whereas the other one is sealed. The syringe is employed not only as a holder but also to add and withdraw the acceptor phase, and directly inject it into the analytical system, thus simplifying the procedure. The length of the fiber is smaller than in the U-shaped configuration, therefore the volume of the acceptor phase is reduced and the EFs are enhanced. On the other

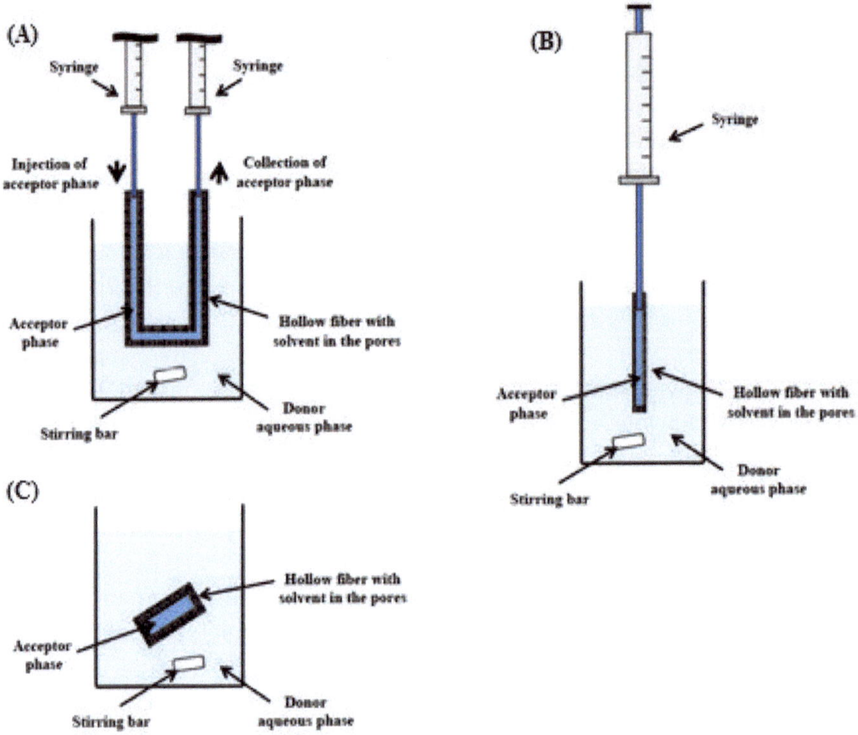

Figure 4.13 Different configurations of HF-LPME: (A) U-shaped; (B) rod-like; and (C) SBME.

hand, the fiber can be directly introduced into the sample with both sealed ends, leading to a methodology called SBME (Jiang & Lee, 2004) (**Figure 4.13C**).

The extraction systems described up to now are all static systems where the donor and acceptor phases are stagnant, with the exception of stirring the sample to enhance mass transfer. HF-LPME can also be performed in a dynamic mode where the lower end of the rod-like fiber configuration is opened. In dynamic two-phase HF-LPME (Zhao & Lee, 2002), the acceptor phase is repeatedly drawn into the barrel and injected into the lumen of the fiber using a syringe. While drawing in sample in the direct immersion mode, aqueous sample fills the lumen and analytes are extracted into a thin film of organic solvents built up in the HF. During injection, sample is driven out of the fiber, the organic film recombines with the organic solvent and analytes distribute into the bulk solution. Dynamic two-phase HF-LPME has also been performed in the headspace mode where the gas phase enters and leaves the lumen of the fiber during sucking and injection cycles (Jiang et al., 2005). In dynamic three-phase HF-LPME (Hou & Lee, 2003), the syringe is first filled with the acceptor aqueous phase and then with an organic solvent. The organic solvent is injected into the HF and withdrawn in the consecutive cycles of sucking and injection, following a similar procedure than in dynamic two-phase, while the final aqueous acceptor phase remains inside the syringe. Dynamic HF-LPME improves extraction speed and efficiency compared with static systems but the extraction setup is more complex. Automated or semi-automated dynamic HF-LPME systems are generally employed due to the obvious tediousness and irreproducibility of manual operation. In these systems, a syringe pump is fixed to the syringe that supports the fiber and is programmed to eject and withdraw the acceptor phase at a selected speed. Related to this, Pezo et al. (2007) devised an automatic multiple dynamic HF-LPME system in which several samples were simultaneously extracted with obvious benefits in throughput and operation time. In spite of reaching different degrees of automation, the developed methods using dynamic HF-LPME are off-line systems since sample preparation and instrumental analysis are carried out separately. Esrafili et al. (2012) have recently proposed an on-line three-phase HF LPME which performed sample extraction and acceptor phase injection into a LC system automatically, without the need of an operator. The method employed a syringe pump for loading solvents, a platform lift for moving the sample vial and a sampling loop for on-line injection of the acceptor phase into the LC. Another novel on-line method for LC, named as push/pull perfusion hollow-fiber liquid-phase microextraction (PPP-HF-LPME), has recently been presented (Chao et al., 2013). PPP-HF-LPME uses a push/pull syringe pump with two opposing syringes as the driving source to perfuse the acceptor phase into the fiber. Therefore, the authors avoided losses or gains of solvent across the porous membrane that normally appear as a consequence of flowing the acceptor phase only by push or pull perfusions. In addition, this method employed US energy to accelerate mass transfer of neutral analytes leading to short extraction times.

HF-LPME has been employed in environmental, bioanalytical and food fields in its different modes and configurations (Lee et al., 2008; Ghambarian et al., 2012). This technique has found its major applications in organic analysis (Ghambarian et al., 2012; Lee et al., 2008), although the determination of metals and their speciation has also been studied in different methods using LC-UV, ETAAS, ICP-MS or ICP-OES systems (Hu et al., 2013). Parameters affecting HF-LPME such as the type of immobilized organic solvent, extraction time, stirring rate and salting out effect are commonly optimized in the different applications of two-phase and three-phase modes. The selection of the appropriate solvent within the pores of the fiber is a critical part of the technique. Some important aspects of the solvent to consider include (Ghambarian et al., 2012): (i) Immiscibility with water to prevent losses by dissolution; (ii) ability to be strongly immobilized in the pores of the fiber; (iii) low volatility to avoid losses by evaporation; (iv) good extractability and high partition coefficients of the target analytes; (v) low viscosity to ensure high diffusion coefficients across the liquid membrane; and (vi) compatibility with the selected analytical technique in two-phase HF-LPME. Organic solvents such as toluene, 1-octanol, undecane or dihexylether fulfill all these requirements and have successfully been employed in both modes of HF-LPME (Ghambarian et al., 2012). The fiber type and its dimensions are also important points to consider. Polypropylene fibers are usually chosen due to their hydrophobicity and compatibility with a broad range of organic solvents. Polypropylene fibers have been modified with carbon nanotubes (CNTs) with the aim of enhancing EE (Es'haghi et al., 2010). CNTs immobilized into the fiber act as solid sorbents, and lead to a methodology called hollow fiber solid–liquid phase microextraction. The dimensions of the fiber have to ensure a good contact surface area, appropriate donor/acceptor volume ratio and mechanical stability. The length of the fiber is chosen by the user and in most cases is conditioned by the employed configuration (i.e., U-shaped, rod-like or SBME), whereas commercially available fibers commonly have an internal diameter of 600 μm and a 200 μm wall thickness. Finally, the pore size (typically of 0.2 μm) should allow the penetration of target analytes while providing good filtration and sample clean-up.

As mention above, organic solvents such as toluene, 1-octanol, undecane and dihexylether have successfully been employed in HF-LPME. However, their losses by evaporation are a disadvantage of the technique and an important source of irreproducibility between extractions. ILs have recently been presented as an interesting and useful alternative to those traditional solvents for the impregnation of the fiber in HF-LPME. Previous studies (Fortunato et al., 2004, 2005) demonstrated the feasibility of using ILs based on 1-*n*-alkyl-3-methylimidazolium cations to obtain stable SLMs, even under vigorous stirring conditions due to properties such as relative high viscosities, interfacial tensions and low solubility in water depending on the anion. The applicability of a SLM based on ILs in HF-LPME was demonstrated by Peng et al. (2007) who employed [C_8mim][PF_6] as intermediary solvent in three-phase HF-LPME for the extraction of chlorophenols from different water samples. Afterwards, Basheer et al. (2008a) presented a three-phase HF-LPME in which [C_4mim][PF_6] IL was employed as intermediary solvent whereas a organic solvent was the final acceptor phase. Thereby, aromatic and aliphatic hydrocarbons could be determined in storm water using a GC-MS system. The use of ILs in two-phase HF-LPME was successfully presented by Abulhassani et al. (2010) who employed [C_6mim][PF_6] for the extraction of ammonium pyrrolidine dithiocarbamate complexes of lead and nickel from water and biological samples, using ETAAS as the detection system.

In general, both organic and inorganic analytes have been extracted from aqueous samples using IL based HF-LPME, although more complex matrices such as biological tissues (Abulhassani et al., 2010), hair and tea (Zeng et al., 2011) have also been studied. The rod-like configuration is most commonly used, but SBME has been chosen by some authors (Zhang et al., 2013).

4.3.2 Electromembrane Extraction

Electromembrane extraction (EME) was presented for the first time in 2006. Pedersen-Bjergaard and Rasmussen (2006) proposed a novel technique based on three-phase HF-LPME in which analytes are extracted from the donor phase through

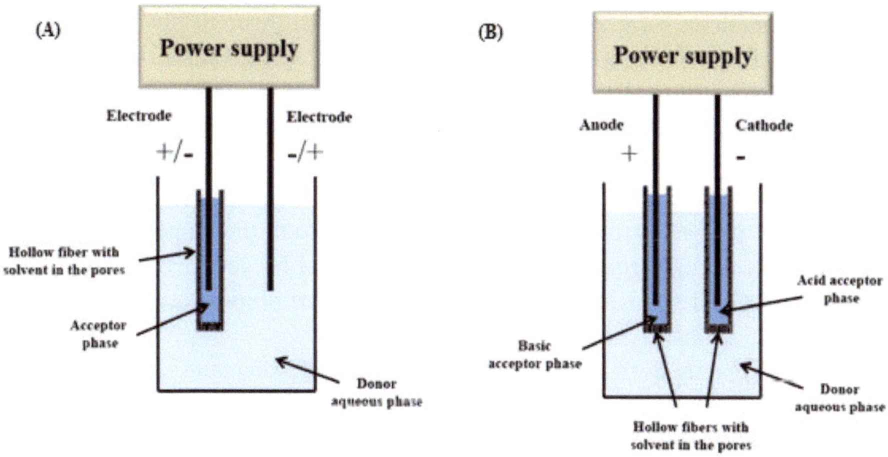

Figure 4.14 Setup of EME: (A) Original configuration; and (B) two hollow fibers for the simultaneous extraction of acid and basic analytes.

the SLM into the acceptor phase using an electrical potential difference as driving force. The authors extracted basic drugs from 300 μL of sample into 30 μL of acceptor phase, both of which were acidic to maintain analyte protonation. The solvent 2-nitrophenyl octyl ether (NPOE) was immobilized in the pores of the fiber, which used the rod-like configuration. A platinum electrode was immersed into the sample and another platinum electrode was placed into the acceptor phase. Both electrodes were connected to a power supply and an electrical field was created across the SLM. Thereby, charged analytes were extracted by electrokinetic migration and not by passive diffusion. A schematic illustration of the general EME setup is shown in **Figure 4.14A**. EME presents the following general advantages (Petersen et al., 2011b; Gjelstad & Pedersen-Bjergaard, 2013): (i) shorter extraction times (1–5 min) than in HF-LPME (typically between 10–60 min) due to the enhancement of mass transport by the force of the electrical potential; (ii) extraction selectivity can easily be modulated by changes in the magnitude and direction of the electrical potential, simply by manipulating the external power supply; (iii) efficient sample clean-up and feasibility of direct extraction from untreated complex matrices; and (iv) possibilities of downscaled format (i.e., microchip devices) and automation. As major drawbacks, the homemade extraction equipment exhibited bubble formation due to electrolysis reactions and sparking at high voltages or during the analysis of real samples with high ion concentrations.

Fundamental theory about EME has been discussed in the literature (Gjelstad et al., 2007). The proposed model describes the influence of the thickness of the fiber, voltage applied and coefficient of diffusion across the SLM over the flux of analytes. As in HF-LPME, commercial fibers with a 200 μm wall thickness are usually employed. The theoretical model predicts an improvement in the flux with thinner fibers but instability problems have to be considered. It has been theoretically proposed and experimentally proven that an increase in the applied voltage produces an increase in the flux across the SLM, thus affecting the recoveries. Potentials between 1 and 300 V are commonly employed. The coefficient of diffusion of analytes across the SLM depends on the viscosity of the immobilized solvent and is higher in those with low viscosity. Apart from these parameters, other considerations such as the pH of donor and acceptor phases, extraction time, the conductance of the immobilized organic solvent or the use of carriers if necessary, are essential for good EME performance. The acceptor phase volume, salt effects and sample stirring speed are also optimized in some applications.

EME has mainly been applied for the extraction of acidic and basic drugs from biological samples (*e.g.*, blood, plasma, urine, breast milk, saliva or amniotic fluids). For the extraction of basic analytes, the anode is placed into an acidic donor phase whereas the cathode is placed into an acidic acceptor solution. SLMs based on NPOE have generally shown the best performance for the extraction of hydrophobic basic analytes. However, the use of ion-pair reagents which act as carriers seems to be necessary for enhancing mass transport across the SLM of more polar basic compounds (Gjelstad et al., 2006). Contrary to the extraction of basic molecules, the extraction of acidic drugs requires the basification of donor and acceptor phases in order to maintain their charge. The direction of the electrical field is reversed with respect to EME of basic compounds, therefore the cathode is placed into the donor phase and the anode is placed into the acceptor solution. SLMs containing 1-octanol have shown the best results for EME of acidic compounds. The simultaneous extraction of acidic and basic drugs has also been presented in the literature using a dual EME with a compartmentalized membrane (Basheer et al., 2010) or two separated hollow fibers (Seidi et al., 2012). The last configuration is simpler and robust and thus has found more use in subsequent works (**Figure 4.14B**). For the extraction of acidic analytes, the lumen of one fiber is filled with a basic solution and the anode is introduced in it. The other fiber is filled with an acid solution and contains the cathode for the extraction of basic analytes. The pH of the donor phase is kept neutral in order to have both acidic and basic compounds ionized. Using this configuration, analytes are simultaneously but separately extracted, therefore a further step of mixturing the two acceptor solutions is necessary for a unique analysis. Peptides (Balchen et al., 2008) and metal ions (Basheer et al., 2008b) have been also target analytes in EME. On the other hand, apart from the excellent performance in biological analysis, EME has also been applied in the environmental field for the extraction of organic compounds (Xu et al., 2008) and metals (Kubáň et al., 2011) from different water samples.

CE and LC are commonly selected as separation techniques after EME. CE has allowed the chiral separation of different drugs in methods where enantiomers were equally extracted (Nojavan & Fakhari, 2010; Fakhari et al., 2013). CE-UV was employed in the first report (Pedersen-Bjergaard & Rasmussen, 2006) and the early years of EME. Later, CE coupled with a contactless conductivity detector (Xu et al.,

2008) appeared as a more universal detection method since it allows the conductometric analysis of all charged species. EME has recently been combined with GC in the classical three-phase mode (Davarani et al., 2012) and in a novel two-phase EME (Davarani et al., 2013) in which the aqueous acceptor phase is replaced by an organic solvent with sufficient electrical conductance (*e.g.*, heptanol, octanol). On the other hand, EME has been combined with SPELs (Ahmar et al., 2013). Electrochemistry offers easy operation, high sensitivity and rapid response. In addition, SPELs present an interesting alternative to traditional bulky and expensive instrumentation and are perfectly compatible with the low-volume acceptor phase in EME. Thus, the use of SPELs in the detection stage could be a powerful tool in future investigations and applications of EME.

Different modifications to the electrical field application have been proposed. Traditionally, a constant voltage was applied across the SLM, however, a recent publication has proposed the use of pulsed voltages (Rezazadeh et al., 2012). In pulsed electromembrane extraction (PEME) the duration of the pulse is long enough for analyte extraction through the SLM, but it is short enough to avoid instability problems, which normally appear in EME at high potentials. Two-way PEME (Rezazadeh et al., 2013) was then introduced for the selective extraction of amino acids (AAs) utilizing their isoelectric pHs in the acceptor phase. Once the extraction of different AAs was achieved, the direction of the electrical field was reversed. The AAs in the zwitteronic form remain in the acceptor phase as they have no mobility in the electrical field, whereas the rest of AAs return to the donor phase. Another publication has suggested the use of voltage steps in PEME (Rezazadeh et al., 2014) and compared it with the application of constant pulses. Similar or higher recoveries were obtained but the step voltage approach required a lower energy supply. In another study, the application of constant voltage was compared with the application of constant current across the SLM (Slampová et al., 2012). The results demonstrated that the use of a stabilized constant current provided better repeatability in the extraction process.

Recent developments in EME membrane composition have also been reported. As in HF-LPME, CNTs have been inserted in the pores of the polypropylene HF in a method called CNTs assisted EME (CNTs-EME) (Hasheminasab & Fakhari, 2013). CNTs possess a high surface area with a high adsorption capacity for organic and inorganic species. In general, when using CNTs-EME, the mechanism of solute transfer involves both liquid and solid phase extractions, acting simultaneously to increase the analyte partition coefficient across the membrane. Therefore, the mass transport across the membrane is improved and better analytical results (lower limit of detection, higher preconcentration factor and recovery) were obtained in comparison with EME (Hasheminasab & Fakhari, 2013; Hasheminasab et al., 2013). CNTs-EME has also been employed in the two-phase mode compatible with GC (Hasheminasab et al., 2014).

EME has been downscaled to a chip format (Petersen et al., 2010). The microfluidic extraction device comprises a SLM with a flat configuration which separates the sample channel and acceptor reservoir. Continuous delivery of fresh sample to the

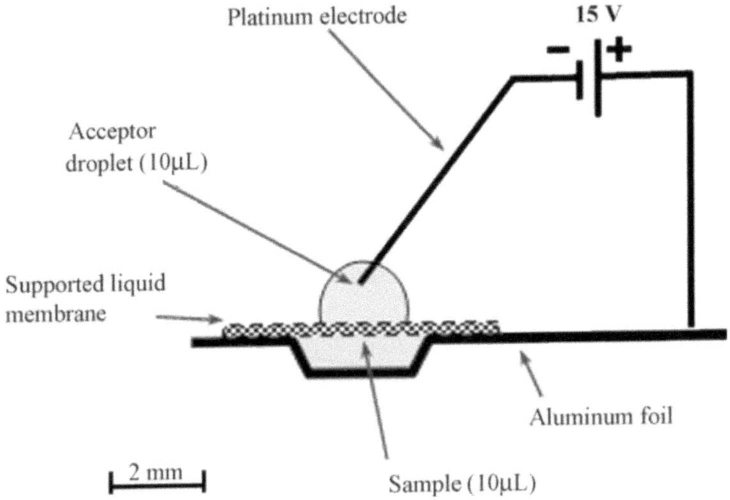

Figure 4.15 Schematic illustration of the set-up for drop-to-drop EME. Reprinted from Petersen et al. (2009), Copyright (2009), with permission from Elsevier.

electromembrane provides higher recoveries compared with EME in which sample solution is stagnant. This miniaturized system was first coupled off-line with CE-UV (Petersen et al., 2010), and thereafter with both CE-UV (Petersen et al., 2011a) and electrospray ionization-MS (Petersen et al., 2011a, 2012) in two on-line systems with a flowing acceptor phase. Another on-line system using CE-UV has been employed in a novel nanoelectromembrane extraction (nano-EME) approach in which analytes were extracted from 200 μL of sample into approximately 8 nL of acceptor solution (Payán et al., 2013). Nano-EME provided a system with an excellent enrichment capacity since the acceptor solution is notably reduced in comparison with EME. Another downscale approach is the drop-to-drop system developed for extraction of basic drugs (Petersen et al., 2009). The extraction was performed through an organic solvent of NPOE immobilized as SLM in a flat polymeric membrane placed between 10 μL of sample and 10 μL of acidic aqueous drop based on a direct current electrical potential of 15 V (**Figure 4.15**). The extraction conditions allowed only cationic compounds with relatively low polarity to cross the SLM whereas neutral and cationic compounds of high polarity could not.

A recent method has introduced the concept of parallel EME (Eibak et al., 2014) in which eight human plasma samples were simultaneously extracted using flat membranes in a multiwell configuration and a single power supply. Thereby, time consumption per sample is reduced and the throughput of the method is notably improved.

4.4 Dispersive Liquid-liquid Microextraction

4.4.1 Classical Dispersive Liquid-liquid Microextraction

Dispersive liquid-liquid microextraction (DLLME) was proposed for the first time in 2006 (Rezaee et al., 2006). Rezaee and co-workers presented a novel microextraction technique based on a ternary component solvent system in which a water-immiscible organic solvent that is denser than water is dispersed in fine drops into the aqueous sample with the aid of an organic disperser agent (Rezaee et al., 2006). A cloudy solution is formed due to the cosolvency of the dispersant with the other two phases and leading to a greater contact surface area. After extraction, phases are separated by centrifugation, and the enriched organic phase at the bottom of the centrifuge tube is collected to be analyzed. A scheme of the described procedure is shown in **Figure 4.16**. This simple methodology presents the characteristics of LPME techniques such as ease of handling, low cost, low solvent and sample consumption, and having a negligible environmental and human health impact due to the reduction of wastes generated. In addition, DLLME offers the following remarkable advantages (Rezaee et al., 2010; Yan & Wang, 2013): (i) fast achievement of the equilibrium state due to the infinitely large contact surface between the fine drops of the extractant solvent and the aqueous phase; (ii) short extraction times; (iii) high EFs; and (iv) problems associated with other LPME techniques such as the dislodgement of the drop in SDME, and the manipulation of the fiber in HF-LPME are avoided.

Parameters affecting DLLME, such as the type and volume of extractant and dispersing solvents, sample volume, salting-out effect, pH, type and amount of chelating agent for metals determination, extraction time and centrifugation time and speed are typically optimized in different applications. Chlorinated solvents (from 20 to 100 μL) (*e.g.*, chlorobenzene, chloroform, tetrachloromethane or tetrachloroethylene) have traditionally been selected as extractants due to their high density, whereas acetone, methanol, ethanol or acetonitrile (from 0.1 to 2 mL) are the most common dispersants.

GC was the analytical technique selected in the first applications of DLLME due to its compatibility with chlorinated solvents. However, polar and ionizable compounds cannot be determined by GC unless a derivatization process converts them into volatile species (Lin et al., 2013). The first use of LC after DLLME was proposed by Farajzadeh et al. (2007). In this work, an additional step of solvent evaporation was included after phase separation, and the solid residue was reconstituted in methanol for injection into the chromatographic system. Afterwards, Wei et al. (2007) demonstrated the feasibility of direct injection of chlorinated solvents in reverse-phase LC columns avoiding the time consuming evaporation steps. For inorganic analysis, ETAAS and FAAS are the most commonly chosen analytical techniques, although fiber optic-linear array detection spectrophotometry, UV-Vis spectrophotometry and ICP-OES or ICP-MS have also been utilized (Andruch et al. 2013a; El-Shahawi & Al-Saidi, 2013). ETAAS requires microvolume samples; therefore, it is perfectly compatible with low

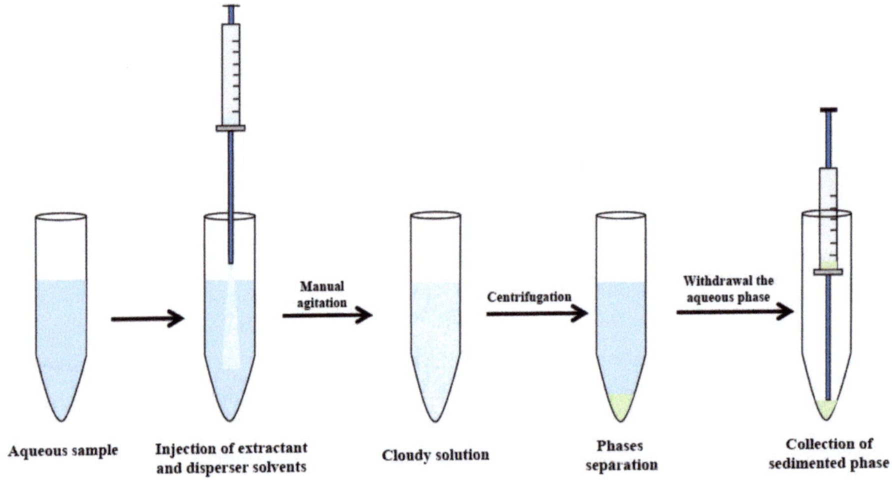

Figure 4.16 General procedure of classical-DLLME.

volume extracts obtained after DLLME. When using FAAS, larger sample volumes are necessary and further steps of dilution of the organic phase or evaporation and replacement of the extractant by a larger volume of other solvent may be necessary (Andruch et al., 2013a). The use of microsample introduction systems have also been reported (Naseri et al., 2008; Baliza et al., 2009). Finally, CE, followed in most cases by UV-Vis detectors, has been used after DLLME (Wen et al., 2012).

DLLME has gained rapid and widespread recognition, attracting the interest of the scientific community and even starting to dominate LPME research publications in recent years (Kokosa, 2013). Despite its wide acceptance, the original configuration of DLLME (Rezaee et al., 2006), termed as classical DLLME, suffers from some drawbacks or limitations that are in continuous revision, namely: (i) environmentally harmful organic solvents denser than water (*e.g.*, chlorinated solvents) are employed as extractants; (ii) emulsification requires a dispersant solvent which competes with the extractant for the analyte, thus reducing EE; and (iii) centrifugation is necessary to separate phases after microextraction. Numerous modifications of classical DLLME have been proposed in order to overcome the above mentioned disadvantages of the technique and to develop efficient, easier and more environmentally friendly approaches. An overview of recent advances is shown in the scheme of **Figure 4.17**.

One of the most representative modifications is the employment of alternative extractant solvents such as those less dense than water (Kocúrová et al., 2012), ILs (Trujillo-Rodríguez et al., 2013) or supramolecular systems (Moradi & Yamini, 2012b). The use of less toxic non-chlorinated solvents that are less dense than water requires developing new strategies to collect the enriched organic phase after phase separation. Almost simultaneously, Leong and Huang (2009) and Xu et al. (2009) combined low-density solvent-based DLLME with the so-called solidification of floating organic

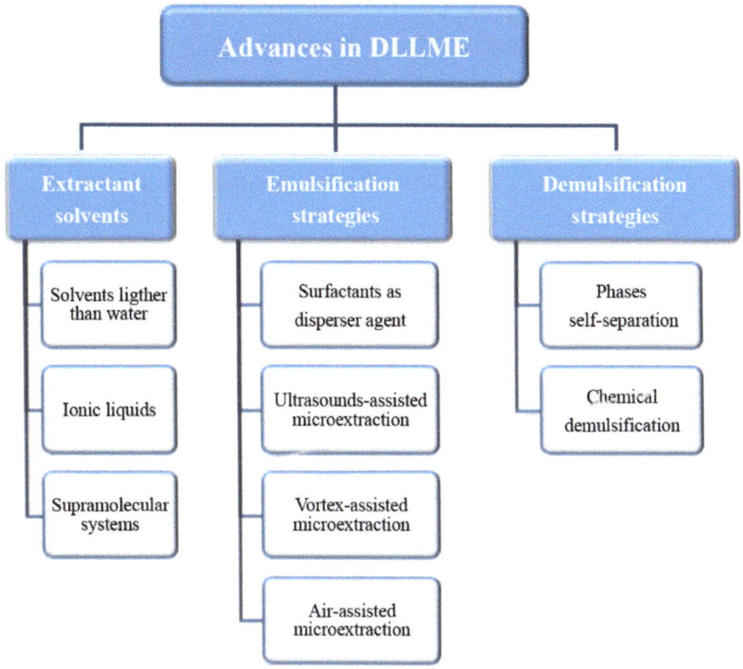

Figure 4.17 Recent advances in DLLME.

drop (SFOD) procedure (Khalili Zanjani et al., 2007). In this technique, the organic solvent, which remains in the upper layer after phase separation, was solidified in an ice-bath, separated from the aqueous phase with a microspatula and then melted at room temperature. Extractant solvents must have a melting point near room temperature (between 10 and 30 °C), be water-immiscible and compatible with the selected analytical technique, have a low vapor pressure to avoid losses by evaporation and a good extractability of the target analytes (Kocúrová et al., 2012). Solvents such as 1-undecanol, 1-dodecanol, 2-dodecanol and hexadecane satisfy these requirements and have successfully been employed in DLLME-SFOD for the determination of both organic and inorganic analytes (Kocúrová et al., 2012). However, the condition of having melting points near room temperature restricts the possibilities of DLLME-SFOD to a few non-chlorinated solvents. Some authors have explored the alternative of using special extraction devices, avoiding the restriction of melting points and the additional freezing step. Different models of homemade vessels have been proposed with the common characteristic of having a narrow upper neck in which the extractant phase is accumulated after centrifugation for its easy collection (Farajzadeh et al., 2009; Hashemi et al., 2009; Saleh et al., 2009). The main limitation of devices is that they are not yet commercially available which hinders their expansion and use in other laboratories. Other original alternatives have been presented in the literature such as the use of disposable polyethylene pipettes (Hu et al., 2010; Guo & Lee, 2011) and pipette tips (Moreno-González et al., 2012) as extraction units, a capillary tube to collect the extractant phase by simply dipping the tube into the floating organic drop (Farajzadeh et al., 2010), or a long-needle syringe to first evacuate the aqueous phase and then collect the organic phase settled in the conical bottom of the test tube (Bidari et al., 2011). Low-density solvents such as cyclohexane (Farajzadeh et al., 2009), toluene (Saleh et al., 2009; Moreno-González et al., 2012), *n*-hexanol (Hashemi et al., 2009), tri-*n*-butylphosphate (Hu et al., 2010), *n*-hexane (Guo & Lee, 2011), octanol (Farajzadeh et al., 2010) and naphthalene (Bidari et al., 2011) have been employed as extractants in the above mentioned approaches.

The use of ILs in classical DLLME (termed classical IL-DLLME) was described for the first time by Y. Liu et al. (2009). ILs present low miscibility with water, good extractability of organic and inorganic compounds and higher density than water, therefore, they appear to be good candidates to act as solvents in DLLME. The authors used [C_6mim][PF_6] as the extractant phase and methanol as the disperser solvent for the determination of heterocyclic insecticides in water samples. Since then, numerous works have employed classical IL-DLLME for the determination of both organic and inorganic analytes (*e.g.*, Trujillo-Rodríguez et al., 2013). In the vast majority of the cases, analytes are extracted from aqueous samples, although more complex matrices such as urine, saliva and cosmetics, have also been explored (Trujillo-Rodríguez et al., 2013).

Classical DLLME requires relatively large amounts of an organic solvent (i.e., acetone, methanol, ethanol or acetonitrile) to disperse the extractant phase. The type and amount of dispersant play a crucial role in the EE. Low volumes prevent the effective formation of the cloudy state whereas high volumes decrease the partitioning of the analyte into the extractant phase, especially for polar compounds. Surfactants have been proposed as alternative disperser agents since they are amphiphilic compounds capable of lowering the surface tension between the two phases and thus facilitating the dispersion of the extractant into the sample (Saraji & Bidgoli, 2010). Cetyltrimethyl ammonium bromide, methyltrialkylammonium chloride, Triton-X or sodium dodecyl sulfate have been employed for the dispersion of both chlorinated (Saraji & Bidgoli, 2010; Deng et al., 2013) and low-density solvents (Moradi et al., 2010; Tehrani et al., 2012) in the so-called surfactant-assisted DLLME. Compared with traditional disperser solvents, lower volumes of surfactant are necessary to generate an effective dispersion and they are more environmentally benign. Nevertheless, the tendency of developing greener methodologies that reduce the use of chemicals and the production of residues, has led to the appearance of new emulsification systems that allow the exclusion of any disperser agent and avoid the competition for the analyte. US energy, vortex agitation, temperature changes, metathesis reactions and air-assisted methodologies have successfully been proposed to assist the dispersion. They are presented and discussed in next sections of this chapter.

The centrifugation step that is needed for phase separation is considered the most time-consuming part and one of the major weaknesses of DLLME. Some authors have attempted to avoid this step by developing alternative strategies to break the emulsion or simply waiting for the self-separation of the phases (Cruz-Vera et al., 2009b; Andruch et al., 2012a; Farajzadeh et al., 2012). Chen et al. (2010) devised a methodology named solvent-terminated DLLME in which an organic solvent was used as a chemical demulsifier. The term solvent-based demulsification DLLME (SD-DLLME) (Zacharis et al., 2010) has been adopted in subsequent works as the name provides more information about the procedure. Briefly, after forming the cloudy state an appropriate volume of an organic solvent, normally the dispersant, is injected into the mixture and the emulsion is rapidly separated in two phases. SD-DLLME has generally been employed with low-density solvents, although a method using chloroform (Liang et al., 2013) has recently been proposed. SD-DLLME notably reduces operation time due to the elimination of the centrifugation step and presents a potential application in field analysis because electrical connections are not necessary. An important drawback to consider is the larger consumption of disperser solvent which can lead to dissolution of certain analytes and therefore decreasing EEs (Zacharis et al., 2010).

Despite of the enormous popularity of DLLME arising from its simplicity and excellent performance, one of the most concerning restrictions of the technique is the difficulty of complete automation, especially when centrifugation is employed for phase separation. Some authors have resolved this issue by developing different methods that attempt to demonstrate a feasible solution. Automated on-line systems based on sequential injection DLLME have been proposed for metals determination by ETAAS (Anthemidis & Ioannou, 2010) and FAAS (Anthemidis & Ioannou, 2009, 2011). In these approaches, phases were separated by the retention of the enriched phase in a packed microcolumn. Then, an appropriate solvent is employed for the elution and transportation of analytes to the detection system. Other methodologies maintaining the original concept of DLLME and avoiding the retention column have recently been introduced (Andruch et al., 2012a; Maya et al., 2012). However, more efforts are needed to address the challenge of an effective, simple and operational automated DLLME.

Finally, it should be mentioned that DLLME has found its major application in environmental water sample analysis (Yan & Wang, 2013). Organic and inorganic analytes have been directly extracted from relatively simple water matrices (i.e., river, tap, lake, sea and wastewaters). In the most complicated cases, simple pre-treatment operations such as filtration have been included to adapt the real-world matrices to the procedure. The interest of expanding the application of DLLME to bioanalytical (Yan & Wang, 2013) and food analysis (Yan & Wang, 2013; Viñas et al., 2014) fields is continuously growing. Sample pretreatments are then essential due to the incompatibility of the direct extraction of analytes from complex matrices and the difficulty of obtaining suitable extracts for analysis. Previous digestion or extraction steps, especially in solid samples, and operations such as dilution, filtration, degassing, centrifugation, protein precipitation or enzymatic hydrolysis are commonly included prior to DLLME of complex biological and food samples.

4.4.2 Ultrasound- and Vortex-assisted Dispersive Liquid-liquid Microextraction

Regueiro et al. (2008) proposed for the first time the use of US energy to assist the dispersion in DLLME. The effect of US energy in a mixture of two immiscible liquids is based on a number of inter-related complex processes which result in the formation of a stable emulsion and increase the mass transfer speed between the two phases (Regueiro et al., 2008). Thus, UA-DLLME allows the elimination of the disperser solvent which enhances EEs. In some studies, the dispersant, in lower concentrations than in classical DLLME, is still included to ensure the effective formation of the cloudy state.

Different sonication devices are employed for US irradiation in UA-DLLME, namely US baths (Regueiro et al., 2008), US probes (Cortada et al., 2011) or sonoreactors (Cabaleiro et al., 2011). US baths are the most employed option due to their lower cost and widespread availability in analytical laboratories. However, well-known disadvantages of US baths can be mentioned such as the lack of uniformity of US energy and the power declining with time, which lead to problems of reproducibility between extractions (Szreniawa-Sztajnert et al., 2013). In addition, US baths provide an indirect US irradiation since waves have to cross the walls of sample vessel. By contrast, US probes are directly immersed into the sample and supply a more powerful, uniform and efficient US energy.

UA-DLLME has been employed to assist the dispersion of chlorinated and low-density solvents and ILs (Andruch et al., 2013b; Picó, 2013; Szreniawa-Sztajnert et al., 2013). When using organic solvents lighter than water, both the use of special extraction devices (Saleh et al., 2009) and the combination of UA-DLLME with SFOD methodology (Mohamadi & Mostafavi, 2010) have been proposed. Surfactants (Wu et al., 2010) or hydrophilic ILs (Gao et al., 2012) have been included as an emulsifier in some UA-DLLME applications with the aim of improving the dispersion of the extractant phase, reducing extraction times and avoiding the use of traditional organic disperser solvents. Organic and inorganic analytes have been determined in a wide range of samples (*e.g.*, water, cosmetics, fruit juices, soil, plant tissues or biological samples) using UA-DLLME (Andruch et al., 2013b; Picó, 2013; Szreniawa-Sztajnert et al., 2013).

US energy offers advantages when improving the speed and efficiency of the extractions in DLLME. However, the above mentioned drawbacks led to a search for alternative mixing modes, such as vortex agitation. Vortex-assisted dispersive liquid-liquid microextraction (VA-DLLME) avoids the generation of free radicals that normally appear when US energy is employed (Andruch et al., 2013b). In addition, the dispersion under vortex-mixing is unstable, therefore phase separation is easily accomplished. Finally, another important point to consider is the fact that vortex methodology is more cost-effective than US baths or US probes (Andruch et al., 2013b).

VA-DLLME was presented for the first time by Yiantzi et al. (2010). The performance of this novel methodology was illustrated with the determination of alkylphenols in water samples, using octanol as extractant phase and omitting the disperser agent. Since then, several approaches have employed vortex methodology with high-density solvents (Zhang & Lee, 2012), low-density solvents (Jia et al., 2010), ILs (Asensio-Ramos et al., 2011a), surfactant-enhanced emulsification (Yang et al., 2011) and solvent demulsification (Seebunrueng et al., 2014). A novel mixed mode of US-vortex-assisted DLLME (Cinelli et al., 2013) has recently appeared in which the mixture of sample, extractant and disperser solvent is vortexed before sonication in a US bath. Organic and inorganic analytes have been determined in samples such as water, sediments, hair, urine, fruit juices, milk, beer and wine using VA-DLLME (Andruch et al., 2013b).

4.4.3 Temperature-assisted Dispersive Liquid-liquid Microextraction

Temperature-controlled IL dispersive liquid-liquid microextraction (TC-IL-DLLME) was presented by Zhou et al. (2008), demonstrating for the first time that ILs could be used in DLLME. This microextraction methodology is based on the use of a temperature increase for the complete dispersion or solubilization of the extractant phase into the sample. Thereafter, when the mixture is cooled in an ice-water or ice bath the solution becomes turbid and phases are finally separated by centrifugation. Temperatures between 50–90 °C are normally required for IL dissolution where the upper limit is established by the boiling point of water. Extraction time is generally considered as the cooling time, being the most time consuming part of the methodology. Cooling times between 10 and 30 min are usually employed, although shorter extraction times have been reported when disperser agents are included.

Baghdadi and Shemirani (2008) proposed, almost at the same time as Zhou et al. (Zhou et al., 2008), an IL-based microextraction technique named cold-induced aggregation microextraction (CIAME). The general procedure and basic principles of CIAME are the same as in TC-IL-DLLME, although the authors named it differently. In both cases, temperature changes are employed to modify IL solubility and a cooling step is needed for the formation of the droplets of the extractant phase from a homogeneous solution.

TC-IL-DLLME and CIAME methodologies have generally been employed in water sample analysis for the determination of both organic and inorganic analytes (Trujillo-Rodríguez et al., 2013). Imidazolium- and hexafluorophosphate-based ILs have been selected as the extractant phase in vast majority of the reported works to date (Trujillo-Rodríguez et al., 2013).

4.4.4 In Situ Ionic Liquid Formation Dispersive Liquid-liquid Microextraction

Simultaneously, Baghdadi and Shemirani (2009) and Jao and Anderson (2009) proposed a novel IL-DLLME methodology based on the *in situ* formation of the extractant phase. Although the first authors named the technique as *in situ* solvent formation microextraction (ISFME) (Baghdadi & Shemirani, 2009), the terms IL-based *in situ* DLLME or *in situ* IL-DLLME have been adopted more frequently. *In situ* IL-DLLME is based on the formation of the extractant phase in the sample solution *via* a metathesis reaction between a water-miscible IL and an ion exchange reagent to form a water-immiscible IL. Briefly, in the general procedure, the water-miscible IL is dissolved into the sample containing the analytes. Then, the ion exchange salt is added, the cloudy solution is immediately formed and the extraction of analytes occurs simultaneously with the ion exchange reaction. Finally, phases are separated by centrifugation and the enriched phase can be analyzed with the selected analytical technique. Imidazolium-based ILs with chloride or tetrafluoroborate anions have been employed as hydrophilic ILs, whereas hexafluorophosphate salts (i.e., $NaPF_6$, KPF_6 and NH_4PF_6) and lithium bis[(trifluoromethyl)sulfonyl]imide ($LiNTf_2$) have acted as ion-pairing agents. Among the advantages that the *in situ* IL-DLLME methodology offers we highlight that the dispersion of the extractant IL takes place *via* a metathesis reaction, therefore, the disperser agent is not needed. Thus, the competition between the IL and the disperser solvent is avoided. Moreover, additional devices such as vortex or US are not necessary to assist the emulsification.

In situ IL-DLLME has mainly been employed for the determination of organic and inorganic analytes in water samples (Trujillo-Rodríguez et al., 2013). More complex matrices such as cereals, marine sediments, fruits, tea and blood have also been explored with the combination of different pre-treatments (Trujillo-Rodríguez et al., 2013). Extraction of DNA from water matrices containing albumin protein or metal ions has been tested using *in situ* DLLME (Li et al., 2013), proving the advantages of this methodology over other DNA extraction protocols in terms of speed, low consumption of solvents and high EE. Recently, a novel methodology combining *in situ* IL-DLLME with electrochemical detection using SPELs has been presented (Fernández et al., 2014). Considering the electrochemical properties of ILs and the low volume of the IL phase formed during *in situ* IL-DLLME, SPELs were perfectly compatible candidates for analyzing the IL drop after microextraction without any further modification. Therefore, the authors employed miniaturized systems both in sample preparation and in the detection stage, and avoided the disadvantages of chromatographic techniques (*e.g.*, special devices or interfaces in GC or shorter column life and resolution problems in LC). The feasibility of the proposed methodology was demonstrated through the determination of 2,4,6-trinitrotoluene in water samples.

4.4.5 Supramolecular-based Dispersive Liquid-liquid Microextraction

As mentioned before, exploring alternative extractant solvents has been a recurring theme in DLLME publications and supramolecular solvents (SUPRASs) have received special attention recently. SUPRASs, also named coacervates, are water-immiscible liquids made of large and three-dimensional assemblies of amphiphilic structures (Yazdi, 2011; Moradi & Yamini, 2012b). Amphiphilic compounds form different kinds of aggregations above a critical concentration (*e.g.*, reverse micelles or vesicles) (Moradi & Yamini, 2012b). SUPRASs are formed by the self-assembly of these aggregates induced by an external stimulus which promotes their separation from the bulk solution (Moradi & Yamini, 2012b). Considering this phenomenon, two microextraction modes based on SUPRASs of alkylcarboxylic acids have been proposed in the literature. In the first methodology, reverse micelles of carboxylic acids are formed in tetrahydrofuran and the subsequent mixture with the aqueous sample induces the formation of the separated SUPRAS phase. The addition of water causes partial desolvation of reverse micelles which facilitates their interaction and the formation of larger water-immiscible structures (Moradi & Yamini, 2012b). In the second methodology, coacervates of carboxylic acid vesicles are formed by the action the of tetrabutylammonium cation (TBA). In the aqueous phase, protonated and deprotonated carboxylic acids form small water-miscible vesicles. The addition of TBA causes the formation of larger water-immiscible vesicles based on hydrophobic interactions between the hydrocarbon chains, hydrogen bonding between carboxylic and carboxylate groups and the electrostatic interactions between the carboxylate and TBA groups (Moradi & Yamini, 2012b). In both described methodologies, extraction of analytes is accomplished after SUPRASs formation and phases are finally separated by centrifugation. Special extraction devices are employed due to the low density of the extractant phase, although an alternative using the SFOD methodology has recently been proposed (Moradi & Yamini, 2012a).

The term supramolecular-based DLLME (SM-DLLME) appeared for the first time in 2011 (Jafarvand & Shemirani, 2011). However, this work followed the same procedure and principles as other previous publications (Ballesteros-Gómez et al., 2007, 2008; García-Prieto et al., 2008a, 2008b; García-Fonseca et al., 2010; López-Jiménez et al., 2010). Terms such as supramolecular solvent-based microextraction (García-Fonseca et al., 2010; López-Jiménez et al., 2010; Caballo et al., 2013), reverse micelle-mediated dispersive liquid-liquid microextraction (Tayyebi et al., 2012) or vesicular coacervate phase microextraction (Moradi & Yamini, 2012a) have been adopted to describe an identical methodology, showing the lack of agreement to establish a unified terminology within the scientific community.

SM-DLLME has successfully been employed for the extraction of organic and inorganic analytes from water, food and biological samples (Yazdi, 2011). SUPRASs possess high concentrations of amphiphilies and regions of different polarities which results in a high number of binding sites with different interactions for the analyte (Yazdi, 2011). In addition, the type of interaction can be tuned by varying the hydrophobic or polar groups, and the most appropriate SUPRAS could be designed for a specific application (Yazdi, 2011). Therefore, SM-DLLME has a promising future for the effective extraction of a wide range of analytes, especially polar and ionizable compounds.

4.4.6 Air-assisted Liquid-liquid Microextraction

Air-assisted liquid-liquid microextraction (AALLME) (Farajzadeh & Mogaddam, 2012) is the most recently developed DLLME methodology, appearing in 2012. Farajzadeh and Moggadam presented a novel approach in which the dispersion of the extractant solvent is obtained by the repeated drawing of sample and extractant mixture into a glass syringe, and subsequent injection into a conical centrifuge tube. The turbidity of the solution continuously increased in every cycle of aspiration and injection and after a certain number of cycles, phases are separated by centrifugation. The effect of the syringe needle dimensions and number of extraction cycles were investigated because were believed to affect the formation of the cloudy state and EE. Results showed that dimensions of syringe needle had negligible effect but EE increased with the number of cycles, remaining constant after a certain number of cycles. This first work also included a comparison with classical DLLME, showing better analytical parameters for AALLME, attributed to the absence of the disperser solvent (Farajzadeh & Mogaddam, 2012).

AALLME has mainly been applied for the determination of organic analytes in water samples (Farajzadeh & Mogaddam, 2012; Farajzadeh & Nouri, 2013; Farajzadeh et al., 2013b), food (Farajzadeh & Khoshmaram, 2013), personal care products (Farajzadeh et al., 2013a) and juices (You et al., 2013). Only one study described the use of AALLME for chromium determination by means of a single-valve sequential injection setup (Alexovič et al., 2013). High-density chlorinated solvents have generally been employed, although low-density solvent-based AALLME has been presented in one publication using a homemade extraction vessel (Farajzadeh & Khoshmaram, 2013).

The most advantageous characteristic of AALLME is the exclusion of the disperser agent without the need for additional devices such as vortex or US, which simplify the procedure and lead to an improvement in the EFs, extraction recoveries and limits of detection (Farajzadeh & Mogaddam, 2012; Farajzadeh et al., 2013b).

4.5 Conclusions

GAC is linked to 12 principles with the aim of reducing the negative impact of chemical analysis on the environment and to enable the implementation of sustainable development principles in analytical laboratories (Gałuszka et al., 2013). Most of the 12 principles negatively affect the quality of analytical parameters (i.e., accuracy, precision, selectivity, sensitivity); therefore, an important challenge is to reach a compromise between increasing the quality of results and improving the environmental friendliness of analytical methods (Gałuszka et al., 2013). Related to this, LPME techniques have strongly helped to achieve this compromise because their numerous advantages fulfill most of the 12 principles without sacrificing the quality of analytical parameters. Therefore, LPME techniques have become widely used since their appearance in 1995. Over the last two decades, many modifications and improvements have been made, converting SDME, HF-LPME and DLLME in the three main LPME groups. For the near future, research should be focused on the use of greener reagents (i.e., those obtained from renewable sources), increasing automation and *in situ* analysis, reducing the use of energy (*e.g.*, more efficient and faster measurement systems) and avoiding toxic reagents that are still in use in order to develop complete sustainable analytical methods.

Abbreviations

$[C_4mim][Cl]$	1-butyl-3-methylimidazolium chloride
$[C_4mim][PF_6]$	1-butyl-3-methylimidazolium hexafluorophosphate
$[C_6mim][PF_6]$	1-hexyl-3-methylimidazolium hexafluorophosphate
$[C_8mim][PF_6]$	1-methyl-3-octylimidazolium hexafluorophosphate
$[C_{10}mim][Br]$	1-decyl-3-methylimidazolium bromide
AALLME	air-assisted liquid-liquid microextraction
AAs	amino acids
CE	capillary electrophoresis
CFME	continuous flow microextraction
CIAME	cold-induced aggregation microextraction
CNTs	carbon nanotubes
DI-SDME	direct immersion single-drop microextraction
DLLME	dispersive liquid-liquid microextraction
dLPME	dynamic liquid-phase microextraction
DSDME	directly suspended droplet microextraction
ECD	electron capture detector
EE	extraction efficiency
EF	enrichment factor
EME	electromembrane extraction
ETAAS	electrothermal atomic absorption spectrometry
FAAS	flame atomic absorption spectrometry
FID	flame ionization detector
GAC	green analytical chemistry
GC	gas chromatography
HF	hollow fiber
HF-LPME	hollow fiber liquid-phase microextraction
HS-LPME	headspace liquid-phase microextraction
HS-SDME	headspace single-drop microextraction
ICP-MS	inductively coupled plasma-mass spectrometry
ICP-OES	inductively coupled plasma-optical emission spectroscopy
IL	ionic liquid

IMS	ion mobility spectrometry
ISFME	*in situ* solvent formation microextraction
LC	liquid chromatography
LLE	liquid-liquid extraction
LLLME	liquid-liquid-liquid microextraction
LPME	liquid-phase microextraction
MS	mass spectrometry
NPOE	2-nitrophenyl octyl ether
PAHs	polycyclic aromatic hydrocarbons
PCR	polychloroprene rubber
PEEK	polyetheretherketone
PEME	pulsed electromembrane extraction
PPP-HF-LPME	push/pull perfusion hollow-fiber liquid-phase microextraction
PTFE	Polytetrafluoroethylene
QDs	quantum dots
RSD	relative standard deviation
SBME	solvent-bar microextraction
SD-DLLME	solvent-based demulsification dispersive liquid-liquid microextraction
SDME	single-drop microextraction
SFOD	solidification of floating organic drop
SFODME	solidification of floating organic drop microextraction
SIA	sequential injection analysis
SLM	supported liquid membrane
SM-DLLME	supramolecular-based dispersive liquid-liquid microextraction
SME	solvent microextraction
SPE	solid-phase extraction
SPELs	screen-printed electrodes
SPME	solid-phase microextraction
SUPRASs	supramolecular solvents
TBA	tetrabutylammonium
TC-IL-DLLME	temperature-controlled IL dispersive liquid-liquid microextraction
UA-DLLME	ultrasound assisted dispersive liquid-liquid microextraction
US	ultrasound
USAEME	ultrasound-assisted emulsification microextraction
VA-DLLME	vortex-assisted dispersive liquid-liquid microextraction
VALLME	vortex-assisted liquid-liquid microextraction
W-coil ETAAS	Tungsten coil electrothermal atomic absorption spectrometer.

Acknowledgements

The authors would like to thank the Spanish Ministry of Science and Innovation (project n. CTQ2011–23968), Generalitat Valenciana (Spain) (projects n. ACOMP/2013/072 and PROMETEO/2013/038) and University of Alicante (Spain) (project n. GRE12–45) for the financial support. E. Fernández also thanks Generalitat Valenciana for her fellowship.

References

Abulhassani, J., Manzoori, J. L., & Amjadi, M. (2010). Hollow fiber based-liquid phase microextraction using ionic liquid solvent for preconcentration of lead and nickel from environmental and biological samples prior to determination by electrothermal atomic absorption spectrometry. *Journal of Hazardous Materials*, *176*, 481–486.

Agrawal, K., & Wu, H.-F. (2007). Drop-to-drop solvent microextraction coupled with gas chromatography/mass spectrometry for rapid determination of trimeprazine in urine and blood of rats: Application to pharmacokinetic studies. *Rapid Communications in Mass Spectrometry*, *21*, 3352–3356.

Aguilera-Herrador, E., Lucena, R., Cardenas, S., & Valcárcel, M. (2008). Direct coupling of ionic liquid based single-drop microextraction and GC/MS. *Analytical Chemistry*, *80*, 793–800.

Aguilera-Herrador, E., Lucena, R., Cárdenas, S., & Valcárcel, M. (2009). Ionic liquid-based single drop microextraction and room-temperature gas chromatography for on-site ion mobility spectrometric analysis. *Journal of Chromatography A*, *1216*, 5580–5587.

Aguilera-Herrador, E., Lucena, R., Cárdenas, S., & Valcárcel, M. (2010). The roles of ionic liquids in sorptive microextraction techniques. *TrAC Trends in Analytical Chemistry*, *29*, 602–616.

Ahmar, H., Fakhari, A. R., Tabani, H., & Shahsavani, A. (2013). Optimization of electromembrane extraction combined with differential pulse voltammetry using modified screen-printed electrode for the determination of sufentanil. *Electrochimica Acta*, *96*, 117–123.

Ahmar, H., Tabani, H., Koruni, M. H., Davarani, S. S. H., & Fakhari, A. R. (2014). A new platform for sensing urinary morphine based on carrier assisted electromembrane extraction followed by adsorptive stripping voltammetric detection on screen-printed electrode. *Biosensors & Bioelectronics*, *54*, 189–194.

Alexovič, M., Andruch, V., Baloghb, I. S., & Šandrejováa, J. (2013). A single-valve sequential injection manifold (SV-SIA) for automation of air-assisted liquid-phase microextraction: Stopped flow spectrophotometric determination of chromium(VI). *Analytical Methods*, *5*, 2497–2502.

ALOthman, Z. A., Dawod, M., Kim, J., & Chung, D. S. (2012). Single-drop microextraction as a powerful pre-treatment tool for capillary electrophoresis: A review. *Analytica Chimica Acta*, *739*, 14–24.

Andruch, V., Acebal, C. C., Škrlíková, J., Sklenářová, H., Solich, P., Balogh, I. S., Billes, F., & Kocúrová, L. (2012a). Automated on-line dispersive liquid–liquid microextraction based on a sequential injection system. *Microchemical Journal*, *100*, 77–82.

Andruch, V., Balogh, I. S., Kocúrová, L., & Šandrejová, J. (2013a). The present state of coupling of dispersive liquid–liquid microextraction with atomic absorption spectrometry. *Journal of Analytical Atomic Spectrometry*, *28*, 19–32.

Andruch, V., Burdel, M., Kocúrová, L., Šandrejová, J., & Balogh, I. S. (2013b). Application of ultrasonic irradiation and vortex agitation in solvent microextraction. *TrAC Trends in Analytical Chemistry*, *49*, 1–19.

Andruch, V., Kocúrová, L., Balogh, I. S., & Škrlíková, J. (2012b). Recent advances in coupling single-drop and dispersive liquid–liquid microextraction with UV–vis spectrophotometry and related detection techniques. *Microchemical Journal, 102*, 1–10.

Anthemidis, A. N., & Adam, I. S. I. (2009). Development of on-line single-drop micro-extraction sequential injection system for electrothermal atomic absorption spectrometric determination of trace metals. *Analytica Chimica Acta, 632*, 216–220.

Anthemidis, A. N., & Ioannou, K.-I. G. (2009). On-line sequential injection dispersive liquid-liquid microextraction system for flame atomic absorption spectrometric determination of copper and lead in water samples. *Talanta, 79*, 86–91.

Anthemidis, A. N., & Ioannou, K.-I. G. (2010). Development of a sequential injection dispersive liquid-liquid microextraction system for electrothermal atomic absorption spectrometry by using a hydrophobic sorbent material: Determination of lead and cadmium in natural waters. *Analytica Chimica Acta, 668*, 35–40.

Anthemidis, A. N., & Ioannou, K.-I. G. (2011). Sequential injection dispersive liquid-liquid microextraction based on fatty alcohols and poly(etheretherketone)-turnings for metal determination by flame atomic absorption spectrometry. *Talanta, 84*, 1215–1220.

Arce, L., Nozal, L., Simonet, B. M., Valcárcel, M., & Ríos, Á. (2009). Liquid-phase microextraction techniques for simplifying sample treatment in capillary electrophoresis. *TrAC Trends in Analytical Chemistry, 28*, 842–853.

Arthur, C. L., & Pawliszyn, J. (1990). Solid phase microextraction with thermal desorption using fused silica optical fibers. *Analytical Chemistry, 62*, 2145–2148.

Asensio-Ramos, M., Hernández-Borges, J., Borges-Miquel, T. M., & Rodríguez-Delgado, M. Á. (2011a). Ionic liquid-dispersive liquid-liquid microextraction for the simultaneous determination of pesticides and metabolites in soils using high-performance liquid chromatography and fluorescence detection. *Journal of Chromatography A, 1218*, 4808–4816.

Asensio-Ramos, M., Ravelo-Pérez, L. M., González-Curbelo, M. Á., & Hernández-Borges, J. (2011b). Liquid phase microextraction applications in food analysis. *Journal of Chromatography A, 1218*, 7415–7437.

Baghdadi, M., & Shemirani, F. (2008). Cold-induced aggregation microextraction: A novel sample preparation technique based on ionic liquids. *Analytica Chimica Acta, 613*, 56–63.

Baghdadi, M., & Shemirani, F. (2009). *In situ* solvent formation microextraction based on ionic liquids: A novel sample preparation technique for determination of inorganic species in saline solutions. *Analytica Chimica Acta, 634*, 186–191.

Balchen, M., Reubsaet, L., & Pedersen-Bjergaard, S. (2008). Electromembrane extraction of peptides. *Journal of Chromatography A, 1194*, 143–149.

Baliza, P. X., Teixeira, L. S. G., & Lemos, V. A. (2009). A procedure for determination of cobalt in water samples after dispersive liquid–liquid microextraction. *Microchemical Journal, 93*, 220–224.

Ballesteros-Gómez, A., Rubio, S., & Pérez-Bendito, D. (2008). Determination of priority carcinogenic polycyclic aromatic hydrocarbons in wastewater and surface water by coacervative extraction and liquid chromatography-fluorimetry. *Journal of Chromatography A, 1203*, 168–176.

Ballesteros-Gómez, A., Ruiz, F.-J., Rubio, S., & Pérez-Bendito, D. (2007). Determination of bisphenols A and F and their diglycidyl ethers in wastewater and river water by coacervative extraction and liquid chromatography-fluorimetry. *Analytica Chimica Acta, 603*, 51–59.

Ballesteros-Gómez, A., Sicilia, M. D., & Rubio, S. (2010). Supramolecular solvents in the extraction of organic compounds. A review. *Analytica Chimica Acta, 677*, 108–130.

Basheer, C., Alnedhary, A. A., Madhava Rao, B. S., Balasubramanian, R., & Lee, H. K. (2008a). Ionic liquid supported three-phase liquid-liquid-liquid microextraction as a sample preparation technique for aliphatic and aromatic hydrocarbons prior to gas chromatography-mass spectrometry. *Journal of Chromatography A, 1210*, 19–24.

Basheer, C., Lee, J., Pedersen-Bjergaard, S., Rasmussen, K. E., & Lee, H. K. (2010). Simultaneous extraction of acidic and basic drugs at neutral sample pH: A novel electro-mediated microextraction approach. *Journal of Chromatography A, 1217*, 6661–6667.

Basheer, C., Tan, S. H., & Lee, H. K. (2008b). Extraction of lead ions by electromembrane isolation. *Journal of Chromatography A, 1213*, 14–18.

Batlle, R., López, P., Nerín, C., & Crescenzi, C. (2008). Active single-drop microextraction for the determination of gaseous diisocyanates. *Journal of Chromatography A, 1185*, 155–160.

Bedendo, G. C., Jardim, I. C. S. F., & Carasek, E. (2012). Multiresidue determination of pesticides in industrial and fresh orange juice by hollow fiber microporous membrane liquid-liquid extraction and detection by liquid chromatography-electrospray-tandem mass spectrometry. *Talanta, 88*, 573–580.

Bidari, A., Ganjali, M. R., Assadi, Y., & Norouzi, P. (2011). A novel methodology based on solvents less dense than water through dispersive liquid–liquid microextraction: Application in quantitation of l-ascorbate in fruit juices and soft drinks by fiber optic-linear array detection spectrophotometry. *Analytical Methods, 3*, 724–730.

Cabaleiro, N., Pena-Pereira, F., de la Calle, I., Bendicho, C., & Lavilla, I. (2011). Determination of triclosan by cuvetteless UV–vis micro-spectrophotometry following simultaneous ultrasound assisted emulsification–microextraction with derivatization: Use of a micellar-ionic liquid as extractant. *Microchemical Journal, 99*, 246–251.

Caballo, C., Sicilia, M. D., & Rubio, S. (2013). Stereoselective quantitation of mecoprop and dichlorprop in natural waters by supramolecular solvent-based microextraction, chiral liquid chromatography and tandem mass spectrometry. *Analytica Chimica Acta, 761*, 102–108.

Cardoso, A. A., & Dasgupta, P. K. (1995). Analytical chemistry in a liquid film/droplet. *Analytical Chemistry, 67*, 2562–2566.

Carletto, J. S., Luciano, R. M., Bedendo, G. C., & Carasek, E. (2009). Simple hollow fiber renewal liquid membrane extraction method for pre-concentration of Cd(II) in environmental samples and detection by flame atomic absorption spectrometry. *Analytica Chimica Acta, 638*, 45–50.

Carmichael, A. J., & Seddon, K. R. (2000). Polarity study of some 1-alkyl-3-methylimidazolium ambient-temperature ionic liquids with the solvatochromic dye, Nile Red. *Journal of Physical Organic Chemistry, 13*, 591–595.

Carrillo-Carrión, C., Simonet, B. M., & Valcárcel, M. (2012). (CdSe/ZnS QDs)-ionic liquid-based headspace single drop microextraction for the fluorimetric determination of trimethylamine in fish. *The Analyst, 137*, 1152–1159.

Chao, Y.-Y., Jian, Z.-X., Tu, Y.-M., Wang, H.-W., & Huang, Y.-L. (2013). An on-line push/pull perfusion-based hollow-fiber liquid-phase microextraction system for high-performance liquid chromatographic determination of alkylphenols in water samples. *The Analyst, 138*, 3271–3279.

Chen, H., Chen, R., & Li, S. (2010). Low-density extraction solvent-based solvent terminated dispersive liquid-liquid microextraction combined with gas chromatography-tandem mass spectrometry for the determination of carbamate pesticides in water samples. *Journal of Chromatography A, 1217*, 1244–1248.

Chisvert, A., Román, I. P., Vidal, L., & Canals, A. (2009). Simple and commercial readily-available approach for the direct use of ionic liquid-based single-drop microextraction prior to gas chromatography determination of chlorobenzenes in real water samples as model analytical application. *Journal of Chromatography A, 1216*, 1290–1295.

Choi, K., Kim, S. J., Jin, Y. G., Jang, Y. O., Kim, J.-S., & Chung, D. S. (2009). Single drop microextraction using commercial capillary electrophoresis instruments. *Analytical Chemistry, 81*, 225–230.

Cinelli, G., Avino, P., Notardonato, I., Centola, A., & Russo, M. V. (2013). Rapid analysis of six phthalate esters in wine by ultrasound-vortex-assisted dispersive liquid-liquid micro-extraction coupled with gas chromatography-flame ionization detector or gas chromatography-ion trap mass spectrometry. *Analytica Chimica Acta, 769*, 72–78.

Cortada, C., Vidal, L., & Canals, A. (2011). Determination of nitroaromatic explosives in water samples by direct ultrasound-assisted dispersive liquid-liquid microextraction followed by gas chromatography-mass spectrometry. *Talanta, 85*, 2546–2552.

Cruz-Vera, M., Lucena, R., Cárdenas, S., & Valcárcel, M. (2008). Ionic liquid-based dynamic liquid-phase microextraction: Application to the determination of anti-inflammatory drugs in urine samples. *Journal of Chromatography A, 1202*, 1–7.

Cruz-Vera, M., Lucena, R., Cárdenas, S., & Valcárcel, M. (2009a). Determination of phenothiazine derivatives in human urine by using ionic liquid-based dynamic liquid-phase microextraction coupled with liquid chromatography. *Journal of Chromatography B: Analytical Technologies in the Biomedical and Life Sciences, 877*, 37–42.

Cruz-Vera, M., Lucena, R., Cárdenas, S., & Valcárcel, M. (2009b). One-step in-syringe ionic liquid-based dispersive liquid-liquid microextraction. *Journal of Chromatography A, 1216*, 6459–6465.

Dadfarnia, S., & Haji Shabani, A. M. (2010). Recent development in liquid phase microextraction for determination of trace level concentration of metals-A review. *Analytica Chimica Acta, 658*, 107–119.

Davarani, S. S. H., Morteza-Najarian, A., Nojavan, S., Pourahadi, A., & Abbassi, M. B. (2013). Two-phase electromembrane extraction followed by gas chromatography-mass spectrometry analysis. *Journal of Separation Science, 36*, 736–743.

Davarani, S. S. H., Morteza-Najarian, A., Nojavan, S., & Tabatabaei, M.-A. (2012). Electromembrane extraction combined with gas chromatography for quantification of tricyclic antidepressants in human body fluids. *Analytica Chimica Acta, 725*, 51–56.

Dehghani, M., Ashraf, N., Chamsaz, M., & Shemirani, F. (2012). An overview of liquid phase microextraction approaches combined with UV-Vis spectrophotometry. *Talanta, 99*, 1–12.

Deng, C., Mao, Y., Hu, F., & Zhang, X. (2007). Development of gas chromatography-mass spectrometry following microwave distillation and simultaneous headspace single-drop microextraction for fast determination of volatile fraction in Chinese herb. *Journal of Chromatography A, 1152*, 193–198.

Deng, Q., Chen, M., Kong, L., Zhao, X., Guo, J., & Wen, X. (2013). Novel coupling of surfactant assisted emulsification dispersive liquid-liquid microextraction with spectrophotometric determination for ultra trace nickel. *Spectrochimica Acta Part A: Molecular and Biomolecular Spectroscopy, 104*, 64–69.

Eibak, L. E. E., Parmer, M. P., Rasmussen, K. E., Pedersen-Bjergaard, S., & Gjelstad, A. (2014). Parallel electromembrane extraction in a multiwell plate. *Analytical and Bioanalytical Chemistry, 406*, 431–440.

Eiceman, Z. K. G. A. (2005). *Ion Mobility Spectrometry*. CRC Press Inc., Boca Raton, USA.

El-Shahawi, M. S., & Al-Saidi, H. M. (2013). Dispersive liquid-liquid microextraction for chemical speciation and determination of ultra-trace concentrations of metal ions. *TrAC Trends in Analytical Chemistry, 44*, 12–24.

Es'haghi, Z., Golsefidi, M. A., Saify, A., Tanha, A. A., Rezaeifar, Z., & Alian-Nezhadi, Z. (2010). Carbon nanotube reinforced hollow fiber solid/liquid phase microextraction: a novel extraction technique for the measurement of caffeic acid in Echinacea purpurea herbal extracts combined with high-performance liquid chromatography. *Journal of Chromatography A, 1217*, 2768–2775.

Escudero, L. B., Castro, A., Martinis, E. M., & Wuilloud, R. G. (2013). Bioanalytical separation and preconcentration using ionic liquids. *Analytical and Bioanalytical Chemistry, 405*, 7597–7613.

Esrafili, A., Yamini, Y., Ghambarian, M., & Ebrahimpour, B. (2012). Automated preconcentration and analysis of organic compounds by on-line hollow fiber liquid-phase microextraction-high performance liquid chromatography. *Journal of Chromatography A, 1262*, 27–33.

Fakhari, A. R., Koruni, M. H., Ahmar, H., Shahsavani, A., & Movahed, S. K. (2014). Electrochemical determination of dextromethorphan on reduced graphene oxide modified screen-printed electrode after electromembrane extraction. *Electroanalysis, 26*, 1–9.

Fakhari, A. R., Tabani, H., Behdad, H., Nojavan, S., & Taghizadeh, M. (2013). Electrically-enhanced microextraction combined with maltodextrin-modified capillary electrophoresis for quantification of tolterodine enantiomers in biological samples. *Microchemical Journal, 106*, 186–193.

Fan, Z., & Liu, X. (2008). Determination of methylmercury and phenylmercury in water samples by liquid-liquid-liquid microextraction coupled with capillary electrophoresis. *Journal of Chromatography A, 1180*, 187–192.

Fang, H., Zeng, Z., Liu, L., & Pang, D. (2006). On-line back-extraction field-amplified sample injection method for directly analyzing cocaine and thebaine in the extractants by solvent microextraction. *Analytical Chemistry, 78*, 1257–1263.

Farajzadeh, M. A., Bahram, M., & Jönsson, J. A. (2007). Dispersive liquid-liquid microextraction followed by high-performance liquid chromatography-diode array detection as an efficient and sensitive technique for determination of antioxidants. *Analytica Chimica Acta, 591*, 69–79.

Farajzadeh, M. A., Djozan, D., & Bakhtiyari, R. F. (2010). Use of a capillary tube for collecting an extraction solvent lighter than water after dispersive liquid-liquid microextraction and its application in the determination of parabens in different samples by gas chromatography-flame ionization detection. *Talanta, 81*, 1360–1367.

Farajzadeh, M. A., Djozan, D., & Khorram, P. (2012). Development of a new dispersive liquid-liquid microextraction method in a narrow-bore tube for preconcentration of triazole pesticides from aqueous samples. *Analytica Chimica Acta, 713*, 70–78.

Farajzadeh, M. A., & Khoshmaram, L. (2013). Air-assisted liquid-liquid microextraction-gas chromatography-flame ionisation detection: A fast and simple method for the assessment of triazole pesticides residues in surface water, cucumber, tomato and grape juices samples. *Food Chemistry, 141*, 1881–1887.

Farajzadeh, M. A., Khosrowshahi, E. M., & Khorram, P. (2013a). Simultaneous derivatization and air-assisted liquid-liquid microextraction of some parabens in personal care products and their determination by GC with flame ionization detection. *Journal of Separation Science, 36*, 3571–3578.

Farajzadeh, M. A., & Mogaddam, M. R. A. (2012). Air-assisted liquid-liquid microextraction method as a novel microextraction technique; application in extraction and preconcentration of phthalate esters in aqueous sample followed by gas chromatography-flame ionization detection. *Analytica Chimica Acta, 728*, 31–38.

Farajzadeh, M. A., Mogaddam, M. R. A., & Aghdam, A. A. (2013b). Comparison of air-agitated liquid-liquid microextraction technique and conventional dispersive liquid-liquid micro-extraction for determination of triazole pesticides in aqueous samples by gas chromatography with flame ionization detection. *Journal of Chromatography A, 1300*, 70–78.

Farajzadeh, M. A., & Nouri, N. (2013). Simultaneous derivatization and air-assisted liquid-liquid microextraction of some aliphatic amines in different aqueous samples followed by gas chromatography-flame ionization detection. *Analytica Chimica Acta, 775*, 50–57.

Farajzadeh, M. A., Seyedi, S. E., Shalamzari, M. S., & Bamorowat, M. (2009). Dispersive liquid-liquid microextraction using extraction solvent lighter than water. *Journal of Separation Science, 32*, 3191–3200.

Fernández, E., Vidal, L., Iniesta, J., Metters, J. P., Banks, C. E., & Canals, A. (2014). Screen-printed electrode-based electrochemical detector coupled with *in-situ* ionic-liquid-assisted dispersive liquid-liquid microextraction for determination of 2,4,6-trinitrotoluene. *Analytical and Bioanalytical Chemistry, 406*, 2197–2204.

Figueroa, R., García, M., Lavilla, I., & Bendicho, C. (2005). Photoassisted vapor generation in the presence of organic acids for ultrasensitive determination of Se by electrothermal-atomic absorption spectrometry following headspace single-drop microextraction. *Spectrochimica Acta Part B: Atomic Spectroscopy*, *60*, 1556–1563.

Fortunato, R., Afonso, C., Benavente, J., Rodriguez-Castellon, E., & Crespo, J. (2005). Stability of supported ionic liquid membranes as studied by X-ray photoelectron spectroscopy. *Journal of Membrane Science*, *256*, 216–223.

Fortunato, R., Afonso, C., Reis, M. A. M., & Crespo, J. (2004). Supported liquid membranes using ionic liquids: study of stability and transport mechanisms. *Journal of Membrane Science*, *242*, 197–209.

Fragueiro, S., Lavilla, I., & Bendicho, C. (2004). Headspace sequestration of arsine onto a Pd(II)-containing aqueous drop as a preconcentration method for electrothermal atomic absorption spectrometry. *Spectrochimica Acta Part B: Atomic Spectroscopy*, *59*, 851–855.

Fragueiro, S., Lavilla, I., & Bendicho, C. (2006). Hydride generation-headspace single-drop microextraction-electrothermal atomic absorption spectrometry method for determination of selenium in waters after photoassisted prereduction. *Talanta*, *68*, 1096–1101.

Gałuszka, A., Migaszewski, Z., & Namieśnik, J. (2013). The 12 principles of green analytical chemistry and the SIGNIFICANCE mnemonic of green analytical practices. *TrAC Trends in Analytical Chemistry*, *50*, 78–84.

Ganjali, M. R., Sobhi, H. R., Farahani, H., Norouzi, P., Dinarvand, R., & Kashtiaray, A. (2010). Solid drop based liquid-phase microextraction. *Journal of Chromatography A*, *1217*, 2337–2341.

Gao, S., Yang, X., Yu, W., Liu, Z., & Zhang, H. (2012). Ultrasound-assisted ionic liquid/ionic liquid-dispersive liquid-liquid microextraction for the determination of sulfonamides in infant formula milk powder using high-performance liquid chromatography. *Talanta*, *99*, 875–882.

Gao, W., Chen, G., Chen, T., Zhang, X., Chen, Y., & Hu, Z. (2011). Directly suspended droplet microextraction combined with single drop back-extraction as a new approach for sample preparation compatible with capillary electrophoresis. *Talanta*, *83*, 1673–1679.

García-Fonseca, S., Ballesteros-Gómez, A., Rubio, S., & Pérez-Bendito, D. (2010). Supramolecular solvent-based microextraction of ochratoxin A in raw wheat prior to liquid chromatography-fluorescence determination. *Journal of Chromatography A*, *1217*, 2376–2382.

García-Prieto, A., Lunar, L., Rubio, S., & Pérez-Bendito, D. (2008a). Decanoic acid reverse micelle-based coacervates for the microextraction of bisphenol A from canned vegetables and fruits. *Analytica Chimica Acta*, *617*, 51–58.

García-Prieto, A., Lunar, M. L., Rubio, S., & Pérez-Bendito, D. (2008b). Determination of urinary bisphenol A by coacervative microextraction and liquid chromatography-fluorescence detection. *Analytica Chimica Acta*, *630*(1), 19–27.

Ghambarian, M., Yamini, Y., & Esrafili, A. (2012). Developments in hollow fiber based liquid-phase microextraction: Principles and applications. *Microchimica Acta*, *177*, 271–294.

Ghambarian, M., Yamini, Y., & Esrafili, A. (2013). Liquid-phase microextraction based on solidified floating drops of organic solvents. *Microchimica Acta*, *180*, 519–535.

Ghambarian, M., Yamini, Y., Esrafili, A., Yazdanfar, N., & Moradi, M. (2010). A new concept of hollow fiber liquid-liquid-liquid microextraction compatible with gas chromatography based on two immiscible organic solvents. *Journal of Chromatography A*, *1217*, 5652–5658.

Gil, S., Fragueiro, S., Lavilla, I., & Bendicho, C. (2005). Determination of methylmercury by electrothermal atomic absorption spectrometry using headspace single-drop microextraction with *in situ* hydride generation. *Spectrochimica Acta Part B: Atomic Spectroscopy*, *60*, 145–150.

Gjelstad, A., Jensen, H., Rasmussen, K. E., & Pedersen-Bjergaard, S. (2012). Kinetic aspects of hollow fiber liquid-phase microextraction and electromembrane extraction. *Analytica Chimica Acta*, *742*, 10–16.

Gjelstad, A., & Pedersen-Bjergaard, S. (2013). Recent developments in electromembrane extraction. *Analytical Methods*, *5*, 4549–4557.

Gjelstad, A., Rasmussen, K. E., & Pedersen-Bjergaard, S. (2006). Electrokinetic migration across artificial liquid membranes tuning the membrane chemistry to different types of drug substances. *Journal of Chromatography A*, *1124*, 29–34.

Gjelstad, A., Rasmussen, K. E., & Pedersen-Bjergaard, S. (2007). Simulation of flux during electro-membrane extraction based on the Nernst-Planck equation. *Journal of Chromatography A*, *1174*, 104–111.

Guo, L., & Lee, H. K. (2011). Low-density solvent-based solvent demulsification dispersive liquid-liquid microextraction for the fast determination of trace levels of sixteen priority polycyclic aromatic hydrocarbons in environmental water samples. *Journal of Chromatography A*, *1218*, 5040–5046.

Han, D., & Row, K. H. (2012). Trends in liquid-phase microextraction, and its application to environmental and biological samples. *Microchimica Acta*, *176*, 1–22.

Han, X., & Armstrong, D. W. (2007). Ionic liquids in separations. *Accounts of Chemical Research*, *40*, 1079–1086.

Hashemi, P., Beyranvand, S., Mansur, R. S., & Ghiasvand, A. R. (2009). Development of a simple device for dispersive liquid-liquid microextraction with lighter than water organic solvents: isolation and enrichment of glycyrrhizic acid from licorice. *Analytica Chimica Acta*, *655*, 60–65.

Hasheminasab, K. S., & Fakhari, A. R. (2013). Development and application of carbon nanotubes assisted electromembrane extraction (CNTs/EME) for the determination of buprenorphine as a model of basic drugs from urine samples. *Analytica Chimica Acta*, *767*, 75–80.

Hasheminasab, K. S., Fakhari, A. R., & Koruni, M. H. (2014). Development of carbon-nanotube-assisted electromembrane extraction in the two-phase mode combined with GC for the determination of basic drugs. *Journal of Separation Science*, *37*, 85–91.

Hasheminasab, K. S., Fakhari, A. R., Shahsavani, A., & Ahmar, H. (2013). A new method for the enhancement of electromembrane extraction efficiency using carbon nanotube reinforced hollow fiber for the determination of acidic drugs in spiked plasma, urine, breast milk and wastewater samples. *Journal of Chromatography A*, *1285*, 1–6.

He, Y., & Kang, Y.-J. (2006). Single drop liquid-liquid-liquid microextraction of methamphetamine and amphetamine in urine. *Journal of Chromatography A*, *1133*, 35–40.

He, Y., & Lee, H. K. (1997). Liquid-phase microextraction in a single drop of organic solvent by using a conventional microsyringe. *Analytical Chemistry*, *69*, 4634–4640.

Ho, T. S., Halvorsen, T. G., Pedersen-Bjergaard, S., & Rasmussen, K. E. (2003). Liquid-phase microextraction of hydrophilic drugs by carrier-mediated transport. *Journal of Chromatography A*, *998*, 61–72.

Hou, L., & Lee, H. K. (2002). Application of static and dynamic liquid-phase microextraction in the determination of polycyclic aromatic hydrocarbons. *Journal of Chromatography A*, *976*, 377–385.

Hou, L., & Lee, H. K. (2003). Dynamic three-phase microextraction as a sample preparation technique prior to capillary electrophoresis. *Analytical Chemistry*, *75*, 2784–2789.

Hu, B., He, M., Chen, B., & Xia, L. (2013). Liquid phase microextraction for the analysis of trace elements and their speciation. *Spectrochimica Acta Part B: Atomic Spectroscopy*, *86*, 14–30.

Hu, X.-Z., Wu, J.-H., & Feng, Y.-Q. (2010). Molecular complex-based dispersive liquid-liquid microextraction: analysis of polar compounds in aqueous solution. *Journal of Chromatography A*, *1217*, 7010–7016.

Huang, K.-J., Wang, H., Ma, M., Sha, M.-L., & Zhang, H.-S. (2006). Ultrasound-assisted liquid-phase microextraction and high-performance liquid chromatographic determination of nitric oxide produced in PC12 cells using 1,3,5,7-tetramethyl-2,6-dicarbethoxy-8-(3',4'-diaminophenyl)-difluoroboradiaza-s-indacene. *Journal of Chromatography A*, *1103*, 193–201.

Jafarvand, S., & Shemirani, F. (2011). Supramolecular-based dispersive liquid-liquid microextraction: A novel sample preparation technique utilizes coacervates and reverse micelles. *Journal of Separation Science*, *34*, 455–461.
Jain, A., & Verma, K. K. (2011). Recent advances in applications of single-drop microextraction: A review. *Analytica Chimica Acta*, *706*, 37–65.
Jeannot, M. A., & Cantwell, F. F. (1996). Solvent microextraction into a single drop. *Analytical Chemistry*, *68*(13), 2236–2240.
Jeannot, M. A., & Cantwell, F. F. (1997). Mass transfer characteristics of solvent extraction into a single drop at the tip of a syringe needle. *Analytical Chemistry*, *69*, 235–239.
Jeannot, M. A., Przyjazny, A., & Kokosa, J. M. (2010). Single drop microextraction-development, applications and future trends. *Journal of Chromatography A*, *1217*, 2326–2336.
Jia, C., Zhu, X., Wang, J., Zhao, E., He, M., Chen, L., & Yu, P. (2010). Extraction of pesticides in water samples using vortex-assisted liquid-liquid microextraction. *Journal of Chromatography A*, *1217*, 5868–5871.
Jiang, X., Basheer, C., Zhang, J., & Lee, H. K. (2005). Dynamic hollow fiber-supported headspace liquid-phase microextraction. *Journal of Chromatography A*, *1087*, 289–294.
Jiang, X., & Lee, H. K. (2004). Solvent bar microextraction. *Analytical Chemistry*, *76*, 5591–5596.
Joshi, M. D., & Anderson, J. L. (2012). Recent advances of ionic liquids in separation science and mass spectrometry. *RSC Advances*, *2*, 5470–5484.
Kanyanee, T., Borst, W. L., Jakmunee, J., Grudpan, K., Li, J., & Dasgupta, P. K. (2006). Soap bubbles in analytical chemistry. Conductometric determination of sub-parts per million levels of sulfur dioxide with a soap bubble. *Analytical Chemistry*, *78*, 2786–2793.
Khalili Zanjani, M. R., Yamini, Y., Shariati, S., & Jönsson, J. A. (2007). A new liquid-phase microextraction method based on solidification of floating organic drop. *Analytica Chimica Acta*, *585*, 286–293.
Kocúrová, L., Balogh, I. S., & Andruch, V. (2013). Solvent microextraction: A review of recent efforts at automation. *Microchemical Journal*, *110*, 599–607.
Kocúrová, L., Balogh, I. S., Šandrejová, J., & Andruch, V. (2012). Recent advances in dispersive liquid–liquid microextraction using organic solvents lighter than water. A review. *Microchemical Journal*, *102*, 11–17.
Koel, M. (2005). Ionic liquids in chemical analysis. *Critical Reviews in Analytical Chemistry*, *35*, 177–192.
Kokosa, J. M. (2007). Automation of liquid pase microextraction, United States Patent US 7,178,414 B1.
Kokosa, J. M. (2013). Advances in solvent-microextraction techniques. *TrAC Trends in Analytical Chemistry*, *43*, 2–13.
Kokosa, J. M., Przyjazny, A., & Jeannot, M. A. (2009). *Solvent Microextraction. Theory and Practice.* Hoboken, New Jersey: John Wiley & Sons, Inc.
Kubáň, P., Strieglerová, L., Gebauer, P., & Boček, P. (2011). Electromembrane extraction of heavy metal cations followed by capillary electrophoresis with capacitively coupled contactless conductivity detection. *Electrophoresis*, *32*, 1025–1032.
Lee, J., Lee, H. K., Rasmussen, K. E., & Pedersen-Bjergaard, S. (2008). Environmental and bioanalytical applications of hollow fiber membrane liquid-phase microextraction: A review. *Analytica Chimica Acta*, *624*, 253–268.
Leong, M.-I., & Huang, S.-D. (2009). Dispersive liquid-liquid microextraction method based on solidification of floating organic drop for extraction of organochlorine pesticides in water samples. *Journal of Chromatography A*, *1216*, 7645–7650.
Li, T., Joshi, M. D., Ronning, D. R., & Anderson, J. L. (2013). Ionic liquids as solvents for *in situ* dispersive liquid-liquid microextraction of DNA. *Journal of Chromatography A*, *1272*, 8–14.

Liang, T.-T., Lv, Z.-H., Jiang, T.-F., & Wang, Y.-H. (2013). High-density extraction solvent-based solvent de-emulsification dispersive liquid-liquid microextraction combined with MEKC for detection of chlorophenols in water samples. *Electrophoresis, 34*, 345–352.

Lin, H., Wang, J., Zeng, L., Li, G., Sha, Y., Wu, D., & Liu, B. (2013). Development of solvent micro-extraction combined with derivatization. *Journal of Chromatography A, 1296*, 235–242.

Liu, H., & Dasgupta, P. K. (1996). Analytical chemistry in a drop. Solvent extraction in a microdrop. *Analytical Chemistry, 68*, 1817–1821.

Liu, J., Jiang, G., Chi, Y., Cai, Y., Zhou, Q., & Hu, J.-T. (2003). Use of ionic liquids for liquid-phase microextraction of polycyclic aromatic hydrocarbons. *Analytical Chemistry, 75*, 5870–5876.

Liu, J., Jönsson, J. Å., & Jiang, G. (2005). Application of ionic liquids in analytical chemistry. *TrAC Trends in Analytical Chemistry, 24*, 20–27.

Liu, R., Liu, J., Yin, Y., Hu, X., & Jiang, G. (2009). Ionic liquids in sample preparation. *Analytical and Bioanalytical Chemistry, 393*, 871–883.

Liu, S., & Dasgupta, P. K. (1995). Liquid droplet. A renewable gas sampling interface. *Analytical Chemistry, 67*, 2042–2049.

Liu, W., & Lee, H. K. (2000). Continous flow microextraction exceeding 1000-fold concentration of dilute analytes. *Analytical Chemistry, 72*, 4462–4467.

Liu, Y., Zhao, E., Zhu, W., Gao, H., & Zhou, Z. (2009). Determination of four heterocyclic insecticides by ionic liquid dispersive liquid-liquid microextraction in water samples. *Journal of Chromatography A, 1216*, 885–891.

López-Jiménez, F. J., Rubio, S., & Pérez-Bendito, D. (2008). Single-drop coacervative microextraction of organic compounds prior to liquid chromatography. Theoretical and practical considerations. *Journal of Chromatography A, 1195*, 25–33.

López-Jiménez, F. J., Rubio, S., & Pérez-Bendito, D. (2010). Supramolecular solvent-based microextraction of Sudan dyes in chilli-containing foodstuffs prior to their liquid chromatography-photodiode array determination. *Food Chemistry, 121*, 763–769.

Ma, M., & Cantwell, F. F. (1998). Solvent microextraction with simultaneous back-extraction for sample cleanup and preconcentration: Quantitative extraction. *Analytical Chemistry, 70*, 3912–3919.

Ma, M., & Cantwell, F. F. (1999). Solvent microextraction with simultaneous back-extraction for sample cleanup and preconcentration: Preconcentration into a single microdrop. *Analytical Chemistry, 71*, 388–393.

Mahugo-Santana, C., Sosa-Ferrera, Z., Torres-Padrón, M. E., & Santana-Rodríguez, J. J. (2011). Application of new approaches to liquid-phase microextraction for the determination of emerging pollutants. *TrAC Trends in Analytical Chemistry, 30*, 731–748.

Márquez-Sillero, I., Aguilera-Herrador, E., Cárdenas, S., & Valcárcel, M. (2011). Determination of 2,4,6-tricholoroanisole in water and wine samples by ionic liquid-based single-drop microextraction and ion mobility spectrometry. *Analytica Chimica Acta, 702*, 199–204.

Martín-Calero, A., Pino, V., & Afonso, A. M. (2011). Ionic liquids as a tool for determination of metals and organic compounds in food analysis. *TrAC Trends in Analytical Chemistry, 30*, 1598–1619.

Martinis, E. M., Berton, P., Monasterio, R. P., & Wuilloud, R. G. (2010). Emerging ionic liquid-based techniques for total-metal and metal-speciation analysis. *TrAC Trends in Analytical Chemistry, 29*, 1184–1201.

Maya, F., Estela, J. M., & Cerdà, V. (2012). Completely automated in-syringe dispersive liquid-liquid microextraction using solvents lighter than water. *Analytical and Bioanalytical Chemistry, 402*, 1383–1388.

Melwanki, M. B., & Huang, S.-D. (2006). Three-phase system in solvent bar microextraction: An approach for the sample preparation of ionizable organic compounds prior to liquid chromatography. *Analytica Chimica Acta, 555*, 139–145.

Metters, J. P., Kadara, R. O., & Banks, C. E. (2011). New directions in screen printed electroanalytical sensors: An overview of recent developments. *The Analyst, 136*, 1067–1076.

Mingyuan, G., Yangcheng, L., & Guangsheng, L. (2009). Directly suspended droplet microextraction in a rotating vial. *Analytica Chimica Acta, 648*, 123–127.

Miró, M., & Hansen, E. H. (2013). On-line sample processing involving microextraction techniques as a front-end to atomic spectrometric detection for trace metal assays: A review. *Analytica Chimica Acta, 782*, 1–11.

Mohamadi, M., & Mostafavi, A. (2010). A novel solidified floating organic drop microextraction based on ultrasound-dispersion for separation and preconcentration of palladium in aqueous samples. *Talanta, 81*, 309–313.

Moradi, M., & Yamini, Y. (2012a). Application of vesicular coacervate phase for microextraction based on solidification of floating drop. *Journal of Chromatography A, 1229*, 30–37.

Moradi, M., & Yamini, Y. (2012b). Surfactant roles in modern sample preparation techniques: A review. *Journal of Separation Science, 35*, 2319–2340.

Moradi, M., Yamini, Y., Esrafili, A., & Seidi, S. (2010). Application of surfactant assisted dispersive liquid-liquid microextraction for sample preparation of chlorophenols in water samples. *Talanta, 82*, 1864–1869.

Moreno-González, D., Gámiz-Gracia, L., Bosque-Sendra, J. M., & García-Campaña, A. M. (2012). Dispersive liquid-liquid microextraction using a low density extraction solvent for the determination of 17 N-methylcarbamates by micellar electrokinetic chromatography-electrospray-mass spectrometry employing a volatile surfactant. *Journal of Chromatography A, 1247*, 26–34.

Myung, S.-W., Yoon, S.-H., & Kim, M. (2003). Analysis of benzene ethylamine derivatives in urine using the programmable dynamic liquid-phase microextraction (LPME) device. *The Analyst, 128*, 1443–1446.

Naseri, M. T., Hemmatkhah, P., Hosseini, M. R. M., & Assadi, Y. (2008). Combination of dispersive liquid-liquid microextraction with flame atomic absorption spectrometry using microsample introduction for determination of lead in water samples. *Analytica Chimica Acta, 610*, 135–141.

Nerín, C., Salafranca, J., Aznar, M., & Batlle, R. (2009). Critical review on recent developments in solventless techniques for extraction of analytes. *Analytical and Bioanalytical Chemistry, 393*, 809–833.

Nojavan, S., & Fakhari, A. R. (2010). Electro membrane extraction combined with capillary electrophoresis for the determination of amlodipine enantiomers in biological samples. *Journal of Separation Science, 33*, 3231–3238.

Ouyang, G., Zhao, W., & Pawliszyn, J. (2005). Kinetic calibration for automated headspace liquid-phase microextraction. *Analytical Chemistry, 77*, 8122–8128.

Ouyang, G., Zhao, W., & Pawliszyn, J. (2007). Automation and optimization of liquid-phase microextraction by gas chromatography. *Journal of Chromatography A, 1138*, 47–54.

Pandey, S. (2006). Analytical applications of room-temperature ionic liquids: A review of recent efforts. *Analytica Chimica Acta, 556*, 38–45.

Patel, D. D., & Lee, J.-M. (2012). Applications of ionic liquids. *Chemical Record (New York, N.Y.), 12*, 329–355.

Pawliszyn, J., & Pedersen-Bjergaard, S. (2006). Analytical microextraction: Current status and future trends. *Journal of Chromatographic Science, 44*, 291–307.

Payán, M. D. R., Li, B., Petersen, N. J., Jensen, H., Hansen, S. H., & Pedersen-Bjergaard, S. (2013). Nano-electromembrane extraction. *Analytica Chimica Acta, 785*, 60–66.

Pedersen-Bjergaard, S., & Rasmussen, K. E. (1999). Liquid-liquid-liquid microextraction for sample preparation of biological fluids prior to capillary electrophoresis. *Analytical Chemistry, 71*, 2650–2656.

Pedersen-Bjergaard, S., & Rasmussen, K. E. (2006). Electrokinetic migration across artificial liquid membranes. New concept for rapid sample preparation of biological fluids. *Journal of Chromatography A, 1109*, 183–190.

Pedersen-Bjergaard, S., & Rasmussen, K. E. (2008). Liquid-phase microextraction with porous hollow fibers, a miniaturized and highly flexible format for liquid-liquid extraction. *Journal of Chromatography A, 1184*, 132–142.

Pena-Pereira, F., Lavilla, I., & Bendicho, C. (2008). Immersed single-drop microextraction interfaced with sequential injection analysis for determination of Cr(VI) in natural waters by electrothermal-atomic absorption spectrometry. *Spectrochimica Acta Part B: Atomic Spectroscopy, 63*, 498–503.

Pena-Pereira, F., Duarte, R. M. B. O., & Duarte, A. C. (2011). Considerations on the application of miniaturized sample preparation approaches for the analysis of organic compounds in environmental matrices. *Central European Journal of Chemistry, 10*, 433–449.

Pena-Pereira, F., Lavilla, I., & Bendicho, C. (2009a). Headspace single-drop microextraction coupled to microvolume UV-vis spectrophotometry for iodine determination. *Analytica Chimica Acta, 631*(2), 223–228.

Pena-Pereira, F., Lavilla, I., & Bendicho, C. (2009b). Miniaturized preconcentration methods based on liquid–liquid extraction and their application in inorganic ultratrace analysis and speciation: A review. *Spectrochimica Acta Part B: Atomic Spectroscopy, 64*, 1–15.

Pena-Pereira, F., Lavilla, I., & Bendicho, C. (2010a). Liquid-phase microextraction approaches combined with atomic detection: A critical review. *Analytica Chimica Acta, 669*, 1–16.

Pena-Pereira, F., Lavilla, I., & Bendicho, C. (2010b). Liquid-phase microextraction techniques within the framework of green chemistry. *TrAC Trends in Analytical Chemistry, 29*, 617–628.

Peng, J.-F., Liu, J.-F., Hu, X.-L., & Jiang, G.-B. (2007). Direct determination of chlorophenols in environmental water samples by hollow fiber supported ionic liquid membrane extraction coupled with high-performance liquid chromatography. *Journal of Chromatography A, 1139*, 165–170.

Peroni, D., van Egmond, W., Kok, W. T., & Janssen, H.-G. (2012). Advancing liquid/liquid extraction through a novel microfluidic device: theory, instrumentation and applications in gas chromatography. *Journal of Chromatography A, 1226*, 77–86.

Petersen, N. J., Foss, S. T., Jensen, H., Hansen, S. H., Skonberg, C., Snakenborg, D., Kutter, J.P., & Pedersen-Bjergaard, S. (2011a). On-chip electro membrane extraction with online ultraviolet and mass spectrometric detection. *Analytical Chemistry, 83*, 44–51.

Petersen, N. J., Jensen, H., Hansen, S. H., Foss, S. T., Snakenborg, D., & Pedersen-Bjergaard, S. (2010). On-chip electro membrane extraction. *Microfluidics and Nanofluidics, 9*, 881–888.

Petersen, N. J., Jensen, H., Hansen, S. H., Rasmussen, K. E., & Pedersen-Bjergaard, S. (2009). Drop-to-drop microextraction across a supported liquid membrane by an electrical field under stagnant conditions. *Journal of Chromatography A, 1216*, 1496–1502.

Petersen, N. J., Pedersen, J. S., Poulsen, N. N., Jensen, H., Skonberg, C., Hansen, S. H., & Pedersen-Bjergaard, S. (2012). On-chip electromembrane extraction for monitoring drug metabolism in real time by electrospray ionization mass spectrometry. *The Analyst, 137*, 3321–3327.

Petersen, N. J., Rasmussen, K. E., Pedersen-Bjergaard, S., & Gjelstad, A. (2011b). Electromembrane extraction from biological fluids. *Analytical Sciences, 27*, 965–972.

Pezo, D., Salafranca, J., & Nerín, C. (2007). Development of an automatic multiple dynamic hollow fibre liquid-phase microextraction procedure for specific migration analysis of new active food packagings containing essential oils. *Journal of Chromatography A, 1174*, 85–94.

Pham, T. P. T., Cho, C.-W., & Yun, Y.-S. (2010). Environmental fate and toxicity of ionic liquids: a review. *Water Research, 44*, 352–372.

Picó, Y. (2013). Ultrasound-assisted extraction for food and environmental samples. *TrAC Trends in Analytical Chemistry, 43*, 84–99.

Poole, C. F., & Poole, S. K. (2010). Extraction of organic compounds with room temperature ionic liquids. *Journal of Chromatography A, 1217*, 2268–2286.

Psillakis, E., & Kalogerakis, N. (2002). Developments in single-drop microextraction. *TrAC Trends in Analytical Chemistry, 21*, 53–63.

Psillakis, E., & Kalogerakis, N. (2003). Developments in liquid-phase microextraction. *TrAC Trends in Analytical Chemistry, 22*, 565–574.

Qian, L., & He, Y. (2006). Funnelform single-drop microextraction for gas chromatography – electron-capture detection. *Journal of Chromatographic Science, 1134*, 32–37.

Ramos, L. (2012). Critical overview of selected contemporary sample preparation techniques. *Journal of Chromatography A, 1221*, 84–98.

Rasmussen, K. E., & Pedersen-Bjergaard, S. (2004). Developments in hollow fibre-based, liquid-phase microextraction. *TrAC Trends in Analytical Chemistry, 23*, 1–10.

Rasmussen, K. E., Pedersen-Bjergaard, S., Krogh, M., Ugland, H. G., & Grønhaug, T. (2000). Development of a simple in-vial liquid-phase microextraction device for drug analysis compatible with capillary gas chromatography, capillary electrophoresis and high-performance liquid chromatography. *Journal of Chromatography A, 873*, 3–11.

Raynie, D. E. (2004). Modern extraction techniques. *Analytical Chemistry, 76*, 4659–4664.

Raynie, D. E. (2006). Modern extraction techniques. *Analytical Chemistry, 78*, 3997–4004.

Regueiro, J., Llompart, M., Garcia-Jares, C., Garcia-Monteagudo, J. C., & Cela, R. (2008). Ultrasound-assisted emulsification – microextraction of emergent contaminants and pesticides in environmental waters. *Journal of Chromatography A, 1190*, 27–38.

Rezaee, M., Assadi, Y., Milani Hosseini, M.-R., Aghaee, E., Ahmadi, F., & Berijani, S. (2006). Determination of organic compounds in water using dispersive liquid-liquid microextraction. *Journal of Chromatography A, 1116*, 1–9.

Rezaee, M., Yamini, Y., & Faraji, M. (2010). Evolution of dispersive liquid-liquid microextraction method. *Journal of Chromatography A, 1217*, 2342–2357.

Rezazadeh, M., Yamini, Y., Seidi, S., & Arjomandi-Behzad, L. (2014). Voltage-step pulsed electro-membrane as a novel view of electrical field-induced liquid-phase microextraction. *Journal of Chromatography A, 1324*, 21–28.

Rezazadeh, M., Yamini, Y., Seidi, S., & Esrafili, A. (2012). Pulsed electromembrane extraction: A new concept of electrically enhanced extraction. *Journal of Chromatography A, 1262*, 214–218.

Rezazadeh, M., Yamini, Y., Seidi, S., & Esrafili, A. (2013). One-way and two-way pulsed electro-membrane extraction for trace analysis of amino acids in foods and biological samples. *Analytica Chimica Acta, 773*, 52–59.

Ruiz-Aceituno, L., Sanz, M. L., & Ramos, L. (2013). Use of ionic liquids in analytical sample preparation of organic compounds from food and environmental samples. *TrAC Trends in Analytical Chemistry, 43*, 121–145.

Sáenz, M., Alvarado, J., Pena-Pereira, F., Senra-Ferreiro, S., Lavilla, I., & Bendicho, C. (2011). Liquid-phase microextraction with in-drop derivatization combined with microvolume fluorospectrometry for free and hydrolyzed formaldehyde determination in textile samples. *Analytica Chimica Acta, 687*, 50–55.

Saleh, A., Yamini, Y., Faraji, M., Rezaee, M., & Ghambarian, M. (2009). Ultrasound-assisted emulsification microextraction method based on applying low density organic solvents followed by gas chromatography analysis for the determination of polycyclic aromatic hydrocarbons in water samples. *Journal of Chromatography A, 1216*, 6673–6679.

Sarafraz-Yazdi, A., & Amiri, A. (2010). Liquid-phase microextraction. *TrAC Trends in Analytical Chemistry, 29*, 1–14.

Sarafraz-Yazdi, A., Beiknejad, D., & Es'haghi, Z. (2005). LC Determination of mono-substituted phenols in water using liquid–liquid–liquid phase microextraction. *Chromatographia, 62*, 49–54.

Sarafraz-Yazdi, A., Mofazzeli, F., & Es'haghi, Z. (2009). Determination of 3-nitroaniline in water samples by directly suspended droplet three-phase liquid-phase microextraction using 18-crown-6 ether and high-performance liquid chromatography. *Journal of Chromatography A, 1216*, 5086–5091.

Saraji, M. (2005). Dynamic headspace liquid-phase microextraction of alcohols. *Journal of Chromatography A, 1062*, 15–21.

Saraji, M., & Bidgoli, A. A. H. (2010). Dispersive liquid-liquid microextraction using a surfactant as disperser agent. *Analytical and Bioanalytical Chemistry, 397*, 3107–3115.

Seebunrueng, K., Santaladchaiyakit, Y., & Srijaranai, S. (2014). Vortex-assisted low density solvent based demulsified dispersive liquid-liquid microextraction and high-performance liquid chromatography for the determination of organophosphorus pesticides in water samples. *Chemosphere, 103*, 51–58.

Seidi, S., Yamini, Y., Rezazadeh, M., & Esrafili, A. (2012). Low-voltage electrically-enhanced microextraction as a novel technique for simultaneous extraction of acidic and basic drugs from biological fluids. *Journal of Chromatography A, 1243*, 6–13.

Shen, G., & Lee, H. K. (2003). Headspace liquid-phase microextraction of chlorobenzenes in soil with gas chromatography-electron capture detection. *Analytical Chemistry, 75*, 98–103.

Shi, Y.-A., Chen, M.-Z., Muniraj, S., & Jen, J.-F. (2008). Microwave-assisted headspace controlled temperature liquid-phase microextraction of chlorophenols from aqueous samples for gas chromatography-electron capture detection. *Journal of Chromatography A, 1207*, 130–135.

Shrivas, K., & Patel, D. K. (2010). Liquid-phase microextraction combined with gas chromatography mass spectrometry for rapid determination of nicotine in one-drop of nightshades vegetables and commercial food products. *Food Chemistry, 122*, 314–318.

Shrivas, K., & Wu, H.-F. (2007). Rapid determination of caffeine in one drop of beverages and foods using drop-to-drop solvent microextraction with gas chromatography/mass spectrometry. *Journal of Chromatography A, 1170*, 9–14.

Slampová, A., Kubáň, P., & Boček, P. (2012). Electromembrane extraction using stabilized constant d.c. electric current--a simple tool for improvement of extraction performance. *Journal of Chromatography A, 1234*, 32–37.

Soukup-Hein, R. J., Warnke, M. M., & Armstrong, D. W. (2009). Ionic liquids in analytical chemistry. *Annual Review of Analytical Chemistry, 2*, 145–168.

Spietelun, A., Marcinkowski, L., de la Guardia, M., & Namieśnik, J. (2014). Green aspects, developments and perspectives of liquid phase microextraction techniques. *Talanta, 119*, 34–45.

Stalikas, C., Fiamegos, Y., Sakkas, V., & Albanis, T. (2009). Developments on chemometric approaches to optimize and evaluate microextraction. *Journal of Chromatography A, 1216*, 175–189.

Sun, P., & Armstrong, D. W. (2010). Ionic liquids in analytical chemistry. *Analytica Chimica Acta, 661*, 1–16.

Szreniawa-Sztajnert, A., Zabiegała, B., & Namieśnik, J. (2013). Developments in ultrasound-assisted microextraction techniques for isolation and preconcentration of organic analytes from aqueous samples. *TrAC Trends in Analytical Chemistry, 49*, 45–54.

Tan, Z., Liu, J., & Pang, L. (2012). Advances in analytical chemistry using the unique properties of ionic liquids. *TrAC Trends in Analytical Chemistry, 39*, 218–227.

Tankeviciute, A., Kazlauskas, R., & Vickackaite, V. (2001). Headspace extraction of alcohols into a single drop. *The Analyst, 126*, 1674–1677.

Tankiewicz, M., Fenik, J., & Biziuk, M. (2011). Solventless and solvent-minimized sample preparation techniques for determining currently used pesticides in water samples: a review. *Talanta, 86*, 8–22.

Tayyebi, M., Yamini, Y., & Moradi, M. (2012). Reverse micelle-mediated dispersive liquid-liquid microextraction of 2,4-dichlorophenoxyacetic acid and 4-chloro-2-methylphenoxyacetic acid. *Journal of Separation Science*, *35*, 2491–2498.

Tehrani, M. S., Givianrad, M. H., & Mahoor, N. (2012). Surfactant-assisted dispersive liquid–liquid microextraction followed by high-performance liquid chromatography for determination of amphetamine and methamphetamine in urine samples. *Analytical Methods*, *4*, 1357–1364.

Theis, A. L., Waldack, A. J., Hansen, S. M., & Jeannot, M. A. (2001). Headspace solvent microextraction. *Analytical Chemistry*, *73*, 5651–5654.

Tobiszewski, M., Mechlińska, A., Zygmunt, B., & Namieśnik, J. (2009). Green analytical chemistry in sample preparation for determination of trace organic pollutants. *TrAC Trends in Analytical Chemistry*, *28*, 943–951.

Trujillo-Rodríguez, M. J., Rocío-Bautista, P., Pino, V., & Afonso, A. M. (2013). Ionic liquids in dispersive liquid-liquid microextraction. *TrAC Trends in Analytical Chemistry*, *51*, 87–106.

Vallecillos, L., Borrull, F., & Pocurull, E. (2012a). Determination of musk fragrances in sewage sludge by pressurized liquid extraction coupled to automated ionic liquid-based headspace single-drop microextraction followed by GC-MS/MS. *Journal of Separation Science*, *35*, 2735–2742.

Vallecillos, L., Pocurull, E., & Borrull, F. (2012b). Fully automated ionic liquid-based headspace single drop microextraction coupled to GC-MS/MS to determine musk fragrances in environmental water samples. *Talanta*, *99*, 824–832.

Vičkačkaitė, V., & Padarauskas, A. (2012). Ionic liquids in microextraction techniques. *Central European Journal of Chemistry*, *10*, 652–674.

Vidal, L., Domini, C. E., Grané, N., Psillakis, E., & Canals, A. (2007a). Microwave-assisted headspace single-drop microextration of chlorobenzenes from water samples. *Analytica Chimica Acta*, *592*, 9–15.

Vidal, L., Psillakis, E., Domini, C. E., Grané, N., Marken, F., & Canals, A. (2007b). An ionic liquid as a solvent for headspace single drop microextraction of chlorobenzenes from water samples. *Analytica Chimica Acta*, *584*, 189–195.

Viñas, P., Campillo, N., López-García, I., & Hernández-Córdoba, M. (2014). Dispersive liquid-liquid microextraction in food analysis. A critical review. *Analytical and Bioanalytical Chemistry*, *406*, 2067–2099.

Wang, Y., Kwok, Y. C., He, Y., & Lee, H. K. (1998). Application of dynamic liquid-phase microextraction to the analysis of chlorobenzenes in water by using a conventional microsyringe. *Analytical Chemistry*, *70*, 4610–4614.

Wei, G., Li, Y., & Wang, X. (2007). Application of dispersive liquid-liquid microextraction combined with high-performance liquid chromatography for the determination of methomyl in natural waters. *Journal of Separation Science*, *30*, 3262–3267.

Wen, X., Deng, Q., Wang, J., Yang, S., & Zhao, X. (2013a). A new coupling of ionic liquid based-single drop microextraction with tungsten coil electrothermal atomic absorption spectrometry. *Spectrochimica Acta Part A: Molecular and Biomolecular Spectroscopy*, *105*, 320–325.

Wen, X., Deng, Q., Wang, J., & Zhao, X. (2013b). Application of portable spectrometer for ultra trace metal analysis after ionic liquid based microextraction. *Analytical Methods*, *5*, 2978–2983.

Wen, Y., Li, J., Ma, J., & Chen, L. (2012). Recent advances in enrichment techniques for trace analysis in capillary electrophoresis. *Electrophoresis*, *33*, 2933–2952.

Wu, H.-F., Yen, J.-H., & Chin, C.-C. (2006). Combining drop-to-drop solvent microextraction with gas chromatography/mass spectrometry using electronic ionization and self-ion/molecule reaction method to determine methoxyacetophenone isomers in one drop of water. *Analytical Chemistry*, *78*, 1707–1712.

Wu, Q., Chang, Q., Wu, C., Rao, H., Zeng, X., Wang, C., & Wang, Z. (2010). Ultrasound-assisted surfactant-enhanced emulsification microextraction for the determination of carbamate

pesticides in water samples by high performance liquid chromatography. *Journal of Chromatography A, 1217*, 1773–1778.
Xia, L., Hu, B., Jiang, Z., Wu, Y., & Liang, Y. (2004). Single-drop microextraction combined with low-temperature electrothermal vaporization ICPMS for the determination of trace Be, Co, Pd, and Cd in biological samples. *Analytical Chemistry, 76*, 2910–2915.
Xia, L., Li, X., Wu, Y., Hu, B., & Chen, R. (2008). Ionic liquids based single drop microextraction combined with electrothermal vaporization inductively coupled plasma mass spectrometry for determination of Co, Hg and Pb in biological and environmental samples. *Spectrochimica Acta Part B: Atomic Spectroscopy, 63*, 1290–1296.
Xiao, Q., Hu, B., Yu, C., Xia, L., & Jiang, Z. (2006). Optimization of a single-drop microextraction procedure for the determination of organophosphorus pesticides in water and fruit juice with gas chromatography-flame photometric detection. *Talanta, 69*, 848–855.
Xie, H.-Y., & He, Y.-Z. (2010). Green analytical methodologies combining liquid-phase microextraction with capillary electrophoresis. *TrAC Trends in Analytical Chemistry, 29*, 629–635.
Xu, H., Ding, Z., Lv, L., Song, D., & Feng, Y.-Q. (2009). A novel dispersive liquid-liquid microextraction based on solidification of floating organic droplet method for determination of polycyclic aromatic hydrocarbons in aqueous samples. *Analytica Chimica Acta, 636*, 28–33.
Xu, H., Liao, Y., & Yao, J. (2007). Development of a novel ultrasound-assisted headspace liquid-phase microextraction and its application to the analysis of chlorophenols in real aqueous samples. *Journal of Chromatography A, 1167*, 1–8.
Xu, L., Basheer, C., & Lee, H. K. (2007). Developments in single-drop microextraction. *Journal of Chromatography A, 1152*, 184–192.
Xu, L., Hauser, P. C., & Lee, H. K. (2008). Electro membrane isolation of nerve agent degradation products across a supported liquid membrane followed by capillary electrophoresis with contactless conductivity detection. *Journal of Chromatography A, 1214*, 17–22.
Yan, H., & Wang, H. (2013). Recent development and applications of dispersive liquid–liquid microextraction. *Journal of Chromatography A, 1295*, 1–15.
Yang, Q., Guo, Y., Wang, L., Liang, S., & Liu, X. (2010). Simultaneous determination of trace benzene and toluene in beverage by ultrasound-enhanced hollow-fiber liquid-phase microextraction coupled with GC. *Chromatographia, 72*, 1157–1161.
Yang, Z.-H., Lu, Y.-L., Liu, Y., Wu, T., Zhou, Z.-Q., & Liu, D.-H. (2011). Vortex-assisted surfactant-enhanced-emulsification liquid-liquid microextraction. *Journal of Chromatography A, 1218*, 7071–7077.
Yangcheng, L., Quan, L., Guangsheng, L., & Youyuan, D. (2006). Directly suspended droplet microextraction. *Analytica Chimica Acta, 566*, 259–264.
Yao, C., & Anderson, J. L. (2009). Dispersive liquid-liquid microextraction using an *in situ* metathesis reaction to form an ionic liquid extraction phase for the preconcentration of aromatic compounds from water. *Analytical and Bioanalytical Chemistry, 395*(5), 1491–1502.
Yao, C., Twu, P., & Anderson, J. L. (2010). Headspace single drop microextraction using micellar ionic liquid extraction solvents. *Chromatographia, 72*, 393–402.
Yazdi, A. S. (2011). Surfactant-based extraction methods. *TrAC Trends in Analytical Chemistry, 30*, 918–929.
Ye, C., Zhou, Q., & Wang, X. (2007). Improved single-drop microextraction for high sensitive analysis. *Journal of Chromatography A, 1139*, 7–13.
Yiantzi, E., Psillakis, E., Tyrovola, K., & Kalogerakis, N. (2010). Vortex-assisted liquid-liquid microextraction of octylphenol, nonylphenol and bisphenol-A. *Talanta, 80*, 2057–2062.
You, X., Xing, Z., Liu, F., & Jiang, N. (2013). Air-assisted liquid-liquid microextraction used for the rapid determination of organophosphorus pesticides in juice samples. *Journal of Chromatography A, 1311*, 41–47.

Zacharis, C. K., Tzanavaras, P. D., Roubos, K., & Dhima, K. (2010). Solvent-based de-emulsification dispersive liquid–liquid microextraction combined with gas chromatography–mass spectrometry for determination of trace organochlorine pesticides in environmental water samples. *Journal of Chromatography A, 1217*, 5896–5900.

Zeng, C., Hu, Y., & Luo, J. (2011). Ionic liquid-based hollow fiber supported liquid membrane extraction combined with thermospray flame furnace AAS for the determination of cadmium. *Microchimica Acta, 177*, 53–58.

Zhang, J., Su, T., & Lee, H. K. (2005). Headspace water-based liquid-phase microextraction. *Analytical Chemistry, 77*, 1988–1992.

Zhang, J., Su, T., & Lee, H. K. (2006). Development and application of microporous hollow fiber protected liquid-phase microextraction *via* gaseous diffusion to the determination of phenols in water. *Journal of Chromatography A, 1121*, 10–15.

Zhang, M., Huang, J., Zheng, X., & Wu, L. (2010). Ultrasonic-assisted drop-to-drop solvent microextraction in a capillary tube for analyzing trace benzene, toluene, xylene in one drop of a water sample. *Chromatographia, 72*, 1163–1168.

Zhang, Y., & Lee, H. K. (2012). Determination of ultraviolet filters in water samples by vortex-assisted dispersive liquid-liquid microextraction followed by gas chromatography-mass spectrometry. *Journal of Chromatography A, 1249*, 25–31.

Zhang, Y., Wang, R., Su, P., & Yang, Y. (2013). Ionic liquid-based solvent bar microextraction for determination of organophosphorus pesticides in water samples. *Analytical Methods, 5*, 5074–5078.

Zhao, D., Liao, Y., & Zhang, Z. (2007). Toxicity of ionic liquids. *CLEAN – Soil, Air, Water, 35*, 42–48.

Zhao, F., Lu, S., Du, W., & Zeng, B. (2008a). Ionic liquid-based headspace single-drop microextraction coupled to gas chromatography for the determination of chlorobenzene derivatives. *Microchimica Acta, 165*, 29–33.

Zhao, F., Li, J., & Zeng, B. (2008b). Coupling of ionic liquid-based headspace single-drop microextraction with GC for sensitive detection of phenols. *Journal of Separation Science, 31*, 3045–3049.

Zhao, H., Xia, S., & Ma, P. (2005). Use of ionic liquids as "green" solvents for extractions. *Journal of Chemical Technology & Biotechnology, 80*, 1089–1096.

Zhao, L., & Lee, H. K. (2002). Liquid-phase microextraction combined with hollow fiber as a sample preparation technique prior to gas chromatography/mass spectrometry. *Analytical Chemistry, 74*, 2486–2492.

Zhao, Q., & Anderson, J. L. (2011). Task-specific microextractions using ionic liquids. *Analytical and Bioanalytical Chemistry, 400*, 1613–1618.

Zheng, C., Zhao, J., Bao, P., Gao, J., & He, J. (2011). Dispersive liquid-liquid microextraction based on solidification of floating organic droplet followed by high-performance liquid chromatography with ultraviolet detection and liquid chromatography-tandem mass spectrometry for the determination of triclosan and 2,4-dichlorophenol in water samples. *Journal of Chromatography A, 1218*, 3830–3836.

Zhou, Q., Bai, H., Xie, G., & Xiao, J. (2008). Temperature-controlled ionic liquid dispersive liquid phase micro-extraction. *Journal of Chromatography A, 1177*, 43–49.

Zhu, L., & Lee, H. K. (2001). Liquid-liquid-liquid microextraction of nitrophenols with a hollow fiber membrane prior to capillary liquid chromatography. *Journal of Chromatography A, 924*, 407–414.

5 Choice of Solvent in Liquid-Phase Microextraction

Shayessteh Dadfarnia* and Ali Mohammad Haji-Shabani
Department of Chemistry, Faculty of Science, Yazd University, Yazd, 89195–741 Iran
*e-mail address: sdadfarnia@yazduni.ac.ir

5.1 Introduction

Liquid-phase microextraction (LPME) is a miniaturized sample preparation technique that emerged in the mid-to-late 1990s (Liu & Dasgupta, 1996; Jeannot & Cantwell, 1996). In LPME, a microliter amount of water-immiscible solvent is used to extract analytes from aqueous samples. The principle of LPME is similar to the traditional liquid-liquid extraction. Thus, the choice of appropriate extracting solvent is an important aspect of a successful LPME. There are several peripheral properties of solvents which are of interest in selecting a solvent (Barwick, 1997) but do not directly affect the separation. These factors are frequently conflicting, and certainly no single substance would ordinary possess every desired characteristic. The final choice should be made after comparing the different physical properties of the solvents available and the required criteria for different modes of LPME to achieve good selectivity, sensitivity and precision. In this chapter, the relative importance of the various factors in choosing a solvent for LPME are considered and briefly discussed.

5.2 Relevance of Physicochemical Properties in Extractant Phase Selection

5.2.1 Solubility

This is the first property that must be considered in deciding the applicability of a solvent, and it refers to the ability of a solvent to dissolve considerable amounts of the analyte. The solute should be very soluble in the extracting phase. Generally, the reason for dissolution is thermodynamic, i.e., when the process is energetically favorable it will occur. However, kinetics can also play a role and solutes that are poorly soluble at room temperature may be dissolved at high temperature. In the qualitative selection of the appropriate solvent, one can use the rule "like dissolves like"; however, there are many parameters that have been used to describe the attractive forces (dispersion, dipole and hydrogen bonding) present within a solvent or liquid. For example, the refractive index value is a good measure of the tendency of compounds to interact by dispersion forces (Barwick 1997), i.e., the greater the refractive index values of the compounds, the stronger the dispersion interaction. Polarity is often used to predict the solubility of a compound, but unfortunately the concept is not straightforward. Among different classifications of solvent polarity, the Hildebrand polarity scale or Hildebrand solubility parameter d is reported as being the most widely applied index of solvent and solute polarity. **Table 5.1** lists d values for a number of solvents along with some of their physical properties such as boiling point, dielectric constant, surface tension, density and vapor pressure. Values of d can be estimated for other compounds by bearing in mind that homologues of polar compounds tend to have similar but slightly lower values of d as molecular weight increases. The solubility of a solute in a solvent is maximized when they have the same d values. In general, two liquids are miscible if the difference in d value is less than 3.4 units. Two solvents whose d values are higher or lower than that of a given solute can be blended to give a mixture with a d value equal to that of solute, thus maximizing the solute solubility. However, there are exceptions to the Hildebrand solubility parameter especially with polar solvents and solutes; therefore, it is often worth testing solubility or solvent miscibility on a small scale even if data are available.

5.2.2 Distribution Coefficient

A distribution coefficient is defined as the ratio of the total concentration of solute in the organic phase to the total concentration of solute in the aqueous phase at equilibrium. It gives a measure of the difference in the solubility of the solute in these two phases. A distribution coefficient value that is greater than unity implies that the solute has a higher affinity for the organic phase than the aqueous phase. The higher the distribution coefficient, the more easily the solute will be extracted into the extracting phase. So, for the extraction of a single solute from the aqueous phase, a large distribution coefficient is desirable to achieve a high enrichment factor (EF) (defined as the ratio of the analyte concentration in the extraction phase to its initial concentration in the sample solution) and extraction efficiency. Thus, the distribution parameter, which depends to a large extent on the type and nature of the extracting phase, is a key variable and it is necessary to emphasize the different extracting phases that can be used.

5.2.3 Selectivity

If the aqueous phase contains more than one solute, then in deciding on the applicability of a solvent the selectivity must be considered. The selectivity refers to the ability of a solvent to extract one solute (x) in preference to another (y) and is defined as the value of the distribution coefficient of x to the distribution coefficient of y. The most suitable solvent from this point of view would dissolve a maximum of one

Table 5.1 Physical properties of the common solvents used in single-drop microextraction (SDME), hollow fiber-liquid phase microextraction (HF-LPME) and dispersive liquid-liquid microextraction (DLLME). These properties were measured at 25 °C unless otherwise stated (Lide,1991–1992, Brawick, 1997, Pena-Pereira et al., 2009).

Solvent	Dielectric constant	Hildebrand solubility parameter (d)	Dipole moment (D)	Surface tension (dyn cm^{-1})	Viscosity (cP)	Density (g cm^{-3})	Water solubility (mg L^{-1})	Vapor pressure (Torr)	Boiling point (°C)
cyclohexane	(20 °C)	8.2	0.00	24.65	0.90	0.78	55	97.8	80.7
n-hexane	1.88	7.24	0.08	17.94	0.29	0.65	1.2	151.3	68.7
octane	(20 °C)	-	0.00	21.18	0.51	0.70	× 10^{-3}	14.0	125.7
iso-octane	(20 °C)	-	0.00	18.8	(20 °C)	0.69	2.4	49	99.2
decane	(20 °C)	-	0.00	23.37	0.86	0.73	0.05	1.3	174.2
benzene	2.27	9.2	0.00	28.2	0.60	0.87	1791	95.2	80.1
toluene	2.38	8.9	0.31	27.92	0.55	0.86	515	28.5	110.6
o-xylene	(20 °C)	8.8	0.45	29.49	0.76	0.88	175	6.6	144.4
m-xylene	(20 °C)	8.8	0.30	28.10	0.58	0.86	146	8.3	139.1
p-xylene	(20 °C)	8.8	0.02	27.76	0.60	0.86	156	8.7	138.4
ethylbenzene	(20 °C)	-	0.37	28.48	0.64	0.86	152	9.6	136.2
methanol	32.66	14.5	2.87	22.30	0.55	0.79	∞	127.0	64.5
ethanol	24.55	12.7	1.66	(20 °C)	1.08	0.79	∞	59.0	78.3
1-octanol	(20 °C)	10.3	1.76	26.92	7.36	0.82	533	0.08	195.2
benzyl alcohol	(20 °C)	-	1.66	39.44	(30 °C)	1.04	800	0.11	205.4

Table 5.1 Physical properties of the common solvents used in single-drop microextraction (SDME), hollow fiber-liquid phase microextraction (HF-LPME) and dispersive liquid-liquid microextraction (DLLME). These properties were measured at 25 °C unless otherwise stated (Lide,1991–1992, Brawick, 1997, Pena-Pereira et al., 2009).

Solvent	Dielectric constant	Hildebrand solubility parameter (d)	Dipole moment (D)	Surface tension (dyn cm^{-1})	Viscosity (cP)	Density (g cm^{-3})	Water solubility (mg L^{-1})	Vapor pressure (Torr)	Boiling point (°C)
ethylene glycol	37.7	17.0	2.31	(20 °C)	13.76	1.11	∞	0.09	197.5
acetone	20.56	9.6	2.69	22.68	0.30	0.78	∞	231.1	56.1
MIBK	(20 °C)	-	-	(20 °C)	0.55	0.80	17000	18.8	117.4
α,α,α-trifluorotoluene	9.03	-	2.56	(20 °C)	(19.8 °C)	1.18	-	38.5	102.0
chlorobenzene	5.62	9.7	1.62	(30 °C)	(30 °C)	(20 °C)	327	11.7	131.7
dichlorometane	8.93	9.6	1.14	(30 °C)	(30 °C)	1.32	1980	435.8	39.6
chloroform	(20 °C)	9.3	1.15	26.53	0.54	1.48	8500	194.8	61.2
tetrachloromethane	2.23	8.6	0.00	26.13	0.90	1.58	770	115.2	76.6
1,2-dichloroethane	10.37	9.76	1.83	(30 °C)	(30 °C)	1.25	8100 (20 °C)	(20 °C)	83.5
1,2-dichlorobenzene	9.93	-	2.14	(20 °C)	1.32	1.30	156	1.3	180.5
tetrachloroethylene	2.28	-	0.00	31.30	(30 °C)	1.61	150	18.5	121.1
bromobenzene	5.40	9.9	1.55	(30 °C)	(30 °C)	1.48	424	4.2	155.9
nitrobenzene	34.78	11.1	4.00	(30 °C)	(30 °C)	1.20	1900 (20 °C)	0.28	210.8
acetonitrile	35.94	12.1	3.53	28.25	0.34	0.78	∞	88.81	81.6
pyridine	12.91	10.6	2.37	36.33	0.88	0.94	∞	20	115.2

Table 5.1 Physical properties of the common solvents used in single-drop microextraction (SDME), hollow fiber-liquid phase microextraction (HF-LPME) and dispersive liquid-liquid microextraction (DLLME). These properties were measured at 25 °C unless otherwise stated (Lide,1991–1992, Brawick, 1997, Pena-Pereira et al., 2009).

Solvent	Dielectric constant	Hildebrand solubility parameter (d)	Dipole moment (D)	Surface tension (dyn cm^{-1})	Viscosity (cP)	Density (g cm^{-3})	Water solubility (mg L^{-1})	Vapor pressure (Torr)	Boiling point (°C)
N',N''-dimethylformamide	36.71	11.8	3.24	36.42	0.80	1.29	∞	3.7	153.0
carbon disulfide	2.64	9.9	0.06	(20 °C)	(20 °C)	1.29	2100 (20 °C)	361.6	46.2
hexyl acetate	-	-	-	-	-	0.8673	(20 °C)	1.32	171.5
[C_4MIM][PF_6]	-	-	-	48.8–49.8	148–450	1.36–1.37	18800	-	-
[C_6MIM][PF_6]	-	-	-	-	560–586	1.29–1.31	7500	-	-
[C_8MIM][PF_6]	-	-	-	34.2–36.5	682–710	1.20–1.23	2000	-	-
[C_6MIM][NTf_2]	-	-	-		-	1.33	3400	-	-
water	78.36	23.5	1.82	71.81	0.89	1.00	∞	23.8	100.0

solute and a minimum of the other. In this regard, the closeness of the polarity of the solvent to the analyte is important. If the polarity of the analyte of interest differs significantly from that of other components, a solvent with polarity close to the analyte will significantly extract the analyte without extraction of other components. In some cases, variation in dispersion interactions may affect the selectivity. Normally dispersion interactions play a minor role in solvent selectivity, but it is worth considering in the separation of solutes with different refractive indices. Thus, replacing a solvent with another solvent of similar polarity but different refractive index may enhance the solvent selectivity. For a useful extraction operation, the selectivity must be greater than unity. A selectivity of unity means that the solvent have no preference for the solutes and no separation is possible.

5.2.4 Immiscibility

For a clean separation, the extraction solvent should have low solubility in the aqueous phase. If the solubility of the organic phase in the aqueous phase is significant, complete recovery of analyte is not possible and an additional separation step is required. One way of arriving at an idea about the degree of immiscibility is through a comparison of the dipole moment of the liquid with that of water (μ= 1.84 Debye). Solvents such as benzene, hexane and carbon tetrachloride with dipole moments of zero have low water solubility (less than 0.1 g/100 g), whereas solvents such as diethyl ether (μ = 1.2 Debye), ethyl acetate (μ = 1.8 Debye), or n-butyl alcohol (μ = 1.7 Debye) have relatively high aqueous solubility (6.90, 7.94, and 7.80 g/100 g, respectively). This generalization has its exceptions, of course; chlorobenzene with a dipole moment of 1.6 Debye has an aqueous solubility less than that of benzene.

5.2.5 Density

Solvent density is critical in conventional liquid-liquid extraction, as well as in many LPME approaches. However, the relevance of the extractant phase density is lower in SDME and HF-LPME techniques. A large difference in densities between the extracting and aqueous phase permits a clean-cut phase boundary between the two phases. It should be noted that it is insufficient to examine only the relative densities of the solution to be extracted and the pure extracting solvent, since on mixing the mutual solubility of the two will alter the individual densities.

5.2.6 Interfacial Tension

Solvent interfacial tension affects the dispersion of one liquid to another and the ease of the separation of two phases. For rapid coalescence of emulsion and phase separation a high interfacial tension between immiscible phases is required. However, too high an interfacial tension may lead to difficulties in the adequate dispersion of two phases and consequently mass transfer efficiency. Thus, a compromise may be necessary in the interfacial tension; it must be high enough to prevent the formation of stable emulsion and low enough to permit adequate dispersion of one liquid in the other to occur. As the information regarding the liquid interfacial tension for complete ternary systems is limited, one should use the differences in the surface tension with air of the contacted liquid as a rough guide to estimate the order of magnitude of interfacial tension.

5.2.7 Chemical Reactivity

The reaction between the solvent and other component of the system is undesirable. Thus, the solvent must be chemically stable and inert toward the sample matrices and materials used in the construction of extraction devices and should not polymerize, condense or decompose under the extraction conditions. However, on occasion, the formation of a chemical compound between the analyte and solvent as part of the extraction process is desirable and the rate and even the extent of extraction may be enhanced.

5.2.8 Corrosiveness

In order to reduce the cost of the extraction equipment, the solvent should be inert and cause no severe corrosion difficulties with common materials of construction. Thus, expensive alloys and other unusual materials should not be required.

5.2.9 Viscosity, Boiling Point and Vapor Pressure

Solvent viscosity also affects the extraction kinetics in LPME. The viscosity of the solvent should be low to permit good contact between the two phases while mixing and to allow rapid settling out of the two liquids after shaking. Thus, a high viscosity leads to difficulties in dispersion and the mass transfer rate.

The solvent should have a sufficiently high boiling point and low vapor pressure so that evaporation of the solvent is not a problem. A solvent with high vapor pressure may cause difficulties in the storage and operation of extraction at atmospheric pressure.

5.2.10 Availability and Cost

The solvent should be readily available in a sufficient state of purity for convenient use. A rare solvent or one that requires long and elaborate purification process before it can be used is not going to become popular and widely used, regardless of its extraction power.

5.2.11 Other Criteria

The freezing point, flammability and toxicity of the solvent must be low. The freezing point of the solvent must be low enough so that it can be conveniently stored. The low flammability and toxicity of the solvent is desirable for occupational health and safety consideration. However, in LPME, in contrast to liquid-liquid extraction, the recoverability of the solvent is not important as the use of solvent is reduced to microliter levels.

5.3 Extracting Solvents for Liquid-phase Microextraction

The extracting solvents used in LPME can be divided into two general groups, namely organic solvents and ionic liquids (ILs).

To date, the most common extracting phases with LPME methodology are organic solvents. The proper choice of the organic solvent should be based on the required criteria for different modes of LPME methods and physical properties of the organic solvents, the aspects of which will be considered later.

ILs are generally composed of bulky, nonsymmetrical organic cations, such as imidazolium, pyrrolidinium, pyridinium, ammonium or phosphonium and many different inorganic or organic anions such as tetrafluoroborate and bromate anions. ILs are gaining widespread recognition as green solvents in chemistry. The unique properties of ILs, such as negligible vapor pressure, good thermal stability, tunable viscosity and miscibility with water and organic solvents and their good extractability for a wide range of organic compounds and metal ions depends mainly on the nature and size of their cationic and anionic constituents. Some ILs are suitable for different modes of LPME due to their immiscibility with water, which permits the formation of two-phase systems, and the high affinity of their organic constituents (Han et al., 2012). However, it should be noted that although room temperature ILs based on alkyl imidazolium hexafluorophosphate have been used by many researchers as green solvents, these compounds may be potentially toxic due to the instability of the $[PF_6]$ anion toward hydrolysis in contact with moisture, which results in forming some volatile species, including HF and POF_3 (Swatloski et al., 2003). In the following section the main required criteria for solvents in different modes of LPME as well as the most common solvents used will be considered.

5.3.1 Extractant Phases for Single-drop Microextraction

The most important parameters that must be considered in the selection of a solvent for direct single-drop microextraction (direct-SDME) and continuous flow microextraction, where the drop is immersed into the aqueous solution, are selectivity, incidence of drop loss, rate of drop dissolution, extraction efficiency and solvent toxicity. Thus, a solvent with relatively high boiling point and high surface tension is required. The high boiling point solvent reduces the possibility of evaporation losses and prevents the possibility of bubble formation inside of the drop whereas a high surface tension increases the cohesive forces at the interface which reduce solvent solubilization (Psillakis & Kalogerakis, 2003).

In the headspace mode of SDME (HS-SDME), where a microdrop of water or water-miscible organic solvent is used for the extraction of volatile or semi-volatile analytes, any solvent with the ability to extract the analytes can be used, but generally, the high boiling point solvents which have low vapor pressure are preferred. The high surface tension of aqueous solutions also allows the use of relatively large drop volumes. HS-SDME is very useful when liquid chromatography or capillary electrophoresis is involved, as the extraction phase can be compatible with the mobile phase or the electrolyte used in these separation techniques (Psillakis & Kalogerakis, 2002).

Ionic analytes can be extracted by SDME into organic phase either by addition of a complexing agent to the sample prior to the extraction or by performing the extraction with a microdrop of organic phase containing a complexing agent (Dadfarnia & Haji Shabani, 2010, Pena-Pereira et al., 2010).

The nature of the analyte and the feasibility of matching between the volume of acceptor phase provided by the SDME technique and the one required for measurement determine which detection system is more suitable. **Figure 5.1** shows the detection techniques used in combination with different mode of SDME methodologies. It can be seen that gas chromatography (GC) is the most common technique combined with SDME, being used in almost 50% of the studies, followed by high performance liquid chromatography (HPLC) and atomic spectroscopy (AS).

As shown in **Figure 5.2**, among the organic solvents used as the extractant in SDME techniques, toluene is the most often used solvent especially when the SDME is combined with GC and HPLC, followed by octanol, chloroform, benzene, *n*-hexane, isooctane and hexyl acetate, respectively. Among AS techniques, the electrothermal atomic absorption spectrometry (ETAAS), where a few microliters are sufficient for measurement and the dilution of the enriched extract is avoided, was most commonly used with SDME techniques (ca. 77 % of them). However, it should be noted that when SDME is combined with ETAAS, chlorinated organic solvents are not recommended as

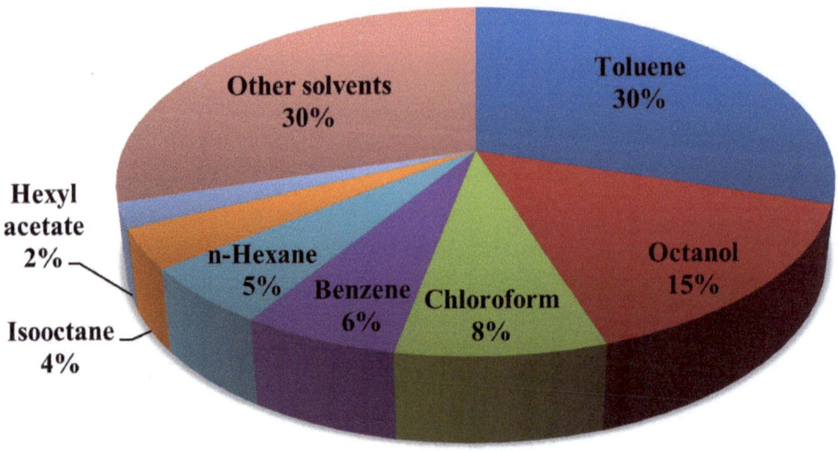

Figure 5.1 Detection techniques used in combination with different modes of SDME methodologies.

extractants due to the possibility of formation of volatile chlorides of analytes in the presence of chlorinated solvent which may result in the significant loss of some analytes. Furthermore, organic solvents may penetrate deeply into the pores of the graphite tube and a high ashing temperature and a long pre-treatment step is required for its removal.

ILs have been used as extracting solvents in both direct-SDME and HS-SDME. Liu et al. reported for the first time the application of ILs in SDME (Liu et al., 2003). Three ILs containing the $[PF_6]$ anion, 1-alkyl-3-methylimidazolium hexafluorophosphate ($[C_nMIM][PF_6]$, n = 4, 6, 8) were used as the extraction phase coupled to HPLC to determine polycyclic aromatic hydrocarbons (PAHs) in water samples. Compared to organic solvents, larger microdroplets were formed using $[C_8MIM][PF_6]$, resulting in several orders of magnitude increase in the EFs. Later, ILs with the form $[C_nMIM][PF_6]$ were used as extractants in both direct-SDME and HS-SDME for the extraction of several compounds, given that they have adequate viscosity and immiscibility in water and nonvolatility (Liu et al., 2004, Peng et al., 2005). The viscosity of ILs plays a critical role in the extraction performance. If the viscosity of ILs is low the microdroplet tends to fall into the sample solution, whereas for more viscous ILs, the microdroplet often traverses up the body of the syringe. Thus, an optimum range of viscosity is needed for a successful extraction.

The main advantages of ILs as extracting solvents for SDME are that they allow the use of larger drop volumes for longer extraction times in comparison with organic solvents, leading to the development of protocols with higher EFs and sensitivity. The negligible volatility and thermal stability of ILs allows their exposure to the headspace of samples heated at high temperature without loss (Peng et al., 2005). In addition to organic compounds, ILs have been also used as extractants in SDME of heavy metal ions from environmental and food samples. In 2009, Manzoori et al. used

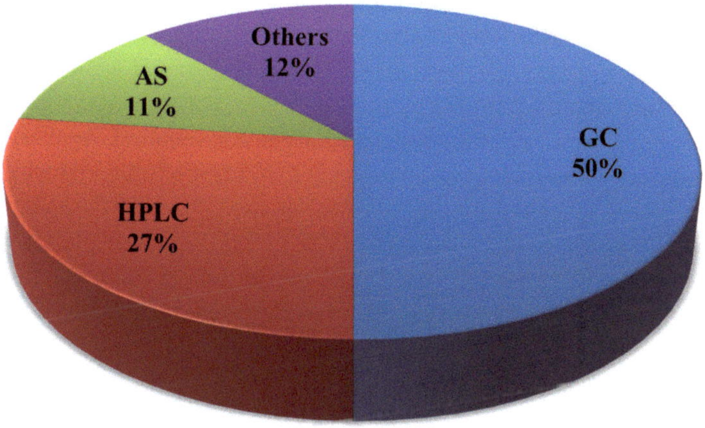

Figure 5.2 Organic solvents used as the extractant in SDME techniques.

IL-SDME for the separation and preconcentration of lead from environmental water samples (Manzoori et al., 2009). Ammonium pyrrolidine dithiocarbamate was used as the complexing agent. The complex was extracted into 7 µL of IL and was directly injected into a graphite furnace for quantification.

ILs based on the $[PF_6]$ anion are the most common ILs used in SDME. However, the limitation of these ILs in SDME is their instability and possibility of dissolution in the sample matrix especially at long extraction times. In 2009, Yao et al. prepared a new class of ILs containing the tris(pentafluoroethyl)trifluorophosphate ([FAP]) anion with imidazolium, phosphonium, and pyrrolidinium cations and used them as extractants in direct immersion SDME for polyaromatic hydrocarbons (PAHs) (Yao, 2009). The extract was analyzed by HPLC. The authors found that the selectivity and sensitivity of the extraction can be monitored by tailoring of ILs using functionalized cations. Thus, with the ILs containing large, hydrophobic cations such as trihexyl(tetradecyl)phosphonium FAP, a high EF can be obtained for extraction of high molecular weight and fused rings PAHs, whereas ILs containing smaller and less hydrophobic cations such as 1-hexyl-3-ethylimidazolium FAP are more suitable for the extraction of smaller and more polar PAHs molecules. The main advantages of these groups of ILs compared to $[PF_6]$- and bis(trifluoromethylsulfonyl)imide ($[NTf_2]$)-based ILs is their strong hydrophobic nature, high thermal stability and better coordination between cations and anion which permit them to be used for longer sampling times in SDME without the dissolution or loss of the IL drop.

Surfactants dissolved in $[C_4MIM][Cl]$ have also been used as extractants in SDME (Yao et al., 2010). The highest sensitivity was obtained with $[C_4MIM][Cl]$-$[C_{12}MIM]Br$ micellar solvent for PAHs with high molecular weights and fused rings, whereas the $[C_4MIM][Cl]$-SDS was more sensitive to smaller and more polar molecules.

For analysis of extracts of IL-based SDME, HPLC is the most often used technique. It should be noted that because of high thermal stability and low pressure, ILs are not compatible with direct injection with GC, the most common technique combined with SDME using organic solvents as the extractant. For direct combination of IL-based SDME with GC, a specially designed interface is required. Aguilera-Herrador et al. (2009) designed a removable interface for GC in which the extracted analytes from the IL microdroplet are transferred to the analytical column but prevent the IL itself from entering the column. The advantages of this design are that the column is not contaminated with IL, the GC system does not require special modification and it is possible to use IL-SDME for extraction and analysis of volatile organic compounds in which analysis by HPLC is not practical.

5.3.2 Extractant Phases for Directly-suspended Droplet Microextraction

Although direct-SDME method has the advantages of simplicity, rapidity and low cost, its main limitation is the instability of the organic drop that is held at the tip of a microsyringe needle due to gravity, shear force and flow-field turbulence. The volume of the microdroplet is also limited to ~5 μL, which results in poor compatibility with some analytical instruments such as HPLC that need larger injection volumes. To overcome these problems, a directly-suspended droplet microextraction (DSDME) method was developed by Yangcheng and co-workers (Yangcheng et al., 2006). In this method, a stirring bar is placed at the bottom of the vial containing the aqueous sample solution and is rotated at a suitable speed to cause a gentle vortex formation. Then, a small volume of the extracting solvent is added to the surface of the sample solution and a single droplet is formed at or near the center of rotation. Compared with SDME, this method has more flexibility in the choice of stirring frequency, stability and the volume of solvent. Sarafraz-Yazdi et al. used a mixture of 90:10 (v/v) toluene/1-octanol as the extraction solvent and combined DSDME with HPLC for the determination of diclofenac (Sarafraz-Yazdi et al., 2008). The same authors also applied DSDME with GC coupled to a flame ionization detector (FID) for the determination of two tricyclic antidepressant drugs, amitriptyline and nortriptyline (Sarafraz-Yazdi et al., 2007), and BTEX compounds (Sarafraz-Yazdi et al., 2009).

The main disadvantage of DSDME is the difficulty of separation of the small volume of the suspended extracting solvent (<5 μL) from the aqueous solution. Exact collection of the extracting solvent with a microsyringe is complicated and some water may be transferred into the syringe that can create difficulties for some instruments (*e.g.*, GC-ECD). To overcome this problem, a new extraction method named solidified floating organic drop microextraction (SFODME) was developed by Khalili-Zanjani et al. (2007). In this method, a small volume of an organic solvent with a melting point near room temperature (in the range of 10–30 °C) is floated on the surface of the aqueous sample. The solution is stirred for a prescribed time and the sample is transferred into an ice bath to solidify the organic droplet. The solidified floating organic drop is then transferred into a small conical vial where it is melted immediately at room temperature. The extracted analyte is then determined by either chromatographic or spectrometric methods.

SFODME has the advantages of simplicity, consumption of a small volume of low-toxicity organic solvent, good reproducibility, low cost, achievement of high preconcentration factors and suitability for the extraction and analysis from samples with complex matrices. In SFODME, the selection of an appropriate extraction solvent is very important to obtain high recovery and EFs. The ideal extraction solvent must have several characteristics: It must be immiscible with water in order to have good extraction efficiency; it should have low volatility to prevent its loss during the extraction; it should have a lower density than water and high affinity for the analytes; it should have a melting point near room temperature (in the range of 10–30 °C); and it must be compatible with the detection system (for example, in combination with chromatographic techniques, its peak must be well-separated from the peaks of analytes). According to these criteria, the organic extraction solvents commonly used in SFODME are presented in **Table 5.2** (Ghambarian et al., 2013). 1-undecanol is frequently the solvent of choice to extract metal ions (Dadfarnia et al., 2008; Shirani Bidabadi et al., 2009), while other solvents such as 1-dodecanol and 2-dodecanol have been used to extract organic compounds (Farahani et al., 2008).

Table 5.2 Common used organic solvents in SFODME.

Extraction solvent	Melting point (°C)	Boiling point (°C)	Density (g cm^{-3})
1-undecanol	13–15	243	0.83
1-dodecanol	22–24	259	0.83
2-dodecanol	17–18	249	0.80
n-hexadecane	18	287	0.77
1-bromohexadecane	17–18	190	0.99
1-chlorooctadecane	20–23	157	0.85
1,10-dichlorodecane	14–16	167	0.99

The volume of the organic extraction solvent directly influences the extraction efficiency and EF of analytes. By increasing the extraction solvent volume to some extent, the extraction efficiency is increased. However, further increases in the extraction solvent volume will cause an increase in the volume of the final floating organic phase, resulting in dilution of the analytes and a decrease in the EF and sensitivity. Therefore, the volume of extraction solvent should be optimized to reach the maximum sensitivity. In other words, the volume of solvent should be sufficient for subsequent analysis and it should also provide a relatively high EF. For experimental purposes, 5–200 μL of extraction solvent is usually chosen.

5.3.3 Extractant Phases for Hollow Fiber Liquid-phase Microextraction

As was discussed earlier, a successful extraction requires a large distribution coefficient for the analyte between the organic and aqueous phase. In the two-phase HF-LPME sampling mode, the analyte *X* is extracted from an aqueous sample (donor phase) through the organic phase which is immobilized in the pores of hollow fiber into the acceptor phase inside the lumen of the hollow fiber. The acceptor phase may be the same organic solvent as the one immobilized in the pores of hollow fiber and the analyte *X* is finally collected in the organic phase. Thus, the extraction process in the two-phase mode can be represented by the following equation (Rasmussen & Pedersen-Bjergaard, 2004):

$$X_{sample} \leftrightarrow X_{organic\ phase} \tag{5.1}$$

The equilibrium process is dependent on the partition coefficient of the analyte between the organic acceptor solution and the donor solution ($K_{a/d}$) which is defined as:

$$K_{a/d} = C_{aqueous}/C_{organic} \tag{5.2}$$

where $C_{aqueous}$ and $C_{organic}$ are the analyte concentration at equilibrium in the donor solution and the acceptor solution, respectively.

In the three-phase mode of HF-LPME, the acceptor solution may be another aqueous phase in which the analyte *X* is extracted from an aqueous sample, through the thin film of organic solvent. The equilibria involved are as follows:

$$X_{sample} \leftrightarrow X_{organic\ phase} \leftrightarrow X_{aqueous\ acceptor\ phase} \tag{5.3}$$

The three phases HF-LPME are characterized by $K_{org/d}$ and $K_{a/org}$ which are the distribution ratio at equilibrium between the organic/donor phases and acceptor solution/organic phase; respectively. The overall distribution ratio $K_{a/d}$ between the acceptor and donor phase can be written as:

$$K_{a/d} = K_{org/d} \times K_{a/org} \tag{5.4}$$

For successful three-phase HF-LPME, the $K_{a/d}$ must be much greater than one. This can be achieved by analyte conversion in the acceptor phase, by exploiting reactions such as complexation or protonation that lead to the formation of analyte species with lower affinity for the organic phase. Thus, in order to achieve a high distribution ratio and develop a sensitive two- or three-phase HF-LPME for efficient analyte separation and preconcentration, the selection of the suitable solvent is a crucial step.

In general, several water-immiscible solvents differing in polarity and water solubility should be tested. It is also possible to use a mixture of organic solvent to obtain the proper solubility parameter (*d*). Care should also be taken to avoid air bubble formation during immobilization of the solvent in the hollow fiber, as bubbles adhering to the surface of the hollow fiber promote solvent evaporation and poor precision. In summary, the final choice of the solvent has to satisfy the following criteria (Rasmussen & Pedersen-Bjergaard, 2004; Pedersen-Bjergaard & Rasmussen, 2008):

1. It should be immiscible or have low solubility in water so as to prevent its dissolution into the aqueous phase.
2. The solubility of the analyte in the immobilized solvent should be higher than its solubility in the donor phase to promote analyte migration through the pores of the hollow fiber.
3. The solvent should have low volatility to prevent its evaporation during extraction.
4. The organic solvent should have a polarity matching that of the polypropylene fiber, so that it can be easily immobilized within the pores of the hollow fiber.
5. The solvent should provide an appropriate extraction selectivity to provide high extraction recovery.
6. It should be compatible with the instrument used for the analysis of the final extract.

The most often used solvents are 1-octanol, toluene and dihexyl ether followed by hexane, octane, nonane, dichloromethane, butyl acetate, 2-octanone and diamyl ether. Typically, two-phase HF-LPME is conducted with either toluene or *n*-octanol as the organic phase, whereas three-phase HF-LPME involves the use of *n*-octanol or dihexyl ether as the support liquid membrane.

ILs have also been used as solvents in HF-LPME. The unique physicochemical properties of ILs such as high stability, high viscosity, tunable polarity, non volatility, good thermal stability, compatibility of its polarity with the polypropylene fiber as well as good extractability for various organic compounds and metal ions make them very useful solvents for HF-LPME. Peng et al. (2007) were the first to support the IL 1-octyl-3-methylimidazolium hexafluorophosphate $[C_8MIM][PF_6]$ on a hollow fiber for the extraction and determination of chlorophenols in environmental water samples. They reported that, in comparison with traditional organic solvents, the HF-LPME based on ILs was more efficient, easier to operate and had good reproducibility and spiked recovery. In addition, supported ILs are quite stable under mild stirring conditions.

In the three-phase mode of HF-LPME, ILs are immobilized within the pores of the membrane, which acts as a barrier between the donor phase and acceptor phase. Furthermore, ILs of $[C_4MIM][PF_6]$ or $[C_6MIM][PF_6]$ have been immobilized in the hollow fiber for the extraction of heavy metal ions such as Pb, Ni and Cd from water samples (Abulhassani et al., 2010).

5.3.4 Extractant Phases for Dispersive Liquid-liquid Microextraction

DLLME was introduced by Rezaee and co-workers in 2006 (Rezaee et al., 2006). This technique uses a few microliters of a binary mixture of disperser and extraction solvents. A cloudy solution is formed when the mixture is rapidly injected into the aqueous sample containing the analytes of interest with a syringe. The hydrophobic solutes are enriched in the fine droplets of extraction solvent and after centrifugation, determination of the analytes can be performed by conventional instrumental analysis. Thus, the selection of an appropriate extraction solvent is an important parameter for DLLME process. To separate the extraction solvent from the aqueous phase by centrifugation, the extraction solvent should have different densities than the aqueous sample solution (usually higher density). The extraction solvent must have good extraction efficiency for compounds of interest, and low solubility in water. It must be miscible with the dispersive solvent and form a stable cloudy solution. Halogenated hydrocarbons with density higher than water, such as chlorobenzene, bromobenzene, dibromobenzene, chloroform, carbon tetrachloride, and tetrachloroethylene, have been used as DLLME extractants. Other non-chlorinated solvents such as 1-undecanol, 1-dodecanol, 2-dodecanol, *n*-hexadecane, carbon disulphide, 1-octanol, toluene and nitrobenzene have also been used as extraction phases. The application of ILs as extraction solvents in DLLME has become a comtemporary research topic (Han et al., 2012). ILs have been dispersed into the aqueous phase either with a common dispersive solvent such as methanol (Liu et al., 2009) or dispersed and dissolved in aqueous sample by temperature control of the mixture (Zhou et al., 2008).

The dispersive solvent also plays a key role in DLLME. It must be miscible with both the aqueous and organic extraction solvents. In practice, the disperser represents 97–99% of the total volume of the extraction mixture and helps to form a cloudy solution of fine droplets of extraction solvent in aqueous samples. In comparison with other LPME methods, there is abundant surface contact between fine droplets and the analytes in DLLME, which speeds up the mass transfer process of analytes from the aqueous phase to the organic phase, and the system quickly reaches equilibrium. Due to low toxicity and low cost, acetone, methanol, ethanol and acetonitrile are generally used as the dispersive solvent, while tetrahydrofuran (THF) has also been used. Although THF is more costly and noxious than other dispersive solvents, it may reduce the volume requirement of chlorinated extraction solvents (Guo et al., 2009).

Unlike SDME, a syringe is not applied to hold the extracting solvent drop during the extraction in DLLME, but used only in the collection and injection of the extract. Thus, problems such as drop dislodgment are avoided.

The main advantages of DLLME are simplicity, rapidity, low sample volume, negligible consumption of extracting and dispersing solvents, low cost, high recovery, high EFs and a very short extraction time. The technique can be applied for the determination of various trace organic pollutants and metal ions in different samples. Factors that affect the process include the type and volume of extraction solvent, type and volume of dispersive solvent, pH of the sample solution and the extraction time. In order to achieve good performance, all of these parameters should be optimized.

Jafarvand and Shemirani developed an alternative DLLME method called supramolecular-based dispersive liquid-liquid microextraction (SM-DLLME) (Jafarvand & Shemirani, 2011). In this method, the analyte is micro-extracted with coacervates composed of reverse micelles made from decanoic acid and dispersed in THF-water mixtures. In comparison with conventional DLLME, SM-DLLME uses decanoic acid, which is a more environmentally friendly solvent.

As stated previously, a suitable disperser solvent has to be miscible with both aqueous and organic phases to ensure the formation of a cloudy solution and to facilitate the extraction process. However, the use of relatively large volume of disperser solvent (mL-level) can be the most significant drawback of DLLME, because it causes a partial dissolution of the analytes in the aqueous phase and decreases the efficiency of extraction. To overcome this drawback, researchers have attempted to perform DLLME without a disperser solvent. In this regard, techniques such as cold induced aggregation microextraction (CIAME), *in situ* solvent-formation microextraction (ISFME), magnetic stirring-assisted DLLME, ultrasound-assisted DLLME and air-assisted DLLME has been developed. Readers are reffered to chapter 4 for further information on these miniaturized techniques.

CIAME is a modified DLLME procedure (Baghdadi & Shemirani, 2008). It involves the addition of an IL as the extractant solvent and a non-ionic surfactant as the anti-sticking agent to an aqueous sample containing the analyte and derivative reagent in a conical-bottom centrifuge tube. IL is dissolved in the aqueous sample by heating the centrifuge tube in a thermostated water bath. After shaking, the centrifuge tube is placed in an ice bath and a cloudy solution like DLLME is formed. Then, the mixture is centrifuged. The fine droplets of ILs are settled in the bottom of the centrifuge tube and are easily separated from the bulk aqueous solution by inverting the tube. The solubility of IL increases with increasing the salt content of the sample. However, according to the common ion effect, the solubility of the IL extractant phase can be decreased by adding another IL with the same ion. Thus, in this method, two different ILs are required to carry out the extraction of analytes in samples with high salt content. Furthermore, the addition of a non-ionic surfactant as anti-sticking agent to the sample prevents the adhesion of ILs to the wall of the tube after centrifugation. Thus, in the presence of a non-ionic surfactant during the phase separation, molecules of the surfactant surround the fine droplets of ILs and decrease their interactions with the wall of the centrifuge tube.

ISFME is another technique derived from DLLME and is based on the *in situ* formation of a hydrophobic IL in the sample (Baghdadi & Shemirani, 2009). In ISFME, a hydrophilic IL (*e.g.*, 1-hexyl-3-methylimidazolium tetrafluorobarate) and an ion-pairing agent (sodium hexafluorophosphate) are added to the aqueous sample. A cloudy solution results from the formation of the fine droplets of a hydrophobic IL (1-hexyl-3-methylimidazolium hexafluorophosphate). In this mode of DLLME, there is

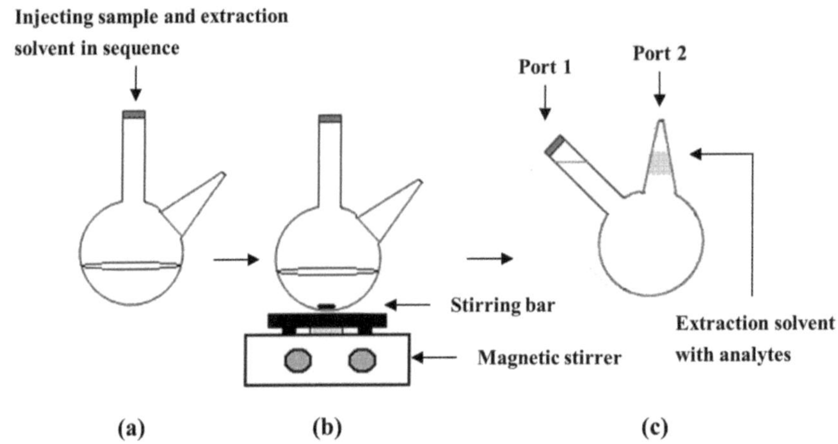

Figure 5.3 Magnetic-assisted DLLME procedure using an extraction solvent that is less dense than water. (A) Extraction device containing aqueous sample and extraction solvent, (B) agitating of the mixture using a magnetic stirrer, (C) elevating the organic phase in the narrow tip of the flask (port 2) by tilting the flask to keep port 2 straight and adding pure water into the flask through port 1.

no interface between the aqueous phase and the extractant phase, and the dispersive solvent is not needed.

Ultrasound radiation in solutions causes acoustic cavitation, which accelerates chemical reactions and mass transfer. Huang et al. first developed an ultrasound-assisted DLLME method to determine NO in cell samples (Huang et al., 2006). In this method, the dispersion of the extractant solvent in the aqueous phase is facilitated by ultrasonic agitation. After the microextraction procedure is complete, a centrifugation step is usually carried out to separate the immiscible phase. This technique is discussed in more detail in chapter 4.

In the magnetic stirring-assisted DLLME method, a binary solvent system of disperser and extraction solvent is avoided. Thus, an extraction solvent having a density lower than water, such as 1-octanol, is added to the aqueous sample solution (Zhang et al., 2011). The extraction phase is dispersed and extraction is accelerated by magnetic agitation of the two phases. After extraction is completed, separation of the organic and aqueous phases is easily obtained by setting aside the extraction system for a period of time. The less dense organic phase, which floats on the surface of the aqueous sample, is then collected *via* the narrow open tip of the flask upon addition of distilled water into the extraction vessel *via* other port to raise the volume level (**Figure 5.3**). The extract is then withdrawn with a microsyringe for the subsequent analysis. The extraction solvent can be easily separated from the aqueous phase without centrifugation, which simplifies the extraction process and make its automation easier.

Air-assisted liquid-liquid microextraction (**Figure 5.4**) is another mode of DLLME which does not require the use of a disperser solvent. In this mode of extraction the

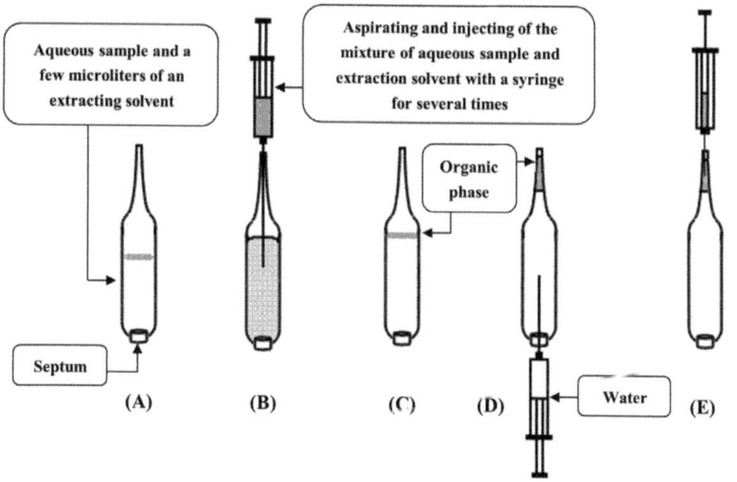

Figure 5.4 Air assisted liquid-liquid microextraction procedure using an extraction solvent less dense than water. (A) Extraction vessel containing aqueous sample and extraction solvent; (B) aspirating and injecting the mixture of aqueous sample solution and extraction solvent with a syringe several times; (C) collection of organic extraction solvent at the top of aqueous phase after centrifuging; (D) elevating the organic phase by injecting de-ionized water through the septum in the bottom of vessel with a syringe; (E) removal of the collected organic phase in the narrow portion of the tube by the use of a syringe.

extraction solvent must also have a density lower than water (Farajzadeh & Khoshmaram, 2013). The aqueous sample solution containing the analytes and appropriate amount of a salt are placed into an extracting vessel and then the mixture is repeatedly drawn into a glass syringe and then injected into the extraction vessel. The solution becomes more and more turbid by this aspiration action. After performing the aspiration and injection cycles for a predetermined number of times, the mixture is centrifuged and the fine droplets of extracting solvent float on top of the aqueous phase. Then, by injection of de-ionized water through the septum at the bottom of extraction vessel, the organic drop containing the extract is elevated and so that it can be collected by a syringe.

The main advantage of these methods over the traditional DLLME technique is that by eliminating the disperser solvent, the EFs, extraction recoveries and limit of detections are significantly improved.

5.4 Conclusions

In this chapter, a strategy for the selection of solvents for different modes of LPME has been discussed. By proper choice of solvent, LPME may provide high EFs, high extraction recoveries and excellent sample clean-up with short extraction times. Of all the desirable properties described, solubility, selectivity, immiscibility, interfacial tension and chemical reactivity are essential for the process to be successfully carried out. The remaining properties that must be given consideration are good engineering work and cost estimation. Furthermore, IL-based LPME methods have become a promising sample preparation technique. The unique and tunable physical and chemical properties of ILs enable the design and synthesis of specific ILs for selective extraction, which expands the application of ILs to LPME technology.

Abbreviations

$[C_nMIM][PF_6]$	n = 4, 6, 8 ,1-alkyl-3-methylimidazolium hexafluorophosphate
AS	atomic spectroscopy
BTEX	benzene toluene ethylbenzene and xylenes
CIAME	cold induced aggregation microextraction
direct-SDME	direct single-drop extraction
DLLME	dispersive liquid-liquid microextraction
DSDME	directly-suspended droplet microextraction
ECD	electron capture detector
EF	enrichment factor
ETAAS	electrothermal atomic absorption spectrometry
FAP	tris(pentafluoroethyl)trifluorophosphate
FID	flame ionization detection
GC	gas chromatography
HF-LPME	hollow fiber-liquid phase microextraction
HPLC	high performance liquid chromatography
HS-SDME	headspace single-drop extraction
ILs	Ionic liquids
ISFME	*in situ* solvent-formation microextraction
LPME	liquid-phase microextraction
NTf_2	bis(trifluoromethylsulfonyl)imide
PAHs	polyaromatic hydrocarbons
SDME	single-drop extraction
SFODME	solidified floating organic drop microextraction
SM-DLLME	supramolecular-based dispersive liquid-liquid microextraction
THF	tetrahydrofuran.

References

Abulhassani, J. Manzoori J. L.,, & Amjadi M. (2010). Hollow fiber based-liquid phase microextraction using ionic liquid solvent for preconcentration of lead and nickel from environmental and biological samples prior to determination by electrothermal atomic absorption spectrometry. *Journal of Hazardous Materials, 176*, 481–486.

Aguilera-Herrador, E., Lucena, R., Cárdenas, S., & Valcárcel, M. (2009). Ionic liquid-based single drop microextraction and room-temperature gas chromatography for on-site ion mobility spectrometric analysis. *Journal of Chromatography A*, 1216, 5580–5587.

Baghdadi, M., & Shemirani, F. (2008). Cold-induced aggregation microextraction: A novel sample preparation technique based on ionic liquids. *Analytica Chimica Acta, 613*, 56–63.

Baghdadi, M., & Shemirani, F. (2009). *In situ* solvent formation microextraction based on ionic liquids: A novel sample technique for determination of inorganic species in saline solutions. *Analytica Chimica Acta, 634*, 186–191.

Barwick, V. J. (1997). Strategies for solvent selection- a literature review. *TrAC Trends in Analytical Chemistry, 16*, 293–309.

Dadfarnia, S., & Haji Shabani, A. M. (2010). Recent development in liquid phase microextraction for determination of trace level concentration of metals—A review, *Analytica Chimica Acta, 658*, 107–119.

Dadfarnia, S., Salmanzadeh, A. M., & Haji Shabani, A. M. (2008). A novel separation/preconcentration system based on solidification of floating organic drop microextraction for determination of lead by graphite furnace atomic absorption spectrometry. *Analytica Chimica Acta, 623*, 163–167.

Farahani, H., Yamini, Y., Shariati, S., Khalili Zanjani, M. R., & Mansour-Baghahi, S. (2008). Development of liquid phase microextraction method based on solidification of floated organochlorine pesticides in water samples. *Analytica Chimica Acta, 626*, 166–173.

Farajzadeh, M. A., & Khoshmaram, L. (2013). Air-assisted liquid-liquid microextraction-gas chromatography-flame ionisation detection: A fast and simple method for the assessment of triazole pesticides residues in surface water, cucumber, tomato and grape juices samples. *Food Chemistry, 141*, 1881–1887.

Ghambarian, M., Yamini, Y., & Esrafili, A. (2013). Liquid-phase microextraction based on solidified floating drops of organic solvents, *Microchimica Acta, 180*, 519–535.

Guo, J. H., Li, X. H., Cao, X. L., Li, Y., Wang, X. Z., & Xu, X. B. (2009). Determination of triclosan, triclocarban and methyl triclosan in aqueous samples by dispersive liquid-liquid microextraction combined with rapid chromatography. *Journal of Chromatography A, 1216*, 3038–3043.

Han, D., Tang, B., Lee, Y. R., & Row K. H. (2012). Application of ionic liquid in liquid phase microextraction technology. *Journal of Separation Science, 35*, 2949 -61.

Huang, K. J. Wang, H., Ma, M., Sha, M. L., & Zhang, H. S. (2006). Ultrasound-assisted liquid phase microextraction and high-performance liquid chromatographic determination of nitric oxide produced in PC12 cells using 1,3,5,7-tetramethyl-2,6-dicarbethoxy-8-(3',4'-diaminophenyl)-difluoroboradiaza-s-indacene. *Journal of Chromatography A, 1103*, 193–201.

Jafarvand, S., & Shemirani, F. (2011). Supramolecular-based dispersive liquid-liquid microextraction: A novel sample preparation technique for determination of inorganic species. *Microchimica Acta, 173*, 353–359.

Jeannot, M. A., & Cantwell F. F. (1996). Solvent microextraction into a single drop. *Analytical Chemistry, 68*, 2236–2240.

Khalili Zanjani, M. R., Yamini, Y., Shariati, S., & Jönsson, A. A. (2007). A new liquid-phase microextraction method based on solidification of floating organic drop. *Analytica Chimica Acta, 585*, 286–293.

Lide, D. R., ed. (1991–1992) CRC Handbook of Chemistry and Physics (1991–1992) 72nd edition, Boca Roaton, Florida, CRC Press.

Liu, J. F., Chi, Y. G., Jiang, G. B., Tai, C., Peng, J. F., & Hu, J. T. (2003). Use of ionic liquids for liquid-phase microextraction of polycyclic aromatic hydrocarbons, *Analytical Chemistry, 75*, 5870–5876.

Liu, H., & Dasgupta P. K. (1996). Analytical chemistry in a drop. Solvent extraction in a microdrop. *Analytical Chemistry, 68*, 1817–1821.

Liu, J. F., Jiang, G. B., Chi, Y. G., Cai, Y. Q., Zhou, O. X., & Hu, J. (2004). Ionic liquid-based liquid phase microextraction, a new sample enrichment procedure for liquid chromatography. *Journal of Chromatography A, 1026*, 143–147.

Liu, Y., Zhao, E., Zhu, W., Gao, H., & Zhou Z. (2009). Determination of four heterocyclic insecticides by ionic liquid dispersive liquid–liquid microextraction in water samples, *Journal of Chromatography A, 1216*, 885–891.

Manzoori, J. L., Amjadi, M., & Abulhassani, J. (2009). Ultra-trace determination of lead in water and food samples by using ionic liquid-based single drop microextraction-electrothermal atomic absorption spectrometry. *Analytica Chimica Acta, 644*, 48–52.

Pedersen-Bjergaard, A., & Rasmussen, K. E. (2008). Liquid-phase microextraction with porous hollow fibers, a miniaturized and highly flexible format for liquid-liquid extraction. *Journal of Chromatography A, 1184*, 132–142.

Pena-Pereira, F., Lavilla, I., & Bendicho C. (2009) Miniaturized preconcentration methods based on liquid-liquid extraction and their application in inorganic ultratrace analysis and speciation. *Spectrochimica Acta Part B: Atomic Spectroscopy, 64*, 1–15.

Pena-Pereira, F., Lavilla, I., & Bendicho C. (2010). Liquid-phase microextraction approaches combined with atomic detection: A critical review. *Analytica Chimica Acta 669*, 1–16.

Peng, J. F., Liu, J. F., Hu, X. L., & Jiang, G. B. (2007). Direct determination of chlorophenols in environmental water samples by hollow fiber supported ionic liquid membrane extraction coupled with high-performance liquid chromatography. *Journal of Chromatography A, 1139*. 165–170.

Peng, J. F., Liu J. F., Jiang, G. B., Tai, C., & Huang, M. J. (2005). Ionic liquid for high temperature headspace liquid-phase microextraction of chlorinated anilines in environmental water samples. *Journal of Chromatography A, 1072*, 3–6.

Psillakis, E., & Kalogerakis, N. (2002). Development in single-drop microextraction. *TrAC Trends in Analytical Chemistry, 21*, 53–63.

Psillakis, E., & Kalogerakis, N. (2003). Development in liquid phase microextraction. *TrAC Trends in Analytical Chemistry, 22*, 565–574.

Rasmussen, K. E., & Pedersen-Bjergaard, A. (2004). Development in hollow fiber-based, liquid-phase microextraction. *TrAC Trends in Analytical Chemistry, 23*, 1–10.

Rezaee, M., Assadi, Y., & Milani Hosseini, M. R. (2006). Determination of organic compounds in water using dispersive liquid-liquid microextraction. *Journal of Chromatography A, 1116*, 1–9.

Sarafraz-Yazdi, A., Amiri, A. H., & Es'haghi, Z. (2009). Separation and determination of benzene, toluene, ethylbenzene and o-xylene compounds in water using directly suspended droplet microextraction coupled with gas chromatography-flame ionization detector. *Talanta, 78*, 936–941.

Sarafraz-Yazdi, A., Mofazzeli, F., & Es'haghi, Z. (2008). Directly suspended droplet three liquid phase microextraction of diclofenac prior to LC. *Chromatographia, 67*, 49–53.

Sarafraz-Yazdi, A., Raouf-Yazdinejad, S., Es'haghi, Z. (2007). Directly suspended droplet microextraction and analysis of amitriptyline and nortriptyline by GC. *Chromatographia, 66*, 613–617.

Shirani Bidabadi, M., Dadfarnia, S., & Haji Shabani, A. M. (2009). Solidified floating organic drop microextraction (SFODME) for simultaneous separation/preconcentration and determination of cobalt and nickel by graphite furnace atomic absorption spectrometry (GFAAS). *Journal of Hazardous Materials, 166*, 291–296.

Swatloski, R. P., Holbrey, J. D., & Rogers, R. D. (2003). Ionic liquids are not always green: hydrolysis of 1-butyl-3-methylimidazolium hexafluouophosphate, *Green Chemistry, 5*, 361–363.

Yao, C., Pitner, W. R, & Anderson, J. L. (2009). Ionic liquids containing the tris(pentafluoroethyl) trifluorophosphate anion: a new class of highly selective and ultra hydrophobic solvents for the extraction of polycyclic aromatic hydrocarbons using single drop microextraction *Analytical Chemistry, 81*, 5054–5063.

Yao, C., Twu, P., & Anderson, J. L. (2010). Headspace single drop microextraction using micellar ionic liquid extraction solvents. *Chromatographia, 72*, 393–402.

Yangcheng, L., Quan, L., Guangsheng, L., & Youyuan, D. (2006). Directly suspended droplet microextraction. *Analytica Chimica Acta, 566*, 259–269.

Zhang, P. P., Shi, Z. G., Yu, Q. W., & Feng, Y. Q. (2011). A new device for magnetic stirring dispersive liquid-liquid microextraction of UV filters in environmental water samples. *Talanta, 83*, 1711–1715.

Zhou, Q. Bai, H., Xie, G., & Xiao J. (2008). Temperature-controlled ionic liquid dispersive liquid phase micro-extraction, *Journal of Chromatography A, 1177*, 43–49.

6 Method Development with Miniaturized Sample Preparation Techniques

Marta Costas-Rodriguez[a]* and Francisco Pena-Pereira[b]
[a] *Ghent University, Department of Analytical Chemistry, Krijgslaan 281-S12, B-9000, Ghent, Belgium*
[b] *Department of Analytical and Food Chemistry, Faculty of Chemistry, University of Vigo, Campus As Lagoas-Marcosende s/n, 36310 Vigo, Spain*
*e-mail: marta.costasrodriguez@ugent.be

6.1 Introduction

Currently, microextraction techniques are established as reliable and environmentally-friendly sample preparation procedures. The success of these techniques comes from their simple operation, rapid sampling, low cost, (virtually) solventless nature, good recoveries and large enrichment factors (EFs). However, they are not free of limitations. Depending on the microextraction mode, difficulties related to fiber breakage, stripping of coatings, instability and losses of organic solvents and porous membranes are still present. The development of more selective, efficient and versatile procedures to overcome these limitations is an expanding field of study. To date, over 20000 papers have been devoted to microextraction techniques, including 750 reviews. During 2013 and early 2014 alone, 70 reviews were published in the literature. A high number of them were focused on important and groundbreaking topics such as i) green aspects, developments and perspectives of microextraction approaches (Spietelun et al., 2014), ii) new sorbent materials as extractant phases (Majors, 2013; Pereira et al., 2013) and iii) application of these approaches in new areas such as bioanalysis (Ahmad, 2013). Innovations and trends are focusing not only on the extraction process itself, and also on automation and the integration of the sample preparation step with analytical instrumentation. Thus, errors associated with manual handling can be minimized, sample and reagent consumption can be reduced, and sensitivity and precision can be improved. Furthermore, the use of ultrasonic (Bendicho et al., 2012) and microwave irradiation (Li et al., 2013), electrochemical supports and autosamplers (commercially available) are being used to improve the capabilities of microextraction approaches (Spietelun et al., 2014).

Molecularly imprinted polymers (MIPs) and nanoparticles are being used as novel fiber coatings for solid phase microextraction (SPME), stir bar sorptive microextraction (SBME), hollow fiber solid phase microextraction (HF-SPME) and liquid-liquid-solid microextraction (LLSME), providing good operation conditions and analytical characteristics (Lasarte-Aragonés et al., 2011; Zhang et al., 2013; Mehdinia & Aziz-Zanjani, 2013). For example, Mosayeb et al. used the pentycaine imprinted polymer to achieve high extraction selectivity of local anesthetics in plasma and urine samples by microextraction by packed sorbent (MEPS) and liquid chromatography-tandem mass spectrometry (LC-MS/MS). The MEPS-MIP-cartridge could be used for 100 extractions (Mosayeb et al., 2013). LLSME based a on porous membrane-protected MIP-coated silica fiber has been used for the determination of triazine herbizides in complex aqueous samples by high performance liquid chromatography (HPLC). LLSME provided the high selectivity of MIP-SPME and the enrichment and sample cleanup capability of HF-LPME in a single device. Moreover, this technique avoids the disturbance from water when the MIP-SPME fiber is directly exposed to aqueous samples (Hu et al., 2009).

Different kinds of nanoparticles are being combined with microextraction approaches. For instance, a liquid phase microextraction (LPME) method based on a dispersion of Pd nanoparticles combined with electrothermal atomic absorption spectrometry (ETAAS) was used for determining Hg concentrations in water samples (Martinis et al. 2013). The use of Pd nanoparticle dispersion in biphasic systems as an alternative to conventional liquid-liquid extraction (LLE) provided high capacity to extract Hg^{2+} without the use of external complexing reagents.

Recently, Lasarte-Aragonés et al. developed an effervescence-assisted dispersive liquid-liquid microextraction method for the determination of herbicides by gas chromatography-mass spectrometry (GC-MS) using Fe_3O_4 magnetic nanoparticles for extractant removal (Lasarte-Aragonés et al., 2014). The extraction solvent dispersion was chemically assisted by an effervescent reaction while its final recovery was performed by means of the nanoparticles. It proved to be a good alternative for *in situ* extraction of aqueous samples.

The numerous modes of microextraction and the novel modifications being adopted require a detailed study of the variables involved and how they affect the enrichment of target analytes. For this, it is crucial to know the fundamental principles governing the mass transfer of analytes in multiphase systems. In general, a high number of variables are involved in all modes of microextraction and the relationships among them often require moving away from the classical approach of optimization, which does not take into account the interactions between the variables. Thus, experimental designs can be used to investigate and optimize the different parameters affecting the response (*e.g.*, analytical signal, extraction efficiency (EE), EF). There has been a noticeable increase of literature pertaining to the use of chemometric experimental designs in combination with microextraction approaches (Stalikas et al., 2009).

6.2 Evaluation of Experimental Parameters

The analytical performance of every miniaturized sample preparation technique is highly dependent on a set of experimental parameters. The influence of each parameter depends on the physicochemical processes involved in a single miniaturized sample pre-treatment approach. Thus, certain experimental parameters show a vital importance in the development of an analytical method and their evaluation is highly recommended for optimal performance. In this section we deal with the variables more commonly evaluated in microextraction techniques.

6.2.1 Type of Miniaturized Sample Preparation Technique

Selection of the most appropriate miniaturized sample pre-treatment technique should take into account several factors, namely physicochemical properties of the analytes (or analyte derivatives if derivatization is required), complexity of the sample matrix, required selectivity and compatibility with available analytical instrumentation.

Microextraction techniques can be used for the extraction of analytes present in solid, liquid and gas samples. However, the selection of the microextraction technique (and more specifically its microextraction mode) is highly dependent on the state of aggregation and complexity of samples. Thus, direct microextraction modes are commonly employed in the preconcentration of neutral (volatile and non volatile) analytes or analyte derivatives from clean samples. Headspace microextraction approaches, such as headspace solid-phase microextraction (HS-SPME) or headspace single-drop microextraction (HS-SDME), enable the extraction and preconcentration of volatile and semi-volatile analytes and analyte derivatives from simple to complex samples. In addition, three-phase liquid-liquid-liquid microextraction (LLLME) systems allow the preconcentration of ionizable compounds. Membrane-based microextraction approaches are employed when dealing with extraction from complex samples. In short, microextraction techniques as a whole can be considered to be complementary tools for multipurpose extraction processes.

The required selectivity can also be tuned by selecting an appropriate microextraction approach. Thus, two-phase (direct) microextraction techniques provide the lowest selectivity possible, especially when non-selective adsorbents are employed. Increased selectivity can be achieved by using three-phase (headspace and liquid-liquid-liquid) microextraction techniques, since non-volatile and non-ionizable compounds are not extracted under experimental conditions, respectively. It is worth noting that higher selectivity can also be achieved by using selective derivatizing agents (Cabaleiro et al., 2012). A high level of selectivity can be obtained when three-phase microextraction is performed with protective membranes which provide efficient microfiltration as a result of their small pore size.

The compatibility of microextraction techniques with analytical instrumentation is another important factor to consider. SPME and related techniques can be coupled with GC, HPLC and capillary electrophoresis (CE) by appropriate selection of desorption conditions. As for LPME, the compatibility with analytical instrumentation is highly dependent on the nature of the extractant phases employed.

6.2.2 Type of Extractant Phase

Selecting the most appropriate extractant phase is of paramount importance in the development of miniaturized sample preparation methodologies. The type of extractant phase will depend on the physicochemical properties of target compounds, the selected microextraction technique and mode, the sample composition and the analytical technique employed. The final selection of the extractant phase for a given application is generally based on a comparison of the EE, selectivity and toxicity.

The basic principle of 'like dissolves like' is generally taken into account when selecting the extractant phase. Three types of intermolecular interactions can take place between target compounds and a given extractant phase, namely 'non polar' interactions (also called as dispersion interactions), δ_d, permanent dipole-permanent dipole interactions, δ_p, and hydrogen bonding interactions, δ_h. The contribution of each of these three types of intermolecular interactions can be used to obtain the corresponding total Hildebrand solubility parameter (δ_t) in accordance with the following equation (Hansen, 2000):

$$\delta_t = \delta_h^2 + \delta_p^2 + \delta_d^2 \tag{6.1}$$

Target compounds will dissolve in solvents whose solubility parameters are not too different from their own (Hansen, 2000). Even though the use of these solubility parameters can be helpful in the selection of candidate extractant phases with potential for the extraction of target analytes, this practice has not been thoroughly followed in analytical method development involving microextraction techniques, probably due to the lack of tabulated solubility parameters for non-conventional extractant phases and target analytes. Interestingly, the applicability of a set of solvents can be predicted by calculating the three individual solubility parameter components by addition of the corresponding functional group contributions on the basis of the chemical structure (Fitzpatrick & Dean, 2002; van Krevelen, 2009).

A variety of extractant phases, namely organic solvents, ionic liquids, supramolecular solvents and aqueous solutions, have been used in LPME approaches (Pena-Pereira et al., 2010). In LPME, extractant phases are initially selected by comparison of their physicochemical properties (mainly dipole moment) with that of the analytes, bearing in mind their compatibility with the analytical technique used for quantification. It is worth noting that in certain LPME modes the selection of extractant phases is first based on specific physicochemical properties that are not directly related to their extraction capability, but with the LPME mode performance, such as density, water solubility, melting point, vapor pressure or boiling point. For instance, low volatility is a must when dealing with extractant phase selection for HS-SDME, while reduced water solubility is needed in two-phase LPME techniques. Furthermore, extractant phases applicable in dispersive liquid-liquid microextraction (DLLME) show a density higher than that of the sample solution, while solvents with a density lower than that of the sample matrix are required in directly suspended droplet microextraction (DSDME). **Table 6.1** shows the relevance of different physicochemical properties when choosing extractant phases to be applied in a variety of LPME approaches.

Apart from their physicochemical properties, extractant phases can be chosen taking into account their potential toxicity. In this way, solvent selection guides in the literature can be consulted with the aim of obtaining relevant information that can also be judged together with the achievable EE and selectivity for final decision-making (Jiménez-González et al., 2005; Alfonsi et al., 2008; Henderson et al., 2011).

Choosing the most appropriate extractant phase is also of great importance when dealing with SPME and related approaches. In fact, this is generally the first experimental parameter to be considered for optimization. Currently, there are several fiber coatings commercially available for SPME that cover a range of polarities, thicknesses of the stationary phase and extraction mechanisms. In addition, novel materials and fiber coating designs are being regularly proposed in the literature with improved thermal, mechanical and chemical resistance that allow analytical challenges such as the extractability of polar compounds from polar matrices to be overcome (Spietelun et al., 2010; Bagheri et al., 2012).

Guides for selecting the most appropriate fiber coating for SPME and related techniques can be found in the literature (Pawliszyn, 2002; Shirey, 2012). Two different retention mechanisms, namely absorption and adsorption, can take place depending on the fiber coating materials employed. In summary, the above mentioned 'like dissolves like' principle is applicable when dealing with liquid coatings such as poly(dimethylsiloxane) (PDMS) or poly(acrylate). However, porous adsorbents are non-selective, and the extraction process in this case is based on the physical trapping of analyte molecules in the pores of the adsorbent. Generally, porous adsorbents such as divinylbenzene or Carboxen are immobilized onto the fiber by using liquid polymers such as polyethylene glycol or PDMS. This results in different selectivity in terms of polarity. Physicochemical properties of the analyte, such as polarity and volatility, should be kept in mind when choosing the most appropriate fiber coating.

6.2.3 Sample and Extractant Phase Volumes

Sample and extractant phase volumes can influence both the EE and the EF that are achievable with microextraction techniques. EE is defined as the percentage of extracted analyte with respect to its initial amount in the sample solution:

$$EE = \left[\frac{K(V_{extr}/V_s)}{1 + K(V_{extr}/V_s)}\right] \times 100 \tag{6.2}$$

Table 6.1 Relevance of physicochemical properties of the extractant phases used in some LPME approaches.

Physicochemical properties	LPME approaches							
	direct-SDME	HS-SDME	LLLME	Two phase HF-LPME	Three phase HF-LPME	DLLME	DSDME	SFODME
miscibility with water	xxx	x	xx[a]	xxx	xx[a]	xxx	xxx	xxx
water solubility	xxx	x	xx[a]	xxx	xx[a]	xxx	xxx	xxx
vapor pressure	x	xxx	x	x	x	x	xx	xx
boiling point	x	xxx	x	x	x	x	xx	xx
melting point	x	x	x	x	x	x	x	xxx
density	x	x	x	x	x	xxx	xxx	xxx
viscosity	xx	xx	xx	xx	xx	xx	xx	xx
polarity	xxx	xxx	xxx	xxx	xxx	xxx	xxx	xxx

x, less important; xx, important; xxx, very important

[a]The physicochemical property is relevant in the case of the extractant phase which is directly in contact with the sample, but not for the acceptor phase.

EF is defined as the concentration of analyte in the extractant phase with respect to its initial concentration in the sample:

$$EF = \frac{K}{1 + K(V_{extr}/V_s)} \tag{6.3}$$

where K is the partition coefficient, and V_{extr} and V_s are the extractant phase and sample volumes, respectively.

According to equations 6.2 and 6.3, at a given K, a reduced V_{extr}/V_s ratio can give rise to non-quantitative EEs even at equilibrium conditions, while increased EFs can be obtained. For instance, EFs in the range 18000–27000 were obtained in the extraction of antidepressant drugs from water samples by three-phase HF-LPME using 1100 mL and 20 µL of donor and acceptor solutions, respectively (Ho et al., 2007). It is important to note that these impressive EFs correspond to EEs in the range 33–49%. Thus, apart from the K values, the selection of an appropriate V_{extr}/V_s ratio can be critical in the preconcentration of target compounds, enabling their determination at trace or ultratrace levels.

In SPME and related techniques, the extractant phase volume is directly related to the film thickness. The higher the film thickness of the coating fiber, the higher the mass of analyte extracted and, therefore, improved EE can be obtained. However, longer enrichment times are needed to reach equilibrium conditions when using thicker film coatings. This fact is exploited in sample preparation techniques related to SPME, such as SBSE, where the use of larger volumes of extractant phase enables the achievement of higher EEs. The influence of the extractant phase volume on the EE values obtained for SPME and SBSE under equilibrium conditions as a function of K is presented in **Figure 6.1**. According to the figure, the use of SBSE would allow the achievement of quantitative extraction of analytes with $K \geq 10^4$, while SPME would give rise to similar results only with the extraction of very hydrophobic analytes. Nevertheless, it can also be deduced from **Figure 6.1** that an SPME fiber would allow the extraction of highly hydrophobic compounds with higher selectivity than SBSE.

6.2.4 Extraction Time

Extraction processes are time-dependent and, in general, the selection of optimum extraction time is critical since sample pre-treatment can be the time-limiting step of the overall analytical method. In any case, a strict control of extraction time is a requirement to obtain good precision when non-equilibrium conditions are used.

Depending on the extraction process and the phases involved, the equation for the concentration of the analyte in the extractant phase ($C^{extr}(t)$), expressed as a function of extraction time, can show different complexity. For the simplest case, i.e., a two-phase microextraction process, this equation is given by (Jeannot & Cantwell, 1996):

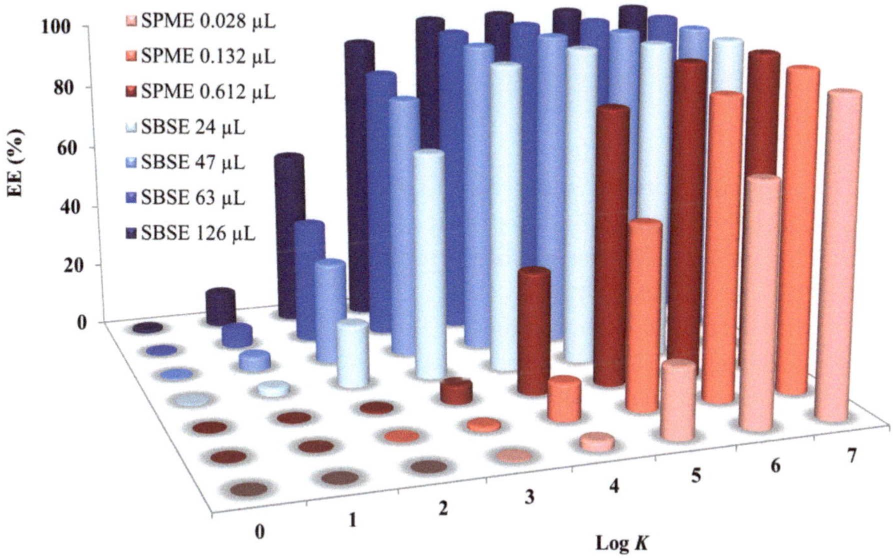

Figure 6.1 Theoretical EEs achievable by commercially available PDMS-based SPME and SBSE coatings under equilibrium conditions as a function of *K*. SPME coatings: 7 µm, 30 µm and 100 µm PDMS; SBSE dimensions: 0.5 mm x 10 mm, 0.5 mm x 20 mm, 1 mm x 10 mm, 1 mm x 20 mm. Other conditions: Sample volume: 10 mL.

$$C^{extr}(t) = C_{eq}^{extr}(1 - e^{-kt}) \tag{6.4}$$

The observed rate constant (*k*) is expressed as:

$$k = \frac{A}{V_{extr}}\bar{\beta}_{extr}\left(1 + K\frac{V_{extr}}{V_s}\right) \tag{6.5}$$

where *A* is the interfacial area, is the overall mass transfer coefficient with respect to the extractant phase, *K* is the distribution coefficient, and V_{extr} and V_s are the extractant phase and sample volumes, respectively.

The extraction of a target compound by a given extractant phase is governed by a mass transfer process that can be explained in accordance with a convective-diffusive mass transfer model such as the Whitman two-film theory. According to this theory, mass transfer is produced by both convection and diffusion in the bulk sample, while steady-state diffusion is produced across layers of thickness δ_{extr} and δ_s at both sides of the interface in the extractant phase and water sample, respectively (Jeannot & Cantwell, 1996). In both the extractant phase and the sample solution, the mass transfer coefficient (β) is directly proportional to the diffusion coefficient (*D*) and inversely proportional to the film thickness in the appropriate phase (δ) (Jeannot

& Cantwell, 1996). If rapid transfer across the liquid-liquid interface is assumed, can be expressed as:

$$\frac{1}{\bar{\beta}_{extr}} = \frac{1}{\beta_{extr}} + \frac{K}{\beta_s} = \frac{\delta_{extr}}{D_{extr}} + \frac{K\delta_s}{D_s} \tag{6.6}$$

When K is large, the first term on the right of equation 6.6 is negligible in comparison with the second term. Thus, the overall mass transfer coefficient and, therefore, the kinetics of extraction greatly depend on the resistance to mass transfer in the sample solution (Jeannot & Cantwell, 1996).

The use of extended extraction times is not commonly recommended in microextraction techniques, since it is usually the limiting step in the whole analytical method. Furthermore, depending on the nature of the extractant phase, extended extraction times could not be practically applicable in certain LPME techniques, since the extractant phase volume can be significantly reduced by evaporation or solubilization. It is worth noting that, according to the above equations, the extraction kinetics are influenced by a variety of experimental parameters that should be evaluated for optimal performance. Furthermore, the use of a different microextraction mode itself can show a significant contribution to the improvement of the extraction kinetics. For instance, DLLME enhances the interfacial area (due to the formation of multiple tiny drops) and higher convection than direct single-drop microextraction (direct-SDME) or HF-LPME, so improvements on the observed rate constant (k) and, as a consequence, on the concentration of the analyte in the extractant phase at a given time can be observed.

6.2.5 Agitation of the Sample

Extraction kinetics can be improved by efficient agitation of the sample, so equilibrium conditions can be achieved in reduced extraction times. Convection can be produced by using an external agitation or pumping device, by using an external energy source or by integrating the extractant phase and an agitation system. Thus, several strategies have been reported to enhance the convective mass transfer in microextraction techniques, namely magnetic agitation, vortex agitation, ultrasound irradiation, needle vibration and flow-through stirring (Pawliszyn, 2002; Lucena, 2012; Andruch et al., 2013). The application of these different strategies has enabled the development of a variety of microextraction modes. In fact, the importance of the strategies employed to enhance convection in different microextraction modes is reflected in their denominations, such as vortex-assisted or ultrasound-assisted microextraction techniques (Andruch et al., 2013; Szreniawa-Sztajnert et al., 2013). In some cases, the extractant phase is immobilized directly on the agitation system in such a way that extraction and agitation of the sample are integrated in the same device. This is the case for SBSE and rotating disk-sorptive extraction (Lucena, 2012).

According to the film theory of convective-diffusive mass transfer, the faster the agitation of the sample the shorter the time needed to reach equilibrium conditions. Fast agitation, however, can be counterproductive when using certain LPME approaches. For instance, the use of a relatively high stirring rate gives rise to the dislodgement of the extractant phase microdrop from the needle of the syringe in direct-SDME. Furthermore, air bubble formation inside the drop is a typical problem observed in direct-SDME and HF-LPME when using fast agitation and/or extended extraction times.

6.2.6 pH

The pH of the sample has an important effect on the EE of ionizable compounds since undissociated forms of analytes are needed for extraction and preconcentration by two-phase microextraction approaches. Thus, the EE of acidic compounds is improved at low pH values, while higher pH values are needed to efficiently extract and preconcentrate alkaline analytes. In SPME and related approaches, however, the adjustment of pH should be done bearing in mind the recommended pH working range for a given fiber coating.

It is worth noting that pH adjustment of both the donor and acceptor phases is of paramount importance in three-phase LPME modes. Thus, high EEs with high selectivity can be obtained by exploiting this parameter in the determination of ionizable analytes by three phase LPME techniques (Zhu et al., 2001). In addition, the adjustment of the pH of the sample can give rise to the formation of a volatile form of the analyte that can be extracted by using a headspace microextraction technique. For instance, extraction and preconcentration of anionic or cationic analytes that have a volatile neutral form can be performed by HS-SDME by appropriate pH adjustment of both the sample and extractant phase bearing in mind the pK_a value of the analytes (Jermak et al., 2007; Pranaitytė et al., 2007).

6.2.7 Ionic Strength

The addition of salt to the sample solution is performed in microextraction techniques to enhance the EE of target compounds by reducing their solubility in the sample solution. The increase in the extractability of the analytes by addition of electrolytes to the sample is commonly known as the “salting-out” effect.

Unlike conventional solvent extraction, the increase in the ionic strength of the sample may give rise to unexpected negative effects on the EE in non-equilibrium microextraction processes.

According to the Stokes-Einstein equation, the diffusion coefficient of the analyte in the sample (D_s) depends on the temperature (T), the viscosity (μ), and analyte radius (R_0) as (Cussler, 1997):

$$D_s = \frac{k_B T}{6\pi\mu R_0} \tag{6.7}$$

where k_B is Boltzmann's constant.

When salt is added to the sample, μ increases and subsequently D_s decreases, according to equation 6.7. Hence, the presence of electrolytes in the sample affects the overall mass transfer of target analytes, in such a way that $C^{extr}(t)$ is decreased in the presence of salts at a fixed microextraction time if equilibrium conditions have not been reached. It can also be deduced that longer extraction times would be required to reach equilibrium conditions when salt-rich samples are subjected to direct-SDME.

Thus, the net effect observed when microextraction is performed in the presence of electrolytes is actually the result of two opposite effects: a) the "salting-out" effect, the magnitude of which depends on the hydrophobicity, diffusion molar volume, molecular diameter and hydrophobic surface area of the analyte (Turner, 2003) and b) the effect due to the decreased diffusion coefficients by modification of the physicochemical properties of the sample (Kokosa et al., 2009). Therefore, positive, negative or neutral effects can be observed depending on the magnitude of each factor, which also depend on the experimental conditions. Hence, Palit et al. demonstrated that an increase on the ionic strength of the sample produced a negative effect on the extraction of chemical warfare agents and related compounds at short direct-SDME times, while a positive effect was observed at longer times using a high concentration of NaCl (Palit et al., 2005). It should also be noted that longer microextraction times were needed to reach equilibrium conditions, as expected from the theoretical aspects described above.

Addition of salt to the sample solution is commonly performed in headspace microextraction to enhance the mass transfer of volatile and semi-volatile compounds to the headspace above the sample. Aqueous solubilities of many compounds are decreased in the presence of large amounts of salt. Thus, improved extractability and, therefore, sensitivity can be obtained by salting-out. Furthermore, adjustment of the salt content in blanks, standards and samples is commonly performed to alleviate for potential matrix effects in the analysis of samples containing high concentrations of salt, such as seawater samples. In certain cases, however, the increase in the ionic strength can be counterproductive, since it can also enhance the mass transfer of interfering compounds. This is especially relevant when non-selective extractant phases, such as porous adsorbents, are used for extraction.

6.2.8 Temperature

Temperature has an important influence on both kinetics and thermodynamics of extraction processes and its control is commonly recommended for optimal performance. Air conditioned and circulating water baths are commonly employed for temperature control. Temperature can give rise to two opposite effects on the extraction process. On the one hand, the increase in the extraction temperature yields increased mass transfer coefficients. This is especially important in headspace microextraction techniques, since temperature affects both sample-headspace and headspace-extractant phase coefficient partitions. Thus, the increase in the extraction temperature would give rise to reduced equilibration times and better precision would be obtained at a given non-equilibrium extraction time. On the other hand, however, the absorption process is exothermic, so the amount of target analytes extracted will decrease with an increase in temperature. It is also important to note that the increase in the extraction temperature also affects the solubility and vapor pressure of the extractant phase. Hence, extractant phases used in two-phase LPME techniques can suffer from increased solubilization and bubble formation, while certain extractant phases employed in HS-SDME can be significantly evaporated during the extraction process when high temperatures are used. Thus, to achieve the highest EE in the shortest extraction times while minimizing such procedural limitations, certain strategies have been reported in the literature. For instance, internally cooled fiber coatings (Zhang & Pawliszyn, 1995), as well as cooled needle and drops (Shariati-Feizabadi et al., 2003) have been reported for headspace microextraction techniques. Nevertheless, a common temperature for both the sample and extractant phase is selected in most reports on analytical method development for simplicity, even though the above mentioned systems enable the achievement of higher EE and improved precision.

Apart from the effect of temperature on the extraction process itself, temperature plays a primary role in the development of certain LPME techniques, such as solidification of floating organic drop microextraction (SFODME) (Khalili Zanjani et al., 2007) or cold-induced aggregation microextraction (Baghdadi & Shemirani, 2008), where the extractant phase is separated from the sample solution at the end of the microextraction process by means of temperature control.

6.2.9 Derivatization

Target analytes do not always show the required physicochemical properties to be efficiently extracted using microextraction techniques and/or to be determined by a given analytical methodology. In these cases, derivatization reactions can be performed with the aim of obtaining a different form of the analytes that can be efficiently and selectively extracted and detected, as well as to avoid interferences from matrix components. In addition, derivatization can be performed to separate analytes with poor chromatographic behavior (Stalikas & Fiamegos, 2008). In general, analyte derivatives should form quickly and quantitatively at room temperature and should exhibit good stability and solubility in the desired solvent or phase.

A variety of strategies can be used to perform derivatization reactions in microextraction techniques. Thus, derivatization reactions can be carried out in the sample (pre-extraction derivatization), in the extractant phase (simultaneous extraction-derivatization or post-extraction derivatization) or in the injection port of a gas chromatograph (post-extraction derivatization).

Direct derivatization in the sample is carried out by addition of a derivatizing agent to the sample. The chemical structure of target analytes is modified by addition of the derivatizing agent to the sample solution to obtain derivatives more amenable for extraction and detection. Extraction of the obtained derivatives is then performed, followed by the introduction of the enriched analyte derivatives in the analytical instrumentation. Pre-extraction derivatization should result in enhanced EE than the original analytes by the corresponding extractant phase. For instance, vapor generation can be performed in the determination of metals and metalloids by headspace microextraction techniques (Mester et al., 2000; Gil et al., 2009). The application of pre-extraction derivatization is simple since widely reported derivatization strategies can be easily implemented.

Derivatization in the extractant phase can be performed simultaneously with the extraction process or after the extraction process. *In situ* extraction, preconcentration and derivatization is convenient, since it enables the direct extraction and simultaneous derivatization in the extractant phase, and reduces the number of unitary steps and time needed to carry out the whole analysis. However, its application depends on the compatibility of the derivatizing agent with the extractant phase and the operational conditions to perform the extraction and the reaction. Furthermore, the optimization of the derivatizing agent composition in the extractant phase is a requirement. When extraction and derivatization cannot be performed in a single step or the partition coefficients of the analyte derivatives are lower than that of the analytes, post-extraction derivatization can be carried out. In this case, the derivatization reaction is performed in the enriched extractant phase by addition of a small amount of derivatizing agent in the case of LPME approaches, or by exposing the enriched SPME fiber to the derivatizing agent solution prior to analysis.

Injection port derivatization is another interesting post-extraction derivatization strategy. It enables the thermal conversion of analytes containing certain polar functional groups such as –OH, –SH, –NH–, and –COOH, into derivatives with improved volatility, thermal stability and detectability (Bizkarguenaga et al., 2013; Wang et al., 2013). Thus, in-port silylation, alkylation and acylation can be efficiently performed by controlling experimental parameters such as the injection port temperature, amount of derivatizing agent and the purge off time (Wang et al., 2013).

The use of derivatization reactions may allow the extraction of analytes that are difficult to extract directly (such as hydrophilic compounds from water samples), the extraction of compounds that do not have the required properties to be extracted by a given microextraction mode (such as non-volatile compounds in headspace microextraction techniques) or enable the determination of extracted compounds that do not show the properties needed (*e.g.*, in-port derivatization of non-volatile extracted compounds). Even though derivatization can be employed to overcome the above aforementioned limitations, it is worth mentioning that derivatizing agents can be a source of interferences and errors. In addition, the use of derivatization reactions should be avoided whenever possible in order to minimize reagent consumption and waste generation, according to the 8th and 6th principles of green chemistry and green analytical chemistry, respectively (Gałuszka et al., 2013). Thus, derivatization processes should be performed only when necessary.

6.2.10 Desorption

Once the microextraction process has finished, target analytes are introduced in the corresponding analytical instrumentation to obtain analytical information. When direct introduction or insertion of the enriched extractant phase is not feasible, desorption can be performed. Desorption conditions must ensure a complete removal of the enriched analytes from the microextraction system with no significant memory effects, damage of the sorbent material or degradation of the analytes. Three possibilities can be considered regarding desorption of analytes when working with SPME and related techniques, namely thermal, liquid and, much less often, laser desorption.

Thermal desorption should enable efficient desorption and rapid transfer of the analytes of interest from the injector to the chromatographic column. In general, high desorption temperatures enable fast desorption. However, high temperatures can affect the stability of the sorbent, and result in bleeding of the polymeric material (Pawliszyn, 2002). Thus, the desorption temperature is commonly determined by the thermal stability of the fiber coating.

Thermal desorption is usually performed by directly inserting the SPME fiber into the inlet of a gas chromatograph at a given temperature for a prescribed time. In thermal desorption, analytes are physically separated from the sorbent phase by using heat to increase the volatility of interest compounds. Thermal desorption is highly dependent on the carrier gas flow rate and the injector temperature (Pawliszyn, 2012).

Related microextraction techniques, such as SBSE, require the employment of a thermal desorption unit to properly perform thermal desorption due to the different dimensions of the specific microextraction system and the injection port of GC. It is worth mentioning that thermal desorption has not only been utilized when microextraction techniques were coupled to GC systems, but also to atomic spectrometry techniques (Mester & Sturgeon, 2005) or ion mobility spectrometry (IMS) (Arce et al., 2008).

When HPLC or CE are used, liquid desorption is the preferred option for analyte introduction. This is also true for desorption of thermally labile compounds. Two options are available: off-line and on-line liquid desorption. Liquid desorption is carried out by using a small volume of an appropriate solvent to quantitatively transfer target analytes to the analytical instrumentation. On-line coupling of SPME and related approaches with liquid phase separation techniques has been reported in the literature by using specially designed desorption interfaces (Saito & Jinno, 2003; Lord, 2007).

As discussed above, analytes can also be desorbed from the SPME fibers by laser desorption. Tong et al. combined SPME with matrix-assisted laser desorption/ionization mass spectrometry for the analysis of large biomolecules (Tong et al., 2002). In this case, analytes were ionized and desorbed from the fiber by using a laser as the energy supply. A series of laser shots are needed to obtain complete ionization and desorption of analyte molecules. Wang et al. employed a SPME/surface enhanced laser desorption/ionization fiber to introduce analytes to an IMS (Wang et al., 2007). Non-volatile and thermally labile compounds were directly desorbed/ionized by a Nd:YAG laser without the addition of a matrix. Laser repetition rate and laser energy were two important experimental parameters to be optimized in the analysis.

6.3 Optimization Strategies For Analytical Method Development

Microextraction techniques require strictly defined experimental conditions. Thus, depending on the mode of microextraction, numerous variables such as extractant phase type and volume, extraction time, agitation, sample volume, ionic strength, temperature and sampling depth should be investigated.

The conventional approach used for the development of the microextraction procedures is 'one-variable-at a-time' (OVAT) where the response of each variable is investigated independently by keeping the remaining variables constant. It requires a high number of experiments, especially if there is an additive effect among the variables, and it does not consider the possible interactions among them (Bianchi & Careri, 2008). OVAT is the most commonly used approach to optimize microextraction procedures even though it is time-consuming and some misinterpretations can occur due to the interactions among the variables.

Currently, chemometric experimental designs are increasingly being used for the optimization of microextraction methods. Using this approach, the variables affecting the response (*e.g.*, the analytical signal, EE, or EF) can be optimized simultaneously, and the interactions among the variables and the significance of the effects of each independent variable can be assessed through statistical analysis (Bianchi, & Careri, 2008; Kokosa et al., 2009). In case of significant interactions among variables, the optimal conditions obtained by OVAT would be different from those obtained by multivariate optimization. This strategy is advantageous compared with OVAT because it considerably reduces the number of experiments, thus reducing time, reagent consumption, experimental work and associated costs.

In general, multivariate optimization entails several steps, including: (i) selection of the appropriate variables (or factors) and responses *via* screening; (ii) choice of the experimental design; (iii) selection of the levels of every variable (domain) and their codification; (iv) mathematical model fitting; (v) evaluation of the model adequacy, i.e., predictability; (vi) analysis of the model and estimation of the effects; (vii) allocation of the optimal values; and (viii) evaluation of the robustness of the analysis (Almeida Bezerra et al., 2008; Stalikas et al., 2009).

The selection of the variables and their levels is crucial to find the conditions which are able to provide the best analytical performance for a target application. As a starting point, they are chosen according to previous experiences and/or reported studies. However, the number of variables under study is generally high and it is not possible to identify the small contributions. Thus, screening (simultaneous) designs are often performed to identify the variables with a significant effect on the response.

The simultaneous designs commonly used for optimization purposes in microextraction methods are the response surface methodologies (RSMs). RSM consists of a group of mathematical and statistical approaches based on the fit of an equation describing the behavior of the experimental data. When there are no interactions among the variables, i.e., the responses do not present any curvature, a first order model (simple linear function) is used to establish which ones are significant:

$$y = \beta_0 \sum_{i=1}^{k} \beta_i x_i + \varepsilon \tag{6.8}$$

where k is the number of variables, β_o is a constant, β_i are the coefficients of the linear parameters, x_i are the variables, and ε the residual of the experiments.

If there are interactions, central points are included in the experimental design to evaluate the curvature and a second order model (equation 6.9) is used to describe the interactions among the variables. The number of terms in the polynomial function is limited to the number of experimental design points. Two-level factorial designs are used in the estimation of first-order effects and three-level factorial design for second-order effects:

$$y = \beta_0 + \sum_{i=1}^{k} \beta_i x_i + \sum_{l \le i \le j}^{k} \beta_{ij} x_i x_j + \varepsilon \tag{6.9}$$

where β_{ij} are the coefficients of the interaction parameters.

Moreover, for determining a critical point (maximum, minimum, or saddle), the polynomial should contain quadratic terms:

$$\sum_{i=1}^{k} \beta_i x_i + \sum_{i=1}^{k} \beta_{ii} x_i^2 + \sum_{l \le i \le j}^{k} \beta_{ij} x_i x_j \varepsilon \tag{6.10}$$

where β_{ii} are the coefficients of the quadratic parameters.

6.3.1 Screening of the Variables

Factorial designs are commonly used for screening purposes. A set of experiments is performed in a systematic way and the equation describing the variation of the responses in the experimental domain is obtained. The equation can be mapped as two-dimensional or three-dimensional (surface response) plots for two and three variables, respectively. To evaluate the statistical significance of these variations, the analysis of the variance (ANOVA) and residual analysis are typically used.

The representation of some common factorial designs used to identify the experimental variables that affect significantly the microextraction methodology is shown in **Figure 6.2**. The two-level factorial design 2^k (where k is number of variables) is one of the most frequently used. In this analysis, each variable adopt two-levels, the maximum (+) and the minimum (–) level.

The levels of variables are chosen according to previous experiences and/or reported studies. The experiments should include all combinations of each level of a variable in all levels of the remaining variables. By contrast, if the optimum region and how far it is from the starting point is unknown, sequential experiments should be performed, i.e., one by one.

Numerous works reported in the literature employed the 2^k design for the evaluation of the significant variables affecting the microextraction methodology. For example, a 2^3 factorial design was used to investigate the significant variables for the determination of pesticide residues in juices by SPME coupled with GC-MS (Bola Abdulra'uf, & Huat Tan, 2013). Three variables (addition of hexane/acetone, salting-out effect, and sample dilution) were studied at two levels and including 3 center points, thus performing 11 experiments. The central points or control runs provide information about the existence of curvature, i.e., rotatability or orthogonality properties of the design. The experiments are carried out randomly to minimize the effects of uncontrolled parameters that may produce bias. However, the central points should be dispersed throughout the matrix design but not randomized. However, the number of experiments increases when k increases (**Figure 6.2C** and **D**). For instance, a three-level factorial design 3^3, resulting in 27 experiments, was performed to study the effects of the parameters affecting the in-tube microextraction of non-polar organic compounds in aqueous samples using multi-walled carbon nanotubes as sorptive material (Hüffer et al., 2013). The authors selected the extraction temperature, number of extraction strokes, and extraction flow based on a previous study. A significant interaction was observed between the extraction temperature and the number of extraction strokes.

To reduce the number of experiments, fractional factorial designs with a particular resolution are highly recommended. In this case, just a fraction of the full factorial design is investigated. The number of experiments is given by $2^{k-p} + C$, where k is the number of variables, C the number of replicates at the center point and p indicates the fractionation of the experimental design. This kind of design is useful for preliminary

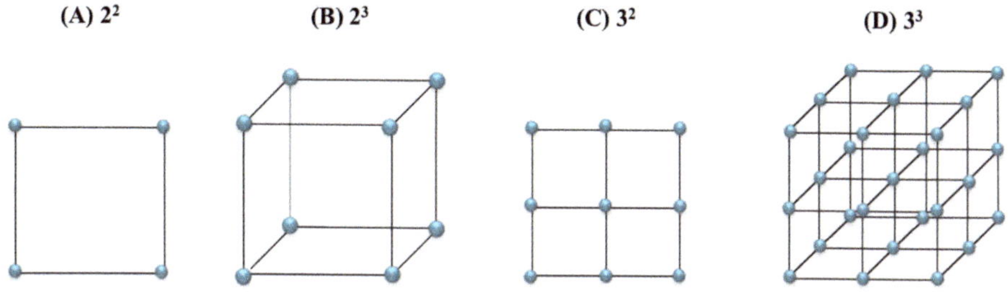

Figure 6.2 Representation of common experimental designs used for screening purposes.

studies or for the initial steps of the optimization due to its simplicity and relatively low cost. As an example, Domeño et al., screened the significant variables affecting the determination of sterols in serum by SPME with on-fiber derivatization combined with GC and flame ionization detection (FID) by means of a 2^{5-2} design (3 central points), performing 8 experiments (Domeño et al., 2005). The screened variables were the sorption time, temperature, amount of salt, sample volume and desorption time. Since each analyte showed its own optimum conditions, a compromise situation was established.

A particular type of two-level fractional factorial design often used is the Plackett-Burman (PB) design. It assumes that the interaction among the variables is negligible, providing information only on the effects of single factors. The PB design involves a reduced number of experiments in comparison with the 2^k design. It involves 4N experiments to study the main effects of 4N-1 factors. N is the number of variables (Montgomeri, 2001). Magiera et al., investigated the variables affecting a technique involving MEPS in combination with reversed phase ultra-high pressure liquid chromatography using a PB design for the determination of drugs in urine (Magiera et al., 2013). The studied variables were nature of the adsorbent, pH and volume of sample, type and pH of the eluting solvent, elution volume, duration of sorbent drying, and the washing solvent. These experimental variables were screened for 4 types of sorbents. The number of experiments performed was 11, including 3 central points, instead of 158 that would be required with a full factorial design.

6.3.2 Optimization

Among the RSMs, Central Composite Design (CCD), Doehlert Design (DD) and Box Behnken Design (BBD) are the most frequently used for optimization of microextraction methods. The selected RSM should guarantee the most precise estimation of the model coefficients and of the estimated response of the experimental conditions when no experiments are performed (Morales et al., 2013). The success of these designs relies on the easy interpretation and the reduced number of experiments required. **Table 6.2** compiles selected applications of different microextraction methods combined with experimental designs for both screening and optimization. Optimum values of the variables obtained after the optimization are also included.

CCD consists of: i) a full or fractional factorial design, usually of two-level; ii) a star design with experimental points on the axis properly distanced from the centre; and iii) an experimental point in the center (Stalikas et al., 2009). For a CCD with a 2^k factorial runs, 2k axial runs and C_0 central point runs, the total number of points (N) are $N=2^k+2k+C_0$ (where k is the number of variables and C_0 is the number of points). CCDs for 2 and 3 variables are shown in **Figure 6.3A** and **B**, repectively. Soares Emídio et al. used the CCD approach to optimize the significant experimental variables of the HF-LPME method for the determination of cannabinoids in hair samples by GC-tandem mass spectrometry (Soares Emídio et al., 2010). After selection of the extraction solvent, the parameters investigated were extraction time, ionic strength, stirring speed, pH and volume of the acceptor (organic) phase. First, the authors performed a 2^{5-1} design (19 randomized experiments) to estimate the experimental variance and check the linearity between the levels of each variable. Extraction time and stirring speed were not statistically significant in the cannabinoid extraction. Afterwards, a 2^3 CCD, i.e., 6 star and 3 center points and thus 17 experiments (2^3 + (2×3) + 3), was applied to optimize the values of the significant variables obtained in the previous design. The analyses of the optimum CCD values were performed using the quadratic model. They observed a significant negative interaction between pH and NaCl concentration.

DD is a highly efficient and economical design. The number of levels is not the same for all variables and the intervals between levels are distributed uniformly. It is a non-routable design with a circular domain for two variables, spherical for three variables, and hyperspherical for more than three variables. A reduced number of experiments (N) is required, given by $N=k^2+k+C_0$, where k is the number of variables and C_0 is the number of center points (Almeida Bezerra et al., 2008; Stalikas et al., 2009). DDs for two and three variables are shown in **Figs. 6.3C** and **D**, respectively. The experimental matrix can also be displayed to another experimental region using previous adjacent points. For instance, to optimize an ultrasound-assisted dispersive liquid-liquid microextraction approach (UA-DLLME) with derivatization integrated in the extraction step, Carro et al., performed: i) an asymmetric screening design $2^2 3^4$//16, i.e., 2 factors at 2 levels and 4 factors at 3 levels, involving 6 parameters (derivatization agent, addition of NaCl, extraction solvent volume, dispersion solvent volume, derivatization agent volume and ultrasonic bath temperature) and 16 runs; ii) DD to optimize 3 parameters (derivatization agent, dispersant volume and NaCl addition) that could not be fixed in the screening step for water samples (16 runs); and iii) a CCD with 8 points of the full factorial design, 6 axial points and 3 central points to investigate the influence of dispersion solvent volume, extraction solvent volume and

Table 6.2 Some recent applications of different microextraction techniques combined with experimental designs.

Analyte	Matrix	Micro-extraction technique	Screening design: variables	Optimization design: variables (optimum conditions)	Separation / Detection technique	Ref.
pesticide residues	apple	HS-SPME	**2^3**: extraction temperature, extraction time and salt addition	-	GC-MS	Bola Abdulra'uf, & Huat Tan, 2013
trihalomethanes	swimming pool waters	HS-SPME	**2^3**: extraction time, extraction temperature and desorption time	**2^3**: extraction time (25 min), extraction temperature (45 °C), and desorption time (5 min)	GC-ECD	Maia et al., 2014
aroma compounds	cooked beef	HS-SPME	-	**CCD**: equilibrium time (10 min), equilibrium temperature (40 °C), and extraction time (25 min)	GC-MS	Ma et al., 2013
PCP	water	alcoholic-DLLE	**PB**: type and volume of extraction and dispersive solvents, salt addition and agitation time	**CCD**: type and volume of extraction solvent (170 µL of octanol), and agitation time (5 min)	HPLC-UV-Vis	Hadjmohammadi et al., 2012
pesticides	water	SPME	**2^{7-3}**: extraction temperature, extraction time, desorption temperature, desorption time, stirring speed, sample pH, salt addition	**CCD**: extraction time (55 min), and NaCl 30% w/v	GC-MS	Passeport et al., 2010
sulfur compounds	white wines	USAE-DLLME	**2^{6-1}**: extracting and dispersing solvents, sample volume, salt addition, cavitation time and centrifugation time	**CCD and multiple response**: extracting volume (150 µL), NaCl 8.75% w/v, cavitation time (20 s), and centrifugation time (50 s)	GC-MS	Jofré et al., 2010
phthalates and polycyclic musks	waters	DLLME	**PB 2^{7-4}**: volume of extraction and disperser solvent, pH, salt addition, extraction time, centrifugation time and centrifugation speed	**CCD**: volume of extraction solvent (250 µL of CCl_4) and disperser solvent (0.62 mL of CH_3OH), pH (natural), centrifugation time (7.5 min), and centrifugation speed (4000 rpm)	GC-MS	Panagiotou et al., 2009

Table 6.2 Some recent applications of different microextraction techniques combined with experimental designs.

Analyte	Matrix	Micro-extraction technique	Screening design: variables	Optimization design: variables (optimum conditions)	Separation / Detection technique	Ref.
ethylene oxide residuals	sterilized medical devices	HS-SPME	-	**DD**: desorption time (0.333 min), extraction temperature (3 °C), GC inlet temperature (174 °C), and extraction time (3min)	GC-FID	DiCicco et al., 2009
volatile compounds	tropical fruits	CF-HS-SPME	**2^4**: extraction/incubation temperature, incubation time, extraction time and sample agitation	**DD**: sample temperature (60 °C), sample agitation (500 rpm), and extraction time (25 min)	GC-MS	Carasek & Pawliszyn, 2006
TCA and TBA	wines	HS-SDME	**2^{5-1}**: extraction time, pH, salt addition, temperature, sample matrix	**BBD**: extraction time (25 min), extraction temperature (48 °C), and NaCl 2.5 M	GC-ECD	Martendal et al., 2007
organophosphorous pesticides	Waters	UA-DLLME	**OVAT**	**BBD**: extraction solvent volume (51 µL of $CHCl_3$), dispersive solvent volume (514 µL of CH_3OH), and sample volume (12 mL)	GC-FID	Khajeh & Musavi Sadeh, 2012
pramipexole drugs	plasma and urine	SBME	**PB**: donor and aceptor pH, extraction time, stirring rate, salt addition, temperature of sample	**BBD**: donor pH 11.5, acceptor pH 3.3 and extraction time 30 min	HPLC-UV-Vis	Ghasemi, 2012
Mn, Cd, Co and Ni	Waters	LL-USAEME	**2^{5-1}**: volume of extraction solvent, pH, sonication time, salt addition, and temperature	**BBD**: extractant volume (30 µL of C_2Cl_4), pH 12, sonication time (5 min), NaCl 5% w/v	ICP-OES	Karimi et al., 2013

CF-HS-SPME, cold fiber headspace solid-phase microextraction; GC-ECD, gas chromatography with electron capture detector; PCP, pentachlorophenol; TBA, 2,4,6-tribromoanisole; TCA, 2,4,6-trichloroanisole; USAE-DLLME, ultrasound assisted-emulsification-dispersive liquid-liquid microextraction.

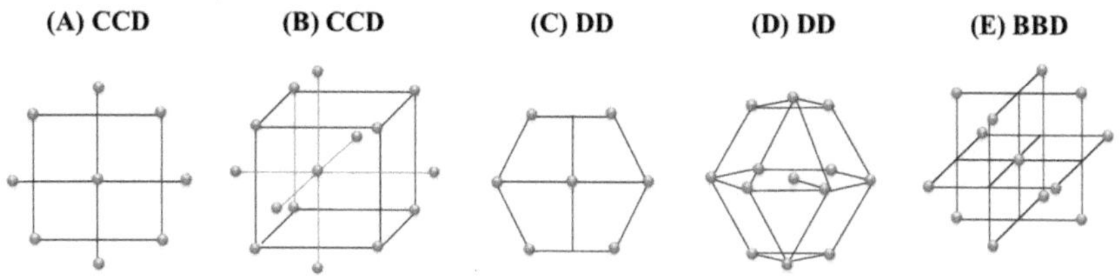

Figure 6.3 Representation of common experimental designs used for optimization purposes. (A) and (C), 2 variables and (B), (D) and (E), 3 variables.

ultrasonic bath temperature in milk samples. UA-DLLME efficiency increased using low levels for the three variables (Carro et al., 2013).

BBD is also an efficient three-level factorial design with the experimental points located on the midpoints of the edges of a cube and at the center (central points), as can be seen in **Figure 6.3E**. This design is rotatable or nearly rotatable and of spherical nature. The estimation of the parameters follows the quadratic model and the design points increase at the same rate as the number of polynomial coefficients. The number of experimental runs required (N) is given by $N=2k(k-1)+C_0$, where k is the number of variables and C_0 is the number of central points. This design is advantageous compared to the three-level full factorial design since BBD does not contain combinations among the parameters at their highest or lowest levels simultaneously. However, this experimental design is not recommended for the assessment of the responses at the vertices of the cube, since BBD does not contain combinations where all factors are at the extremes. In terms of efficiency, it has been proved that: BBD ~ DD > CCD > 3^k factorial design, defining the efficiency as the number of coefficients in the estimated model divided by the number of experiments (Ferreira et al., 2007). BBD was used to optimize the volume of extraction solvent, pH, salt addition and sonication time for the simultaneous determination of heavy metals in water samples by ligandless-ultrasound-assisted emulsification-microextraction (LL-USA-EME) combined with inductively coupled plasma-optical emission spectrometry (ICP-OES). Before the BBD, a 2^{5-1} design was used for the screening of the variables. To interpret the interactions and visualize the relationship between the recovery of the target analytes and the level of each variable, two three-dimensional plots were mapped: i) extractant volume and pH and ii) extractant volume and sonication time *vs* recoveries. Optimum conditions were 190 µL of extractant volume, pH 11, 15 % w/v salt and 5 min of sonication time (Sereshti et al., 2001).

Another advantageous multivariate approach is Artificial Neural Networks (ANNs), although it has not yet been extended for the optimization of microextraction methods. ANNs provides a non-linear modeling of the data, in contrast to regression methods, therefore, more complex interactions are considered. ANN analysis is quite flexible with regard to the number and form of the experimental data and all parameters can be included in the model. The RSMs mentioned above are dependent on the statistical significance of the considered levels, and only the significant terms are included in the model (Almeida Bezerra et al., 2008). Recently, Khakjeh et al. developed a reliable modeling method to predict the efficiency of the homogenous LLME of zinc in flour samples and to optimize the EE of the target analyte using the ANN and a particle swarm optimization method (Khakjeh et al., 2014).

6.4 Validation of Microextraction Methodologies

Small changes in the variables involved in the microextraction procedure can considerably affect the final result, and thus the performance under the optimum conditions is a requisite to ensure the quality of the results. Under these conditions, the developed methodology must be validated in order to demonstrate its suitability for the intended application. Internationally accepted protocols and guidelines are available for the validation of analytical methods, most notably the "*Harmonized guidelines for single laboratory validation of methods of analysis*" reported by the International Union of Pure and Applied Chemistry (IUPAC) (Thompson et al., 2006). AOAC International, the International Organization for Standardization (ISO) and IUPAC co-operated to produce this protocol. According to the harmonized protocol 'full' validation comprises the examination of the characteristics of the method *via* an inter-laboratory exercise. However, the method should be validated in-house first. For that, analytical characteristics such as sensitivity, EF, precision, accuracy, selectivity and specificity are determined for the microextraction approach combined with the separation/detection technique.

The sensitivity is an arbitrary characteristic that is dependent on the instrumental settings (slope of the calibration curve) and it is usually evaluated as the limit of detection (LOD) and the limit of quantification (LOQ). Several definitions of LOD and LOQ can be used depending on the organization protocol/guidance, including IUPAC (Thompson et al., 2006), Commission Regulation (EC) No 333/2007, US Environmental Protection Agency (EPA) or Food and Drug Administration (FDA, 2001) and thus, to specify how they have been calculated is highly recommended during publication.

In general, microextraction techniques provide good sensitivity, precision and selectivity mainly due to the large EFs that can be achieved. In SPME, EFs ranging between 2 and 27000 have been reported (Kokosa et al., 2009). Microextraction techniques performed under equilibrium conditions provide larger EFs, and thus, better results in terms of sensitivity and precision. DSDME and, to a lesser extension, HF-LPME, give rise to larger EFs than direct-SDME, which commonly operates under non-equilibrium conditions. Nevertheless, complete equilibrium is not necessary to attain accurate and precise results. SPME combined with chemical derivatization improves the sensitivity for many compounds, such as aliphatic amines, compared with direct-SPME (Pan et al., 1997) even though it depends on the structure of the derivatized compound. Herráez-Hérnandez et al. compared different strategies to couple SPME and chemical derivatization for the determination of short-chain aliphatic amines using 9-fluorenylmethyl chloroformate (FMOC) as a derivatizing agent: (i) derivatization of the analytes in solution followed by the extraction of the analyte derivatives; (ii) extraction of the analytes and derivatization by immersing the SPME fiber into the reagent solution; and (iii) extraction/derivatization of the analytes using fibers previously coated with the reagent. The third option provided the best sensitivity, being 126 and 223 times higher than i) and ii), respectively (Herráez-Hérnandez et al., 2006).

Relative standard deviations (RSDs) obtained for SPME and LPME procedures are around 10% (Kokosa et al., 2009) and 5%, respectively (Pena-Pereira et al., 2010; Pinto et al., 2010). In spite of equilibrium methods which provide better precision and sensitivity than pre-equilibrium methods, the two can be comparable when automated systems are used. For instance, in-tube SPME and multi-fiber SPME methods give rise to better precision than that of manual SPME methods due to the automation and the reproducibility of timing and/or fiber positioning (Vas & Vékey, 2004; Kokúrová et al., 2013). SPME/derivatization often yields higher RSDs due to the additional steps and factors such as reaction yields involved in the sample preparation procedure. RSDs of up to 24% have been reported for a SPME on-fiber derivatization method coupled with LC-fluorescence detection (Herráez-Hernández et al., 2006). As HS-SPME is based on the equilibrium of analytes among three phases (fiber coating, headspace and sample), precision is often higher compared to conventional SPME. Nevertheless, RSDs lower than 11% were obtained for furanic compounds in coated deep-fried products by HS-SPME-GC-MS (Pérez-Palacios et al., 2012). Regarding LPME techniques, DSDME, DLLME and SFODME, display better results in terms of precision because equilibrium conditions are commonly attained.

Different tests are used to assess the accuracy of microextraction procedures for the intended purpose, mainly *via*: i) recovery studies of samples spiked with a known amount of the analyte; ii) analysis of certified reference materials (CRMs); and iii) comparison with alternative/standard methodologies. An adequate calibration method is crucial to obtain good recoveries, especially for complex matrixes. Even though microextraction approaches provide excellent sample clean-up, matrix effects can be present. The absolute matrix effects are typically evaluated by comparison of the signals/concentrations of a procedural blank, sample and sample spiked with the target analyte. Traditional calibration methods, i.e., external standard, internal standard, standard addition and isotope dilution are used for quantification purposes. For on-site *in vivo* sampling, calibration is performed by means of equilibrium extraction, exhaustive extraction and diffusion-based calibration (Pawliszyn, 2012).

Selectivity is usually evaluated by comparing i) the slope of the calibration curve and the slope of the response independently produced by a potential interference; ii) a sample/matrix blank and the same solution spiked with the interfering compound at one appropriate concentration. In SPME, the selectivity is mainly controlled by the physicochemical properties of the coating. The selectivity can also be enhanced by simultaneous extraction and on-fiber derivatization. Thus, compounds with poor chromatographic behaviour, high reactivity or thermal instability can be quantified using this approach (Kokosa et al., 2009). Headspace sampling, used for more volatile compounds, also provides a higher degree of selectivity as a result of the phase separation involved. The use of membranes coated or placed around a sorbent can add a certain degree of selectivity to the extraction process.

As for LPME, three-phase LPME methodologies, such as liquid-liquid-liquid microextraction (LLLME), three-phase HF-LPME and HS-SDME provide a higher degree of selectivity. Ionizable molecules are only extracted in LLLME and three-phase HF-LPME, while volatile or semi-volatile compounds are extracted in HS-SDME. Hollow fiber-based LPME methodologies, such as two- and three-phase HF-LPME and SBME, ensure an efficient microfiltration since only the small molecules penetrate through the pores of the hollow fiber. In two-phase LPME approaches, such as direct-SDME, continuous-flow microextraction, DSDME, DLLME and SFODME, the selectivity of the sample preparation method depends exclusively on the partition coefficients of analyte and potential interferences between the sample solution and the extractant phase (Pena-Pereira et al., 2010).

6.5 Conclusions

Numerous advantages of the microextraction techniques have been documented in both analytical and green chemistry contexts. The analytical performance of every technique is highly dependent on a set of experimental parameters and thus, their evaluation is highly recommended for optimal performance. By means of chemometric approaches, such as RSM, more information is obtained by performing a reduced number of experiments which reduces the experimental work and time, reagent consumption, and costs. In general, microextraction approaches have proven to provide enough characteristics for a wide number of applications and even to fulfill regulatory requirements. Moreover, accurate and precise results have been obtained after inter-laboratory exercises.

Abbreviations

ANN	artificial neural networks
ANOVA	analysis of the variance
BBD	Box Behnken design
CCD	central composite design

CE	capillary electrophoresis
CRM	certified reference material
DD	Doehlert design
direct-SDME	direct-single-drop microextraction
DLLME	dispersive liquid-liquid microextraction
DSDME	directly duspended droplet microextraction
EE	extraction efficiency
EF	enrichment factor
ETAAS	electrothermal atomic absorption spectrometry
FDA	Food and Drug Administration
FMOC	9-fluorenylmethyl chloroformate
GC	gas chromatography
GC-MS	gas chromatography-mass spectrometry
HF-LPME	hollow fiber liquid phase microextraction
HF-SPME	hollow fiber solid phase microextraction
HPLC	high performance liquid chromatography
HS-SDME	headspace single-drop microextraction
HS-SPME	headspace solid-phase microextraction
ICP-OES	inductively coupled plasma optical emission spectrometry
IMS	ion mobility spectrometry
ISO	International Organization for Standardization
IUPAC	International Union of Pure and Applied Chemistry
LC-MS/MS	liquid chromatography-tandem mass spectrometry
LLE	liquid-liquid extraction
LLLME	liquid-liquid-liquid microextraction
LLSME	liquid-liquid-solid microextraction
LL-USAEME	ligandless ultrasound assisted emulsification microextraction
LOD	limit of detection
LOQ	limit of quantification
LPME	liquid phase microextraction
MEPS	microextraction by packed sorbent
MIP	molecularly imprinted polymer
OVAT	one-variable-at a-time
PB	Plackett-Burman
PDMS	poly(dimethylsiloxane)
RSD	relative standard deviation
RSM	response surface methodology
SBME	stir bar sorptive microextraction
SFODME	solidification of floating organic drop microextraction
SPME	solid phase microextraction
UA-DLLME	ultrasound assisted dispersive liquid liquid microextraction
US-EPA	United States - Environmental Protection Agency

Acknowledgements

F. Pena-Pereira thanks Xunta de Galicia for financial support as a post-doctoral researcher of the I2C program.

References

Ahmad, S. (2013). Is SPME a destination or just another station for bioanalysis? *Bioanalysis, 5*, 2897–2901.

Alfonsi, K., Colberg, J., Dunn, P. J., Fevig, T., Jennings, S., Johnson, T. A., et al. (2008). Green chemistry tools to influence a medicinal chemistry and research chemistry based organisation. *Green Chemistry, 10*, 31–36.

Almeida Bezerra, M., Erthal Santelli R., Padua Oliveira, E., Silveira Villar, L., & Amélia Escaleira., L. (2008). Response surface methodology (RSM) as a tool for optimization in analytical chemistry. *Talanta, 76*, 965–977.

Andruch, V., Burdel, M., Kocúrová, L., Šandrejová, J., & Balogh, I. S. (2013). Application of ultrasonic irradiation and vortex agitation in solvent microextraction. *TrAC Trends in Analytical Chemistry, 49*, 1–19.

Arce, L., Menéndez, M., Garrido-Delgado, R., & Valcárcel, M. (2008). Sample-introduction systems coupled to ion-mobility spectrometry equipment for determining compounds present in gaseous, liquid and solid samples. *TrAC Trends in Analytical Chemistry, 27*, 139–150.

Baghdadi, M., & Shemirani, F. (2008). Cold-induced aggregation microextraction: A novel sample preparation technique based on ionic liquids. *Analytica Chimica Acta, 613*, 56–63.

Bagheri, H., Piri-Moghadam, H., & Naderi, M. (2012). Towards greater mechanical, thermal and chemical stability in solid-phase microextraction. *TrAC Trends in Analytical Chemistry, 34*, 126–139.

Bendicho, C., de la Calle, I., Pena, F., Costas, M., Cabaleiro, N., & Lavilla, I. (2012). Ultrasound-assisted pretreatment of solid samples in the context of green analytical chemistry. *TrAC Trends in Analytical Chemistry, 31*, 50–60.

Bianchi, F., & Careri, M. (2008). Experimental design techniques for optimization of analytical methods. Part I: Separation and sample preparation techniques. *Current Analytical Chemistry, 4*, 55–74.

Bizkarguenaga, E., Iparragirre, A., Navarro, P., Olivares, M., Prieto, A., Vallejo, A., et al. (2013). In-port derivatization after sorptive extractions. *Journal of Chromatography A, 1296*, 36–46.

Bola Abdulra'uf, L, & Huat Tan, G. (2013). Multivariate study of parameters in the determination of pesticide residues in apple by headspace solid phase microextraction coupled to gas chromatography–mass spectrometry using experimental factorial design. *Food Chemistry, 141*, 4344–4348.

Cabaleiro, N., de la Calle, I., Bendicho, C., & Lavilla, I. (2012). Enzymatic single-drop microextraction for the assay of ethanol in alcohol-free cosmetics using microvolume fluorospectrometry detection. *Analytica Chimica Acta, 733*, 28–33.

Carasek, E., & Pawliszyn, J. (2006). Screening of tropical fruit volatile compounds using solid-phase microextraction (SPME) fibers and internally cooled SPME fiber. *Journal of Agricultural and Food Chemistry, 54*, 8688–8696.

Carro, A. M., González, P., & Lorenzo R. A. (2013). Simultaneous derivatization and ultrasound-assisted dispersive liquid–liquid microextraction of chloropropanols in soy milk and other aqueous matrices combined with gas-chromatography–mass spectrometry. *Journal of Chromatography A, 1319*, 35–45.

Commission Regulation (EC) No 333/2007 of 28 March.

Cussler, E. (1997). *Diffusion: Mass Transfer in Fluid Systems.* Cambridge, U.K.: Cambridge University Press.

DiCicco, M. P., Lang, B., & Harper, T. I. (2009). Optimization of a novel headspace–solid phase microextraction–gas chromatographic method by means of a Doehlert uniform shell design for the analysis of trace level ethylene oxide residuals in sterilized medical devices. *Biomedical Chromatography, 23*, 647–657.

Domeño, C., Ruiz, B., & Nerín, C. (2005). Determination of sterols in biological samples by SPME with on-fiber derivatization and GC/FID. *Analytical and Bioanalytical Chemistry, 381*, 1576–1583.

Ferreira, S. L. C., Bruns, R. E., Ferreira, H. S., Matos, G. D., David, J. M., Brandão, G. C., da Silva, E. G. P., Portugal, L. A., dos Reis, P. S., Souza, A. S., & dos Santos W. N. L. (2007). Box-Behnken design: An alternative for the optimization of analytical methods. *Analytica Chimica Acta, 597*, 179–186.

Fitzpatrick, L. J., & Dean, J. R. (2002). Extraction solvent selection in environmental analysis. *Analytical Chemistry, 74*, 74–79.

Food and Drug Administration. Guidance for Industry. Bioanalytical Method Validation. (2001). http://www.fda.gov/downloads/Drugs/GuidanceComplianceRegulatory Information/ Guidances/ucm070107.pdf (Accessed March 2014).

Gałuszka, A., Migaszewski, Z., & Namieśnik, J. (2013). The 12 principles of green analytical chemistry and the SIGNIFICANCE mnemonic of green analytical practices. *TrAC Trends in Analytical Chemistry, 50*, 78–84.

Ghasemi, E. (2012). Optimization of solvent bar microextraction combined with gas chromatography mass spectrometry for preconcentration and determination of tramadol in biological samples. *Journal of Chromatography A, 1251*, 48–53.

Gil, S., de Loos-Vollebregt, M. T. C., & Bendicho, C. (2009). Optimization of a single-drop microextraction method for multielemental determination by electrothermal vaporization inductively coupled plasma mass spectrometry following *in situ* vapor generation. *Spectrochimica Acta Part B: Atomic Spectroscopy, 64*, 208–214.

Hadjmohammadi, M. R., Fatemi, M. H., & Shakeri, P. (2012). Developing an alcoholic-assisted dispersive liquid-liquid microextraction for extraction of pentachlorophenol in water. *Journal of Separation Science, 35*, 3375–3380.

Hansen, C. M. (2000). *Solubility Parameters: A User's Handbook* (2nd ed.). Boca Raton, FL: CRC Press.

Henderson, R. K., Jiménez-González, C., Constable, D. J. C., Alston, S. R., Inglis, G. G. A., Fisher, G., et al. (2011). Expanding GSK's solvent selection guide – embedding sustainability into solvent selection starting at medicinal chemistry. *Green Chemistry, 13*, 854–862.

Herráez-Hernández, R., Cháfer-Pericás, C., Verdú-Andrés, J., & Campíns-Falcó, P. (2006). An evaluation of solid phase microextraction for aliphatic amines using derivatization with 9-fluorenylmethyl chloroformate and liquid chromatography. *Journal of Chromatography A, 1104*, 40–46.

Ho, T. S., Vasskog, T., Anderssen, T., Jensen, E., Rasmussen, K. E., & Pedersen-Bjergaard, S. (2007). 25,000-fold pre-concentration in a single step with liquid-phase microextraction. *Analytica Chimica Acta, 592*, 1–8.

Hu, Y., Wang, Y., Hu, Y. Li, G. (2009). Liquid-liquid-solid microextraction based on membrane-protected molecularly imprinted polymer fiber for trace analysis of triazines in complex aqueous samples. *Journal of Chromatography A, 1216*, 8304–8311.

Hüffer, T., Osorio, X. L., Jochmann, M. A., Schilling, B., & Schmidt T. C. (2013). Multi-walled carbon nanotubes as sorptive material for solventless in-tube microextraction (ITEX2)-a factorial design study. *Analytical and Bioanalytical Chemistry, 405*, 8387–8395.

Jeannot, M. A., & Cantwell, F. F. (1996). Solvent microextraction into a single drop. *Analytical Chemistry, 68*, 2236–2240.

Jermak, S., Pranaityte, B., & Padarauskas, A. (2007). Ligand displacement, headspace single-drop microextraction, and capillary electrophoresis for the determination of weak acid dissociable cyanide. *Journal of Chromatography A, 1148*, 123–127.

Jiménez-González, C., Curzons, A. D., Constable, D. J. C., & Cunningham, V. L. (2005). Expanding GSK's Solvent Selection Guide-application of life cycle assessment to enhance solvent selections. *Clean Technologies and Environmental Policy, 7*, 42–50.

Jofré, V. P., Assof, M. V., Fanzone, M. L., Goicoechea, H. C., Martínez, L. D., & Silva, M. F. (2010). Optimization of ultrasound assisted–emulsification–dispersive liquid–liquid microextraction by experimental design methodologies for the determination of sulfur compounds in wines by gas chromatography–mass spectrometry. *Analytica Chimica Acta, 683*, 126–135.

Karimi, M., Sereshti, H., Khojeh, V & Samadi, S. (2013). Ligandless-ultrasound-assisted emulsification microextraction followed by inductively coupled plasma-optical emission spectrometry for simultaneous determination of heavy metals in water samples. *International Journal of Environmental Analytical Chemistry, 93*, 401–415.

Khajeh, M., & Musavi Sadeh, F. (2012). Response surface modeling of ultrasound-assisted dispersive liquid–liquid microextraction for determination of benzene, toluene and xylenes in water samples: Box–Behnken Design. *Bulletin of Environmental Contamination and Toxicology, 89*, 38–43.

Khajeh, M., Kaykhaii, M., Hashemi, H. S., & Shakeri, M. (2014). Particle swarm optimization–artificial neural network modeling and optimization of leachable zinc from flour samples by miniaturized homogenous liquid–liquid microextraction. *Journal of Food Composition and Analysis, 33*, 32–38.

Khalili Zanjani, M. R., Yamini, Y., Shariati, S., & Jonsson, J. A. (2007). A new liquid-phase microextraction method based on solidification of floating organic drop. *Analytica Chimica Acta, 585*, 286–293.

Kocúrová, L., Baloghb, I.S., & Andruch, V. (2013). Solvent microextraction: A review of recent efforts at automation. *Microchemical Journal, 110*, 599–607.

Kokosa, J. M., Przyjazny, A., & Jeannot, M. A. (2009). *Solvent microextraction-Theory and practice.* New Jersey.: John Wiley & Sons.

Lasarte-Aragonés, G., Lucena, R., Cárdenas, S., & Valcárcel, M. (2014). Effervescence assisted dispersive liquid–liquid microextraction with extractant removal by magnetic nanoparticles. *Analytica Chimica Acta, 807*, 61–66.

Lasarte-Aragonés, G.·, Lucena, R., Cárdenas, S., & Valcárcel, M. (2011). Nanoparticle-based microextraction techniques in bioanalysis. *Bioanalysis, 3*, 2533–2548.

Li, Y., Fabiano-Tixier, A. S., Vian, M. A., & Chemat, F. (2013). Solvent-free microwave extraction of bioactive compounds provides a tool for green analytical chemistry. *TrAC Trends in Analytical Chemistry, 47*, 1–11.

Lord, H. L. (2007). Strategies for interfacing solid-phase microextraction with liquid chromatography. *Journal of Chromatography A, 1152*, 2–13.

Lucena, R. (2012). Extraction and stirring integrated techniques: examples and recent advances. *Analytical and Bioanalytical Chemistry, 403*, 2213–2223.

Ma, Q. L., Hamid, N., Bekhit, A. E. D., Robertson, J., & Law T. F. (2013). Optimization of headspace solid phase microextraction (HS-SPME) for gas chromatography mass spectrometry (GC–MS) analysis of aroma compounds in cooked beef using response surface methodology. *Microchemical Journal, 111*, 16–24.

Magiera, S., Gülmez, S., Michalik, A., & Baranowska, I. (2013). Application of statistical experimental design to the optimisation of microextraction by packed sorbent for the analysis of nonsteroidal anti-inflammatory drugs in human urine by ultra-high pressure liquid chromatography. *Journal of Chromatography A, 1304*, 1–9.

Maia, R., Correia, M., Brás Pereira, I.M., & Beleza, V. M. (2014). Optimization of HS-SPME analytical conditions using factorial design for trihalomethanes determination in swimming pool water samples. *Microchemical Journal, 112*, 164–171.

Majors, R. E. (2013). Novel sorbents for solid-liquid extraction. *LCGC North America 31*, 984–991.

Martendal, E., Budziak, D., & Carasek, E. (2007). Application of fractional factorial experimental and Box-Behnken designs for optimization of single–drop microextraction of 2,4,6-trichloroanisole and 2,4,6-tribromoanisole from wine samples, *Journal of Chromatography A, 1148*, 131–136.

Martinis, E.M., Escudero, L.B., Salvarezza, R., Calderón, M.F., Ibáñez, F.J. & Wuilloud, R. (2013). Liquid–liquid microextraction based on a dispersion of Pd nanoparticles combined with ETAAS for sensitive Hg determination in water samples. *Talanta, 108*, 46–52.

Mehdinia, A., & Aziz-Zanjani, M. O. (2013). Advances for sensitive, rapid and selective extraction in different configurations of solid-phase microextraction. *TrAC Trends in Analytical Chemistry, 51*, 13–22.

Mester, Z., & Sturgeon, R. (2005). Trace element speciation using solid phase microextraction. *Spectrochimica Acta Part B: Atomic Spectroscopy, 60*, 1243–1269.

Mester, Z., Sturgeon, R. E., & Lam, J. W. (2000). Sampling and determination of metal hydrides by solid phase microextraction thermal desorption inductively coupled plasma mass spectrometry. *Journal of Analytical Atomic Spectrometry, 15*, 1461–1465.

Montgomeri, D. C, (2001). *Design and Analysis of experiments* (5th ed.). New Jersey: John Wiley & Sons.

Morales, R., Sarabia, L. A., Sánchez, M. S., & Ortiz, M. C. (2013). Experimental design for the optimization of the derivatization reaction in determining chlorophenols and chloroanisoles by headspace-solid-phase microextraction–gas chromatography/mass spectrometry. *Journal of Chromatography A, 1296*, 179–195.

Mosayeb Daryanavard, S. Jeppsson-Dadoun, A., Andersson, L. I., Hashemi, M., Colmsjö, A., & Abdel-Rehim, M. (2013). Molecularly imprinted polymer in microextraction by packed sorbent for the simultaneous determination of local anesthetics: lidocaine, ropivacaine, mepivacaine and bupivacaine in plasma and urine samples. *Biomedical Chromatography, 27*, 1481–1488.

Palit, M., Pardasani, D., Gupta, A. K., & Dubey, D. K. (2005). Application of single drop microextraction for analysis of chemical warfare agents and related compounds in water by gas chromatography/mass spectrometry. *Analytical Chemistry, 77*, 711–717.

Pan, L., Chong, J. M., & Pawliszyn J. (1997). Determination of amines in air and water using derivatization combined with solid-phase microextraction. *Journal of Chromatography A, 773*, 249–260.

Panagiotou, A. N., Sakkas, V. A., & Albanis, A. T. (2009). Application of chemometric assisted dispersive liquid–liquid microextraction to the determination of personal care products in natural waters. *Analytica Chimica Acta, 649*, 135–140.

Passeport, E., Guenne, A., Culhaoglu, T., Moreau, S., Bouyé, J. M., & Tournebize, J. (2010). Design of experiments and detailed uncertainty analysis to develop and validate a solid-phase microextraction/gas chromatography–mass spectrometry method for the simultaneous analysis of 16 pesticides in water. *Journal of Chromatography A, 1217*, 5317–5327.

Pawliszyn, J. (2002). *Solid phase microextraction*. In J. Pawliszyn (Ed.), Sampling and Sample Preparation in Field and Laboratory (pp. 389–477). Amsterdam: Elsevier.

Pawliszyn, J. (2012). *Handbook of solid phase microextration*. Waltham: Elsevier.

Pena-Pereira, F., Lavilla, I., & Bendicho, C. (2010). Liquid-phase microextraction techniques within the framework of green chemistry. *TrAC Trends in Analytical Chemistry, 29*, 617–628.

Pereira, J., Goncalves, J., Silva, C.L., Mendes, B. Silva, P., Alves, V., Figueira, J., & Câmara, J.S. (2013). Metabolomic applications of liquid chromatography: from food bioactive metabolites to disease biomarkers research. *Liquid Chromatography*, 269–289.

Pérez-Palacios, T., Petisca, C., Melo, A., & Ferreira, I. M. (2012). Quantification of furanic compounds in coated deep-fried products simulating normal preparation and consumption: Optimisation of HS-SPME analytical conditions by response surface methodology. *Food Chemistry, 135*, 1337–1343.

Pinto, M. I., Sontang, G., Bernardino, T. J., & Noronha, J. P. (2010). Pesticides in water and the performance of the liquid-phase microextraction based techniques. A review. *Microchemical Journal, 96*, 226–237.

Pranaitytė, B., Jermak, S., Naujalis, E., & Padarauskas, A. (2007). Capillary electrophoretic determination of ammonia using headspace single-drop microextraction. *Microchemical Journal, 86*, 48–52.

Saito, Y., & Jinno, K. (2003). Miniaturized sample preparation combined with liquid phase separations. *Journal of Chromatography A, 1000*, 53–67.

Sereshti, H., Khojeh, V., Karimi, M., Samadi, S. (2012). Ligandless-ultrasound-assisted emulsification-microextraction combined with inductively coupled plasma-optical emission spectrometry for simultaneous determination of heavy metals in water samples. *Analytical Methods, 4*, 236–241.

Shariati-Feizabadi, S., Yamini, Y., & Bahramifar, N. (2003). Headspace solvent microextraction and gas chromatographic determination of some polycyclic aromatic hydrocarbons in water samples. *Analytica Chimica Acta, 489*, 21–31.

Shirey, R. E. (2012). *SPME commercial devices and fibre coatings*. In J. Pawliszyn (Ed.), Handbook of solid-phase microextraction (pp. 99–133). Waltham, MA, USA: Elsevier.

Soares Emídio, E., de Menezes Prata, V, Malagueño de Santaña F. J., & Silveira Dórea, H. (2010). Hollow fiber-based liquid phase microextraction with factorial design optimization and gas chromatography – tandem mass pectrometry for determination of cannabinoids in human hair. *Journal of Chromatography B: Analytical Technologies in the Biomedical and Life Sciences, 878*, 2175–2183.

Spietelun, A., Marcinkowski, L., de la Guardia, M., & Namieśnik, J. (2014). Green aspects, developments and perspectives of liquid phase microextraction techniques. *Talanta*, 119, 34–45.

Spietelun, A., Pilarczyk, M., Kloskowski, A., & Namieśnik, J. (2010). Current trends in solid-phase microextraction (SPME) fibre coatings. *Chemical Society Reviews, 39*, 4524–4537.

Stalikas, C. D., & Fiamegos, Y. C. (2008). Microextraction combined with derivatization. *TrAC Trends in Analytical Chemistry, 27*, 533–542.

Stalikas, C., Fiamegos, Y., Sakkas, V., & Albanis, T. (2009). Developments on chemometric approaches to optimize and evaluate microextraction. *Journal of Chromatography A*, 1216, 175–189.

Szreniawa-Sztajnert, A., Zabiegała, B., & Namieśnik, J. (2013). Developments in ultrasound-assisted microextraction techniques for isolation and preconcentration of organic analytes from aqueous samples. *TrAC Trends in Analytical Chemistry, 49*, 45–54.

Thompson, M., Ellison, S. L. R., & Wood, R. (2006). The international harmonized protocol for the proficiency testing of analytical chemistry laboratories (IUPAC technical report). *Pure and Applied Chemistry, 78*, 135–196.

Tong, H., Sze, N., Thomson, B., Nacson, S., & Pawliszyn, J. (2002). Solid phase microextraction with matrix assisted laser desorption/ionization introduction to mass spectrometry and ion mobility spectrometry. *The Analyst, 127*, 1207–1210.

Turner, A. (2003). Salting out of chemicals in estuaries: implications for contaminant partitioning and modelling. *Science of the Total Environment, 314–316*, 599–612.

van Krevelen, D. W. (2009). *Properties of polymers, their correlation with chemical structure; their numerical estimation and prediction from additive group contributions* (4th ed.). Amsterdam: Elsevier.

Vas, G., & Vékey, K. (2004). Solid-phase microextraction: a powerfull sample preparation tool prior to mass spectrometric analysis. *Journal of Mass Spectrometry, 39*, 233–354.

Wang, Q., Ma, L., Yin, C. R., & Xu, L. (2013). Developments in injection port derivatization. *Journal of Chromatography A, 1296*, 25–35.

Wang, Y., Nacson, S., & Pawliszyn, J. (2007). The coupling of solid-phase microextraction/surface enhanced laser desorption/ionization to ion mobility spectrometry for drug analysis. *Analytica Chimica Acta, 582*, 50–54.

Zhang, M, Zeng, J., Wang, Y., & Chen, X. (2013). Developments and trends of molecularly imprinted solid-phase microextraction. *Journal of Chromatography Science, 51*, 577–586.

Zhang, Z., & Pawliszyn, J. (1995). Quantitative extraction using an internally cooled solid phase microextraction device. *Analytical Chemistry, 67*, 34–43.

Zhu, L., Zhu, L., & Lee, H. K. (2001). Liquid–liquid–liquid microextraction of nitrophenols with a hollow fiber membrane prior to capillary liquid chromatography. *Journal of Chromatography A, 924*, 407–414.

7 Miniaturized Alternatives to Conventional Sample Preparation Techniques for Solid Samples

Noelia Cabaleiro and Inmaculada de la Calle*
Department of Analytical and Food Chemistry, Faculty of Chemistry, University of Vigo, Campus As Lagoas-Marcosende s/n, 36310 Vigo, Spain
*e-mail: incalle@uvigo.es

7.1 Introduction

In recent years, several efforts were made in relation to the miniaturization of conventional sample preparation procedures. Analytical chemistry researchers are concerned about the use of conventional protocols, which are generally large-scale, tedious, time-consuming, require manual labor and involve the use of large quantities of hazardous reagents. Thus, the development of new miniaturized procedures has become necessary due to the increasing demand for environmentally friendly, green, fast and alternative approaches.

Today, there is continuous research focused on improving sample preparation procedures. Miniaturization is one of the main trends in analytical chemistry in addition to simplification and automation (Valcárcel & Cárcenas, 2000; Ríos et al., 2009). The main principles of miniaturization approaches focus on downsizing the methods by reduction of the amount of sample, reduction or elimination of chemical reagents and the development of one-step treatments, on-line procedures and in-the-field analysis. It is also desirable to design automated and unattended procedures (Halls, 1995; Ramos et al., 2005). **Figure 7.1** shows the number of publications per year devoted to miniaturization of sample preparation procedures. As can be seen, there is a marked increase since 2000.

A special challenge regarding miniaturization is the analysis of solid samples, which often require advanced sample pre-treatment in order to be suitable for a specific type of analysis. Different instrumental techniques are applied for the analysis of organic and inorganic analytes and each presents unique challenges.

One option for the analysis of metal ions in solid samples is the direct solid analysis (Bendicho & de Loos-Vollebregt, 1991; Stoeppler, 1997; Potts & Robinson, 2003). However, sample pre-treatment may be necessary prior to the analysis with some instruments, if the solid sample is not the appropriate form. The limitations of direct solid analysis include: (i) reduced representativeness of subsamples; (ii) lack of appropriate calibration standards; (iii) time-consuming steps for multielemental analysis; (iv) high number of interferences; (v) difficult sample introduction (non automated) and (vi) low precision (Hoenig & de Kersabiec, 1996; Ebdon et al., 2003). Due to these limitations, sample preparation of solids has received tremendous research attention during the last few years.

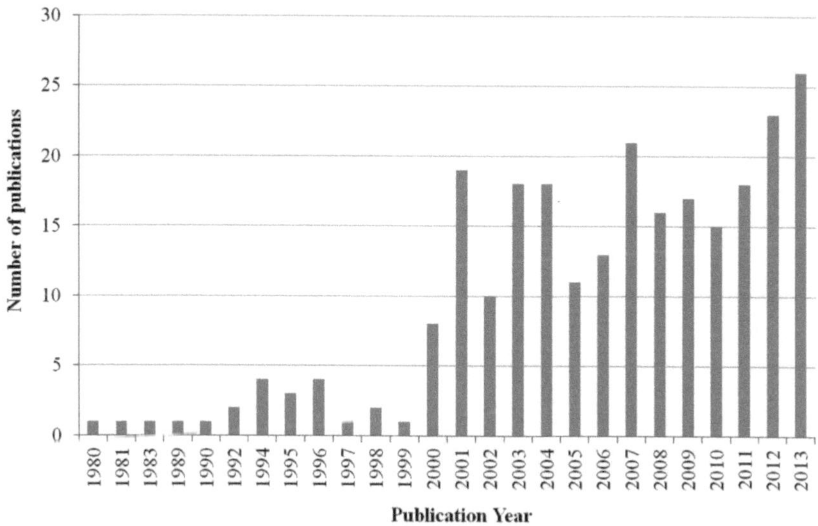

Figure 7.1 Number of publications devoted to miniaturized sample preparation procedures since 1980 (source: ISI Web of knowledge (Web of Science)-Thomson Reuters).

In this chapter, examples of conventional or established procedures for solid sample preparation and the miniaturized alternative approaches evolved from the previous protocols will be discussed. A scheme of the evolution from conventional sample preparation procedures to miniaturized versions can be seen in **Figure 7.2.** Moreover, a detailed description of the year of appearance of some conventional and miniaturized procedures is presented in **Figure 7.3.**

Conventional or established sample preparation procedures encompass the commonly applied procedures and reference methods used for validation purposes. These classical procedures involve several approaches that can be classified as: procedures involving no sample decomposition (slurry sampling, SS), procedures involving sample decomposition (such as acid digestion (AD) and dry ashing) and extraction procedures (such as liquid-liquid extraction (LLE), Soxhlet extraction and solid-phase extraction, SPE). However, these protocols can entail several drawbacks, such as the amount of sample required, the volume of organic and/or toxic reagents and solvents needed, or the extensive use of energy. Thus, these conventional sample preparation approaches evolved toward miniaturized and alternative approaches in order to solve these inconveniences.

In this chapter, miniaturized and alternative approaches will be discussed, including, but not limited to, matrix solid-phase dispersion (MSPD), solid-phase microextraction (SPME), micro-solid-phase extraction (μ-SPE), liquid-phase microextraction (LPME), small-volume microwave-assisted extraction (MAE), vapor-phase acid digestion (VPAD), vapor-phase microwave-assisted digestion (VPMAD), miniaturized ultrasonic SS, and miniaturized ultrasound-assisted extraction (UAE) and digestion (UAD).

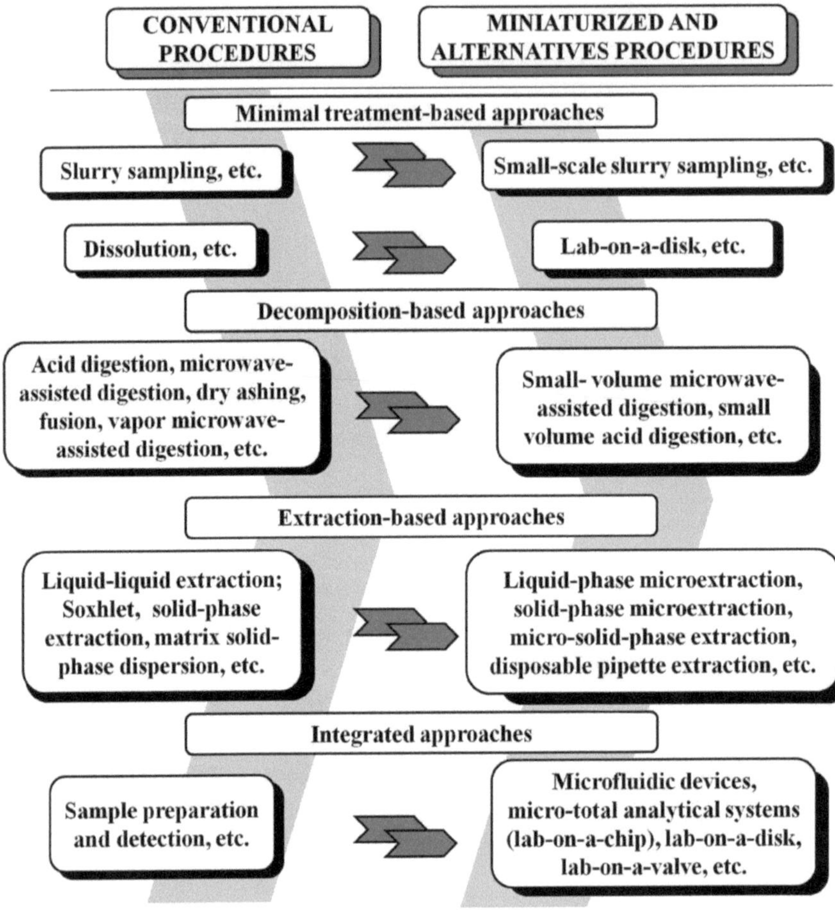

Figure 7.2 Evolution of sample preparation procedures from conventional to miniaturized approaches.

7.2 Objectives and Benefits of Miniaturized Sample Preparation Procedures

Recently, miniaturization of classical analytical techniques and procedures has emerged as a new branch of scientific research. Likewise, the trend toward small, autonomous approaches that consume smaller amounts of reagents, require less sample and take less time to process is clear (Pawliszyn, 2003; Felton, 2003). The general objective of miniaturized approaches lies in improving the sample preparation procedures by developing integrated and automated systems, simplified protocols using less amount of sample in a short time and with less energy and reagent consumption. The main principles of miniaturization approaches are summarized in **Figure 7.4**.

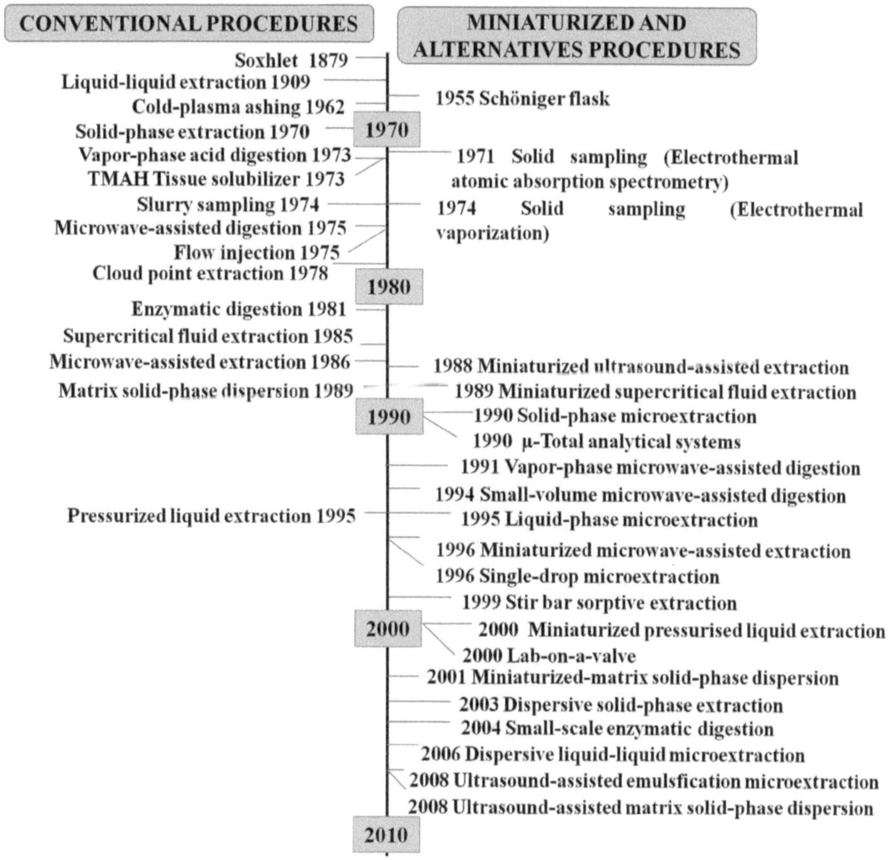

Figure 7.3 Chronological order of appearance of different miniaturized sample preparation procedures compared to conventional methods.

7.3 Challenges of Solid Sample Analysis

Sample preparation continues to be a critical step in the analytical process and the most important source of error. In fact, due to the number of steps that are often involved, sample preparation usually takes approximately two thirds of the total analysis time (50–70%) (Hyötyläinen, 2009) and tends to be slow and labor-intensive. The appropriate choice of the sample preparation procedure should be done according to the main matrix and the analyte of interest. For example, environmental and biota samples are complex matrices, and generally contain a wide variety of compounds in addition to the target analytes (Clement & Hao, 2010). Thus, sample preparation steps are required to bring the sample into solution, to extract the analyte from the matrix, to avoid matrix interferences and/or to preconcentrate the analyte prior to the analysis.

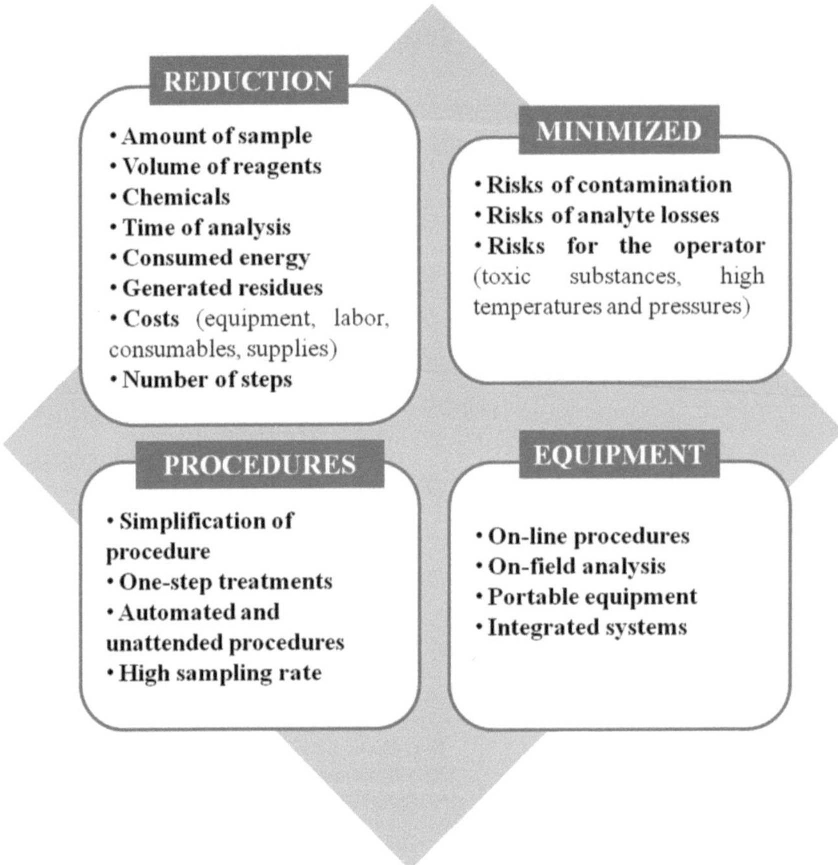

Figure 7.4 Main principles of miniaturized sample preparation procedures.

Solid sample analysis is itself a challenge to both routine analysis and analytical method development due to the complexity of matrices and the very low concentration of target analytes. The main challenges found during sample preparation could be: i) the difficulty of having homogeneous and representative samples, ii) losses of elements by volatilization, iii) contamination of the sample during the processing or storage of the sample, iv) time-consuming and labor-intensive clean-up protocols, v) extensive operator manipulation, vi) difficculty of automation, vii) incomplete dissolution of inorganic samples, and viii) the presence of interferences.

For instance, several errors can occur in the determination of elements in solid samples due to the inherent difficulty in bringing the sample into a solution, including contamination risks, analyte looses and incomplete dissolution (Hoening, 2001). Since solid samples are generally heterogeneous, preliminary treatments are required to obtain a more representative sample with less particle size (Hoenig, 2001). Besides, several metals could be retained into the matrix of samples with high silicon content, such as plants, soils and sediments, due to insolubilization phenomena (Hoenig et al., 1998; Hoenig, 2001), being necessary more stringent conditions for sample digestion. Another difficulty consists of deciding if dilution is required, because the approximate concentrations of the analyte and the main matrix components in the sample is usually unknown (Hoenig, 2001).

The determination of organic compounds in biological tissues requires carrying out the disruption of cells and, as a consequence, high concentrations of lipids and proteins are often co-extracted. Therefore, analysis of biological matrices is a tedious process which often gives rise to low recovery and poor reproducibility (Petrovic & Barceló, 2004). This is the case for Soxhlet-based methods, in which further manipulation can be required. For instance, dirty extracts produced may require the implementation of an extra clean-up step. Besides, the need of large volumes of organic solvents for extraction may require solvent evaporation in order to concentrate the analytes before analysis. In addition, a considerable quantity of co-extracted lipids (levels of lipids can vary widely depending on the tissue) can damage the chromatographic column, particularly in gas chromatography (GC). In fact, several pretreatment steps, such as AD using HCl, are needed to remove lipids prior to analysis (Clement & Hao, 2010).

7.3.1 Types and Composition of Solid Samples

The target analytes to quantify in solid samples can be very different, including volatile and semi-volatile organic compounds, metals, organometallics and inorganic ions. Moreover, the purpose of the analysis varies from one case to another. For example, analyses can be performed to determine levels of pollutants in an area (*e.g.*, a road, an industry) or a specific sample (*e.g.*, wastes), to evaluate the possibility of reusing wastes for agricultural applications (Osberghaus et al., 1997), to search for bioindicators of pollution, to assess the nutritional composition of food, to evaluate impurities of industrial samples, or to analyze biomolecules related with a disease.

In general, solid matrices are complex and very different in nature and composition. Likewise, each type of sample entails the use of a specific sample preparation procedure. Solid matrices can usually be classified as organic or inorganic regarding their main composition, but in most cases they are a mixture of both types of components. There are samples with high proportion of organic matter (biological tissues, food, plastics) while others are predominantly inorganic in composition (soils, sediments, metals, alloys, fly ash, rocks, sludge, etc.). Other samples that are commonly analyzed include food, clinical, pharmaceutical, industrial and waste samples (Hoenig de Kersabiec, 1996; Hoenig, 2003).

Biological tissues may include human, animal, fish and plant tissues. Biological samples also encompass food, feedstocks, clinical and agricultural samples because they consist basically of animal and plant-based materials. A major component of biological tissues is water (*e.g.*, 70 % in mammals). However, water is often removed by lyophilization or drying of the sample in the first steps of sample preparation. According to the different organization levels, biological tissues (from the smallest to the highest level) in increasing size order are composed of atoms (*e.g.*, C, H, O, N, P, S, inorganic ions) < molecules and macromolecules (*e.g.*, proteins, lipids, carbohydrates and nucleic acids) < organelles (*e.g.*, mitochondria, liposome, Golgi apparatus) < cells < tissues (*e.g.*,neural tissue, epithelial tissue, muscle tissue, connective and supportive tissue) < organs and organ systems (*e.g.*, lung and respiratory system, etc.) < organisms (*e.g.*, humans, fish, animals, plants) (Müller-Esterl, 2008). The main differences between animal and plant tissues are the presence of the cell wall, chloroplasts and silica in plant tissues. The cell wall is mainly composed of cellulose, which is the most abundant organic compound in plants. Moreover, silicon in plants provides structural support and improves tolerance to diseases, drought and metal toxicity (Hodson et al., 2005). Siliceous residues can be found in some plants and plant-based foods. Clinical samples include other tissues such as teeth and bones that present high inorganic content apart from the organic components. The organic matrix of these samples consists of collagen and a small fraction of non-collagen proteins, lipids, citrates and sugars. Biological hard tissues (enamel, dentin and bone) consist of a mineral matrix (hydroxyapatite crystals ($Ca_5(PO_4)_3(OH)$), water and an organic matrix (Bachmann et al., 2006). Clinical or biomedical samples may include both solid (bone, teeth, hair, nails, tissue) and semi-solid samples (blood, faeces, viscera) (Sansoni & Panday, 1994) and may also contain organic biomolecules.

Geological samples, soils and sediments are also widely analyzed. These type of samples present a variable organic matter content (from less than 1 % to 40 % of C). Soils and sediments are composed of different phases such as organic matter (humic and fulvic acids, organism residues) and minerals (oxides of iron, aluminum and manganese, phyllosilicate minerals, carbonates and sulfides) (Filgueiras et al., 2002a). Silicate rocks comprise a wide range of different mineral species. Moreover, soils without pre-treatment contain a biotic portion that consists of a diverse array of bacteria, fungi, worms and insects (Del Castilho & Breder, 1997).

On the contrary, solid waste is a comprehensive term and generally implies heterogeneous material, including *e.g.*, industrial waste, domestic refuse, excavated material, wrecked cars, used tires, sewage sludge and numerous other materials discarded by human society (Osberghaus & Helmers, 1997). It should be pointed out that sewage sludge constitutes a relatively homogenous matrix due to the long deposition time of sludges in sewage treatment plants (Schladot & Backhaus, 1997). Other waste samples include fly ash and other combustion products, such as industrial powders, solid sewage and industrial wastes (Clement & Hao, 2010). These industrial solid materials, as well as cements and ceramics, contain refractory compounds such as Al_2O_3, SiO_2, MgO, CaO, Cr_2O_3 and ZrO_2 and can be also complicated matrices (Rechenberg, 1997).

Apart from the target samples, certified reference materials (CRMs) of different origins are often purchased and analyzed along with unknown samples. These reference materials are used to test and validate newly-developed procedures as they help to establish the accuracy and precision of a methodology. Moreover, CRMs can be used to calibrate instruments and assure analytical quality control. CRMs of most of the previously mentioned types of samples are available in different sources, such as National Research Council of Canada (NRCC, Canada), Community Bureau of Reference (BCR, Brussels, Belgium), National Research Center for Certified Reference Materials (China) and National Institute of Standards and Technology (NIST, USA).

7.3.2 Pre-treatment of Solid Samples

The common steps involved in almost every analytical process include sampling, sample transport and storage, sample preservation, sample pre-treatment and sample preparation (or sample treatment) and analysis. These processes are highly linked to the nature of the sample and the analyte and are aimed at minimizing sample heterogeneity and ensuring sample representativeness. Herein, several main types of solid sample pre-treatment will be discussed.

Sample pre-treatment can include processes such as drying, homogenization, extraction, concentration and/or clean-up depending on the sample, the analyte and the analytical technique to be employed (Mitra & Brukh, 2003). Preservation of samples should avoid physical (*e.g.*, volatilization, diffusion, adsorption) and chemical changes (*e.g.*, photochemical reactions, oxidation, precipitation), as well as biological degradation, with the goal of preventing or minimizing any changes in sample composition (Mitra & Brukh, 2003). A table of suitable conditions of storage for different analytes can be found in the literature (Mitra & Brukh, 2003).

In some cases, the analysis of solids may require the initial removal of extraneous matter. For instance, removal of shells of some seafood can be required in the case of analysis of marine organisms when only soft tissue is to be analyzed. In the case of analysis of internal organs, marine or terrestrial animals must be eviscerated. With regard to plants, depending on the anatomical part to be analyzed (i.e., flower, stem, leaf) structures that are not being analyzed must be removed.

Once the specific matrix of interest is selected, another stage of the pre-treatment process includes, in many cases, drying the sample to prevent it from having a high concentration of moisture, which could disturb the analysis or encourage biodegradation (Petrovic & Barceló, 2004), as is often the case for soil and sediment analysis. The type of drying conditions and the temperature are parameters to take into account at this stage. For example, the ammonium and nitrate content of soil samples can vary significantly depending on whether the sample has been open air-dried or dried in an oven.

After drying, the sample may need to be reduced in size and mass. Reduction of particle size can be achieved by several processes including, but not limited to, chopping, grinding, mincing, pulverization, blending or mixing. Similar to other cases, care has to be taken to avoid loss of homogeneity or even analyte losses during these processes. For instance, volatile compounds are likely to be released from the solid matrix throughout these stages. This step is especially critical when it comes to the analysis of low mass samples, since sample representativeness could be lost during the process. In this regard, the minimum sample mass intake that ensures homogeneity should be determined.

Storage and sample preservation is also an important issue, especially for those solid samples that will not be immediately analyzed after preparation. Temperature (and therefore stability), oxygen/inert atmosphere, light and storage material conditions have to be established to ensure sample integrity. In all cases, sample headspace should be minimized to avoid degradation. In general, a fast transportation to the analytical laboratory is of high importance, especially for the analysis of fresh products such as edible matrices or biological tissues, whose molecular degradation could lead to a loss of sample integrity (Pawliszyn, 2002).

7.3.3 Extraction Mechanisms

The mechanisms that govern the extraction of analytes from the matrix are much more complicated when the matrix is a solid. Since the analyte has to be extracted from the matrix and then transported into different phases, numerous processes take part in the extraction mechanism. In this regard, a comprehensive chapter on the theory of extraction has been published by Pawliszyn as a part of a book aimed at sample preparation. All these mass transfer processes can be expressed through mass balance equations, which are modified for those energy-, pressure- and temperature-assisted techniques for the preparation of solid samples (Veggi et al., 2013).

Basically, diffusion of the extractant through the matrix, solvation and interactions with the matrix components, diffusion of the analyte out of the solid and migration from the surface matrix towards the bulk extractant are the main processes occurring during the extraction of the analyte from the solid sample matrix. In this process, partitioning of the analyte depends on its affinity towards the surrounding media and, therefore, the extractant must be able to establish strong interactions with the analyte, enough to overcome those forces keeping the analyte bound to the matrix. In this sense, both physical adsorption to the matrix and solid matrix-analyte chemical interactions (*e.g.*, van der Waals forces, electrostatic interactions and hydrogen bondings) play an important role.

In addition, extractions using sorbent-based techniques must consider the specific sorption and desorption mechanisms between the analyte and the sorbent. The characteristics of the solid sample, the physico-chemical interactions of the analyte (inorganic or organic) with the matrix, along with the characteristics of the phase used for extraction also influence the mechanism. In most cases, the extractant phase to sample ratio should be high to ensure complete extraction, especially in the case of exhaustive extraction techniques (*e.g.*, Soxhlet extraction).

7.4 Sample Preparation Techniques for Solid Samples: from Conventional to Miniaturized Alternatives

Preparation of samples for metal and organic compound determination at trace levels is applied to i) degrade or solubilize the matrix, ii) dilute the matrix and iii) extract, separate or concentrate the analyte (Mitra & Brukh, 2003). Sometimes, it is necessary to destroy the organic matter (including proteins, lipids and carbohydrates) in order to determine inorganic compounds. Usually, sample preparation for trace analysis requires more sophisticated procedures than major component analysis. Whereas carrier gas hot extraction or combustion can be enough for the analysis of major components, additional clean-up and preconcentration techniques (i.e., SPE of LLE-based techniques) are required in many cases for trace analysis, some of them even using specific instrumentation as in the case of pressurized liquid extraction (PLE) or supercritical fluid extraction based approaches (Mitra & Brukh, 2003).

As discussed above, one of the trends in sample preparation is the downsizing of analytical methodologies. Miniaturization of analytical systems is performed at small and very small scales. A possible classification of the different devices into three levels could be acheived based on the size of the devices and the volume of sample used: mini- (from cm to 1 mm and µL), micro- (lower than 1 mm and nL) and nano- (lower than 1 µm and pL, fL, aL) devices or systems (Ríos et al., 2009).

Some problems inherent to miniaturization include sampling, sample preparation and sample introduction, especially in the case of solid samples (Rios et al., 2009). More research is needed in these fields of study.

In this section, different conventional and miniaturized sample preparation procedures for solid and semi-solid samples are discussed. Different classifications have been compiled for organic and inorganic analytes. In **Table 7.1**, an assessment of different strategies for miniaturized sample preparation according to their 'miniaturization profile' for metals, organometallics and organics analytes is provided.

7.4.1 Trace Elemental and Organometallic Analysis

Several procedures mainly employed for trace elemental and organometallic analysis are described in this section. The evolution from conventional (more established procedures) to miniaturized procedures is also discussed. **Table 7.2** shows selected applications of miniaturized sample preparation for metals and organometallics.

Table 7.1 Miniaturized sample preparation procedures for trace metals, organometal and organic compounds analysis.

Miniaturized sample preparation	Sample mass	Treatment time	Volume of reagents/solvents	Auto-mation	Simultaneous multitreatment	Miniaturiza-tion profile
DSS	few µg to mg	few seconds	none	no	1 sample	high
miniaturized-SS	5 - 200 mg	20 s - 30 min	1–5 mL	yes	depending on the agitation system	medium
miniaturized sample emulsification	15 mg	1 min	1 mL	yes	depending on the agitation system	medium
Schöniger flask digestion (dry ashing)	3–10 mg	20 min	-	no	1 sample	low
cold plasma ashing (CPA)	100 mg	several hours	0.5–2 mL	no	1 sample	medium
miniaturized *in situ* fusion (on the boat-type graphite platform)	few µg (200 µg)	few seconds	10 µL	no	1 sample	very high
low-volume acid digestion	10–100 mg	5–15 min	200–700 µL	no	2 or 3 samples for digestion vessel (and 6–12 digestion vessels per microwave oven)	high
VPAD	< 250 mg	5–15 min	200 - 700 µL (directly on the sample) 3–15 mL acid for the vapor	no	several samples (1–3 sample per digestion vessel, that is, 6–12 samples per microwave oven; 4 samples using a focused-microwave oven)	medium
miniaturized ultrasound-assisted digestion (miniaturized-UAD)	few mg to g	20 s - 30 min	1–5 mL	no	several samples using an ultrasonic bath	medium
miniaturized ultrasonic enzymatic digestion	15 mg	few seconds	1 mL	no	depending on the ultrasonic device. (ultrasonic probe: 1 sample, ultrasonic bath: several samples, cup-horn sonoreactor: 6 samples)	medium
miniaturized-MSPD	few milligrams to 0.2 g	2 mL	up to 3 mL	no	1 sample	medium

continued **Table 7.1** Miniaturized sample preparation procedures for trace metals, organometal and organic compounds analysis.

Miniaturized sample preparation	Sample mass	Treatment time	Volume of reagents/solvents	Automation	Simultaneous multitreatment	Miniaturization profile
miniaturized-UAE	few mg to g	20 s-30 min	1–5 mL	no	depending on the ultrasonic device. (ultrasonic probe: 1 sample, ultrasonic bath: several samples, cup-horn sonoreactor: 6 samples)	medium
miniaturized-MAE	0.15- 5g	5–15 min	up to 10 mL	no	several samples depending on the microwave oven (6–12 samples)	low
miniaturized supercritical fluid extraction (miniaturized-SFE)	g	30 min	-	no	1 sample	medium
miniaturized accelerated solvent extraction (miniaturized-ASE)	few mg to 0.3 mg	5–30 min	up to 6 mL	no	1 sample	medium
μ-SPE	few mg of sample- 3 g	few minutes-hour	1–8 mL	no	1 sample	high
SPME	Few mg of sample- 3 g	3min-180 min	none- 10 mL	yes	1 sample	high
LPME	few mg of sample- 5 g	30s-60 min	1–100 μL extractant	yes	1 sample	high
micro-total analytical systems (μ-TAS) or lab-on-a-chip (LOC)	few mg	few seconds-few minutes	few μL	yes	1 sample	very high
lab-on-a-valve (LOV)	(sample preparation off-valve)	few seconds	few μL or sub- μL	yes	1 sample	very high
magnetically driven solid sample preparation	few mg to 0.1 g	few seconds	up to 1 mL	yes	6 samples	very high

Table 7.2 Selected applications of miniaturized sample preparation procedures for trace metals and organometallics determination.

Miniaturized Sample preparation	Matrix	Analyte	Description	Detection limits	Recovery (%)	Calibration	Detection	Ref.
DSS	metallic and non-metallic samples	Fe, Cr	sample, mounted in a perpendicular position in the carrier, was ablated by a Nd:YAG laser. The ablated sample was collected directly on a TXRF sample carrier, which has a mass of approx. 10 ng.	5–6 mg g^{-1}	-	addition of an internal standard (Ge)	LA-TXRF	(Spanke et al., 2000)
ultrasonic slurry sampling (USS)	fertilizer sample	Cd	samples (0.4–5.5 mg) were suspended in 1.0 mL of 5% (v/v) HNO_3, 0.05% (m/v) Triton X-100 and 10% (v/v) ethanol and sonicated in an ultrasonic bath for 30 min. Prior to analysis, the slurry was homogenized manually for 5 s.	ng g^{-1},	98 %	external calibration (aqueous standard solutions)	HR-CS-ETAAS	(Borges et al., 2011)
sample emulsification	cosmetic samples	Cd, Cr, Ni	15 mg of sample was dispersed in 1 mL volume of a mixture 0.5% (m/v) of SDS + 3% (v/v) of HNO_3 or HCl, followed by 1 min of sonication using an ultrasonic probe.	0.95–11 µg g^{-1}	93–102%	external calibration (aqueous standard solutions)	ETAAS	(Lavilla et al., 2009)
dry ashing	airborne dust sample (collected on filters)	As, Ca, Co,Cd, Cr, Fe, Ga, Ni, Pb, Pt, Ti, V, Zn	filter disks of a 6 mm diameter were fixed on a sample carrier with 6 mL of a solution of polyethylene glycol:water (1:1, v/v). After freeze drying for 15 min the samples were treated in the plasma asher for 50 min.	-	89–98%	addition of an internal standard (Ga)	CPA-TXRF	(Theissen & Niessner, 1999)

continued **Table 7.2** Selected applications of miniaturized sample preparation procedures for trace metals and organometallics determination.

Miniaturized Sample preparation	Matrix	Analyte	Description	Detection limits	Recovery (%)	Calibration	Detection	Ref.
in situ fusion on the boat-type graphite platform	cement samples	Cr, Co, Mn	200 µg of sample was placed in a boat-type graphite platform. Then, 10 µL of a flux mixture (4 % (m/v) Na_2CO_3 + 4 % (m/v) ZnO + 0.1% (m/v) Triton X-100) were added over the sample and heated at 800 °C for 20 s in the ETAAS platform. After the fusion step, the heating is stopped and an aliquot of 10 µL of 0.1 % (m/v) HNO_3 was added to the sample and the heating program was continued to the end.	-1–9 µg g^{-1}	96- 120%	external calibration (aqueous standard solutions)	ETAAS	(Intima & Väisänen, 2009)
low-volume microwave-assisted digestion (LVMAD)	biological samples (human breast biopsies and CRMs)	Ag, As, Cd, Co, Cu, Cr, Mn, Mo, Ni, Pb, Se, Zn	20–30 mg of sample was digested in a microwave using 3 vials of 6 mL placed in a 100 mL PTFE conventional vessel. Heating program: 15 min at 250 W.	0.04–19.5 ng g^{-1}	92–111 %	external calibration (aqueous standard solutions)	ICP-MS	(Millos et al., 2008)
vapor-phase-microwave-assisted digestion (VPMAD) with focused-MW	biological samples	Co, Fe	30 mg of samples were placed in PTFE cups and digested by HNO_3 vapor formed at 115 °C, 10 min for Co and 60 min for Fe. 15 mL of HNO_3 was placed in the bottom of the vessel (without contact with the sample) for the vapor steaming. 150 µL of H_2O_2 was added to the samples.	-	82–99 %		ETAAS	(Araújo et al., 2000)

continued **Table 7.2** Selected applications of miniaturized sample preparation procedures for trace metals and organometallics determination.

Miniaturized Sample preparation	Matrix	Analyte	Description	Detection limits	Recovery (%)	Calibration	Detection	Ref.
miniaturized-UAD	environmental and biological samples	Cd, Cr, Ni, Pb	0.1 g of sample was extracted in 1 mL of extractant phase. The extract and the sonication time depend on the type of sample. Sonication times of 2–20 min and a mixture of HNO_3:H_2O_2 (2:1, v/v) and HNO_3 for biological samples and sonication time of 5–50 min and a mixture of concentrated HNO_3:$HClO_4$:HF (2:1:1, v/v/v), HNO_3:HCl (1:3, v/v), HNO_3:H_2O_2 (2:1, v/v) and HNO_3 were tested.	0.05- 0.55 $\mu g\ L^{-1}$ (2.5- 27.5 $ng\ g^{-1}$)	96–102%	external calibration (aqueous standard solutions)	ETAAS	(Kazi et al., 2009b)
enzymatic probe sonication (EPS)	yeast, oyster and mussel tissue	Se	10 mg of sample, 1 mg of enzyme were mixed and diluted in 1 mL of water. The mixture was sonicated for 5 s at 20 W using an ultrasonic probe.	-	96–105-%		ICP-MS	(Capelo et al., 2004)
tetramethyl ammonium hydroxide (TMAH) tissue solubilization	human hair	As, Cd, Ni, Pb	125 mg of sample was solubilized in 0.5 mL of 25 % (m/v) TMAH at room temperature for 2 h with occasional shaking or alternatively under heating in a water bath at 60–70 °C, 10 min.)	0.02–8.8 $\mu g\ L^{-1}$ (0.004–1.8 $\mu g\ g^{-1}$)	92–107 %	addition of TMAH to the calibration solutions.	ETAAS	(Ribeiro et al., 2000)
magnetic agitation acid extraction	soil samples	Cr, Mn, Fe, Ni, Cu, Zn, Pb	for EDTA extraction, 0.1 g of sample was extracted in 1 mL of a solution of 0.5 mol L^{-1} EDTA for 1 h. for acetic acid extraction, 0.025 g of sample was extracted in 1 mL of a solution of 0.43 mol L^{-1} acetic acid for 16 h.	0.4–1.9 $\mu g\ g^{-1}$	81–122%	addition of an internal standard (Ga)	TXRF	(De la Calle et al., 2013a)

continued **Table 7.2** Selected applications of miniaturized sample preparation procedures for trace metals and organometallics determination.

Miniaturized Sample preparation	Matrix	Analyte	Description	Detection limits	Recovery (%)	Calibration	Detection	Ref.
miniaturized-UAE with cup-horn sonoreactor	biological (animal and plant tissue) and environmental samples (soils, sediments, fly ash)	Cd, Pb, Cr, Mn, Ni	3–25 mg of sample was extracted in 1 mL of diluted HNO_3, HCl, H_2O_2, or HF (0.5–50 % v/v). Sonication time: 3–40 min using the cup-horn sonoreactor.	0.0045–0.034 μg g^{-1}	80–120 %	external calibration (aqueous standard solutions)	ETAAS	(De la Calle et al., 2009)
MAE	flour samples	Organotin	0.5 g of sample was placed into a PTFE tube and 4.5 mL of a mixture hexane-acetic acid (80/20 v/v) was added. The sample was digested in a microwave oven at 100 °C for 3 min. The extract is centrifuged and an aliquot of the supernatant is diluted.	0.14–0.31 μg mL^{-1}	73–101 %	comparison of retention times with organotin standards	HPLC-UV and HPLC-MS	(Wang et al., 2006)
miniaturized-ASE	hair samples	As(III), As(V), As total	sample (50 mg) was mixed with 350 mg of dispersant (PTFE balls). This mixture was placed in the extraction cell of a laboratory-made miniaturized device which was employed for PLE. optimum conditions: extraction solvent, mixture of 1 % (m/v) SDS and 5 % (v/v) isopropanol; temperature, 125 °C; pressure, 6 MPa (for two cycles of 5 min each). Finally, 1.5 mL of extract was collected and analyzed.	14–20 ng L^{-1}	70 %	standard additions method to correct any matrix effect	HPLC-ICP-MS	(Sanz et al., 2007)

continued **Table 7.2** Selected applications of miniaturized sample preparation procedures for trace metals and organometallics determination.

Miniaturized Sample preparation	Matrix	Analyte	Description	Detection limits	Recovery (%)	Calibration	Detection	Ref.
supercritical fluid extraction (SFE) (fractionation BCR studies)	aquatic sediments and biofilm samples	Zn, Pb, Ni and Cu	sample (0.5 g) was mixed with SiO_2 in mass ratio of 5:95. Stainless steel extraction columns were filled with 1 g SiO_2, then 10 g of sample–SiO_2 mixture, and the remaining volume was filled with SiO_2 again. 1st step: using supercritical CO_2 (CO_2-soluble organic fraction and sulfides), 2nd step: subcritical H_2O (water-soluble fraction). 3rd step: subcritical H_2O+CO_2 (fraction bound to carbonates). each step was performed at 80 °C and 27 MPa for 30 min.	-	-	calibration was performed by MERCK multielement standard with matrix-matching according to the applied extractant	ICP-OES	(Horváth et al., 2013)
MSPD	seafood products	Speciation of As	0.25 g of sample was blended with diatomaceous earth (1.75 g) for 5 min and the mixture packed into a cartridge containing 2 g of C18 as co-sorbent. after that, arsenic species were eluted with 10 mL of 1:1 methanol/water, concentration by rotary evaporation to aprox 2 mL and analyzed.	6.4–23 ng g^{-1}	98%	different aqueous calibration using Ge as an internal standard.	HPLC-ICP-MS	(Moreda-Piñeiro et al., 2008)

continued **Table 7.2** Selected applications of miniaturized sample preparation procedures for trace metals and organometallics determination.

Miniaturized Sample preparation	Matrix	Analyte	Description	Detection limits	Recovery (%)	Calibration	Detection	Ref.
SPME with thermal desorption	marine sediment and biological tissue	Species of As, Se, Hg, Pb, Sn	sample (1 g) was placed in a vial on a hot plate heated to a certain temperature 70–150 °C for a period of 1–5 min.	1.8–121 ng g^{-1}	105–109%	matrix matched standard is required (using CRMs as external calibration standards)	ICP-MS	(Mester, 2002)
lab on a chip (LOC) (chip-based LPME)	human hair and cell samples	Cu. Zn, Cd, Hg, Pb and Bi	sodium diethyldithiocarbamate (DDTC) was used both as the chemical modifier and as extracting reagent for the chip-based LPME. the extraction took 3.5 min and 7 µL of the organic phase was collected and introduced using a micropipette into the graphite furnace.	6.6–89.3 pg mL^{-1}	87–120%		ETV-ICP-MS	(Wang et al., 2013a)
LOV	ash and river sediment samples	Ni, Bi	samples were digested in a PTFE beaker using HNO_3 + HF. LOV was coupled to ICP-MS by a special valve that allowed aspiration of only 15 µL.	4–15 ng L^{-1}	93 %	calibration standard of periodically analysed.	ICP-MS	(Wang & Hansen, 2001b)
LOD	potassium ferrocyanide	-	complete dissolution of 0.1g of sample in 1 mL of water in only 3 s. Magnetically driven solid sample preparation for centrifugal microfluidic devices.	-	-		-	(Duford et al., 2009)

7.4.1.1 Minimal Treatment-based Techniques

As discussed above, sample treatment is one of the most energy- and reagent- consuming steps of the analytical process, and the one with the greatest risk of sample contamination and analyte losses. Non-intensive procedures avoid: i) time-consuming steps of decomposition or extraction of the solid sample, ii) the use of large volumes of acids and organic solvents, iii) generation of additional wastes and iv) risks or hazards to the operator. Here, procedures without treatment (direct solid sampling, DSS) and with minimal sample treatment (SS and emulsification) are presented.

7.4.1.1.1 Direct Solid Sampling

As discussed in the introduction section, there are several analytical techniques that can directly analyze a solid sample. Classically, electrothermal atomic absorption spectrometry (ETAAS) using specially-designed graphite tubes (Bendicho & de Loos-Vollebregt, 1991) and laser-induced breakdown spectroscopy (Yamamoto et al., 1996; Capitelli et al., 2002) were applied for the direct analysis of solid samples. More recently, glow discharge optical emission spectrometry (Bengtson & Lundholm, 1988; Pisonero et al., 2006), high resolution continuum source ETAAS (HR-CS-ETAAS) (Welz et al., 2007), laser ablation (LA)-ICP-MS (Günther, 2005; Arroyo et al., 2009) and electrothermal vaporization with inductively coupled plasma mass spectrometry (ETV-ICP-MS) (Vanhaecke et al., 2002), and inductively-coupled plasma optical emission spectrometry (ETV-ICP-OES) (Matschat et al., 2005) have gained more attention. Other possibilities for direct solid analysis are instrumental neutron activation analysis (Kolmogorov et al., 2009), X-ray fluorescence spectrometry (XRF) (Khuder et al., 2009) or total reflection X-ray fluorescence spectrometry (TXRF) (Magalhaes et al., 2010; Vázquez et al., 2010).

Conversely, direct solid sample analysis is not usually combined with flame atomic absorption spectrometry (FAAS), ICP-OES and ICP-MS, since it is necessary to prepare (pre-treat and treat) the sample prior to the analysis. In the case of FAAS, this occurs due to the insufficient dissociation of solid particles in the relatively cold flame; while in the case of ICP-OES and ICP-MS, it is due to physical and chemical interferences (Hoenig & de Kersabiec, 1996).

Some of the advantages of DSS analysis are i) higher sensitivity (no dilution), ii) faster analysis without sample preparation, iii) reduced risks of contamination and analyte loss, iv) hazardous reagents are not required, v) reduced generation of wastes and vi) reduced hazards for the operator. However, some of the limitations are: i) inhomogeneity of solid samples (due to the small sample size), ii) less precise results and iii) the need of CRMs (Bendicho & de Loos-Vollebregt, 1991; Welz, 2007).

From the point of view of miniaturization, ETAAS, ETV-ICP-OES or ETV-ICP-MS, LA-ICP-MS and TXRF are microanalysis techniques since only a very small amount of solid sample is employed. ETV consists of a simple combination of a graphite tube as an ETV device with ICP-MS or ICP-OES. The aim of using ETV consists of achieving selective atomization/vaporization of the analyte (Belarra et al., 2002). LA generates a vapor-phase aerosol using a pulsed laser beam focused on a sample surface (Russo & Baldwin, 2003; Potts & Robinson, 2003; De la Guardia & Armenta, 2011). Thus, LA offers a simple form to deal with materials that are difficult to digest or dissolve such as alloys and refractory materials. Another miniaturized approach was applied using LA directly onto quartz sample carriers of TXRF (Bredendiek-Kamper et al., 1996; Spanke et al., 2000) (**Figure 7.5A**). The main advantages involved in this last procedure are no sample preparation, minimum risks of contamination and the possibility of local and micro-distribution analysis.

While techniques such as XRF require high amount of sample (10 g), lower amounts of samples are enough in some other techniques: 0.5–3 mg (typically 1 mg) in ETV-ICP-MS (Vanhaecke et al., 2002) and ETV-ICP-OES (Matschat et al., 2005), 0.5 mg of medicinal plants by HR-CS-ETAAS (Figueredo-Rego et al., 2012), a few µg of tissue or powder sample by TXRF (Magalhaes et al., 2010; Vázquez et al., 2010), 0.1–10 µg in LA-ICP-MS or LA-ICP-OES (De la Guardia & Armenta, 2011) and a few ng of steel and ceramic samples in LA-TXRF (Bredendiek-Kamper et al., 1996; Spanke et al., 2000).

7.4.1.1.2 Slurry Sampling

SS is an alternative for direct solid analysis and provides a faster and cheaper option for sample preparation than decomposition approaches. Several problems derived from the use of DSS, such as the inhomogeneity, can be alleviated using SS (Hsu et al., 2013). The main advantages of SS include: i) ease of sample preparation, ii) use of non-corrosive reagents, iii) reduced risks of loss of volatile elements and iv) reduced risks of contamination. Additionally, it does not require complete dissolution of the sample, which is a great advantage in case of samples that are difficult to dissolve.

SS involves the preparation of a suspension of powdered sample in a liquid medium and several reviews on this topic have been published (Bendicho & de Loos-Vollebregt, 1991; Cal-Prieto et al., 2002; Ferreira et al., 2010). Preparation of slurries can involve partial decomposition and extraction, particularly if acids are present in the suspension medium (Ihnat, 2003).

Several parameters should be studied when SS is applied, including the i) type of diluents, ii) stabilizing agents, iii) homogenization systems, iv) particle size and v) sample mass-to-volume ratio.

Usually, the liquid medium consists of a dispersant or surfactant (*e.g.*, Triton X-100, Triton X-114) in the presence of a highly diluted acid to avoid flocculation of particles. In other instances, reagents such as diluted glycerol, ethanol, hexametaphosphate, or TMAH were used as stabilizing agents.

Apart from manual agitation (Ferreira-Damin et al., 2011), different systems are commonly used in order to obtain a homogeneous suspension, such as magnetic agitation, vortex, Ultraturrax and gas mixing, or a combination of systems. For instance, the application of Ultra-turrax homogenizer (5 min) and vortex agitation (1 min) was performed for analysis of baby food samples (Ozbek & Akman, 2012). More recently,

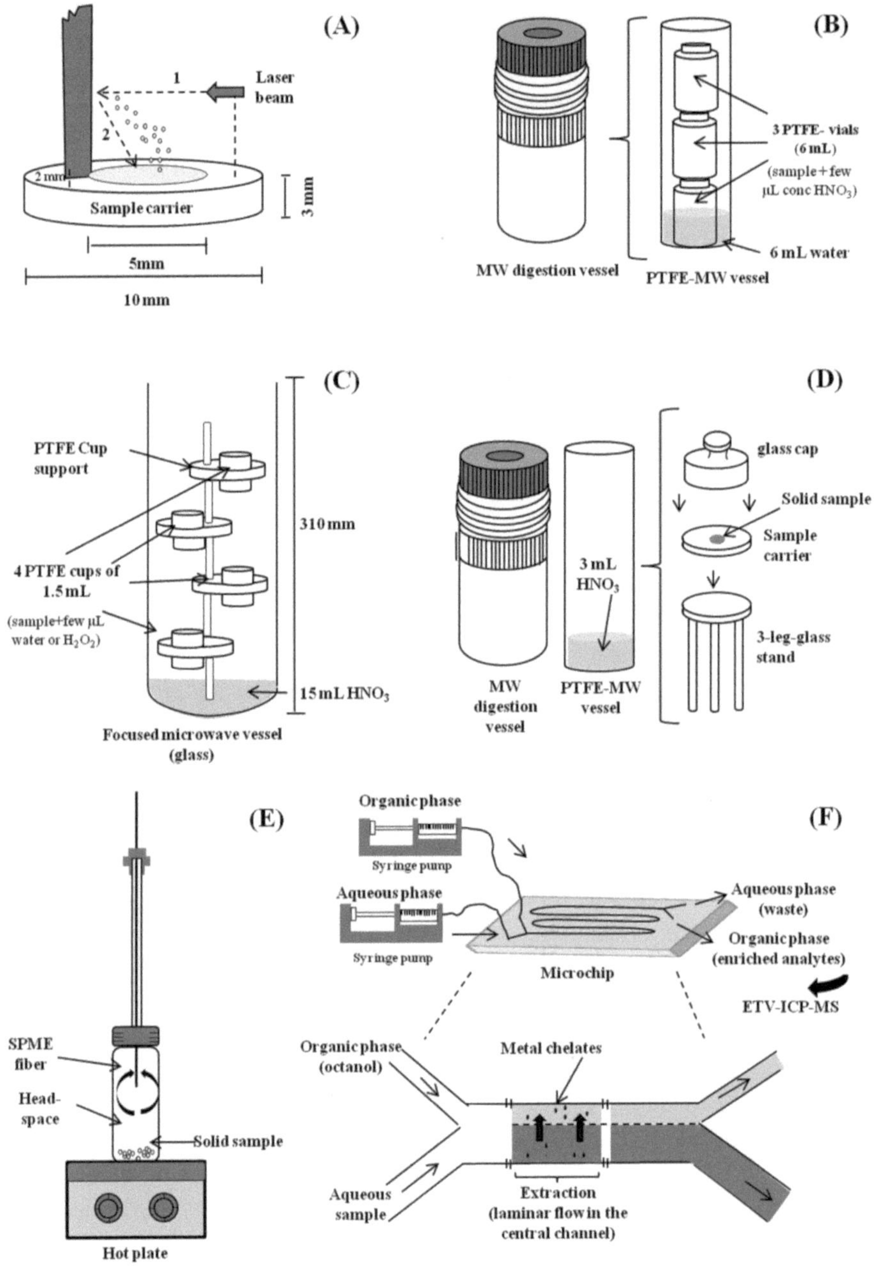

Figure 7.5 Examples of miniaturized sample preparation procedures for determining metals in solid samples. (A) DSS (1. Laser ablation and 2. Deposition onto the sample carrier). (B) Small-vessel microwave-assisted digestion. (C) VPMAD in a focused microwave (FMW) vessel. (D) VPMAD in TXRF sample carriers. (E) Headspace extraction of solid on a SPME fiber. (F) Chip-based LPME.

ultrasound energy was applied for this purpose since this type of energy can drive slurry mixing processes. Generally, ultrasonic probes are preferred due to the high intensity and low time (few seconds) that is required to form a homogenous slurry. On the contrary, when an ultrasonic bath is applied, higher sonication times (15–30 min) are required due to the lower intensity, (Vignola et al., 2010; Kadenkin et al., 2011; De Jesus et al., 2013).

Particle size is also a key parameter for slurry stability and homogeneity. High particle sizes (≥300 μm) provide less reproducibility and sensitivity while a particle size of <50 μm seems to be adequate for SS (Cal-Prieto et al., 2002). A grinding step is usually required to obtain a homogeneously-distributed slurry. Generally, the appropriate sample mass-to-volume ratio is less than 0.25 (Miller-Ihli, 1993). Typically, a sample mass in the range of 5–200 mg and a volume of diluents in the range of 1–100 mL are employed.

In spite of being a classical sample preparation approach (Brady et al., 1974), SS procedures have been reported very recently (De la Calle et al., 2012, 2013a; Hsu et al., 2013). As early as 1974, Brady et al. (1974) developed a very fast approach using small amounts of sample and reagents (*e.g.*, 3 mg of sediment in 10 mL of water, vortex-mixed for 20 s).

The main trends in SS are focused on miniaturization and acceleration by application of ultrasound energy. Miniaturized SS has been applied based on reducing the mass of sample and the volume of reagents. Recent publications include the trace elemental analysis of biological tissues (De la Calle et al., 2012) and soils (De la Calle et al., 2013a). The amount of sample employed in the most recent publications range from 0.4 to 50 mg, while the volume of suspension ranged from 700 μL to 3 mL (Sánchez-Moreno et al., 2009; Vignola et al., 2010; Borges et al., 2011; Ferreira-Damin et al., 2011; De la Calle et al., 2012, 2013a; Scaccia & Mecozzi, 2012; Dobrowolski et al., 2013a). The agitation systems generally used were manual agitation (Ferreria-Damin et al., 2011), magnetic agitation (De la Calle et al., 2012, 2013a), vortex (Dobrowolski et al., 2013), ultrasonic bath (Vignola et al., 2010; Borges et al., 2011; Scaccia & Mecozzi, 2012) and ultrasonic probe (De la Calle et al., 2012, 2013a). The agitation time depends on the agitation system employed, and it ranges from 20–30 s (Sánchez-Moreno, 2009; De la Calle et al., 2012, 2013a), to 30 min (Vignola et al., 2010; Borges et al., 2011; Scaccia & Mecozzi, 2012).

Other improvements on SS are related with the combination with instrumentation, automation, use of permanent modifiers and preconcentration. First, SS has been combined with multiple analytical techniques (especially multi-elemental techniques). Some examples of these combinations include USS with ETV-ICP-MS (Hsu et al., 2013; Lin & Jiang, 2013), cold-vapor atomic absorption spectrometry (CVAAS) (De Jesus et al., 2013), ETAAS (Sánchez-Moreno et al., 2009; Ferreira-Damin et al., 2011; De Paula et al., 2012), USS-HR-CS-ETAAS (Vignola et al., 2010; Borges et al., 2011), ETV-ICP-OES (Amberger, 2010), USS-FAAS (Ozbek & Akman, 2012), TXRF (Kadenkin et al., 2011; De la Calle et al., 2012, 2013a). As for automation, sample introduction systems have included autosamplers equipped with ultrasonic probe slurry samplers since the 1990s (*e.g.*, the USS-100 Ultrasonic slurry sampler from Perkin Elmer) (Amoedo et al., 1999; Santos et al., 2000; Bendicho et al., 2012b). Some examples can also be found in recent publications (Hsu et al., 2013; Lin & Jiang, 2013). Another specific improvement of SS analysis for ETAAS and ETV-ICP-MS consists of the use of permanent modifiers in the graphite tube, such as Nb and Ir (Dobrowolski et al., 2013) for V determination and Pd nanoparticles for Zn, As, Cd, Sb, Hg and Pb determination by ETV-ICP-MS (Yi et al., 2012). Finally, another interesting strategy that consists of using SS for preconcentration by means of solid sorbents (Dobrowolski et al., 2012) will be described in section 7.4.1.4.1.

7.4.1.1.3 Sample Emulsification

Sample emulsification was proposed as a simple sample treatment for semi-solid liposoluble samples prior to atomic spectrometry (Burguera & Burguera, 2012). Target samples included oils (De Jesus et al., 2008) and cosmetics (such as shampoo, gel and cream) (Salvador et al., 2000). Sample emulsification is based on forming an emulsion of the semi-solid sample. Therefore, this process provides the sample with similar properties to an aqueous solution (Carballo et al., 2013), being more suitable to be analyzed directly or after dilution. This facilitates the use of aqueous standard solutions for calibration purposes. Emulsification avoids the use of strong acids and large volumes of organic solvents. Moreover, it is a rapid and simple sample preparation procedure with low manipulation of the sample and low generation of wastes compared to other procedures that are commonly employed for these type of samples (*e.g.*, acid dissolution and/or digestion using heating (Adeyemo et al., 2004), ultrasound irradiation (Bangroo et al., 1995; Sanz-Segundo et al., 1999) and microwave energy (Bellido-Mila et al., 2002)).

Similar to SS, several parameters should be optimized, including: i) the type of dispersion medium (surfactants, diluents), ii) the sample mass-to-dispersion medium and iii) homogenization systems.

Conventional approaches involving magnetic stirring and heating have been reported in the literature for Na, K, Ca, Mg, Zn and Fe determination in eggs by FAAS (Ieggli et al., 2010) and for oils, margarine and butter analysis by FAAS (Na, K, Ca, Mg, Zn and Fe) and ETAAS (Cr, Ni, As, Pb, Cd, Cu, Mn) (Ieggli et al., 2010).

Mechanical and manual agitation are generally used for emulsion formation and stabilization (Salvador et al., 2000). However, the use of ultrasound energy allows one to obtain more stable and homogeneous emulsions in a faster way than conventional shaking (Luque de Castro & Priego-Capote, 2007). The use of ultrasound energy accelerates the process of mass transfer between the two immiscible phases thus generating smaller sized droplets of the dispersed medium. This phenomenon occurs due to processes such as interfacial instability of the oil–water interface and transient cavitation bubbles that generate micro streaming, high-pressure shockwave and high local temperature during their collapse (Cucheval & Chow, 2008). When ultrasound energy is employed, the approach is named ultrasound-assisted emulsification. In addition to the above mentioned parameters, sonication amplitude and sonication time should be studied. Ultrasonic probes are generally used instead of ultrasonic baths due to the higher intensity of the former.

Ultrasound-assisted emulsification was proposed as a simple sample treatment for semi-solid cosmetic samples (*e.g.*, shampoo, gel, cream and oils) prior to atomic spectrometry and trace elemental analysis (As, Cd, Cr, Cu, Hg, Mg, Mn, Ni, Sr and Zn) (Lavilla et al., 2009). Lavilla et al. employed two different procedures (large and small scale) depending on the analytical technique used. In the large scale procedure, 0.05–0.2 g of sample was dispersed in a 15 mL volume for FAAS, CVAAS and ICP-OES analysis. On the contrary, in the small scale procedure, 15 mg of sample was dispersed in a 1 mL volume for ETAAS analysis. The dispersion medium consisted of a mixture of 0.5% (m/v) of SDS + 3% (v/v) of HNO_3 or HCl which was sonicated for 1 min using an ultrasonic probe, which produced a stable emulsion.

Recent approaches applied to the analysis of lubricating oils also used a miniaturized procedure after dissolution of the sample in toluene (0.25–0.35 g of sample in 10 mL of toluene). The emulsions were prepared directly in an autosampler cup (1 mL) for ETAAS analysis, adding an aliquot of the dissolved sample and HNO_3, followed by manual agitation (1 min), ultrasonic bath agitation (5 min) and subsequent addition of xylene + Triton X-100 with additional manual agitation (1 min) (Carballo et al., 2013). Another experiment involved the preparation of a three-component solution directly in an autosampler cup (1 mL). A small aliquot of dissolved sample was placed into the autosampler cup followed by 200 µL of methyl isobutyl ketone (MIBK) and HNO_3 each, and 100 µL HCl were added. The mixture was sonicated in an ultrasonic bath (15 min) and after adding propan-1-ol, the mixture was sonicated in an ultrasonic probe (10 s) just before the analysis (Carballo et al., 2013).

7.4.1.2 Decomposition-based Techniques

In this section, different procedures to bring a solid sample into solution will be described. Since these procedures are undesirable from an environmentally-friendly point of view, several procedures evolved toward miniaturization with subsequent savings in energy, reagents and time. Evolution from the point of view of miniaturization will be taken into account. These procedures include dry ashing, AD, and digestion accelerated by means of microwave and ultrasound energy, enzymatic digestion and tissue solubilization.

7.4.1.2.1 Dry Ashing

Dry ashing and dissolution of the residue with dilute inorganic acids (usually HCl) is a common way to remove organic matter (Hoenig & de Kersabiec, 1996; Hoenig, 2001, 2003). However, the general procedure requires the use of a high amount of solid sample (0.2–2 g) (Hoenig, 2003; Soylak et al., 2004).

Dry ashing in muffle furnace consists of subjecting the solid sample located in a suitable vessel (usually a porcelain crucible) to controlled high temperature (450 °C) until constant weight of the sample is obtained (1–48 h). Usually, these procedures are performed at atmospheric pressure in programmable muffle furnaces. After ashing, the resulting inorganic residue (ash) is dissolved with mineral acids (Hoenig & de Kersabiec, 1996; Heonig, 2001).

Conventional dry ashing leads to complete removal of the organic matrix and to accurate analytical results for commonly determined elements (Hoenig, 2003). It ensures the quantitative decomposition and elimination of organic matter and an efficient release of elements initially associated with it. Due to the high temperatures involved in the procedure, there are associated risks of loss of elements by volatilization. The elements with highest loss risks are Se, As, Sb, Ge, Hg, Cd, Pb and Zn. In addition to volatilization problems, other drawbacks could be the insolubilization that occurs when several elements are associated with the silica matrix. Thus, the addition of HF may be required to solubilize the residue (Hoenig & de Kersabiec, 1996; Hoenig, 2001).

In addition to conventional dry ashing in a muffle furnace, other systems are available. While dry ashing and low temperature ashing (LTA) are carried out in open vessels, combustion bomb, oxygen flask (Schöniger flask) and microwave-induced combustion (MIC) are performed in closed systems. Flores et al. (2007) reviewed sample preparation techniques based on combustion reactions in closed vessels.

In combustion bombs made of stainless steel, a sample mass higher than 0.5 g is prepared as a pellet and positioned in a metallic cup. A volume of absorbing solution (5–10 mL) is added to the vessel. Then, the system is closed, pressurized with oxygen and ignition takes place.

A miniaturized approach derived from the previous one consists of the use of a Schöniger flask that employs a small amount of sample (MacDonald, 1961). It is a very classical approach developed in 1955 by W. Schöniger as a pioneer for microanalysis (Schöniger, 1955; Schöniger, 1956). The combustion of the sample takes place in a closed Pyrex conical flask filled with oxygen for approximately 20 min. The sample (3–10 mg) is wrapped in a piece of paper, linked to a Pt wire, and burnt in the flask containing the appropriate absorption solution (H_2O, H_2O_2, NaOH). This strategy is more commonly employed for non-metals and metalloids. The two main limitations observed for this approach are incomplete oxidation and incomplete absorption by the solution (MacDonald, 1971). Usually, a sample with mass smaller than 100 mg is burned in glass vessels of up to 1 L, but using a volume of reagents in the range of 1–20 mL. Other applications of the oxygen flask were applied for determination of Se (Sun & Li, 2000) and As (Puttermans et al., 1983) in biological tissues (hair, kidney and liver tissue) using an amount of sample in the range of 40–200 mg. Oxygen flask combustion offers a faster and simpler method than AD in the case of Se and does not suffer from losses of Se.

MIC combines the advantages of microwave-assisted digestion (MAD) and dry ashing. MIC involves the combustion of organic samples in closed quartz vessels containing pressurized oxygen but using microwave radiation in the ignition step (Flores, 2007). After combustion, the gases are absorbed in a solution of diluted acid. In MIC, a quartz device is inserted into a quartz vessel and it acts as a sample holder. The sample (less than 0.5 g) is prepared as a pellet and temperatures of 1300 °C are achieved. MIC applications were developed for samples which were difficult to digest by MAD, such as coal (Antes et al., 2010) and biological samples (Duarte et al., 2009; Mesko et al., 2010). Quantification is usually achieved by ICP-MS, or ICP-OES. A recent application of MIC was employed for seafood digestion and As determination by hydride generation coupled to atomic spectrometry (HG-AAS, HG-ICP-MS and ICP-MS) (Duarte et al., 2009). Pellets of sample mass between 100–500 mg and 6 mL of 0.1 mol L^{-1} HNO_3 as the absorbing solution were used. Up to eight samples could be processed simultaneously in 25 min. The residual carbon content was very low (<0.5 %).

Another possibility is the use of oxygen plasma ashing (i.e., CPA, LTA) which is a low temperature (100 °C) and low pressure (100 Pa) ashing techniques employing a cold plasma of activated oxygen (Hoenig & de Kersabiec, 1996; Flores, 2007). The sample is contained in a quartz test tube equipped with a cooling finger and the oxygen plasma is generated at high frequency. Volatile elements are retained on the cooling finger during the ashing procedure, avoiding element losses. The temperature reaches 150 °C and the required volume of acid is only 1–2 mL. However, this method is unsuitable for routine analysis due to the long mineralization time. CPA tends to be slow, taking several hours per sample, for instance, 1.5 h for muscle tissue, fat, fecal matter filter paper and rat tissue (Gleit & Holland, 1962), whereas 12–18 h for dentin samples, 48 h for bones, or even 60 h for complete destruction of organic matter in non-powdered samples (Sansoni & Panday, 1994).

Several miniaturized procedures of CPA were applied prior to TXRF analysis in two ways. First, deposition of the sample (a few mg) onto the quartz sample carrier after ashing (Reus, 1991; Schemling et al., 1997) and secondly, application of ashing directly to the quartz sample carrier of TXRF (Theisen & Niessner, 1999). The last option is of interest from the point of view of miniaturization and is called *in situ* CPA (De la Calle et al., 2013b). Moreover, AD or dry ashing were combined with *in situ* CPA (Woelfl et al., 2003, 2004; Mages et al., 2008; Wagner & Mages, 2010; Savoly et al., 2012). In these examples, 2 h of CPA was required for nematodes and microcrustaceans (Woelfl, 2003; Savoly et al., 2012), and between 1 and 7 h for filters (Wagner and Mages, 2010).

7.4.1.2.2 Fusion

Some samples such as ceramics, aluminosilicates and alloys are difficult to decompose by wet digestion. Fusion offers an alternative for these type of samples. For fusion procedures, basic (NaOH, Na_2CO_3), acidic ($K_2S_2O_7$), oxidizing (Na_2O_2) and reducing (carbide) fluxes are used. Both powdered samples and fluxes are mixed in a crucible and this mixture is heated to a temperature above the melting point of the flux in a furnace or burner and agitated until total dissolution occurs (Claisse, 2003). Approximate proportions of sample/flux are 1 g of sample per 10 g of flux. Fusion processes usually last between 1–3 h at a temperature in the range of 300–1000 °C. The high concentration of added fluxes and the aggressive media increase the risks of contamination and the amount of added salts may produce matrix interferences. In addition, the loss of volatile components has to be carefully controlled.

Based on a large scale procedure using 0.5 g of sample and 2 g of flux (Na_2CO_3: ZnO, 1:1 m/m) loaded into a platinum crucible and heated in a microwave muffle furnace for 1 h (including cooling time), a miniaturized procedure was developed (Associaçao Brasileira de Cimento Portland, 2003). *In situ* fusion on the boat-type graphite platform for the direct determination of elements (Cr, Co and Mn) by ETAAS in cement samples by solid sampling was proposed as a miniaturized approach (Intima & Väisänen, 2009). This strategy works with up to 200 µg of sample. It consists of using 10 µL of a flux mixture (consisting of 4 % (m/v) Na_2CO_3 + 4 % (m/v) ZnO + 0.1% (m/v) Triton X-100) added over the cement sample and heated at 800 °C for 20 s in the ETAAS platform. The addition of Triton X-100 was necessary to facilitate the interaction between the flux and the sample. After the fusion step, the heating is stopped, the platform is removed and an aliquot of 10 µL of 0.1 % (m/v) HNO_3 is added to the resulting mold which has been completely dissolved. Then, the platform was re-introduced into the graphite tube and the heating program was continued to the end. This method is faster than the normal-scale fusion and allows direct determination on the boat-type graphite platform (Intima & Väisänen, 2009).

7.4.1.2.3 Acid Digestion and Microwave-assisted Digestion

Conventional AD, also called wet digestion, involves the total dissolution of the sample matrix, which ensures the complete availability of analytes for the analysis (Mitra & Brukh, 2003). This is probably the most common sample preparation procedure for solid samples.

Conventional AD generally encompasses the use of acids (HNO_3, HCl, H_2SO_4, or HF) and oxidants (H_2O_2). Furthermore, traditional heating systems such as a sand bath, hot plate, furnace, or Bunsen burner are used for AD (Bendicho et al., 2011). The simplest configuration consists of placing the sample and the acids in a glass beaker covered with a watch glass while heating. When the sample is digested, the acid is evaporated to near dryness and diluted in acid solution (Mitra & Brukh, 2003). Conventional AD generally uses a large amount of sample (0.3–5 g) and 10–100 mL of concentrated reagents while the heating is performed at elevated temperatures depending on the employed acids (120–330 °C) for long periods of time (4–48 h) (Soylak et al., 2004).

Faster and more focused digestion was developed using microwave energy since 1975 (Abu-Samra et al., 1975), giving rise to MAD. The principles, equipment and applications of microwave-assisted techniques have been described in book chapters (Luque de Castro & Luque-García, 2002; Matusiewicz, 2003) and many review articles

(Smith & Arsenaut, 1996; Luque-García & Luque de Castro, 2003). This approach has been widely applied and established as a reference sample preparation procedure for trace element analysis in routine analysis and research laboratories. There are different schemes for AD methods (Matusiewicz, 2003), including open systems (conventional heating, microwave heating and ultraviolet digestion), closed systems (conventional heating, microwave heating), and flow systems (conventional heating, UV on-line decomposition and microwave heating). Recently, a combined procedure of exposure to infrared radiation (IR) for 10 min followed by cavity microwave-assisted digestion was proposed for the digestion of a food complement which includes flaxseed, wheat bran, wheat germ, cashew nuts, soybeans, sesame seeds, brown sugar and oats (Dantas et al., 2013). Moreover, several pieces of equipment were developed to improve its safety and efficiency. Both low and high pressure digestion can be employed in MAD. In addition, focused-microwave and cavity-microwave digestion have also been reported (De Oliveira, 2003).

MAD reduces the time of operation from 1–2 h in conventional AD to 5–15 min (Matusiewicz, 2003). Conventional MAD requires 50–120 mL vessels, the addition of 5–10 mL of reagents (usually acids and H_2O_2) and a sample mass of up to 0.5 g (Soylak et al., 2004; Borgese et al., 2009).

In order to reduce the amount of sample and necessary reagents, several miniaturized strategies have been tested by researchers using electric ovens, hot plates, domestic microwave ovens, microwave systems, focused-microwaves and cavity-microwaves (Luque-García & Luque de Castro, 2003).

One possibility of miniaturization is the use of 7 mL vials as sample holders and carrying out the sample digestion by means of an electric oven (W. Zhang et al., 2012) and/or a hot plate (Yu et al., 2001; W. Zhang et al., 2012). As an example, sample mass ranges from 50 to 100 mg of geological material and 4.5 mL of digestion reagents (NH_4HF_2 (W. Zhang et al., 2012) and HNO_3+HF (Yu et al., 2001)) have been reported. Similarly, 2 mL polypropylene vials were used for the digestion of hair samples and bovine liver (15 mg) using a conduction oven (20 h at 70 °C) or a conventional microwave oven (7 min). Only 50 µL of H_2SO_4 + 150 µL of HNO_3 were needed for the digestion (De Oliveira, 2003; Flores, 2007). Another recent option consists of using a domestic microwave oven for the partial digestion of bovine tissue prior to ICP-OES (Matos et al., 2009). In this case, a sample mass of 50 mg was placed in a 4 mL laboratory-made Polytetrafluoroethylene (PTFE) microvessel inside a baby-bottle sterilizer. Very low volumes of HNO_3 (50 µL) and H_2O_2 (100 µL) were added to the uncovered microvessels and positioned inside the sterilizer, containing 500 mL of water and exposed to microwave radiation for 7 min (Matos et al., 2009).

More attention has been paid to the miniaturization of MAD approaches since the mid-1990s (Baldwin et al., 1994) by using microwave energy and significantly reducing the amount of sample (0.5 mg - 0.1 g) and the volume of acids (0.1–1 mL). These approaches are commonly named small-volume microwave-assisted digestion or LVMAD.

First, smaller vessels (20 mL) were developed for the analysis of organic and biological microsamples (Müller, 1998). These modified vessels allow digestion of 50–100 mg of sample using only 0.7 mL of HNO_3 (Müller, 1998). Moreover, 15 mL polystyrene vessel liners were used for analysis of 2–10 mg of biopsy samples after adding 0.5 mL of HNO_3. However, specific microwave ovens and rotors, which are not available in all labs, are needed in this approach (Bocca et al., 2007). Several examples of small-volume MAD can be found for the analysis of different elements combined with FAAS (Baldwin et al., 1994), ETAAS (Deaker & Maher, 1997, 1999; Iavicoli et al., 2001), TXRF (Varga et al., 2005), HG-AFS (Zhao et al., 2010), ICP-OES (Millos et al., 2009) and ICP-MS (Esslemont et al., 2000; Maher et al., 2001; Varga et al., 2005; Millos et al., 2008). Diverse configurations were developed in the literature for small-volume MAD with the goal of increasing the sample throughput: i) 1 vial of 7 mL (Maher et al., 2001; Iavicoli et al., 2001; Varga et al., 2005), ii) 2 vials of 7 mL placed one over the other (Baldwin et al., 1994; Deaker & Maher, 1997, 1999; Esslemont et al., 2000), iii) 3 vials of 6 mL placed one over the other (**Figure 7.5B**) (Millos et al., 2008, 2009), iv) 3 vials of 6 mL positioned in triangle (Zhao et al., 2010) or v) 4 vials of 2 mL-polypropylene minivials positioned in circle (Brancalion et al., 2005), but all placed in larger PTFE vessels of 100–120 mL. As can be noted, high throughput can be achieved when multiple small vials are placed inside a large PTFE vessel, and placed in a rotor of 6–12 digestion vessels.

Moreover, miniaturization was also extended to focused-MW (FMW). In one study, 4 vials of 4 mL or 5 vials of 2 mL were placed inside a glass tube of a FMW for the determination of metals in hair samples (Tan, 2001; De Oliveira, 2003). The mass of sample employed was in the range of 30–60 mg. Similarly, 4 polypropylene minivials of 2 mL positioned in circle were placed in a support fitted in a glass vessel for the digestion of medicinal plants and determination of Cd (Brancalion et al., 2005). The digestion of 5 mg of sample was carried out by adding 0.2 mL of HNO_3 + 0.15 mL 30 % (v/v) H_2O_2 for 4–10 min.

There is even a more miniaturized approach that it is carried out directly for a specific analytical technique called TXRF due to the type, form and configuration of the sample carrier that is used. The sample carrier in TXRF is generally a disc of quartz with a diameter of 30 mm and a thickness of 3 mm (De la Calle et al., 2013b). These miniaturized sample preparation procedures are called *in situ* AD and *in situ* MAD, depending on the heating source, usually a hot plate (Woelfl et al., 2003; Sávoly et al., 2012) or a domestic microwave oven (Marcó et al., 2001). These approaches consist of placing a very small amount of sample (μg-mg) directly onto the sample carrier and adding a very small volume of reagents (5–20 μL of HNO_3 or HNO_3+H_2O_2). Then, the digestion occurs at 90–100 °C over 10 min using a hot plate (Woelfl et al., 2003; Sávoly et al., 2012) and 8 min using a domestic microwave oven (Marcó et al., 2001). Several examples of this approach are reported in the literature for the analysis of nematodes (Sávoly et al., 2012), freshwater microcrustaceans, (Woelfl et al., 2003; Mages et al., 2004), fish embryos (Mages et al., 2008), bovine liver tissue (Marcó et al., 2001) and human hair (Borgese et al., 2010).

7.4.1.2.4 Vapor-phase Acid Digestion and Vapor-phase Microwave-assisted Digestion

VPAD is an alternative to AD in which the attack of the sample occurs from the acid-vapor phase rather than the solution. The most important advantages of VPAD are the decreased acid concentration of the digestates, the lower dilution and, as a consequence, the improvement of detection limits (Matusiewicz, 1991; Matusiewicz et al., 1991). Moreover, this approach allows the simultaneous purification of the acid, prevents the introduction of impurities and improves blank samples (Araújo et al., 2004; Matos et al., 2009). This procedure has proved useful for biological tissues since both organic and inorganic components are completely solubilized by vapor phase attack (Matusiewicz et al., 1991).

Vapor-phase digestion has been employed for sample preparation using different inorganic acids by conventional heating (Thomas & Smythe, 1973; Klitenick et al., 1983), PTFE pressure bombs (Czégény et al., 1998; Matusiewicz, 1989), a high-pressure asher (Amarasiriwardena et al., 1994) and a high-pressure digestion vessel (Lásztity et al., 1995).

Furthermore, microwave energy was employed for accelerating VPAD, giving rise to VPMAD using closed-vessel microwave systems (Matusiewicz et al., 1991), and, more recently, FMW ovens (Araújo et al., 2000; Trevizan et al., 2003; Araújo et al., 2004).

In 1991, Matusiewicz et al. developed VPMAD using a laboratory-made high pressure-digestion bomb (Matusiewicz et al., 1991). A PTFE microsampling cup with one leg (to lift the vial up off the reagents) was used to place the sample (250 mg) and 0.6 mL of acid was added to the vial. Then, the unit was placed inside a 100 mL perfluoroalkoxy-PTFE vessel containing 6 mL of acids which are not in direct contact with the sample and thus only the acid vapors are able to digest the sample.

VPAD and VPMAD can be considered miniaturized approaches since very small amounts of sample are generally used. Usually, a small amount of sample is used (less than 250 mg) and a small volume of H_2O_2 or HNO_3 (150–600 μL) is added to the sample to prevent charring of organic matter (Matusiewicz et al., 1991). Usually, micro-vessels are employed for sample digestion of small masses (Matusiewicz et al., 1991; Araújo et al., 2004), which are advantageous in these procedures because they improve the interaction of the vapor with sample particles (Araújo et al., 2004; Matos et al., 2009). Moreover, this characteristic is attractive when the sample size is limited, such as in the case of toxicological analysis of biological tissues. Since then, different configurations were developed in order to increase the sample throughput (Araújo et al., 2000; Eilola & Perämäki, 2001).

A VPMAD was developed for the digestion of organic samples (polymers, biological and drug samples) using quartz inserts inside the digestion vessels (3–12 vessels) placed in a carrousel in a microwave oven (Eilola & Perämäki, 2001, 2003, 2009; Niemelä et al., 2004). The sample mass was in the range of 0.1–0.2 g (Eilola & Perämäki, 2001, 2003; Niemelä et al., 2004) and 0.5 g (Eilola & Perämäki, 2009). Two sample digestion arrangements were used, consisting of placing a quartz insert without the holder or a quartz insert with a three-legged holder inside a digestion vessel. A holder was employed in order to segregate the sample vial away from the reagents, since conductive heating could enhance the digestion if hot acids are in contact with the insert wall. Several modifications were made in a similar system depending on the difficulty of sample digestion:

1. Two-step procedure VPMAD+MAD (Eilola & Perämäki, 2001). First, in the VPMAD phase the sample was placed in the quartz insert (without direct contact with the acid) and the digestion vessel contained 3 mL of HNO_3 + 0.5 mL of H_2O_2 and vapor phase digestion took place for 80 min. After that, since a small residue remained, 5 mL of diluted HNO_3 or HF were added to the sample which was digested for 15 min in a microwave oven.
2. Simultaneous VPMAD and MAD (Eilola & Perämäki, 2003; Niemelä et al., 2004). The sample was placed in the quartz insert and 1 mL of H_2SO_4 + 0.5 mL of HNO_3 was directly added to the sample and the digestion vessel contained 3 mL of HNO_3 (+ 0.5 mL of H_2O_2). Two MW programs of heating were performed while allowing the vials to cool down between programs.
3. VPMAD using a quartz insert and a holding insert (Eilola & Perämäki, 2009). The sample was placed in the quartz insert (without direct contact with the acid), 2 mL H_2SO_4 were added to the holding insert and 4.5 mL of HNO_3 + 4.5 mL of H_2SO_4 were added to the digestion vessel. A MW program of heating was performed and, in the case where a solid residue remained, 5 mL of hot HNO_3 was added to complete digestion.

In the case of FMW, a homemade PTFE support was designed by placing 4 cups containing the sample (30 mg) one over another at different heights from the base of the support. Then, the support was adapted to fit on the microwave glass vessel of FMW containing 15 mL of HNO_3 in the bottom part. As a consequence of MW heating, the acid vapor partially condensed in the upper part of the flask and was partially collected in each sample cup, thus the samples were directly exposed to acid steam (**Figure 7.5C**) (Araújo et al., 2000, 2004). This procedure was applied for the determination of Fe and Co from biological samples by ETAAS. It was observed that the distance of the sample cup to the support is a critical parameter.

In a similar way, a VPMAD procedure was adapted for TXRF analysis (Varga et al., 1999). In this procedure, freeze-dried algae samples (6–21 mg) were placed in 2 mL open quartz microvials and 75 µL of 30 % (m/v) HNO_3 was added to the microvials. The vials were placed in 120 mL PTFE microwave vessels and the vapor-digestion took place for 90 min at 180 °C (Varga et al., 1999).

Likewise, VPMAD was extended for the direct analysis of cancer cells in the TXRF sample carrier for the determination of Cu, Fe and Zn (**Figure 7.5D**) (Szosboszlai et al., 2008). Cancer cells were placed directly into the quartz sample carrier. Then, it was covered with a cap and assembled onto a three-leg glass support inside the digestion vessel. Finally, 3 mL of HNO_3 were added to the digestion vessel and the HNO_3 vapor attacked and digested the cancer cell sample. However, it should be noted that in this approach of vapor phase digestion developed for TXRF, higher concentrations of organic and inorganic matter remained in the sample than in conventional AD or MAD due to incomplete decomposition.

7.4.1.2.5 Ultrasound-assisted Digestion and Pseudodigestion

Another possibility for AD is the application of ultrasound energy, which gives rise to ultrasound-assisted digestion (UAD). UAD encompasses total or partial matrix decomposition (digestion or pseudodigestion). UAD only provides information of the total or pseudo-total metal content present in the solid sample. Generally, concentrated acids and oxidants are used, even HF which is used to solubilize the silicate matrix. When HF is employed, mostly in UAD but also in UAE (discussed in section 7.4.1.3.2.), the use of ultrasonic probe should be avoided because of damage to the probe tip (De la Calle et al., 2009). As a consequence, higher sonication times are usually applied, from 1 min up to 2 h in an ultrasonic bath, depending on the type of sample (*e.g.*, 1–5 min for biological tissue samples (Arain et al., 2007), 5–10 min for plant tissue (Kazi et al., 2009a) and 18–30 min for soils and sediments (Brunori et al., 2004; Ilander & Väisänen, 2007, 2009; Pontes et al., 2010).

Several applications of UAD using the ultrasonic bath were reported for a wide variety of samples, including muscle tissue (Arain et al., 2007), cigarette (Kazi et al., 2009a), sediments (Brunori et al., 2004), street dust samples (Elik, 2005) and ash (Ilander & Väisänen, 2007, 2009; Pontes et al., 2009). Typical mixtures include only HNO_3 (Kazi et al., 2009a), HNO_3:H_2O_2 (1:1, v/v) (Brunori et al., 2004; Arain et al., 2007), HNO_3:$HClO_4$:HF (2:1:1, v/v/v) (Elik, 2005), or *aqua regia* + HF (Ilander & Väisänen, 2007, 2009; Pontes et al., 2009). Variable volumes of reagents can be employed in an ultrasonic bath, usually ranging from 1–10 mL (Ilander & Väisänen, 2007, 2009; Kazi et al., 2009b). However, there is a growing trend toward the reduction of the volume of reagents and mass of sample. For instance, Kazi et al. (2009b) employed an ultrasonic bath for the UAD of heavy metals (Cd, Cr, Ni and Pb) from biological and environmental samples. They used a mass of sample of 0.1 g and 1 mL of extractant phase. The digestion reagent composition and the sonication time depended on the type of sample. Thus, sonication times of 2–20 min and a mixture of HNO_3:H_2O_2 (2:1, v/v) and HNO_3 were used for biological samples, while a sonication time of 5–50 min and a mixture of concentrated HNO_3:$HClO_4$:HF (2:1:1, v/v/v), HNO_3:HCl (1:3, v/v), HNO_3:H_2O_2 (2:1, v/v) and HNO_3 was used for soils and sediments.

In the case of UAD, the cup-horn sonoreactor has not yet been used. Also, it is sometimes difficult to distinguish between UAD and UAE with the different ultrasonic devices since partial digestion of the sample occurs in both techniques (De la Calle et al., 2009, 2011a). However, the main differences are that more dilute acids and shorter sonication times are employed in UAE.

7.4.1.2.6 Enzymatic digestion

Enzymatic digestion is mainly used for drug analysis and is less used for matrix degradation prior to total trace element analysis and chemical speciation due to the long operation time and low analyte recovery (Bermejo et al., 2004). Generally, hydrolytic enzymes (i.e., lipases, amylases, proteases) are used. These enzymes catalyze the introduction of water at a specific bond of the substrate (Bermejo et al., 2004), which provokes protein destruction and tissue degradation (Carpenter, 1981) and, as a consequence, metals associated with hydrolyzable proteins are released to the aqueous medium (Bermejo et al., 1999). Moreover, using hydrolytic enzymes it is possible to differentiate between fractions of elements associated to diverse components of the sample matrix, maintaining the chemical form of a species after the enzymatic digestion and thus making speciation analysis possible (Bermejo et al., 1999). The procedure for total metal analysis by enzymatic digestion usually involves the mixture of the sample with the enzyme and a buffer medium and incubation at 37–55 °C for a certain time, which usually ranges from 4 to 48 h.

In 1981, Carpenter published the first work on proteolytic enzymes for the digestion of kidney and liver tissue and the determination of Cd, Cu, Pb and Tl by FAAS (Carpenter, 1981). In this work, a very high amount of tissue (5 g) was used, and *crystalline subtilisin A* (5 mg) was employed as a non-specific protease (a protein-digesting enzyme). The tissue and the enzyme were mixed in 30 mL of buffer and incubated at 55 °C for 60 min.

In addition to biological tissues (Tan & Marshall, 1997; Chen & Marshall, 1999; Bermejo et al., 1999), enzymatic digestion was also applied to sediments (Turner & Olsen, 2000; Turner et al., 2001; Turner, 2006). In these examples, and as a first step towards miniaturization, a sample mass lower than that employed by Carpenter (0.2 g) and a lower volume of buffer (3–10 mL) were used. The samples were incubated for 4–8 h at 37 °C in presence of different enzymes. In some papers, the enzymatic digestion was accelerated by using high-pressure homogenization, which applied gentle agitation prior to introducing the slurry into the instrumentation (Tan & Marshall, 1997; Chen & Marshall, 1999).

As a second step toward minaturization, these procedures were accelerated by the use of ultrasound energy, giving rise to ultrasound-assisted enzymatic digestion (Vale et al., 2008). Both ultrasonic baths (Peña-Farfal et al., 2004) and probes (Capelo et al., 2004; Cabañero et al., 2005; Maduro et al., 2006; Vale et al., 2007) were used in order to speed up enzymatic digestion. Although different publications addressed the increase of enzymatic kinetics under ultrasound energy (Capelo et al., 2004), its effects are still controversial. Some authors confirm that cavitation boosts enzyme-substrate kinetics (Bermejo et al., 2004; Vale et al., 2008), while others maintain that denaturation of the enzymes took place and, as a consequence, their activity declined (Talukder et al., 2006). In the case of using an ultrasonic bath, the incubation time was reduced to 30 min when sonicated (Peña-Farfal et al., 2004), but similar mass sample and volume of reagents to the previously mentioned publications (0.2 g of sample, 7 mL of buffer and 20 mg of enzyme) were employed. On the contrary, the EPS approach is more appealing since the operation time was reduced to just a few seconds (Capelo et al., 2004; Maduro et al., 2006; Vale et al., 2007). The amount of sample used in these cases was reduced to 10 mg, the amount of enzyme was in the range of 1- 10 mg and the volume of water or buffer was 1 mL. The sonication time varied from 5 s in the case of Se in yeast (Capelo et al., 2004), to 15–300 s for Cd and Pb in BCR-101 spruce needles, BCR-414 plankton, BCR-679 white cabbage, BCR-710 oyster tissue and IAEA-0392 algae (Maduro et al., 2006), and 120 s for Se in oyster tissue, BCR-414 plankton and ERM-CE 278 mussel tissue (Vale, 2007).

Although enzymatic digestion does not compete with conventional approaches for trace metal analysis from the point of view of operation time, these procedures are generally carried out under mild conditions of temperature and pressure due to the temperature-dependent activity of enzymes. As a result, there are low losses by volatilization and reduced risks of contamination (Bermejo et al., 2004). When using ultrasound power, this methodology meet the requirements of miniaturization, because the reactions are generally carried out in Eppendorf vials of 1 mL with low amount of samples (10 mg) and in a short period of time (few seconds). Moreover, this procedure offers low cost, low sample requirements and low generation of wastes (Bermejo et al., 2004).

Enzymes have also been involved in sample pre-treatment for speciation analysis (Vale et al., 2008; Mesko et al., 2011). Several examples were found for Se speciation (organic and inorganic species) in animal tissue and oyster (Quijano et al., 2000; Moreno et al., 2001). These conventional approaches required 24 h of enzyme-sample incubation. Similar approaches were accelerated by reducing the incubation time in presence of ultrasonic waves to only 1–15 min. EPS applications were reported for Se speciation in plants (Montes-Bayón et al., 2006) and krills (Siwek et al., 2006) and As speciation in rice and chicken samples (Sanz et al., 2005a, 2005b). However, the most miniaturized and accelerated approaches for speciation involve the use of lower amounts of sample (around 10 mg instead of 0.2–3 g) and reduced sonication time (around 0.5–10 min). Some examples of miniaturized approaches were found for Se speciation in plants (Pedrero et al., 2007), yeast, oyster tissue and mussel tissue (Capelo et al., 2004), and As speciation in human hair (Sanz et al., 2007).

7.4.1.2.7 Tissue Solubilization

In the case of solubilization procedures, the matrix is not significantly altered but completely dissolved (i.e., the solid sample is converted into a liquid). These procedures can be divided according to the reagent used: buffer salts, surfactants, reducing agents, TMAH, and acids (Mesko et al., 2011).

Buffer salts, such as tris(hydroxymethyl)aminomethane (Tris), are commonly used to obtain the necessary ionic strength for protein solubilization of the sample. Tris buffer can be used in combination with HCl so that the free and weakly bound analyte species or metallo-amino acids are released (Mesko et al., 2011). The general trend toward miniaturization lies on the reduction of volumes (1 mL) and sample masses (20 mg) (Tran et al., 2010; Lavilla et al., 2012) and the speeding up of the procedure by applying ultrasound energy (3 min) (Lavilla et al., 2012).

Surfactants such as sodium dodecyl sulfate (SDS) are recommended for element speciation. SDS provokes the disruption of the cell membranes, breaking the interactions between lipids and proteins. As a consequence, the metal-binding proteins are solubilized and hydrophobic interactions are prevented (Szpunar et al., 2003; Mesko et al., 2011). As a miniaturized approach, 20 mg of fungi separated from maize sample was solubilized using 1 mL of Tris-HCl containing SDS and boiling in a water bath for 5 min (Muñoz et al., 2005a).

TMAH is a strong alkaline reagent that helps tissue solubilization. TMAH provokes hydrolytic scission, methylation and breaks sulfide chemical bonds in proteins (Nóbrega et al., 2006). The use of TMAH was also reported for Hg and As speciation (Sanz et al., 2007; Reyes et al., 2008). Some of the drawbacks of TMAH solubilization include limited time-stability of the obtained solution and the prolonged treatment time that is necessary (Subramanian, 1996). Conventional approaches involved the solubilization of 0.1–1 g of sample in 1–5 mL of 25 % (m/v) TMAH over 8–48 h at room temperature or 2–8 h at 50–70 °C (Subramanian, 1996). Reduction of volume of TMAH solution up to 0.5 mL with occasional shaking was performed at room temperature for 2 h for the determination of As, Cd, Ni and Pb in human hair (Ribeiro et al., 2000). Recently, the solubilization of tissues using TMAH in MeOH was also accelerated by using ultrasound energy, either by using an ultrasonic bath (Silva et al., 1999; Reyes et al., 2008) (30–45 min) or probe (2–4 min) (Matusiewicz &Golik, 2004; Sanz et al., 2007). The main advantage of these procedures is the low losses of metals by volatilization. Different techniques were employed for the analysis of samples after treatment with TMAH, such as ICP-OES (Matusiewicz & Golik, 2004), ICP-MS (Reyes et al., 2008) or ETAAS (Silva et al., 1999).

Ethylenediaminetetraacetic acid (EDTA) in alkaline medium has been used for solubilization of plant tissue in combination with probe-sonication (for only 3 min) followed by FAAS (Ca, Fe, Mg, Mn and Zn) and ETAAS (Pb and Cd) determination (Filgueiras et al., 2001). Moreover, microwave energy was also employed for the acceleration (30 min digestion + 10 min cooling) of biological sample solubilization using TMAH in combination with EDTA (Zhou et al., 1996b).

Several reducing agents including dithiothreitol (DTT), dithioerythritol (DTE) and β-mercaptoethanol were employed with the aim of i) reducing the disulfide bonds, ii) helping protein denaturation, iii) avoiding protein oxidation and iv) improving the solubilization of proteins (Mesko et al., 2011), all of which can assist with tissue solubilization. A volume of about 10 mL of a 0.01 % DTT solution was used in some applications (Goenaga-Infante et al., 2004, 2005).

7.4.1.3 Extraction-based Techniques

Extraction is usually a time-consuming step that involves the transfer of analytes of interest from the solid sample to another phase. Solid-liquid extraction (SLE) and MSPD are some options for the analysis of solid samples without previous treatment.

SLE of metals and organometallics from environmental matrices has been commonly called leaching (Camel, 2000). SLE can also be accelerated by means of: i) acids, ii) ultrasound energy, iii) microwave energy, iv) solvents and v) supercritical fluids.

For the extraction of metals from solid samples, the most extended approach is based on acid extraction and accelerated by heating or ultrasound irradiation. MAE and supercritical fluid extraction (SFE) are mostly employed for organic compounds, but several applications to metal speciation and fractionation procedures exist.

7.4.1.3.1 Acid Extraction (leaching)

Numerous elements can be extracted from a solid matrix by adding diluted acids under soft conditions. Agitation and temperature can accelerate this process. This scenario is not considered a proper acid digestion because the organic matrix is not destroyed, but dissolution of the rest of the sample and release of the metals does occur. In some case, the metal-matrix bond breaks down without alteration of the matrix. This process is called acid extraction or acid leaching (Welz & Sperling, 1999).

Diluted acids such as HNO_3, HCl, $HClO_4$, H_2O_2 or mixtures such as HNO_3 + HCl, acetic acid, and even HF (for plant and rock analysis) were used. Leaching could be performed from room temperature to 105 °C (Welz & Sperling, 1999). The time necessary to perform the extraction process varies from 5 min to 24 h. Several representative examples found in the literature used 30 min for vegetables (Wieteska et al., 1996), 1 h for mussel tissue (Solchaga et al., 1986), 4 h for bovine liver (Steiner et al., 1987), 24 h for eggs or fih tissue (Benson et al., 1983) and overnight for lake sediments (De Vevey et al., 1993). In some of these examples, the approximate sample mass was about 0.5 g and the volume of acid used was 5 mL.

This procedure avoids the organic matrix destruction step and results in a simple, versatile and inexpensive method with reduced hazards and low risks of loss and contamination. However, it can only be applied to limited matrices and specific elements (Welz & Sperling, 1999).

Fractionation studies have also been subjected to miniaturization by magnetic agitation extraction (Filgueiras et al., 2002b; De la Calle et al., 2013a). Specifically, they consisted of a scale reduction, but maintained the same mass-to-volume ratio of the operationally-defined protocols of BCR (Quevauviller et al., 1993) and EDTA and acetic acid extractable metal content from soils (Quevauviller et al., 1997). In the case of BCR, the sample mass was reduced from 0.5 g to 25 mg and the volume of reagents from 5–20 mL to 1 mL or 250 µL, respectively (depending on the step). In the case of EDTA and acetic acid extraction, the sample mass was reduced from 5 g to 0.1 g for the EDTA-extractable fraction and to 25 mg for the acetic acid-extractable fraction.

Moreover, the extractant volume was reduced from 50 mL for extractions with EDTA and 200 mL for extractions with acetic acid to just 1 mL. However, the same long agitation times in the operationally-defined procedures were used for the BCR (48 h) (Filgueiras et al., 2002b), EDTA (1 h) and acetic acid extractions (16 h) (De la Calle et al., 2013a).

7.4.1.3.2 Ultrasound-assisted Extraction

Recently, ultrasound energy has been expanded to sample pre-treatment due to the enhancement of diverse processes (Capelo et al., 2005; Santos-Junior et al., 2006; Priego-Capote & Luque de Castro, 2007; Bendicho et al., 2012a). One of these procedures, UAE, has gained increased attention during recent years due to its simplicity and reliability. UAE was developed in the 1990s, when Miller-Ihli (1989) realized that partial extraction of several elements took place when using USS. The first study involved a sample mass of 3–5 mg and reagent volumes of 2.5–5 mL (Mierzwa et al., 1997), while another used 0.2 g of sample in 10 mL of solvent (El-Azouzi et al., 1998). Some discrepancies were found in the sonication time employed and the recoveries achieved by different authors using different ultrasonic devices and types of samples.

Power ultrasound provokes particle disruption and, as a consequence, facilitates penetration of the extraction solvent inside the particles and the release of metals (Mason & Lorimer, 2002). Ultrasound energy can speed up SLE of elements from different matrices for different purposes, including the determination of total-element contents, speciation analysis and fractionation studies (*e.g.*, metal bound to organic matter, oxides, or carbonates) (Bendicho et al., 2012a).

For the determination of total elements (Cd, Pb, Cr, Zn, Cu, Hg, Se, and rare earth elements), diluted acids are usually employed in combination with sonication by means of ultrasonic baths, probes and cup-horn sonoreactors. While the ultrasonic bath allows the use of a variety of vial sizes (*e.g.*, floating supports for Eppendorf vials), there are also ultrasonic probe devices with different tip diameters depending on the volume to be sonicated (i.e., 2 mm tip diameter for 150 µL–5 mL, 3 mm for 250 µL–10 mL, 6 mm for 10–50 mL and 13 mm for 50–150 mL). Probe tips with small diameter produce greater intensity of cavitation, but the energy released is restricted to a narrower area. On the contrary, probe tips with large diameters produce lower intensity but over a greater area, thus permitting larger volume to be sonicated. Finally, the cup-horn sonoreactor is provided with a special adaptor that allows multitreatment for up to 6 Eppendorf vials.

Several protocols using relatively high amounts of sample (0.1–5 g) and high volumes of diluted acids in the range of 3–100 mL were employed, leading to a reduction of time *via* the use of ultrasound energy. A few examples regarding the use of the ultrasonic bath (for 3–30 min) include the extraction of elements from plant samples (Álvarez et al., 2003; Arruda et al., 2003), soils (Väisänen et al., 2002; Lukkari et al., 2004), biological samples (Liva et al., 2000; Santos et al., 2000; Afridi et al., 2007), hair samples (Bermejo-Barrera et al., 2000) and baby food (Kazi et al., 2010). Other examples report the use of the ultrasonic probe (3–20 min) for the extraction of elements from biological samples (Afridi et al., 2007; Neves et al., 2009), marine tissues (Costas et al., 2010), leaves (Caballo-López & Luque de Castro, 2007), plants (Filgueiras et al., 2000), sewage sludge (Hristozov et al., 2004), and rocks (Pumure et al., 2009). It should be pointed out that matrices such as soils, sewage sludge and rocks required higher sonication time than biological tissues for complete element extraction due to their complexity.

The most miniaturized UAE protocols involved a sample mass in the range of 1–30 mg and an extractant volume of 1–2 mL. Several examples were detailed with the three ultrasonic devices mentioned above. The extraction with the ultrasonic bath generally takes 10–60 min for aquatic plant, sewage sludge, mussel tissue, river sediment and tea leaves (Amoedo et al., 1999) and seaweed (Domínguez-Gonzalez et al., 2005). Conversely, the ultrasonic probe requires only 2–15 min for element extraction from animal tissue (Capelo et al., 1998, 1999; Amoedo et al., 1999; Lima et al., 2000; Maduro et al., 2006), hair (Batista et al., 2009), seafood samples (Lavilla et al., 2006b, 2008), biopsy tissues (Lavilla et al., 2006a), sediment samples (Lima et al., 2000), and soils (López-García et al., 2005). Last but not least, the extraction by means of the cup-horn sonoreactor takes 1–45 min for biological tissues (De la Calle et al., 2009, 2012, 2013c) and environmental solid samples (De la Calle, 2011a, 2011b, 2013a).

Although UAE is mostly applied in combination with ETAAS (Capelo et al., 1998; De la Calle et al., 2009), several applications were developed for FAAS (Filgueiras et al., 2000), TXRF (De la Calle et al., 2012, 2013a, 2013c), ICP-OES (Väisänen et al., 2002; Balarama-Krishna & Arunachalam, 2004; Carvalho-Santos et al., 2009) and ICP-MS (Balarama-Krishna & Arunachalam, 2004; Batista et al., 2009; Costas et al., 2010). It is worth noting that, in the case of ICP-MS, 5 mL of extractant (Balarama-Krishna & Arunachalam, 2004; Costas et al., 2010) or a significant dilution with water after carrying out the extraction (Batista et al., 2009) is required.

Speciation analysis of As, Hg and Se using suitable extractants are not truly miniaturized. Usually, they involve extraction using a soft medium (water or diluted H_3PO_4) and then speciation is performed by HPLC-ICP-MS (Hirata & Toshimitsu, 2005; Sanz et al., 2005a) or HPLC-HG-AFS (Huerga et al., 2005). When using an ultrasonic probe, lower sonication times (1–5 min) (Huerga et al., 2005; Sanz et al., 2005a) than using the ultrasonic bath (15 min) (Hirata & Toshimitsu, 2005) were reported. Balarama-Krishna et al. (2005) developed a procedure for Hg and methylmercury extraction from animal and plant tissue and coal fly ash using an ultrasonic probe for 3–4 min. The sample mass employed was 0.1–0.2 g of sample and 5 mL of extractant (HNO_3+thioruea). Sonication should be followed by selective reduction of species to Hg(0) prior to analysis, since $SnCl_2$ can reduce only Hg(II) and $NaBH_4$ can reduce both (Hg(II) and methylmercury).

Several protocols have been designed to evaluate the metal content bound to different geochemical phases in soils and sediments, such as single and sequential extraction schemes from the Standards, Measurements and Testing program from the European Commission (SM&T) (Quevauviller et al., 1993) and the Tessier protocol (Tessier et al., 1979). These schemes allow the separation of the total metal content in different fractions according to their potential mobility and bioavailability. The procedures consist of the dissolution of different solid phases in environmental samples (*e.g.*, organic matter, oxides, carbonates) using specific reagents solutions as extractants for each phase (Filgueiras et al., 2002a). Generally, BCR, Tessier and modified procedures employ large amounts of sample (1 g), large volumes of reagents (50–200 mL × 3 or 5 repetitions) and long agitation times (1–48 h) (Tessier et al., 1979; Quevauviller et al., 1993).

Two trends were identified in recent publications of fractionation studies: i) acceleration of the extraction by ultrasound energy but using the same or similar volumes recommended by the standardized protocols and ii) acceleration by ultrasound energy and minimization of volume of reagents and sample mass employed. First, the sample mass used was reduced from 1 g to 0.5 or 0.25 g and the volume of reagents to 20 mL each step. Significant shortening of the extraction times was achieved for BCR and Tessier protocols from several hours to 9–15 min (each step) using an ultrasonic probe (Pérez-Cid et al., 1998) for soils (Pérez et al., 2008), compost (Greenway & Song, 2002), mud and soils (Marín, 2001), and sediments (Davidson & Delevoye, 2001), or to 15–30 min using the ultrasonic bath from sewage sludge (Kazi et al., 2006), contaminated soils (Väisänen & Kiljunen, 2005), or sediments (Davidson & Delevoye, 2001; Arain et al., 2008). In the case of acetic acid and EDTA extraction, a sample mass of 0.5–3 g and a reagent volume of up to 75 mL were used. Moreover, the use of the ultrasonic bath allows one to reduce the operation time from 16 h to 40 min for the acetic acid extraction (Krasnodebska-Ostrega et al., 2006), while the use of the ultrasonic probe enabled a processing time reduction from 1 h to 6–12 min for the EDTA extraction (Hwang et al., 2007; Remeteiova et al., 2008).

Secondly, the small-scale and accelerated UAE was also developed for single extraction using the ultrasonic probe (Filgueiras et al., 2002b; De la Calle et al., 2013a) and the cup-horn sonoreactor (De la Calle et al., 2013a). The sample mass used was reduced to 25 mg in the BCR protocol, to 0.1 g in the EDTA extraction and to 25 mg in the acetic acid extraction, with an extraction volume of 1 mL. Using the ultrasonic probe, the time of the procedures was reduced to 0.5–1 min (each step) in the BCR protocol, to 2 min for EDTA and 6 min for the acetic acid extraction. On the contrary, when the cup-horn sonoreactor was used for EDTA and acetic acid extractions, higher sonication time (10 and 30 min, respectively) was required due to the lower intensity of the sonoreactor. Nevertheless, the cup-horn sonoreactor allows for the extraction of up to six Eppendorf vials at a time.

7.4.1.3.3 Microwave-assisted Extraction

Microwave energy was first applied in sample preparation for the mineralization or digestion of samples. The matrix is not completely decomposed in MAE, since lower time and softer conditions than in MAD are selected. Microwaves accelerate and enhance the extraction by simultanously heating the mixture of extractant (usually an organic solvent) and sample. This is done very fast and without heating the vessel. Choosing a high temperature and an appropriate organic solvent favors the extraction of the analyte in a very short time (Camel, 2000). In fact, MAE provides several advantages over classical techniques, such as i) lower solvent consumption, ii) lower sample usage, iii) reduction in analysis time iv) reduction of contamination, v) analyte loss prevention, vi) more controlled and reproducible results and vii) enhanced operator safety (Camel, 2000). MAE is generally used for organometallic compound extraction from environmental solid samples (Camel, 2000), but several applications for total trace metal analysis, speciation and fractionation in a variety of solid samples have been reported.

MAE has been used for total or pseudo-total metal analysis for diverse environmental samples (including rocks, atmospheric particulate matter, soils, sediments, spruce needles and apple leaves) using nonionic surfactants (Akinlua et al., 2013), mixture of acids (Nieuwenhuize et al., 1991; Zhou et al., 1996a; Borkowska-Burnecka, 2000; Lesniewicz & Zyrnicki, 2003; Landajo et al., 2004; Kulkarni et al., 2007; Bettiol et al., 2008; Balarama-Krishna et al., 2013), EDTA (Borkowska-Burnecka, 2000; Öztan & Düring, 2012) and *aqua regia* (Muránszky et al., 2011; Horváth et al., 2013). Generally, a sample of mass of 0.1–1 g, a volume of 5–40 mL of acids and an extraction time of 20–45 min were needed for total metal content.

As commented before, MAE has been employed for metal speciation. It should be pointed out that mild extraction conditions should be used to preserve inorganic As species in solution from plankton samples (with water as extractant, for 50 min) (Lehman et al., 2013), or diluted $HCl+H_2O_2$ in food and feed of marine origin (25 min) (Rasmussen et al., 2012). The sample mass ranged from 0.1 to 0.2 g and the volume of extractants is about 10 mL.

Microwave energy has been applied for the extraction of organometallic compounds from industrial and environmental samples such as textiles, plastics, biological tissues, including such species as organotins (Donard et al. 1995; Szpunar, 1996; Rodríguez et al., 1997; Wang et al., 2006, 2008), methylmercury (Vázquez et al., 1997; Ramalhosa et al., 2001), or arsenic species (Vilanó & Rubio, 2001) in just 2–10 min. Open vessels are generally preferred for the extraction of organometallic compounds, in order to control the applied microwave energy and prevent the destruction of the carbon–metal bonds (Camel, 2000). Most of the MAE analyses were carried out on a large scale, using sample amounts of 0.15 g to 5 g and reagent volumes of 4–40 mL for organometallics. Usually, low-power microwaves are applied in combination with diverse extractants, *e.g.*, methanol:water (1:1, v/v) (Vilanó & Rubio, 2001), acetic acid:hexane (1:4, v/v) (Wang et al., 2006), methanol:water (3:2, v/v) (Wang et al., 2008), methanolic potassium hydroxide (Ramalhosa et al., 2001) or methanol (Yang et al., 2010).

Moreover, MAE was used for single extraction tests which involve the use of extractants such as EDTA and acetic acid to obtain the mobilizable and bioavailable fractions from river sediments by FAAS (Cu, Cr, Ni, Pb, and Zn) (Pérez-Cid & Boia, 2001). Furthermore, MAE was used for metal fractionation. In fact, different sequential extraction schemes (SES) such as Tessier and BCR protocols have been enhanced and accelerated by MW energy, giving rise to what have been termed microwave sequential extraction procedures (MSE). MSE generally consists of using the same reagents as in the BCR procedure, but replacing the magnetic agitation by applying MW energy. Several examples can be found in the literature for the extraction of metals from sediments by modified BCR (Ipolyi et al., 2002; Canepari et al., 2005; Arain et al., 2008; Jamali et al., 2009; Alonso-Castillo et al., 2011; Reid, 2011) and Tessier protocols (Perez-Cid et al., 1999; Reid et al., 2011). These approaches were coupled with different analytical techniques such as FAAS (Arain et al., 2008), ETAAS (Arain et al., 2008), ICP-OES (Reid et al., 2011), and ICP-MS (Alonso-Castillo et al., 2011) for the determination of different metals including Cu, Ni, Cr, Pb, Zn and Cd. A fractionation method was also performed by continuous-flow sequential extraction assisted by FMW (Nakazato et al., 2006). Most of the BCR and Tessier protocols assisted by microwave energy were carried out on a large scale using sample amounts of 0.25–5 g and reagent volumes of 8–40 mL. Remarkably, more recent approaches were tested to decrease both sample mass and volume of conventional protocols while maintaining the sample mass-to volume ratio unmodified, following the trend towards miniaturization. For example, the sample mass was reduced from 1 g to 0.25 g and the volume of reagents from 40 to 10 mL in a BCR protocol (Arain et al., 2008; Jamali et al., 2009; Alonso-Castillo et al., 2011) in which the time was reduced from 32 h to 6 min (Alonso-Castillo et al., 2011) or to 2 min (Arain et al., 2008; Jamali et al., 2009) with the aid of MW energy.

7.4.1.3.4 Accelerated Solvent Extraction

Accelerated solvent extraction (ASE), also called pressurized solvent extraction (PSE), PLE, high-pressure solvent extraction (HPSE), high-pressure, high temperature solvent extraction (HPHTSE), pressurized hot solvent extraction (PHSE), and subcritical solvent extraction (SSE) involves analyte extraction from solid and semi-solid sample matrices under elevated temperature (50–200 °C) and pressure (500–3000 psi) for short periods of time (5–10 min), using a compressed gas to purge the sample extract to a vessel (Richter et al., 1996; Mesko et al., 2011). Static extractions can be performed at elevated temperatures and pressures or only at elevated pressures in order to avoid decomposition of thermally labile compounds (Szpunar, 2003). Some advantages related to ASE are its high sample throughput, low consumption of solvents, fast extraction cycles, automation with reduced sample handling and suitability for solid and semi-solid samples (Goenaga-Infante et al., 2004).

Although ASE is generally applied for organic compounds, several examples were reported for organometallics and metals (Szpunar, 2003; Mesko et al., 2011), such as As in fish tissue (Mckiernan et al., 1999) and seaweed (Gallagher et al., 2001), Se species in yeast supplements (Goenaga-Infante et al., 2004, 2005) and multielemental analysis of lubricating oil (Carballo-Paradelo et al., 2012). Methanol and water are the usual extractant phases.

In most of cases, the sample mass used is 0.1–0.7 g and the volume of extractant added is 10 mL (McKiernan et al., 1999; Goenaga-Infante et al., 2004, 2005; Carballo-Paradelo et al., 2012). However, a 3 mL ASE cell, smaller than the conventional one, allowed the miniaturization of the ASE approach (Gallagher et al., 2001). The solvent involved in the miniaturized ASE approach was MeOH:H_2O (1:1, v/v) (Gallagher et al., 2001). Usually, 3 or 6 cycles of 5 min extraction time were performed at 1500 psi (McKiernan et al., 1999; Gallagher et al., 2001; Goenaga-Infante et al., 2004, 2005). The temperature was fixed at 100 °C and diatomaceous earth was added as dispersing agent (Goenaga-Infante et al., 2004, 2005).

Recently, a laboratory-made miniaturized device for PLE was patented for the extraction of As(III), As(V) and total As from hair samples (Sanz et al., 2007). About 50 mg of hair sample was mixed with 350 mg of dispersant medium. The homogenized mixture was then introduced into the extraction cell, which consisted of a stainless steel liquid chromatographic column. The PLE device was equipped with an homemade oven whose temperature was controlled by a digital electronic controller. The extraction cell was mounted between two valves. One valve was connected to the HPLC pump that introduced the extraction solvent in the cell and controlled the pressure in the system, while the second valve was connected to the collection vial (Sanz et al., 2007).

7.4.1.3.5 Supercritical Fluid Extraction

Supercritical fluid extraction (SFE) consists of the extraction of analytes from a matrix using a supercritical fluid as the extraction solvent. While in off-line SFE hundreds of mg to g of sample are used, in on-line SFE only a few mg of sample are extracted. A complexing agent is required in order to form neutral metal complexes and facilitate the extraction, usually dithiocarbamate. In general, CO_2 is utilized as supercritical fluid extractant because it is chemically inert, non-toxic, non flammable and non-corrosive. Moreover, a modifier could be used, for example 1–10 % (v/v) MeOH, acetic acid or formic acid.

Apart from avoiding the use of organic solvents, SFE avoids or reduces the formation of wastes. However, SFE is not widely applied and the energy cost needed to maintain a substance in the supercritical state is very high (Bendicho et al., 2011).

Even though SFE is more commonly used for organic compounds (*e.g.*, polycyclic aromatic hydrocarbons (PAHs) and polychlorinated biphenyls), different elements have been extracted from wood (Cr, Cu, As), fly ash (Zn, Pb, Mn, Cd, Cu, V, Sb, Ni, Mo, Cr and Co) (Lin et al., 2014) and sand (S. Wang et al., 2003, Yabalak & Gizir, 2013) by SFE using diverse chelating agents at 100–200 bar and 40–130 °C. Furthermore, the SFE of organometallic species of Sn, Pb, Hg and As after derivatization was also reported in the literature (Alzaga et al., 2003). The application of supercritical CO_2 in heavy metal extraction has been recently reviewed (Sunarson & Ismadji, 2009; Umale & Mahanwar, 2010; Lin et al., 2014).

As a novel approach, SFE was adapted for improving fractionation studies of a three-step BCR sequential extraction procedure (SEP) of lake and river sediments (Horváth et al., 2013). The tested procedure consisted of a three-step method for the extraction of different fractions. Thus, CO_2-soluble organic fraction and sulfides were first extracted by means of supercritical CO_2, while water-soluble fraction and a fraction bound to carbonates where subsequently obtained by using subcritical H_2O and subcritical H_2O+CO_2, respectively. Each step was performed at 80 °C and 27 MPa for 30 min, but without the addition of organic solvents, thus involving a solventless procedure. The sample mass used was 0.5 g.

7.4.1.3.6 Matrix Solid-phase Dispersion

MSPD consists of the disruption of a sample by mechanical blending using a solid support or a dispersing agent (silica-based support) bonded phase which is suitable for extraction. After blending, the mixture is placed inside a cartridge or column. A new sample matrix-solid support phase is formed and analytes tend to be more weakly bonded to it. Then, the analytes are eluted using different solvents by gravity or by applying pressure or vacuum (Kristenson et al., 2006). This approach was developed as an alternative to conventional techniques such as Soxhlet extraction or LLE that require higher amounts of sample, volume of eluents and time (Moliner-Martinez et al., 2009). Moreover, MSPD avoids the problem of emulsion formation in LLE (Ramos, 2012). MSPD is, in general, less time consuming than Soxhlet extraction, and shows higher extraction efficiencies (EEs) than achieved by conventional LLE (Capriotti et al., 2013).

Several reviews are focused on the MSPD technique (Barker, 2000; Ramos et al., 2005; Kristenson et al., 2006; Barker, 2007; Moreda-Piñeiro et al., 2009; Ramos, 2012; Caprioti et al., 2010, 2013). A scheme of MSPD is shown in **Figure 7.6.**

Several parameters should be taken into account when optimizing a MSPD method, including the nature of the solid support, the bonded phase, the components of the matrix, and the polarity of the eluting solvents (Moreda-Piñeiro et al., 2009). Several sorbents have been described in the literature, including molecularly imprinted polymers (MIPs) and carbon nanotubes (CNTs) (Caprioti et al., 2013). Other options that increase the analyte recovery are the use of ultrasound energy for improving the EE (Capriotti et al., 2010) and the use of high-temperature and high-pressure solvents that facilitate the automation of MSPD (Kristenson et al., 2006). Moreover, interfering components of the matrix could be selectively retained on the column by interactions with the stationary phase (Ramos, 2012).

Although MSPD has been designed for organic compound purification and extraction (Ramos et al., 2005, 2009; Ramos, 2012), it is also a promising and appealing sample pre-treatment technique for trace elemental analysis. Several applications have been reported in the literature for inorganic and organometallic compounds (Moreda-Piñeiro et al., 2008, 2009; Martínez-Fernández et al., 2011), and they will be described in detail below.

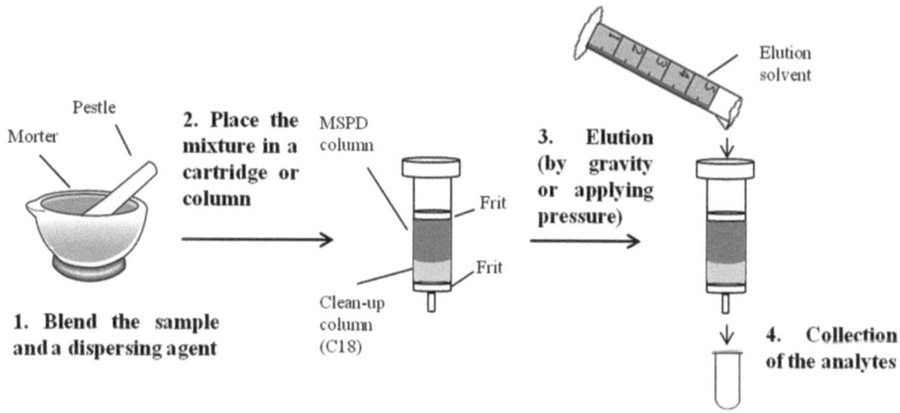

Figure 7.6 Scheme of MSPD including the different steps.

MSPD is considered advantageous from different points of view. Firstly, MSPD requires a small amount of sample (0.1–0.25 g) versus conventional techniques (10 g) (Moliner-Martínez et al., 2009), an important consideration when only small amount of sample is available (Kristenson et al., 2006). Secondly, a relatively small volume of solvent (2 mL) is needed (Moliner-Martínez et al., 2009), which is generally lower than reported in classical methods (hundreds of mL) (Ramos et al., 2005; Moliner-Martínez et al., 2009). In fact, it has been reported that MSPD requires 95% less solvent compared to classical approaches (Moreda-Piñeiro et al., 2009). Additionally, MSPD is shorter (less than 1 h) than classical methods (several hours) (Moliner-Martínez et al., 2009), with some researchers reporting that MSPD requires 90% less time than classical methods (Moreda-Piñeiro et al., 2009). Moreover, MSPD offers the possibility of performing several processes in one step, involving extraction and clean-up, resulting in a faster approach (Kristenson et al., 2006). MSPD is also a low cost, simple technique that does not require any special instrumentation (Ramos et al., 2005).

The applications dealing with MSPD for the determination of metals and organometallics (Moreda-Piñeiro et al., 2009; Capriotti et al., 2013) include Cd, Co, Cr, Mn, Ni, Sr and Zn in lake and estuarine sediments (Martínez-Fernández et al., 2011), Sb(III) and Sb(V) in marine sediment (Moreda-Piñeiro et al., 2009), As(III) and As(V) in seafood (Moreda-Piñeiro et al., 2009, 2012) and organometallics such as arsenobetaine (AsB), arsenocholine (AsC), monomethylarsonic acid (MMA), and dimethylarsinic acid (DMA) in seafood samples (Moreda-Piñeiro et al., 2009, 2012). For the MSPD extraction of inorganic and organic species of As in seafood samples (Moreda-Piñeiro et al., 2009, 2012), 0.25 g of sample was blended with diatomaceous earth (1.75 g) and the mixture packed into a cartridge containing 2 g of C_{18} as a co-sorbent. After that, arsenic species were eluted with 10 mL of methanol:water (1:1, v/v), concentrated by rotary evaporation to approx 2 mL and determined by anion-exchange HPLC-ICP-MS.

As discussed in previous sections, SES or SEP have been accelerated by ultrasound and microwave energy (Sutherland, 2010). MSPD has also been applied for evaluating the partitioning of metals (Cd, Co, Cr, Mn, Ni, Sr and Zn) in lake and estuarine sediments prior to ICP-MS analysis (Martínez-Fernández et al., 2011). The SES approach begins by mixing 0.5 g of sample with 1.0 g of a dispersing agent (such as alumina, silicates, or diatomaceous earth) and blending the mixture with a mortar and pestle for 5 min. Then, the dispersed sample is packaged inside a disposable syringe and placed in a vacuum manifold station. The extractants conventionally used in the BCR protocol (0.11 mol L^{-1} acetic acid, 0.1 mol L^{-1} hydroxylammonium chloride and 8.8 mol L^{-1} hydrogen peroxide + 1 mol L^{-1} ammonium acetate) are then passed through the sample. The time of the procedure and the volume of reagents were reduced in relation with the conventional approach (Sutherland, 2010) from 51 h to 3–4 h the whole procedure and from 40 mL to 10 mL and 25 mL in the first and second steps, respectively. The main disadvantage of this procedure (Martínez-Fernández et al., 2011) is the washing (or rinsing) step required after each extraction which involves passing 10 mL of water through the column and takes 20 min. Moreover, inaccuracy was observed for the reducible fraction, as occurs in the conventional, ultrasound and microwave-assisted BCR protocols.

7.4.1.3.7 Solid-phase Microextraction

SPME is a miniaturized sample preparation technique first introduced by Pawliszyn and co-workers in the early 1990s. Generally, SPME is mainly used to extract organic compounds from solutions as well as headspace. In the case of headspace enrichment, samples can be of solid or liquid nature. SPME is a solventless sample preparation technique, since no addition of solvent is required in headspace solid-phase microextraction (HS-SPME). The procedure involves two integrated steps: i) mass tranfer of the analytes from the solid sample to the headspace by heating and ii) preconcentration of the analytes in the SPME fiber by exposing it to the headspace. After that, sample introduction is generally performed by thermal desorption of analytes into an analytical instrument such as a GC. A detailed description of SPME techniques can be found in chapter 2 of this book.

HS-SPME has been applied to solid samples, such as marine sediment, dogfish muscle and liver tissue and the volatile metals and organometallics can be collected onto the fiber (**Figure 7.5E**). It is necessary to heat the solid sample to promote the volatilization of elements of interest. This approach was tested for the analysis of As (as AsB), Se, Hg (as methylmercury), Pb (as trimethyllead) and Sn (as butyltin species, mono-, di- and tributyl tin) by ICP-MS. 1 g of solid sample was placed in a vial and heated on a hot plate to a temperature between 70–150 °C for a period of 1–5 min (Mester, 2002). A similar approach was employed for the analysis of organotin compounds by SPME-ICP-TOF-MS from sediment and urban dust by simply increasing the temperature to about 40 °C (Mester & Sturgeon, 2002). This approach is completely green, since no chemicals were used.

7.4.1.4 Combined Techniques

A wide range of miniaturized sample preparation approaches that are generally applied to liquid samples cannot be directly used for the analysis of solid and semi-solid samples. Thus, solid samples have to be converted into liquid samples by diverse procedures such as digestion, extraction or ashing prior to another sample preparation procedure such as SPE, SPME, LPME, or prior new generation miniaturized analytical systems including μ-TAS or LOC.

7.4.1.4.1 Solid-phase Extraction

SPE is frequently used for the preconcentration or clean-up of target analytes from aqueous samples. SPE is an alternative sample preparation technique to LLE which can reduce the volume of solvents needed. SPE involves the extraction of target compounds from an aqueous sample, extract or a digested solid sample (Camel, 2003). Conventional SPE uses solid phases such as C_{18}-bonded silica and the required volume of solvents ranges from 1 mL to 1 L depending on the scale of the system.

Generally, solid samples are first digested or dissolved and then subjected to SPE using cartridges, syringe barrels, microcolumns or disks (Camel, 2003). Several examples in the literature involve the analysis of geological materials (Pu et al., 1998) and biological tissues (Yin et al., 1998).

CNTs that were previously oxidized were used as sorbents in SPE for the analysis of metals and organometallics in solid samples (Muñoz et al., 2005b; Ravelo-Pérez et al., 2010; Herrero-Latorre et al., 2012). Some of the samples (i.e., cigarettes, biological samples, garlic, canned fish, leaves, human hair, soils) were digested by MAD (Teixeira-Tarley et al., 2006; Barbosa et al., 2007; Tuzen et al., 2008), while soils, human hair, tea and rice were extracted by sonication (H_2SO_4:HNO_3 (3:1, v/v) for 2 h and stirred at 55 °C for 7 h prior to SPE (Liu, 2008). The amount of CNTs in the different procedures was in the range of 10–300 mg, and the volume of sample after treatment was in the range of 25–400 mL. The enrichment factors (EFs) achieved varied from 30 to 200. Usually, the elution of analytes is carried out by adding diluted HNO_3. The analytical technique most commonly employed was FAAS (Teixeira-Tarley et al., 2006; Barbosa et al., 2007; Liu et al., 2008, 2009; Chen et al., 2009), although a few applications using ICP-OES (Zang et al., 2009) and ICP-MS (Chen et al., 2009) were also published.

Another related technique is dispersive micro solid-phase extraction (DMSPE). The use of multiwalled carbon nanotubes (MWCNTs) as a solid sorbent for the DMSPE preconcentration of Se from biological tissues was recently reported (Skorek et al., 2012). First, 250 mg of solid sample was digested in 5 mL of HNO_3 by MAD. HCl was added to reduce Se(VI) to Se(IV) and DMSPE was performed prior to XRF analysis. For

DMSPE, 100 mL of the digested sample was mixed with 100 µL of a solution of APDC and 200 µL of 5 mg L^{-1} dispersed oxidized MWCNTs, and stirred for 5 min at 700 rpm. After that, the sample was filtered and MWCNTs containing the adsorbed metal chelates that remained in the filter were dried under IR radiation and directly analyzed.

As a modification of SPE, another interesting improvement is the use of SS for preconcentration analysis using solid phases as sorbents. However, solid samples had to be previously digested by MAD. Several recent studies reported the use of activated carbon as a sorbent for the extraction and preconcentration of Au from geological samples (Dobrowolski et al., 2012), Ni and Co from plant materials (Dobrowolski & Otto, 2012), and Se in complementary feed (Dobrowolski & Otto, 2013). Separation and enrichment of metals consisted of the adsorption of metal ions onto activated carbon from aqueous solutions. After adsorption, SS introduction replaced the step of elution of analytes, thus avoiding sample dilution and, as a consequence, decreased EFs. Powdered activated carbon was chemically modified, followed by drying and washing steps. Different protocols can be employed for the modification of activated carbon by: i) treatment with H_2O_2 and HNO_3 (Dobrowolski et al., 2012), ii) treatment with HNO_3 and outgassing in the argon atmosphere at high temperature (1100 ºC) followed by impregnation with dimethylglyoxime solution (Dobrowolski & Otto, 2012), and iii) loading with Fe species by impregnation method and thermo-chemical reactions using different concentrations of $Fe(NO_3)_3$ (Dobrowolski & Otto, 2013). Thus, modified activated carbon was used for adsorption of the target elements. Firstly, previously digested samples were evaporated almost to dryness. Then, 0.1–0.2 g of the modified activated carbon was mixed with the sample. The suspension was filtered and dried at 120 ºC. Finally, after preconcentration of metals onto modified carbon, the determination of metals is performed in a suspension of carbon. Thus, the carbon suspension was prepared in a range of 1–50 mg mL^{-1} in 5 % v/v HNO_3 as a liquid medium, homogenized by Vortex agitator and introduced in a graphite furnace by GFAAS (Dobrowolski et al., 2012; Dobrowolski & Otto, 2012, 2013).

7.4.1.4.2 Headspace Solid-phase Microextraction

Different protocols have been designed for the enrichment of metals and organometallic compounds by HS-SPME. Although the extraction of elements from the sample to the aqueous phase is not miniaturized, the preconcentration process involves a miniaturized procedure. Derivatization reactions are typically used to obtain volatiles to be extracted by the SPME fiber when dealing with the enrichment of metals and organometallic compounds. For example, SPME has been used for the determination of methylmercury in biological samples. Tissue samples (0.25 g) were extracted using 5 mL of 3 mol L^{-1} HCl and diluted to 30 mL. Then, the SPME fiber was exposed to the headspace of these extracts without the need of centrifugation step to capture methylmercury for subsequent ICP-MS analysis (Mester et al., 2000). In another study, a digestion step was incorporated prior to HS-SPME for Hg determination in biological tissues using a Pd wire instead of a common fiber (Romero et al., 2010). 200 mg of sample were digested using 2 mL HNO_3 + 1 mL H_2O_2 by MAD. Then, the Pd wire was placed in the headspace for 30 min and analyzed by CVAAS. Another study analyzed butyltin in sediment samples by SPME combined with GC-atomic emission detection (Carpinteiro et al., 2004). 0.1–0.5 g of sample was sonicated with 5 mL of acetic acid and neutralized with ammonium hydroxide. After derivatization, SPME was carried out in the headspace mode for 10 min at room temperature.

Determination of methylmercury and mercury (II) in sediment and biota (fish, crab, prawn and bivalves) samples was also performed by SPME-GC-MS (Mishra et al., 2005). Here, 0.5 g of sediment sample was extracted using 15 mL ($HCl:CH_3OH$, 1:1 (v/v) and sonicated for 2 h. Similarly, for biota samples, 0.3–0.8 g of edible tissue were first hydrolyzed using 25 % (m/v) KOH (10–20 mL) for 2 h under ultrasonic treatment. After the extraction, the preconcentration of analytes was performed by exposing the SPME fiber to the headspace for 15 min at 40 °C after alkylation of mercury species. Another example of mercury speciation analysis involving SPME was based on the combination of HS-SPME with furnace atomization plasma emission spectrometry in biological samples (Gringberg et al., 2003). A 0.25 g sample of biological tissue was mixed with 25 mL of 25 % (m/v) methanolic KOH solution for 4 h. For HS-SPME sampling, 12.5 mL of acetate buffer (pH = 5) was added to 200–1000 µL of the dissolved sample along with 1 mL of derivatizing agent. The mixture was magnetically stirred to ensure the proper mixing of the sample solution and to enhance transfer of the analytes from the solution to the headspace. After a predetermined time (20–200 s), the SPME fiber was introduced into the vial headspace (at a temperature in the range of 22–55 °C depending on the mercury species) for enrichment. Organomercury and organotin compounds were also determined in sediment and biota samples by GC with microwave-induced plasma atomic emission detection using HS-SPME for preconcentration (Delgado et al., 2008). In this case, the extraction of these compounds from the sample involved the use of 0.2–1 g of sediment extracted under sonication in 10 mL of diluted HCl in an ultrasonic bath for 1 h. Conversely, the extraction from the biological tissue used a sample mass of 0.1–0.3 g and extraction in an ultrasonic bath for 1 h with a diluted KOH in MeOH solution. SPME extraction involved a 3–24 min headspace exposure time, temperatures between 20–90 °C, 1–10 min desorption time, 150–260° C desorption temperature and sample volumes between 5–22 mL. A fiber with a 30 µm coating of polydimethylsiloxane (PDMS) was used in this case.

7.4.1.4.3 Liquid-phase Microextraction

LPME, also named solvent microextraction (SME), involves the miniaturization of conventional LLE (Pena-Pereira et al., 2009, 2010a, 2010b). LPME is most frequently uses to analyzed aqueous solutions due to their relative simplicity and low matrix effects. More efforts are needed to extend the applicability of LPME to solid matrices. Digestion of the solid or extraction to an aqueous phase is usually performed prior to LPME. A detailed description of LPME techniques can be found in chapter 4 of this book. In this section, the treatment step prior to different modes of LPME of solid samples will be discussed.

One mode of LPME, headspace single-drop microextraction (HS-SDME), was applied to extract methylmercury from fish tissue by *in situ* hydride generation and ETAAS (Gil et al., 2005). First, methylmercury was selectively extracted from fish tissue samples (0.4 g) by means of an ultrasonic probe for 2–5 min using 2 mol L^{-1} HCl. Generation of methylmercury hydride was performed in a closed vial containing 5 mL of 0.1 mol L^{-1} NaOAc/HOAc buffer. An aqueous drop of 3 μL containing Pd(II) or Pt(IV) was selected as trapping agent. After 3 min of extraction and with agitation of the sample, methylmercury was preconcentrated in the drop. A similar approach was reported for the GC analysis of organotin species from extracts of sediment samples using an organic drop containing 2 μL of decane (Colombini et al., 2004).

Dispersive liquid-liquid microextraction (DLLME) was also applied to solid samples. A pre-treatment step, ranging from conventional AD to small-scale UAE, was required. In the first case, acid digestion of soil samples (1 g) was carried out with 2 × 10 mL of *aqua regia* for 1 h followed by the addition of 1 mL of HF + 1 mL of HCl, and diluted to 25 mL, whereas only 1 mL was used for the preconcentration of Pd by DLLME (Shamsipur et al., 2009). In addition, a miniaturized procedure consisting of the combination of UAE and DLLME for determination of Au in soils and sediments by ETAAS was recently reported (De la Calle et al., 2011b). Specifically, a portion of solid sample of 3–30 mg was weighed into an Eppendorf vial and 1 mL of a 25 % (v/v) HNO_3 + 25 % (v/v) HCl mixture was added to extract Au. Ultrasound energy was applied for 20 min by means of a cup-horn sonoreactor. After sonication, centrifugation was carried out for 2 min at 5000 rpm. The preconcentration step was then performed using 900 μL of the extract.

Another miniaturized sample preparation technique generally applied for water samples is ultrasound-assisted emulsification microextraction (USAEME). USAEME has been used to extract previously digested solid samples such as plant tissues followed by ICP-OES determination of Al, Cu, Fe and Zn (Sereshti et al., 2011). In addition, soil and sediment samples can be directly extracted using ultrasound energy and used to quantify Au by ETAAS (De la Calle et al., 2011b). Hollow-fiber liquid-phase microextraction (HF-LPME) was also applied for the preconcentration of trace elements (Cu, Zn, Pd, Cd, Hg, Pb, Bi) from environmental and biological samples (peach leaves) after pre-treatment of the samples coupled to ETV-ICP-MS (Xia et al., 2007).

7.4.1.4.4 Fully Miniaturized Analytical Systems

Several reviews (Miró & Hansen, 2007; Mark et al., 2010) and a book (Ríos et al., 2009) are focused on fully miniaturized analytical systems. All of these advanced miniaturized devices share common characteristics such as the very low volume of reagents (μL, nL, pL) used. In addition to miniaturization, integration and automation are other key working principles. This section includes applications of μ-TAS, LOC, LOV and lab-on-disk (LOD) devices. It should be highlighted that fully miniaturized analytical systems are generally applied for aqueous samples and, in the case of solid samples, an off-line sample preparation is required.

μ-TAS, also named LOC, have been applied to trace metals analysis from aqueous samples such as river and drinking water samples (Xue et al., 2013; Ibarlucea et al., 2013) and fruit juices (Chailapakul et al., 2008). LOC was also combined with ICP-MS (Al-Suhaimi, 2011). In this case, a downscaled SPE device was coupled to ICP-MS using a PTFE connector and coupled with a flow injection manifold for determination of Cd, Co and Ni in seawater samples. Some examples found in the literature for LOC analysis of solid samples (*e.g.*, environmental and food samples), are summarized here. However, off-chip sample preparation is usually performed due to the difficulty to directly incorporate a complex matrix into a microchip (Ríos et al., 2009). More research is highly required in relation with the on-chip treatment of solid samples.

Chang et al. (2005) designed a miniaturized sensor for monitoring Pb(II) in environmental samples. In this study, electroplating sludge samples were analyzed according to US EPA Method 3050B (AD of sediments, sludges and soils. Method 3050B, 1996), involving off-chip extraction, centrifugation and filtration. Then, the sample was diluted in a solution of ammonium hydroxide before injection on the microchip. This microdevice (50 μm width, 30 μm depth and 14 mm length) made of PDMS, contained a network of microfluidic channels that were fluidically coupled *via* a nanocapillary array interconnect. The selectivity of the sensor for Pb(II) was obtained using a lead-specific DNAzyme followed by laser-induced fluorescence as the detection mode. The limit of detection achieved for Pb(II) was 11 nM. It should be pointed out that despite the very small volume (<100 pL) still allowed efficent molecular recognition reactions to occur.

A capillary electrophoresis (CE) microchip was fabricated for the separation and detection of Cd, Pb, Cu, Co, Ni and Hg (Deng & Collins, 2003). Firstly, a spiked sample was deposited onto a Plexiglas surface. After evaporation, the surface was washed with 0.1 mol L^{-1} HNO_3 and different reagents including an special chelating agent were added for the formation of metal chelates. The resulting sample can be directly introduced into the CE microchip, or preconcentrated by off-line SPE (using a C18 silica gel microcolumn) and subsequently separated by CE microchip. The combination of surface sampling and SPE coupling with CE allows the improvement of detection limits by 100-fold.

He et al. (2007) developed a method to determine nitrite in different food samples (*e.g.*, vegetables, fruits and meats) using a poly(methylmethacrylate) (50 ×40 ×5 mm) substrate. For the treatment of solid samples, off-chip extraction with a 732-Cation Resin column and filtration was employed. A microflow injection analysis (μ-FIA) system was used on a chip with chemiluminescence detection using the luminol-ferricyanide system. The volume of reaction area was about 1.8 μL.

An electrochemical sensor with a microfabricated on-chip Bi electrode was designed for *in situ* heavy metal ion (Cd and Pb) determination in soil pore, ground water and cell culture media (Zou et al., 2008). In this work, soil pore and ground water samples were taken from a lab-scale reactor for *in situ* measurements to simulate the pollution of the soil pore and ground water of the heavy metal wastes disposable sites. Thus, a solution of Cd(II) was flowed through the soil and the groundwater was collected. The system consisted of a microfluidic chip with microchannels, syringe pumps (for sample loading), Bi working electrode, Ag/AgCl reference electrode and gold counter electrode (each electrode was 3 mm in length). Moreover, this sensor is disposable and easy to use, has very low cost, is able to be mass produced, involves very small analyte consumption and generates a very small amount of waste.

Another procedure was reported for the determination of nitrite in food samples, namely sausages and ham (Shiddikiy et al., 2009). In this case, off-line sample preparation encompassed two possibilities - sonication (i.e., 2.5 g of sample in 50 mL extractant, sonication time of 5–10 min) and MAE (1 g of sample in 2 mL extractant, heated in a microwave oven for 1–3 min). The extracted samples were analyzed using microchip electrophoresis with electrochemical detection (MCE-ED) using a modified carbon paste electrode. The time required for analysis was less than 200 s.

Recently, a chip-based LPME was fabricated and combined with ETV-ICP-MS for the determination of trace metals (Cu. Zn, Cd, Hg, Pb and Bi) in human hair and cell samples (Wang et al., 2013a). DDTC was used both as the chemical modifier for low-temperature ETV-ICP-MS and as a chelating reagent using octanol as the organic phase for the chip-based LPME. Double "Y" type microchannels were designed. The chip had two inlet branch channels (one for aqueous phase and another for organic phase), one central channel for LPME (where the aqueous and organic phase combine in a laminar flow) and two outlet branch channels (one for aqueous phase and another for organic phase). Syringe pumps were applied for the introduction of two phases. These two phases formed laminar flow in the central channel, increasing the interfacial contact between the phases, favoring the entrance of metal chelates into the organic phase (**Figure 7.5F**). The extraction took 3.5 min and 7 μL of the organic phase was collected and subsequently introduced into the graphite furnace for ETV-ICP-MS analysis. Limits of detection (LODs) were in the range of 6.6–89.3 pg mL^{-1} while sample recoveries in the range of 87–119% were achieved. The developed method was fast, suitable for microsamples and exhibited low reagent consumption.

LOV constitutes the third generation of flow injection techniques, showing the significant progress that has been made in miniaturization, automation and integration of on-line sample pre-treatment (Yu et al., 2011). Several reviews related to LOV can be found in the literature (J. Wang et al., 2003; Chen & Wang, 2007; Yu et al., 2011).

LOV was combined with instruments like ETAAS (Wang & Hansen, 2001a; Long et al., 2005) and ICP-MS (Wang & Hansen, 2001b) for total metal detection (Ni, Bi) and metal speciation (Cr(III), Cr(VI)) in order to automate the process and downscale sample and reagent consumption. This procedure can be applied straightforwardly for aqueous samples (Long et al., 2005) but digestion with HNO_3 + HF is required for solid samples, such as ash and river sediment (Wang & Hansen, 2001a, 2001b). LOV facilitates processes such as mixing, dilution, etc. It is worth noting that volumes involved in LOV are in the range of microliter to sub-microliter. Thus, its combination with ETAAS is suitable, but it involves a challenge in the case of ICP-MS, which requires larger volumes (mL) for analysis. Wang and Hansen (2001b) developed a homemade direct injection high efficiency nebulizer to solve this problem which allowed the ICP-MS to detect metals after aspirating only 15 μL of sample.

The lab-on-a-valve bead-injection (LOV-BI) sample pre-treatment emerged to solve several inconveniences of SPE (*e.g.*, long operation time, clogging of the column and deactivation of the SPE material) and to facilitate the separation and preconcentration of metal ions using different sorbent materials. In the BI technique a packed microcolumn is generated *in situ* by aspirating into the system a suspended solution of microcarrier beads from a peripherical port of the valve. Two channels act as microcolumns positions connected with a small tubing. This tubing present two functions: a) holding the beads inside the column reactor avoiding escaping of the beads and b) automatically transporting of the beads between different column positions. Advantages of LOV-BI include: i) sorbent materials which are renewed after each cycle, ii) matrix interferences are avoided, iii) flow resistance is eliminated and iv) analytical performance is improved. LOV-BI has been coupled to different atomic spectrometric techniques for metal analysis, as reported in a recent review (Yu et al., 2011).

A new generation of centrifugal devices has also emerged. As in total analysis systems, several operations can be integrated, including valving, siphoning, liquid mixing and volume metering and splitting. In centrifugal microfluidics devices, sometimes called LOD, all processes are controlled by centrifugal force by means of a rotating microstructured substrate to provide a pumping action for liquids (Ducree et al., 2007; Ríos et al., 2009; Mark et al., 2011). A centrifugal microfluidic device has been designed for solid sample preparation. This procedure could be extended as an alternative to large scale protocols for the dissolution of solid samples prior to analyses by other LOC or LOV protocols. It consists of a magnetically driven device, which allows the pulverization of solid samples by mechanical movement (Duford et al., 2009). In this work, a 0.1 g sample of potassium ferrocyanide was completely dissolved in 1.0 mL of water in 3 s while rotating at 1000 rpm. This system allows complete sample dissolution with up to 70% reduction in time with respect to static dissolution (Duford et al., 2009). The system also demonstrated the ability to pulverize a hard crystal, suggesting that a solid sample preparation step may be integrated on centrifugal microfluidic devices in the future.

7.4.2 Organic Compound Analysis

In this section, different miniaturized sample preparation procedures for the analysis of organic compounds are described. A wide range of organic pollutants, drugs and naturally-occurring compounds have been determined in a variety of matrices, including food and feedstuffs, biological and environmental solid samples. **Table 7.3** shows selected applications of miniaturized sample preparation for organic compounds.

7.4.2.1 Minimal Treatment-based Techniques

In certain cases, very simple sample treatments such as dilution, dissolution or centrifugation can be carried out for the preparation of samples. In those cases, miniaturization has been overcome through important reductions in the volume of solvents used. Furthermore, additional automation of these procedures and direct solid analysis by different instrumental (sometimes miniaturized) techniques is also possible.

SS can be considered as a very simple alternative for sample preparation of solids. In the case organic compounds are analyzed, the solid sample is dissolved in organic solvents such as tetrahydrofuran, dimethylformamide, toluene, chlorinated solvents, alcohol-based solutions or even water, depending on the characteristics of the analyte (Self, 2005; Pawliszyn & Lord, 2010). Volumes of solvents do not exceed 10 mL, especially in the miniaturized format.

In some works, ultrasound irradiation has been applied to facilitate and improve sample dispersion since slurry homogeneity is an essential parameter affecting the sample representativeness. In this regard, it has to be considered that all of the solid matrix is introduced in the analytical instrument, and therefore, significant background effects and poorer detection and quantification limits are obtained. Additional effects related to organic compound degradation caused by ultrasound irradiation must be also taken into account.

In some cases, direct solid analysis is possible and techniques such as nuclear magnetic resonance (NMR), near infrared spectrometry (NIR), Raman spectroscopy, ion mobility spectrometry (IMS), direct analysis in real time mass spectrometry (DART-MS), and atmospheric pressure ionization mass spectrometry (API-MS) have been also employed for determination of organic analytes (Garrigues & De la Guardia, 2013). In the case of mass spectrometric detection, the ionization source has to be selected depending on the type of analyte and nature of the sample. For example, in the case of IMS, laser sources for ionization allow direct analysis of organic compounds in solid environmental matrices with improved sensitivity in relation with other sources. This is the case, for example, of the determination of PAHs in petroleum-derived products, or the detection of explosives and drugs in different solid matrices by laser source-based IMS, for which any sample preparation is required (Márquez-Sillero et al., 2011).

GC techniques have been used for direct solid analysis, in most cases using the headspace sampling mode. New advances in this field include the coupling of GC–IMS in portable instruments for *in situ* detection of volatiles from solid samples such as plants (Reyes-Garcés et al., 2013).

Table 7.3 Selected applications of miniaturized sample preparation procedures for organic compound analysis.

Miniaturized sample preparation	Matrix	Analytes	Description	Detection limits	Recovery (%)	Calibration	Detection	Ref.
MSPD	orange, apple, pear and grape	diazinon, methylparathion, fenitrothion, malathion, fenthion, ethylchlorpyriphos, methylbromophos, methidathion, methylazinphos, permethrin	25 mg of sample are blended with 25 mg C_8 bonded silica sorbent. The mixture is packed in a 10×4 mm holder. Washing cycle with water at 1 mL/min (8 min), Drying with N_2 (30min). Desorption with 0.1 mL ethyl acetate.	10–115 μg kg^{-1}	43–118	standard addition	GC-MS	(Kristenson et al., 2001)
UAE	zebrafish	chlorpyrifos, atrazine, dicofol	addition of 40 μL of internal standard and 0.150 mL 1:1 (v/v) *n*-hexane-ethyl acetate to 50 mg sample. Extraction under sonication for 40 s using a 3mm sonication probe. Centrifugation (2 min). Supernatant purification with C_8 packed in a 1-mL glass column using 4 aliquots of 0. 25 mL of *n*-hexane-ethyl acetate (1:1). Dry under nitrogen stream. Reconstitution with 40 μL of iso-octane.	3- 45 ng g^{-1}	88–95	external calibration	GC-μECD	(El-Amrani et al., 2012)

continued **Table 7.3** Selected applications of miniaturized sample preparation procedures for organic compound analysis.

Miniaturized sample preparation	Matrix	Analytes	Description	Detection limits	Recovery (%)	Calibration	Detection	Ref.
MAE	raw propolis	caffeic acid, *p*-coumaric acid, ferulic acid, quercetin, cinnamic acid, apigenin, kaempferol, isorhamnetin, luteolin, chrysin, pinocembrin, galangin	0.50 g was extracted for 15 min with 5 ml EtOH–H_2O (80:20, v/v), in a 10 mL glass vessel using a closed-vessel FMW system at 106 °C. Centrifugation (5 min). Filtration.	0.5–3 μg mL^{-1}	94.1–101.3	external calibration	HPLC-DAD	(Pellati et al., 2013)
PLE	soil, sediments	naphthalene, acenaphthylene, acenaphthene, fluorene. phenanthrene, anthracene, fluoranthene, pyrene, benzoanthrazene, chrysene, benzo[*b*] fluoranthene, benzo[*k*] fluoranthene, benzo[*a*] pyrene, indeno[1,2,3-*cd*]pyrene, benzo[*ghi*] perylene, dibenzo[*a,h*] anthracene, [$^2H_{10}$]phenanthrene	50 mg of sample is extracted with 0.1 mL toluene. 10 min static-dynamic PLE. A SPE drying heatable 10×3.0 mm internal diameter stainless-steel cartridge holder is used as extraction cell.	0.040.5 ng mL^{-1}	94–104	standard addition	large-volume injection (LVI)–GC–MS	(Ramos et al., 2000)

continued **Table 7.3** Selected applications of miniaturized sample preparation procedures for organic compound analysis.

Miniaturized sample preparation	Matrix	Analytes	Description	Detection limits	Recovery (%)	Calibration	Detection	Ref.
SFE	pig tissues	enrofloxacin, danofloxacin, ciprofloxacin	2 g of pig tissue are mixed with 7 g sand and 0.2 g Na_4-EDTA and subjected to SFE for 40 min using 30% v/v methanol at 80 °C. Extract is evaporated and redissolved in 10 mL phosphate-acetonitrile buffer.	0.002–0.01 mg kg^{-1}	87–113	external calibration	HPLC-FLD	(Choi et al., 2009)
multiple-headspace sampling-single-drop microextraction (MHS-SDME,direct solid analysis)	pharmaceutical powders	methanol, ethanol	10 mg of sample is incubated for 3 h at 140 °C before the initial extraction and thermostated for 15 min at 140 °C) between each extraction. SDME exposing 2 µL drop of dimethyl sulfoxide (DMSO) to the headspace for 5 min.	µg	-	external calibration	GC-FID	(Yu et al., 2010)
in vivo-SPME and *in vivo*-thin film microextraction (TFME)	rainbow trout	fluoxetine, diazepam, nordiazepam, gemfibrozil, atorvastatin, ibuprofen, carbamazepine, diclofenac, naproxen, bisphenol-A, paroxetine, sertraline, atrazine	exposure of the C_{18} fibers / C_{18} thin film phase to the dorsal-epaxial muscle for 30 min. Washing of the fiber with water and desorption for 90 min in 60 % (v/v) methanol.	0.08–0.21 ng g^{-1}	-	kinetic calibration	LC-MS/MS	(Togunde et al., 2012)

continued **Table 7.3** Selected applications of miniaturized sample preparation procedures for organic compound analysis.

Miniaturized sample preparation	Matrix	Analytes	Description	Detection limits	Recovery (%)	Calibration	Detection	Ref.
one step-MAE-μ-SPE	pak choi, tomato, apple, grape	ethoprophos, monocrotophos, phorate, terbufos, fonofos, diazinon, chlorpyrifos-methyl, malathion, chlorpyrifos, phenthoate	1 g sample, 8 mL hexane and 3 polypropylene membrane envelopes (containing 0.1 g activated carbon) are placed in a PTFE MW closed-vessel. MW is applied (400W) at 60 °C for 10 min. After extraction, enriched envelopes are washed with hexane and dried in filter paper. Desorption in 2.5 mL ethyl acetate under sonication (15 min). Filtration, evaporation to dryness and reconstitution in 0.1 mL hexane.	0.07–0.23 μg kg^{-1}	93.5–104.6	standard addition	GC-MS	(Wang et al., 2013b)
HF-SPME	apple	carbofuran, carbaryl, isoprocarb, diethofencarb, methiocarb, phenol	MWCNT-reinforced hollow fibre. 2 g of pulp sample is mixed with 9 mL of 16 % (w/v) NaCl. The MWCNT-HF impregnated in 1-octanol is placed in the sample solution for 60 min, at 950 rpm. Then it is washed with water, dried with filter paper. Placed in an end-sealed pipette tip containing 0.025 mL methanol. Sonication for 25 min for desorption.	0.09–6ng g^{-1}	92–113	external calibration	HPLC-DAD	(Song et al., 2013)

continued **Table 7.3** Selected applications of miniaturized sample preparation procedures for organic compound analysis.

Miniaturized sample preparation	**Matrix**	**Analytes**	**Description**	**Detection limits**	**Recovery (%)**	**Calibration**	**Detection**	**Ref.**
HS-SDME	dried fruit *Fructus forsythia*	essential oils	a flask with 0.7 g sample, 1.7 g 1-ethyl-3-methylimidazolium acetate is introduced in the microwave system. The flask is connected to a condenser through an open hole in the system. The needle of the microsyringe is introduced through one of the holes of the condenser. For SDME, 2.0 µL of *n*-heptadecane is exposed to the headspace of the system for 3.4 min, at 78 °C and 300W.	-	-	external calibration	GC-MS	(Jiao et al., 2013)
magnetically-driven centrifugal micro-fluidic	soil	pyrene	30 mg of sample is placed in each sample unit (10 total) along with a magnet. 0.2 mL of hexane added. 10 min of rotation cycles of 50, 400 and 1200 rpm for extraction, filtration and transferral to detection unit, respectively.	µg	78–84	external calibration	UV-Vis spectrophotometry	(Xi et al., 2010)

Improvements in solid sample analysis using these techniques has evolved through the development of miniaturized and, in some cases, hand-held devices. Electrochemical techniques, mainly coulometric- and voltammetric-based, have also been used for direct solid analysis, although to a lesser extent for organic compounds in comparison with inorganic analytes (Doménech-Carbó et al., 2013).

7.4.2.2 Extraction-based Techniques

Extraction-based techniques are frequently used for the analysis of organic compounds since analytes of interest can be directly extracted from the solid sample with no previous sample preparation step. In this sense, conventional and energy-assisted extractions, namely SFE, ASE, MAE or UAE, and MSPD have evolved towards their miniaturized modes or towards more environmentally friendly miniaturized alternatives such as LPME and SPME. In general, all of these techniques require a lower sample mass and/or reagent consumption. Experimentally, the volumes used can vary from few microliters to ten milliliters. In addition, low volume extraction vessels are required for accomplishing the solid sample preparation in miniaturized conditions. However, the implementation of new miniaturized procedures for the analysis of solid samples has been more limited in comparison with those for liquid samples (Ramos et al., 2005).

In most cases, extraction-based sample preparation techniques are essential tools for transferring analytes into a liquid phase that is suitable to be analyzed, or even suitable to be subjected to another separation/preconcentration sample preparation technique prior to analysis. In the vast majority of cases, separation-based techniques such as HPLC, GC or electrophoretic techniques coupled to different detectors are used.

7.4.2.2.1 Direct Solid Treatment by Miniaturized Matrix Solid-phase Dispersion

MSPD is one of the few extraction techniques that can be directly applied to solid, semi-solid and high-viscosity matrices. In fact, MSPD has been mainly used for the analysis of animal tissues, plants and foods, as well as environmental and human biological samples (Capriotti et al., 2013). A vast number of MSPD-based analytical methodologies have been developed for the determination of drugs, natural compounds and contaminants (Kristenson et al., 2001; Moliner-Martínez et al., 2009; Gutiérrez Valencia & García de Llasera, 2011; Lu et al., 2011; Muñoz-Ortuño et al., 2012). However, publications concerning the use of MSPD for the determination of inorganic analytes are scarce (Capriotti et al., 2013).

In MSPD, the sample is homogeneously blended with a solid support (i.e., C_8, C_{18}, or florisil, among others, for the determination of organic compounds). The mixture is packed into an analytical column for further elution of the analytes. Miniaturized versions of this technique, in terms of reduced amounts of sample, inert support or elution solvent, have been successfully applied to the analysis of different samples, especially fatty biological tissues or environmental samples. Sample masses in the range 0.025–0.1 g, inert support masses between 0.2–0.4 g and elution volumes between 0.1–3 mL have been used in miniaturized MSPD. Solvents such as acetonitrile, methanol, ethyl acetate, hexane, or their mixtures with other solvents in a much lower proportion, have primarily been used for elution.

For example, Kristenson et al. developed an automated miniaturized-MSPD using only 0.025 g of sample for the extraction of different pesticides from fruits, a C_8 bonded silica sorbent and 0.1 mL ethyl acetate for elution. Reasonable recoveries from samples with a high reproducibility, automation and high sample throughput were emphasized by the authors (Kristenson et al., 2001). In another example, Muñoz-Ortuño et al. (2012) determined di(2-ethylhexyl) phthalate in fatty biological samples such as mussels following a miniaturized MSPD technique. 0.1 g of bivalve sample was mixed with only 0.4 g of C_{18} sorbent. Elution was carried out with 1.2 mL of acetonitrile.

The extraction of organic compounds can be challenging depending on the sample, as in the case of environmental solid samples. Strong interactions between the target molecule and the structure of the solid may lead to a more difficult extraction of the analyte from the matrix (Capriotti et al., 2013). In addition, miniaturized-MSPD may lead to matrix effects, such as the co-extraction of fatty acids in the case of biological tissues, thus necessitating different calibration strategies such as matrix-matching (Covaci et al., 2010) or more frequently, the standard addition method (Kristenson et al., 2001; Gutiérrez Valencia & García de Llasera, 2011). In some cases, the use of an extra solid inert phase (co-sorbent) such as florisil allows for improved retention of the lipid phase of the sample, thus minimizing effects from matrix components (Moliner-Martinez et al., 2009; Gutiérrez Valencia & García de Llasera, 2011; Lu et al., 2011; Muñoz-Ortuño et al., 2012).

In recent years, different sorbents have been introduced in miniaturized-MSPD to circumvent the potential lack of selectivity and matrix effects in the analyses. In this sense, innovations have been mainly introduced through the use of MIPs (Qiao & Sun, 2010; Qiao & Yan, 2010; Yan et al., 2011, 2012a, 2012b; Hong & Chen, 2013). For example, Qiao and Yan et al. (2010) determined fluoroquinones and xanthine in serum using an MIP of ofloxacin and theophylline as a mixed template for selective recognition and extraction. Only 0.2 g of MIP particles along with 0.2 g of sample were used for MSPD and 4 mL acetonitrile-trifluoroacetic acid (99.5/0.5 (v/v)) for elution. Using standard addition to minimize potential matrix effects, recoveries were in the range of 90–104 % for all samples (Qiao & Yan, 2010).

Although emerging novel sorbents have still found scarce application in miniaturized-MSPD, novel solid blending materials such as graphene or MWCNTs have also been used (Q. Liu et al., 2011; Su et al., 2011). For example, Liu et al. used graphene for the extraction of polybrominated diphenyl ethers (PBDEs) and their analogs from environmental samples. 0.1 g of sample was mixed with 0.01 g of graphene sorbent. 0.05 g of florisil was used as a co-sorbent and 1 mL hexane/dichloromethane or 1 mL acetone was used for elution. In spite of the enhancement in selectivity as a result of the capability of graphene to retain co-eluting matrix components, a co-sorbent was required to retain polar interferences and lipids. Standard addition was used to minimize matrix effects (Q. Liu et al., 2011).

New trends in miniaturized-MSPD are the use of energy-assisted procedures (*e.g.*, ultrasound irradiation) or the combination with other clean-up steps. Recently, ultrasounds have been applied to sample preparation technique to improve EE as a result of improved contact between the solid adsorbent and the sample (Ramos et al., 2008; Karageorgou & Samanidou, 2010, 2011; Rezaei & Hosseini, 2011; Karageorgou et al., 2012, 2013, 2014; Barfi et al., 2013). In these cases, after mixing the sample and the sorbent, the mixture is transferred to a support (*e.g.*, a syringe barrel or cartridge) and the extraction solvent is added. Both ends of the support are then closed, and the column is immersed in a sonicating system. Finally, the extract is collected from the support by gravity elution.

In the majority of these cases, ultrasonic baths are used to disperse the mixture for 5–30 min at a thermostated temperature of around 35–40 °C. In the case of a highly effective extraction system for organophosphorus pesticides and triazines in fruits, a sonoreactor system was used which reduced the sonication time to only 1 min (Ramos et al., 2008).The application of ultrasounds requires one to take into account potential degradation of the target analytes. For example, Rezaei and Hosseini (2011) found that an increase in sonication time gave rise to degradation products of organochlorine pesticides.

Along with conventional MSPD sorbents such as C_{18}, C_8 or florisil, other inert blending materials such as commercial Oasis HLB sorbent or Nexus polymeric sorbent have been used when ultrasound-assisted miniaturized-MSPD is used (Ramos et al., 2008; Karageorgou & Samanidou, 2010, 2011; Rezaei & Hosseini, 2011; Karageorgou et al., 2012, 2013, 2014; Barfi et al., 2013;).

7.4.2.2.2 Direct Solid Treatment by Miniaturized Ultrasound- and Microwave-assisted Extraction, Supercritical Fluid Extraction and Pressurized Liquid Extraction

Energy-assisted extractions (*e.g.*, UAE, MAE, SFE and PLE) involving the transfer of analytes to an aqueous or organic phase have been widely used for the extraction of organic compounds from solid samples. Energy-assisted extractions not only can enhance analyte extraction, but also shorten extraction times. Miniaturization in these sample preparation techniques arise from a drastic reduction in the volumes used for extraction. However, scarce miniaturized applications aimed at the analysis of solids can be found in the literature.

Although UAE can be considered a well-established sample preparation technique, few ultrasound-assisted extractions in miniaturized format have been developed for the analysis of organic compounds. Exemplary applications have been devoted to the determination of PBDEs, PAHs, colistin and pesticides in different food, feedstuffs and environmental and biological tissue samples (Morales-Muñoz & Luque de Castro, 2005; Aydin et al., 2006; Domeño et al., 2006; El-Amrani et al., 2012;

Pena-Abaurrea et al., 2013a; Bizkarguenaga et al., 2014). The solvent-to-sample mass ratio is one of the most important parameters to take into account, apart from those related to the ultrasonic irradiation. Thus, the volumes of solvents have been reduced to 0.15–10 mL, in comparison with traditional procedures which use up to 300 mL of solvents and no less than 1 g of sample for extraction (Bagherian et al., 2011). For example, El-Amrami et al. extracted chlorpyrifos, atrazine and dicofol from zebrafish using only 50 mg of sample and 0.150 mL of hexane-ethyl acetate 1:1 (v/v). The low volumes allow shortened extraction times of just 40 s with good recoveries by using a high intensity sonication system with a 3 mm sonication probe (El-Amrani et al., 2012).

Although the use of conventional solvents such as hexane, ethyl acetate, methanol or acetone is common, new solvents have been also tested for miniaturized-format UAE. This includes natural deep eutectic solvents, which have the advantage of biodegradability and low toxicity. These solvents, formed by heating different alcohols and choline chloride at 80 °C, have been used for the extraction of flavonoids from plants by ultrasound irradiation. For this purpose, 0.2 g of powdered plant tissue was mixed with just 2 mL of the eutectic solvent and subjected to ultrasound irradiation for 40 min (Bi et al., 2013).

The use of sonication probes not only allows reduced solvent consumption, but also enables automation (Morales-Muñoz & Luque de Castro, 2005; Domeño et al., 2006). Morales-Muñoz & Luque de Castro (2005) employed this approach for colistin analysis in animal feed by extracting 0.1 g of sample in a closed chamber with a flow injection manifold. A 1 mm sonication probe was inserted into the chamber for extraction using water as the solvent.

The incomplete removal of the matrix prompted different authors to consider the implementation of an extra clean-up step after miniaturized-UAE. As a result, co-extracted compounds are eliminated prior to the analytical determination. SPE has been considered for this purpose as a suitable technique after miniaturized-UAE of different organic compounds (Aydin et al., 2006; El-Amrani et al., 2012; Pena-Abaurrea et al., 2013a).

Pena-Abaurrea et al. (2013a) achieved an extra clean-up by subjecting the extract to a disposable pipette SPE procedure. To this end, the low volume UAE extracted supernatant was taken with a micropipette into a 5 mL pipette tip containing 0.8 g acidic silica as a clean-up adsorbent to eliminate lipids from biological tissues for polychlorobiphenyls (PCB) determination.

A similar number of procedures based on MAE in miniaturized format (i.e., using less than 10 mL solvent volume) exist for the determination of organic compounds in solid matrices. Although plant and environmental materials have been mostly exensively analyzed, other applications for the quantification of contaminants in food can be also found in the literature. Volumes between 0.5 and 10 mL of organic extractants such as heptane, methanol, ethanol or ethyl acetate have been used for this purpose (Ramalhosa et al., 2012; Agudelo-Mesa et al., 2013; Azzouz & Ballesteros, 2013; Pellati et al., 2013; Barrera-Vázquez et al., 2014; Song et al., 2014). Barrera-Vázquez et al. (2014) treated 0.3 g of sample extract, previously subjected to miniaturized-UAE, to miniaturized-MAE using 6 mL of ethyl acetate for 6 min. This procedure showed higher EEs of anthraquinones from plants than the procedure using miniaturized-UAE alone.

As is the case for UAE-based procedures, co-extraction can be a common problem in procedures assisted by microwave irradiation. Several researchers have tried to circumvent this problem through the use of different strategies for calibration (Azzouz, 2012; Song, 2014), and/or applying a further extraction or clean-up step, either in miniaturized format or not (Azzouz & Ballesteros, 2012; Barrera-Vázquez et al., 2014; Song et al., 2014). A different strategy was followed by Agudelo-Mesa et al. for determining aromatic amines in beef burgers. The authors used MAE as a clean-up technique for extraction (three consecutive extraction steps with heptane, 3 min in duration each). In this case, miniaturized-MAE was used to eliminate fatty material and soluble organic compounds from the matrix. Non-polar heterocyclic aromatic amines remaining in the settled phase were further solubilized in alkaline media to desorb the analytes from the matrix and to precipitate remaining proteins. A miniaturized preconcentration step based on DLLME was applied afterwards. After the whole procedure, however, matrix effects were found to affect the methodology, thus standard addition was used for calibration (Agudelo-Mesa et al., 2013).

The use of MW energy in closed vessels has also been reported not only to reduce extraction time, but also to reduce solvent volume in comparison with conventional maceration and heat reflux extraction for the extraction of plant-based materials. Phenolics and flavonoids have been extracted from plant materials using this approach (Pellati et al., 2013).

On the other hand, fewer publications concerning miniaturized versions of PLE for the analysis of solid samples are reported. The use of commercial PLE equipment limits the applicability of miniaturized formats, since these are usually developed for extractions in higher volume format. Solutions come through the use of smaller extraction cells or the construction of homemade miniaturized PLE equipment. Between 0.01–0.2 g of solid samples (foodstuffs, feedstuffs and soils) have been analyzed for to determine PCB, PBDE and PAH concentrations. Solvents used are typically hexane, dichloromethane or toluene, in volumes varying from only 0.1 mL to 6 mL. As in the case of conventional PLE, both static and dynamic extraction steps are necessary for extraction (Crescenzi et al., 2000; Hyötyläinen et al., 2000; Ramos et al., 2000, 2007; Damm & Kappe, 2011; Pena-Abaurrea et al., 2013b).

In most cases, miniaturized PLE is carried out by mixing the solid sample with an inert material such as silica or sand for retaining lipids or other interfering compounds, thus increasing the selectivity (Crescenzi et al., 2000; Ramos et al., 2007; Pena-Abaurrea et al., 2013b). For example, Pena-Abaurrea et al. developed an approach combining the use of a silica-based inert material and a homemade PLE device for improving selectivity in the extraction of PCBs and PBDEs from feedstuffs.

The sample was mixed with Na_2SO_4 and acidified silica and placed in the extraction cell. For increased clean-up and selectivity, different layers of both acidic and neutral silica were also packed in the cell. Extraction was carried out in two static PLE cycles (7 min each) using 3 mL of hexane and hexane-dichloromethane and a final dynamic extraction (10 min) using 3 mL of hexane-dichloromethane (Pena-Abaurrea et al., 2013b).

Other relatively-miniaturized extractions based on the SFE have also found applications in the analysis of organic compounds in solid matrices (Pourmortazavi et al., 2014). These techniques have been used for the extraction of PAHs, pesticides and other organics from a variety of solids such as foods, diesel particulates, soils or biological and plant tissues (Starr & Selim, 2008; Choi et al., 2009; Portet-Koltalo et al., 2009; Matsubara et al., 2012). Automation is possible when using this technique, and extraction in low volume cells (2.5–10 mL) can be achieved when an extra-miniaturization step is added (Matsubara et al., 2012). In most cases, a supercritical fluid is the only requirement for extraction, usually CO_2 due to its greener analytical characteristics, with extraction times ranging from 20 to 150 min. Sample masses from 10–500 mg have been used. Sometimes, an organic modifier (i.e., methanol, ethanol, hexane, acetonitrile, etc.) is used to improve the extractability, especially for more polar analytes in complex environmental samples such as soils or sediments. After extraction, analytes can been collected in solvents (up to 10 mL) that in most cases, but not always, are the same as that the organic modifier.

Due to the nature of solid samples, selective extraction is not always possible, thus leading in some cases to interferences. This is the case for lipids that are co-extracted from biological tissues. In order to circumvent this potential problem, addition of solid materials such as florisil, silica gel or sand to the SFE extraction cells has been carried out. In a similar approach, Choi et al. added sand and 0.2 g of (ethylenedinitrile)-tetraacetic acid tetrasodium salt into 10 mL extraction cells for the SFE of fluoroquinones from 2 g pig tissues (Choi et al., 2009). Extraction was carried out using 30 % v/v methanol as an organic modifier for 40 min. The extraction of the most polar and hydrophilic compounds was significantly improved (87–113 % recovery) when the extra solids were added to the sample.

7.4.2.2.3 Direct Solid Treatment by Microextraction Techniques

Several examples can be found in literature in which solid samples are directly subjected to new generation miniaturized sample preparation techniques based on microextraction. However, it should be highlighted that only some formats of LPME and SPME are suitable for the direct analysis of solid samples with no previous treatment. This is the case of HS-SDME and DLLME modes, and both headspace and direct immersion-SPME.

Solid pharmaceuticals, plastics, vegetables and plants have been directly subjected to HS-SDME using sample masses in the range 0.01–5 g. Obviously, volatility of the selected analyte is a must, and therefore volatile residual organic solvents, volatile aldehydes, terpenes and styrenes have been selected as target compounds (Kim et al., 2005; Hansson & Hakkarainen, 2006; Ligor & Buszewski, 2008; Yu et al., 2010). In many cases, incubation at high temperatures (60–150 °C) is required to improve the transfer of the analyte to the headspace before HS-SDME (Kim et al., 2005; Hansson & Hakkarainen, 2006; Yu et al., 2010). Solvents such as pentanol, butyl acetate or DMSO have been used for the extractions in volumes ranging from 0.5–2 µL.

Although HS-SDME is not an exhaustive sample preparation technique, complete direct extractions from solids can be achieved by exposing the extractant drop to the headspace during multiple consecutive cycles (Hansson & Hakkarainen, 2006; Yu et al., 2010). Quantitative microextraction of residual solvents in pharmaceutical samples has been achieved following this technique. In this case, a long incubation step of 3 h at 140 °C was initially required, as well as additional steps of thermostation (15 min at 140 °C) between extractions in order to keep the headspace saturated with the analyte prior to extraction (Yu et al., 2010).

Owing to the difficulty of directly analyzing solids, only semi-solid samples have so far been directly subjected to DLLME procedures. Thus, formaldehyde and triclosan have been directly quantified in cosmetics following this sample preparation technique using only 20–80 µL of dichloromethane and ionic liquids, respectively, for microextraction (Lavilla et al., 2010; Cabaleiro et al., 2011). The nature of cosmetic samples, which often contain high amounts of surfactants, makes a disperser solvent unnecessary, thus reducing the amount of reagents required.

SPME-based approaches have accounted for many more applications in the direct analysis of solid samples. The wide applicability of direct SPME methods lies in the fact that the volume of sample does not influence the extraction, when the sample volume is much higher than the volume of the SPME fiber (Müller, 1999). As such, applications for environmental, plant and biota analysis are numerous, although some other applications to the analysis of pharmaceuticals and packagings can be found (X. Zhang et al., 2012; Souza Silva et al., 2013; Zhu et al., 2013). Detection of antioxidant and aromatic compounds in plants, pharmaceuticals in living biological systems and volatiles in packaging are just some examples of applications of SPME to solid samples (Bicchi et al., 2000; Sides et al., 2001; Baták et al., 2003; Carrillo & Tena, 2006; Togunde et al., 2012). In these cases, the SPME procedures typically consist of the direct insertion of the fiber into the living organism (*in vivo*-SPME) or, alternatively, exposing the SPME fiber to the headspace of a vial containing the solid sample (HS-SPME), which makes this technique completely free of extractant solvents. In the first case, biocompatibility of the fiber is essential in order to avoid adverse effects when sampling is performed with live animals. In this regard, PDMS, polypyrrole and C_{18} commercial fibers have been used with success. In the case of plant materials, being biocompatible is not a critical issue, thus PDMS-based fibers (PDMS, PDMS/Divinylbenzene (DVB), PDMS/Carboxen (CAR)), or polyacrylate (PA) fibers have been mostly used. For example, trace pharmaceuticals such as gemfibrozil, atorvastatin, ibuprofen, carbamazepine, diclofenac, naproxen, bisphenol A, paroxetine and sertraline have been determined in live marine organisms such as rainbow trout and fathead minnow. In these experiments, C_{18} fibers were directly inserted into the dorsal-epaxial muscle of the fish, and microextraction was carried out over 30 min. Acceptable reproducibility (4–22% relative standard deviation, RSD) was achieved (Togunde et al., 2012).

HS-SPME mode has demonstrated its usefulness for the analysis of diverse volatile compounds in plants, mostly for analyzing aromatic profiles and scents (Bicchi et al., 2000; Barták et al., 2003; Carrillo & Tena, 2006; Beck et al., 2008; Stashenko et al., 2009), either in the static or dynamic mode. Given the variety of possible aromatic compounds in plants, Bicchi et al. compared the effectiveness of PDMS, PDMS/DVB, Carbowax (CW)/DVB, CAR/PDMS and CAR/DVB/PDMS fibers for the HS-SPME analysis of aromatic organic analytes from rosemary, sage, thyme and valerian. For extraction, 0.6 g of each plant was thermostated at 60 °C for 1 h in a closed vial prior to extraction. Vials were shaken for 10 s at 10 min intervals and the extraction was carried out for 60 min. The authors concluded that, although the fiber selection is strongly influenced by the polarity and volatility of each aromatic organic compound of the plant, PDMS-based fibers combined with porous solids provided, in general, good results (Bicchi et al., 2000).

7.4.2.3 Combined Techniques

A wide range of miniaturized sample preparation approaches are applied to liquid samples and cannot be directly used, in general, for the analysis of organics in solid and semisolid samples. In spite of the important advances in sample preparation, basic sample preparation techniques (i.e., homogenization, extraction, etc.) are still required for converting the solid into a suitable physical state (mostly liquid) prior to new generation miniaturized sample preparation techniques. This previous step can be also miniaturized or not. As discussed in the case of inorganic analysis, miniaturized techniques such as µ-SPE, SPME, LPME and micro-total analysis systems (µ-TAS), among others, are applied. These techniques are in general characterized not only by the low volumes of reagents (from µL to nL) and liquid-state sample (from few mL to nL) required, but also by a high degree of sample clean-up, and, in most cases, analyte preconcentration.

7.4.2.3.1 Solid-phase Extraction in Miniaturized Format

Different formats of miniaturized SPE have been developed in recent years. Since sorbents control the selectivity and EE, efforts have been focused in this area, using polymers, MIPs, zeolites or activated carbon, among others. Examples include the determination of PBDEs in soils using copper (II) isonicotinate-based sorbents, ochratoxin A in coffee and cereals using Zeolite Linde Type L sorbent, pesticides in tissues using HayeSep A/C_{18} sorbents and dicyandiamide in milk using a functionalized magnetic polymer sorbent (Basheer et al., 2008; Lee et al., 2012; Zhou et al., 2012; Chen et al., 2014).

In most cases, miniaturized formats of SPE have evolved towards procedures based on the so called μ-SPE. This mode is based on the use of a porous polypropylene membrane that protects the adsorbent from direct contact with the solid matrix, thus minimizing matrix effects. In general, the envelope-like polypropylene membrane containing the sorbent (around 20–50 mg) has dimensions of 1–2 cm in length and 0.5–1 cm in width. In general, analytes are first dissolved or solvent-extracted from the solid sample prior to be subjected to the μ-SPE techniques.

Poly(methacrylic acid-ethylene glycol dimethacrylate) has been used for the extraction of sulfamerazine, sulfamethazine, sulfathiazole and sulfadiazine from chicken muscle. In order to transfer the analytes to a liquid phase, 1 g of sample was first extracted with 10 mL acetonitrile 10 % (v/v). For μ-SPE, just 20 mg of sorbent in the polypropylene membrane were used for the extraction of 10 mL sample solution for 30 min under sonication. As in conventional SPE, steps of sorbent conditioning, extraction, washing and desorption are required. In this case, desorption was carried out with 0.4 mL of acetone containing 1 % (v/v) ammonia under sonication for 30 min (Huang et al., 2012).

Interestingly, energy-assisted μ-SPE procedures are common and ultrasound or microwave irradiation is frequently applied simultaneously to enhance the extraction (Basheer et al., 2008; Huang et al., 2012, 2013; Wang et al., 2013b). A novel approach based on miniaturized one-step MAE-μ-SPE for the analysis of solid samples was recently reported. Wang et al. introduced three envelope-like polypropylene membranes containing 20 mg of activated carbon into a PTFE microwave vessel which had been previously filled with 1 g of vegetable sample dissolved in 8 mL hexane. Extraction was carried out for 10 min at 400 W and 60 °C. Good recoveries (94–105%) and reproducibility was obtained (Wang et al., 2013b).

Another miniaturized-SPE format developed for solid sample analysis is pipette tip SPE which has been applied to the analysis of pesticides, picrosides, toxins or contaminants in plant materials and other biological tissues. 0.1–0.2 mL volume pipette tips containing 2–60 mg of sorbent are commonly used. As in other modes, different sorbents can be selected depending on the characteristics of the analyte. Conventional sorbents such as styrene divinylbenzene, C_{18} and graphene can be used. As the sample has to be aspirated into liquid form, a previous solid sample pre-treatment technique is required. Solvent extraction with organic solvents such as acetonitrile (Guan et al., 2010) or, in many cases, UAE with alcohol-based solvents (Pena-Abaurrea et al., 2013a ; Shen et al., 2013a, 2013b) has been reported. In addition, direct analysis of herbal powders has also been carried out (Wang et al., 2014). In this format, the extraction recovery is critically influenced by the number of aspirating and dispensing cycles. In one study, 25 consecutive aspirating/dispensing cycles were used for the extraction of pricosides from plants using a conventional C_{18} sorbent packed into a 0.2 mL pipette tip (Shen et al., 2013a).

7.4.2.3.2 Solid-phase Microextraction

SPME techniques have found extensive use in the analysis of solid samples such as biological tissues, food, pharmaceuticals, plants and environmental samples and a vast number of different analytes have been determined, including organic contaminants, pesticides, herbicides and biologically active molecules. Except for those cases of direct solid analysis discussed in previous sections, samples have to be adapted to a liquid state containing the target analytes. Many different modes such as direct immersion-SPME, HS-SPME, in-tube-SPME, microextraction in a packed syringe or stir bar sorptive extraction (SBSE) have been used, all of them based on adsorption phenomena and characterized by minimal consumption of solvents, which in many cases are not required for extraction.

As in other techniques, solid samples must first be treated following procedures such as dissolution or homogenization in water (Domeño et al., 2005; Bali-Prasad et al., 2010) or organic solvents as acetone, acetonitrile or methanol. Other techniques for adapting the solid sample to a liquid state prior to SPME include extraction procedures such as miniaturized-MSPD (Campíns-Falcó et al., 2008; Moliner-Martinez et al., 2009), UAE (Ke et al., 2014), MAE (Ramil-Criado et al., 2004), SPE (Matin et al., 2013), subcritical water extraction (Hawthorne et al., 2000) or SFE (W. Liu et al., 2011).

Apart from the classical PDMS fiber coating, PA fibers, commercial mixed coatings such as PDMS/CAR, PDMS– DVB, CW/DVB and CW/templated resin, other solid sorbent phases based on antibodies, aptamers, polymers, nanoparticulated or carbon-based phases can be used in different SPME-based modes. In fact, new technologies come not only through the development of new SPME-based configurations, but also through the implementation of new adsorbent phases. In all cases, the polarity of the analyte determines the nature and characteristics of the sorbent used. In the case of direct immersion-SPME and HS-SPME, typical fiber lengths are in the range of 1–2 cm. Fiber exposure times ranging from just a few min up to 60 min may be required for extractions. The use of high temperatures is also possible for speeding up the microextraction process, although this is more common in headspace mode.

Examples of SPME-based methods for the analysis of organics in solid phase samples can be found in the literature, such as the extraction of hydrocarbons in environmental samples (Matin et al., 2013), microextraction in packed syringe of clenbuterol from pork samples (Du et al., 2014), rotating disk sorptive extraction of PAHs from soils and sediments, (Hawthorne et al., 2000), SBSE of biologically active compounds from biological tissues and feed samples (Xu et al., 2010) or the evaluation of biotic exposure to contaminants by thin-film microextraction (Meloche et al., 2009).

The robustness of the supporting fiber and SPME coatings is a much more critical parameter when placed in direct contact with the sample (Mehdinia & Aziz-Zanjani, 2013). In order to circumvent potential problems related to fiber robustness and carry-over effects, novel strategies based not only on the development of more robust fibers, but also on new reinforced ones can be found in the literature (Ahmadi-Golsefidi et al., 2012; Song et al., 2013). The reinforcement of a hollow fiber with CNTs and

MIPs combined with CNTs are emerging trends. For example, Song et al. developed a CNT-reinforced hollow fiber for the SPME of pesticides from apples. CNTs were immobilized in the pores of the hollow fiber and used for selective extraction of carbofuran, carbaryl, isoporcarb, diethofencarb and methiocarb. At the same time, the hollow fiber provided an extra sample clean-up step, since it was in direct contact with the sample. The only sample pre-treatment required was solubilization in 9 mL of 16 % (v/v) NaCl aqueous solution (Song et al., 2013).

Extractions are usually carried out with stirring and the time needed may be reduced by as adding energy in the form of microwaves. Several one-step microwave-assisted sorbent-based microextraction procedures have been developed in this regard and a few of them have been applied to analyze solid samples such as oysters, sediments and wood by SPME (Pino et al., 2007; Wu & Ding, 2010; Wu et al., 2012). In these cases, samples are only dispersed in water, and HS-SPME is carried out by inserting the needle through the top of the microwave system, exposing it to the headspace of the vial containing the slurried sample. The assistance of MW energy allowed equilibrium to be reached in only 5 min with good extraction recoveries. To the best of our knowledge no ultrasound-assisted-SPME based procedures have been published so far.

7.4.2.3.3 Liquid-phase Microextraction

A vast number of LPME techniques as SDME, HF-LPME, DLLME and their different modes have also been applied for the analysis of organic compounds in solid samples such as soils, foodstuffs, pharmaceuticals, biological tissues, plant materials, plastics and textiles (Pena-Pereira et al., 2010c; Chao et al., 2013; Pirsaheb et al., 2013; Viñas et al., 2013; Caballo et al., 2014; Lu & Zhu, 2014). However, applications to solid sample analysis by LPME-based techniques only represent around 10 % of the total (Kokosa et al., 2009).

Unlike SPME-based methods, all LPME modes require solvents for extractions, although at very low volumes varying from 0.5 to 100 μL, the highest volumes being required in the solidified organic droplet microextraction and DLLME modes and the lower volume in SDME-based modes.

Organic solvents and ionic liquids are commonly used to extract solid samples, depending mainly on the polarity of the organic analyte. In recent years, new aqueous extraction solvents have been gradually introduced, including aqueous cyclodextrins, surfactant-based extractants, coacervates and supramolecular solvents. Extraction times can vary from only 30 s to 40 min depending on the LPME technique and mode used. As in many other sample preparation techniques, temperature, ultrasound and microwave irradiation and vortexing have been used to acheive faster kinetics and to improve microextraction effectiveness from the solid matrix (Zhang & Shi, 2010; Wang et al., 2012; Guo & Kee Lee, 2013; Jiao et al., 2013; Lana et al., 2013; Abu-Bakar et al., 2014). For example, microwave energy has been used for the distillation and simultaneous HS-SDME of essential oils from the dried fruit *Fructus forsythia*. 0.7 g of sample, placed in a reaction flask containing 1.7 g of 1-ethyl-3-methylimidazolium acetate, was introduced in the cavity of the microwave apparatus. The flask was connected to a condenser through an open hole in the upper part of the microwave system, and to a microsyringe. For microextraction, a 2.0 μL drop of *n*-heptadecane was exposed to the headspace of the system for 3.4 min, at 78 °C and at an irradiation power of 300 W. In this case, the solvent was chosen on the basis of high boiling point, good EE, and also non-absorption of microwave energy. Precision, expressed as RSD, was found to be about 8 % (Jiao et al., 2013).

Again, the physical state of the sample, with the exceptions of those modes discussed previously, must meet the requirement of being liquid, and thus, different sample pre-treatment strategies are applied in the case of solid samples. More specifically, processes based on solubilization (Seidi et al., 2011), UAE (González-Curbelo et al., 2013; Sereshti et al., 2014) and MAE have been mainly used (Campillo et al., 2011; Agudelo Mesa et al., 2013). Other pre-treatment techniques such as MSDP (Yan et al., 2011), PLE (Lu & Zhu, 2013) or SFE (Naeeni et al., 2011) have been less used.

Lu and Zhu (2013) determined chlorobenzenes in polyester fabrics by pressurized hot water extraction followed by vortex-assisted liquid-liquid microextraction (VALLME). First, the low polar analytes were selectively extracted from a 2 g sample by a static cycle of extraction at 160 °C and 1500 psi for 10 min and using 20 % (v/v) acetonitrile. A 5 mL aliquot of the diluted extract was subjected to VALLME using 60 μL of carbon tetrachloride. After vortexing for 3 min and centrifugation, the enriched sedimented phase was used for analysis. Sample recoveries were obtained in the range of 68–92%.

As mentioned before, not all pre-treatment techniques applied before LPME are miniaturized. For example, in the extraction of organophosphorous pesticides from cereals and flour, 20 mL of acetonitrile was required for UAE (5 min) portion of the analysis. After centrifugation and filtration of the supernatant, the UAE was repeated again. Finally, pesticides were microextracted and preconcentrated by HF-LPME using 20 μL of 1-octanol for 45 min (González-Curbelo et al., 2013).

7.4.2.3.4 Fully Miniaturized Analytical Systems

The use of micro-total analysis systems (μ-TAS or LOC) is undoubtedly one of the most challenging and, at the same time, promising developments in the analysis of organic compounds. The integration of the entire analytical process into a single miniaturized system will provide a step forward towards total miniaturization, automation and greening of analytical chemistry. In this sense, the use of volumes of both sample and reagents in the pL-nL range be highlighted as one of its several advantages. Nevertheless, one of its biggest pitfalls still lies in the achievement of adequate selectivity and sensitivity when it comes to trace analysis in highly complicated samples such as solid matrices. In addition, channels of the miniaturized analytical systems are likely to clog as a result of the introduction of suspended particles from solids.

In recent years, the interest in the possibilities of μ-TAS for organic analysis has increased and is reflected in the growing number of μ-TAS applications in analytical chemistry (Ríos & Zougagh, 2013). However, publications of these systems to the analysis of solids as biological, clinical, environmental and food samples are still scarce (Ríos et al., 2012).

Applications of LOC to solid analysis found in the literature were performed on soils, tablets, vegetables, fruits, and infant formula powder. The target analytes were antioxidants, vitamins, isoflavones, zearalenone, folic acid or formaldehyde (Miró et al., 2006; Lee et al., 2008; Plata et al., 2008; Atalay et al., 2011). In almost all of these cases, the most critical and error-prone step of sample pre-treatment is still carried out 'off-chip'. In off-chip sample preparation, steps such as pulverization, extraction, dilution and filtration are commonly carried out, not necessarily miniaturized at the same scale as the chip. For example, conventional UAE and subsequent SPE with C_{18} sorbent and filtration have been employed for the analysis of food dyes in fish, noodles and pepper powders. The combination of two different clean-up extraction steps and filtration was required in order to obtain extracts clean enough to flow through the microfluidic channels of the proposed microfluidic system. Analyte preconcentration, separation and chronoamperometric detection was then carried out on-chip (Lee et al., 2008).

By incorporating in the miniaturized system three different channels of field-amplified sample stacking (FASS), field-amplified sample injection (FASI) and micellar electrokinetic chromatography, Lee et al. achieved a 10800-times higher sensitivity compared with conventional micellar electrokinetic chromatography with electrochemical detection (MEKC-ED). In spite of the clean-up steps carried out for sample preparation, salts and organic compounds from the matrix were found to affect the electrical signal and electrophoretic migration of the analytes (Lee et al., 2008).

Without a doubt, μ-TAS will expand in the coming years, with much research being done in the field of solid sample analysis (Ríos & Zougagh, 2013) and new advances based on both miniaturization and automation are continually being published. Centrifugal microfluidic systems have been used recently for automated integration of extraction, filtration and detection in soil analysis (Duford et al., 2009; Xi et al., 2010). Only Xi et al. used a magnetically-driven centrifugal microfluidic system for the direct determination of pyrene in soils. A detailed scheme on the proposed system is depicted in **Figure 7.7**. 30 mg of soil was placed in each of the ten extraction chambers containing a magnet and 0.2 mL of hexane was added. Each chamber comprised stages of magnetically driven SLE, filtration and spectrometric detection, respectively. By rotating the systems at different speeds, the different steps took place. Good extraction yields and reproducibility was obtained, along with a very important reduction in sample mass and extractant volume in comparison with conventional techniques (Xi et al., 2010).

The combination of miniaturized sample preparation techniques such as μ-SPE or LPME with μ-TAS are also promising, and the first steps in this direction have already

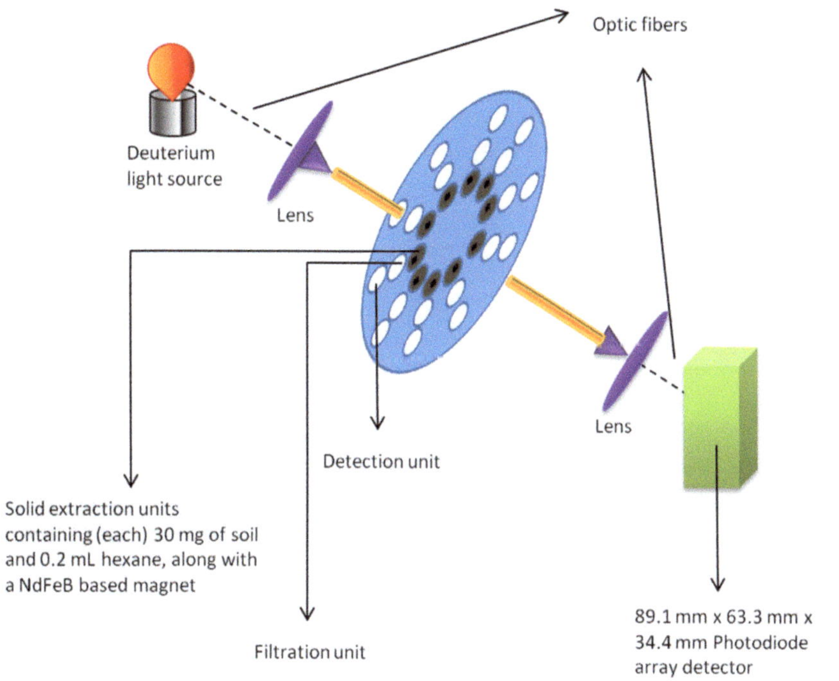

Figure 7.7 Schematic of a magnetically-driven centrifugal microfluidic system that includes automated integration of extraction, filtration and detection.

been achieved for the analysis of solid samples. For example, by combining HS-SDME with microfluidic chip electrophoresis, aliphatic amines have been determined in seafood. Sample pre-treatment was not completely eliminated however, since 0.5 g of sample was initially homogenized with 2 mL of aqueous buffer and NaOH prior to microextraction (Mark et al., 2011).

The LOV concept has also found application for the determination of organic compounds in solids. These miniaturized flow systems, which enable complete system automation and handling of volumes in the nanoliter range, are implemented using optical and electrochemical detection. Less frequently, separation-based instrumental detection techniques are used. However, important steps have already been taken in this direction with the aim of achieving analytical methodologies fully adapted to routine analysis laboratories, in which mainly chromatographic systems are used. In this regard, one of the bottlenecks lies in the hyphenation of the low sample flows with chromatographic systems and in potential column deterioration. Very recently, PCBs were quantified in solid waste leachates by coupling online LOV-BI and GC. In this system, PCBs were preconcentrated in the BI system (3 mg of commercial reversed-phase copolymeric sorbent) which served as an SPE-based device that allowed on-line sample pre-treatment and avoided further column damage. Injection of the different solutions was done using an automated multisyringe flow injection system. Even in these cases, sample pre-treatment is necessary and thus, previous filtration and dilution of samples are required. 4 mL of sample solution were finally used for analysis and only 80 μL of ethyl acetate for elution. Recoveries in the range 87–116% were obtained. The authors also indicated a 25-fold reduction of sorbent consumption with regard to traditional SPE procedures (Quintana et al., 2009).

7.5 Future Trends

Future trends in sample preparation should be oriented toward the increase of productivity through automation and the simultaneous direct treatment of several samples, but also considering the miniaturized procedures and the use of portable instrumentation for on-site determinations. In many cases, this will also mean the development of new interfaces to directly couple new sample preparation techniques with the detection unit of the portable instrumentation.

Research is ongoing to broaden the applications of virtually solventless techniques. The use of new extractant phases with improved sorption and robustness characteristics in the case of sorbent-based techniques (*e.g.*, solid nanoparticle materials, graphene, PDMS-based modified coatings, magnetic nanocomposites and polymeric ionic liquids) is likely to improve extractions from highly complex solid matrices. Regarding LPME, new formats that are able to directly treat the solid sample and overcome potential matrix effects are required. The implementation of new extractant phases, based on nanoparticle materials, polymeric ionic liquids, or supramolecular solvents can also be anticipated.

Miniaturized microfluidic devices are very interesting for environmental and clinical monitoring. However, further developments are needed to link solid sample preparation to these miniaturized devices. In this sense, the implementation of new functionalized nanomaterials and new multifunctional particles (i.e., multifunctional protein particles encapsulating biologically active molecules such as enzymes and antibodies) within these techniques reveals promising routes for improving selectivity and sensitivity in the analysis of solid samples. This means the integration of all steps of the analytical process and avoiding off-chip sample preparation, especially in the case of solid samples.

Research efforts are also aimed at developing new sensors for on-site, real-time pollutant and biologically active molecule measurements. Special attention should be paid to the robustness, selectivity and precision of these systems. Electrochemical detection is expected to increase its popularity for analytical microchips due to: i) high sensitivity and selectivity, ii) low cost and iii) the possibility of miniaturization, portability and design of disposable devices.

Magnetic beads are suitable and have increasing applications in microfluidic devices. Thus, magnetic nano- and microparticles could be used in SPE microchips and LOV systems due to the easy manipulation (separation) of magnetic particles using magnets and the possibility of automation. Associated advantages such as sample clean-up, mixing, analyte adsorption, transport and separation will also be considered for boosting their analytical applications. The coupling and automation of miniaturized systems using magnetic beads with chromatographic and other separation-based instrumental systems is still a challenge, although steps have been taken in this direction.

The development of new materials and nanomaterials could help several extraction and microextraction approaches. Moreover, the use of new solvents, like ionic liquids, or eutectic solvents should be continued.

With regard to increasing sample throughput, more research should be focused, for example, on centrifugal devices (6–10 sample throughput), microextraction in packed sorbent (with 96-well plates in polypropylene tips) and similar techniques. In addition, there are commercially available ultrasonic dual-probe and multi-probe systems (containing several tips) that presumably will play an important role in upcoming analytical method development and applications.

To summarize, fast, cheap, easy, integrated and automated procedures will be the future of miniaturized solid sample preparation procedures.

7.6 Conclusions

In recent decades, the pressure to decrease the usage of chemical reagents in the analytical laboratory has increased due to their potential toxicity and environmental impacts. Furthermore, there is a growing tendency towards the reduction of sample size, the time of operation and associated costs, along with the reduction of toxic solvents and wasteful residues. Moreover, the risks for the operator and analyte losses or contamination should be avoided. Thus, miniaturization and improvement of sample preparation alternatives is a focus of research. In addition, the possibility of integrating several analytical steps into only one miniaturized step, automation of these procedures with the possibility of direct on-site solid sample preparation, and enhancement of sample throughput are other state-of-the-art branches of research.

As reviewed in this chapter, different procedures could be applied for the analysis of metals, organometallics and organic compounds. In addition to miniaturization, acceleration of sample preparation procedures from several hours to a few minutes or seconds is carried out by means of ultrasound and microwave energy, the use of supercritical fluids and solvents at elevated temperature and pressure. Furthermore, several small vials, microvessels and cells were specially designed for different procedures, such as ASE, VPMAD, VPAD, AD, MAD, and PLE involving low-volume sample preparation procedures derived from the conventional ones. Generally, low amounts of sample (*e.g.*, a few mg), and reduced volumes of reagents (*e.g.*, µL-mL) are used in these procedures, thus minimizing the generation of wastes or residues.

Some procedures can be applied directly to solid samples while in other cases sample digestion or extraction of analytes is required prior to sample preparation. In the last case, preconcentration of analytes, with an associated elimination of potential matrix interferences, is generally performed.

As for the required volume of reagents and solvents, it can be stated that different levels of miniaturization have been achieved. Thus, some developed approaches require 5–10 mL (*e.g.*, MAE, ASE, µ-SPE), others only 1 mL (*e.g.*, SS, UAE, UAD, LVMAD), while the most miniaturized techniques (*e.g.*, LPME, LOC, LOV, LOD) only require a few µL, nL or pL.

Abbreviations

µ-FIA	microflow injection analysis
µ-SPE	micro-solid-phase extraction
µ-TAS	micro-total analysis systems
AD	acid digestion
API-MS	atmospheric pressure ionization mass spectrometry
ASE	accelerated solvent extraction
AsB	arsenobetaine
AsC	arsenocholine
BCR	Community Bureau of Reference (Brussels Belgium)
BI	bead injection
CAR	Carboxen
CNTs	carbon nanotubes
CPA	cold plasma ashing
CRMs	certified reference materials
CVAAS	cold-vapor atomic absorption spectrometry
CW	Carbowax
DART-MS	direct analysis in real time mass spectrometry
DDTC	diethyldithiocarbamate
DLLME	dispersive liquid-liquid microextraction
DMA	dimethylarsinic acid
DMSPE	dispersive micro solid-phase extraction
DMSO	dimethyl sulfoxide
DSS	direct solid sampling
DTE	dithioerythritol
DTT	dithiothreitol
DVB	divinylbenzene
DPX	disposable pipette extraction

EDTA	ethylenediaminetetraacetic acid
EE	extraction efficiency
EF	enrichment factor
EPS	enzymatic probe sonication
ETAAS	electrothermal atomic absorption spectrometry
ETV	electrothermal vaporization
ETV-ICP-OES	electrothermal vaporization with inductively-coupled plasma optical emission spectrometry
ETV-ICP-MS	electrothermal vaporization with inductively coupled plasma mass spectrometry
FAAS	flame atomic absorption spectrometry
FASS	field-amplified sample stacking
FASI	field-amplified sample injection
FMW	focused microwave
GC	gas chromatography
GC-FID	gas chromatography flame ionization detector
GC-MS	gas chromatography mass spectrometry
GC-μECD	gas chromatography with micro electron capture detector
HF-LPME	hollow-fiber liquid-phase microextraction
HG-AAS	hydride generation atomic absorption spectrometry
HG-AFS	hydride generation atomic fluorescence spectrometry
HG-ICP-MS	hydride generation coupled to inductively coupled plasma mass spectrometry
HPHTSE	high-pressure, high temperature solvent extraction
HPLC-DAD	high pressure liquid chromatography diode array detection
HPLC-FLD	high pressure liquid chromatography with fluorimetric detector
HPLC-HG-AFS	high pressure liquid chromatography hydride generation atomic fluorescence spectrometry
HPLC-ICP-MS	high pressure liquid chromatography coupled to inductively coupled plasma mass spectrometry
HPLC-MS	high pressure liquid chromatography mass spectrometry detection
HPLC-UV	high pressure liquid chromatography ultraviolet detection
HPSE	high-pressure solvent extraction
HR-CS-ETAAS	high resolution continuum source electrothermal atomic absorption spectrometry
HS-SDME	headspace single-drop microextraction
HS-SPME	headspace solid-phase microextraction
ICP-OES	inductively-coupled plasma optical emission spectrometry
ICP-MS	inductively coupled plasma mass spectrometry
IR	infrared radiation
IMS	ion mobility spectrometry
LA	laser ablation

LA-ICP-MS	laser ablation inductively coupled plasma mass spectrometry
LC-MS/MS	liquid chromatography tandem mass spectrometry
LLE	liquid-liquid extraction
LOC	lab-on-a-chip
LOD	lab-on-disk
LODs	limits of detection
LOV	lab-on-a-valve
LOV-BI	lab-on-a-valve bead-injection
LPME	liquid-phase microextraction
LTA	low temperature ashing
LVMAD	low-volume microwave-assisted digestion
MAD	microwave-assisted digestion
MAE	microwave-assisted extraction
MCE-ED	microchip electrophoresis with electrochemical detection
MEKC-ED	micellar electrokinetic chromatography with electrochemical detection
MIBK	methyl isobutyl ketone
MIC	microwave-induced combustion
MIPs	molecularly imprinted polymers
NIR	near infrared spectrometry
MMA	monomethylarsonic acid
NMR	nuclear magnetic resonance
MSE	microwave sequential extraction procedures
MSPD	matrix solid-phase dispersion
MWCNTs	multiwalled carbon nanotubes
NIST	National Institute of Standards and Technology
NRCC	National Research Council of Canada
PA	polyacrylate
PAHs	polycyclic aromatic hydrocarbons
PBDEs	polybrominated diphenyl ethers
PCBs	polychlorobiphenyls
PDMS	polydimethylsiloxane
PHSE	pressurized hot solvent extraction
PLE	pressurized liquid extraction
PSE	pressurized solvent extraction
PTFE	Polytetrafluoroethylene
RSD	relative standard deviation
SBSE	stir bar sorptive extraction
SDS	sodium dodecyl sulfate
SEP	sequential extraction procedure
SES	sequential extraction schemes
SDME	single-drop microextraction

SFE	supercritical fluid extraction
SLE	solid-liquid extraction
SM&T	Standards Measurements and Testing program from the European Commission
SPE	solid-phase extraction
SPME	solid-phase microextraction
SS	slurry sampling
SSE	subcritical solvent extraction
TMAH	tetramethyl ammonium hydroxide
Tris	tris(hydroxymethyl)aminomethane
TXRF	total reflection X-ray fluorescence spectrometry
UAD	ultrasound-assisted digestion
UAE	ultrasound-assisted extraction
USS	ultrasonic slurry sampling
USAEME	ultrasound-assisted emulsification microextraction
USEPA	United States Environmental Protection Agency
VALLME	vortex-assisted liquid-liquid microextraction
VPAD	vapor-phase acid digestion
VPMAD	vapor-phase microwave-assisted digestion
XRF	X-ray fluorescence spectrometry.

Acknowledgements

I. de la Calle thanks Xunta de Galicia for financial support as a post-doctoral researcher of the I2C program and cofinancing by the European Social Fundings P.P. 0000 421S 140.08.

References

Abu-Bakar, N., Makahleh, A. & Saad, B. (2014). Vortex-assisted liquid–liquid microextraction coupled with high performance liquid chromatography for the determination of furfurals and patulin in fruit juices. *Talanta, 120,* 47–54.

Abu-Samra, A., Morris, J. S. & Koirtyohann, S. R. (1975). Wet ashing of some biological samples in a microwave oven. *Analytical Chemistry, 47,* 1475–1477.

Acid digestion of sediments, sludges and soils; Method 3050B *(1996)*. Retrieved March, 11, 2014 from http://www.epa.gov/sw-846/pdfs/3050b.pdf.

Adeyemo, D. J., Umar, I. M., Thomas, S. A. & Agbaji, E. B. (2004). Analysis of lubricating oil for Cu, Fe, Mn, Ni and Zn using electrothermal atomic absorption spectrometry. *Instrumentation Science & Technology, 32,* 689–694.

Afridi, H. I., Kazi, T. G., Arain, M. B., Jamali, M. K., Kazi, G. H. & Jalbani, N. (2007). Determination of cadmium and lead in biological samples by three ultrasonic-based samples treatment procedures followed by electrothermal atomic absorption spectrometry. *Journal of AOAC International, 90,* 470–478.

Agudelo-Mesa, L. B., Padró, J. M. & Reta, M. (2013). Analysis of non-polar heterocyclic aromatic amines in beefburguers by using microwave-assisted extraction and dispersive liquid–ionic liquid microextraction. *Food Chemistry, 141,* 1694–1701.

Ahmadi-Golsefidi, M., Es'haghi, Z. & Sarafraz-Yazdi, A. (2012). Design, synthesis and evaluation of a molecularly imprinted polymer for hollow fiber–solid phase microextraction of chlorogenic acid in medicinal plants. *Journal of Chromatography A, 1229,* 24–29.

Akinlua, A., Torto, N. & McCrindle, R. I. (2013). A new approach to sample preparation for the determination of trace metals in petroleum source rocks. *Analytical Methods, 5,* 4929–4934.

Alonso-Castillo, M. L., Vereda-Alonso, E., Siles-Cordero, M. T., Cano-Pavón J. M. & García de Torres, A. (2011). Fractionation of heavy metals in sediment by using microwave assisted sequential extraction procedure and determination by inductively coupled plasma mass spectrometry. *Microchemical Journal, 98,* 234–239.

Al-Suhaimi, A. O. & McCreedy, T. (2011). Microchip based sample treatment device interfaced with ICP-MS for the analysis of transition metals from environmental samples. *Arabian Journal of Chemistry, 4,* 195–203.

Alvarez, J., Marcó, L. M., Arroyo, J., Greaves, E. D. & Rivas, R. (2003). Determination of calcium, potassium, manganese, iron, copper and zinc levels in representative samples of two onion cultivars using total reflection X-ray fluorescence and ultrasound extraction procedure. *Spectrochimica Acta Part B: Atomic Spectroscopy, 58,* 2183–2189.

Alzaga, R. Díez, S. & Bayona, J. M. (2003). Supercritical fluid extraction. In Z. Mester & R. Sturgeon (Eds.), *Sample preparation for trace element analysis* (pp. 313–341). The Netherlands: Elsevier.

Amarasiriwardena, D. Krushevska, A., Argentine, M. & Barnes, R. M. (1994). Vapour-phase acid digestion of micro samples of biological material in a high-temperature, high-pressure asher for inductively coupled plasma atomic emission spectrometry. *The Analyst, 119,* 1017–1021.

Amberger, M. A. & Broekaert, J. A. C. (2010). Direct multielement determination of trace elements in boron carbide powders by slurry sampling ETV-ICP-OES. *Journal of Analytical Atomic Spectrometry, 25,* 1308–1315.

Amoedo, L., Capelo, J. L., Lavilla, I. & Bendicho, C. (1999). Ultrasound-assisted extraction of lead from solid samples: a new perspective on the slurry-based sample preparation methods for electrothermal atomic absorption spectrometry. *Journal of Analytical Atomic Spectrometry, 14,* 1221–1226.

Antes, F. G., Duarte, F. A., Mesko, M. F., Nunes, M. A. G., Pereira, V. A., Müller, E. I., Dressler, V. L. & Flores, E. M. M. (2010). Determination of toxic elements in coal by ICP-MS after digestion using microwave-induced combustión. *Talanta, 83,* 364–369.

Arain, M. B., Kazi, T. G., Jamali, M. K., Jalbani, N., Afridi, H. I., Sarfraz, R. A. & Shah, A. Q. (2007). Determination of toxic elements in muscle tissues of five fish species using ultrasound-assisted pseudodigestion by electrothermal atomic absorption spectrophotometry: optimization study. *Spectroscopy Letters, 40,* 861–878.

Arain, M. B., Kazi, T. G., Jamali, M. K., Jalbani, N., Afridi, H. I. & Baig, J. A. (2008). Speciation of heavy metals in sediment by conventional, ultrasound and microwave assisted single extraction methods: A comparison with modified sequential extraction procedure, *Journal of Hazardous Materials, 154,* 998–1006.

Araújo, G. C. L., Nogueira, A. R. A. & Nóbrega, J. A. (2004). Microwave single vessel acid-vapor extraction: effect of experimental parameters on Co and Fe determination in biological samples. *Microchimica Acta, 144,* 81–85.

Araújo, G. C. L., Nogueira, A. R. A. & Nóbrega, J. A., (2000). Single vessel procedure for acid-vapour partial digestion in a focused microwave: Fe and Co determination in biological samples by ETAAS. *The Analyst, 125,* 1861–1864.

Arroyo, L., Trejos, T., Gardinali, P. R. & Almirall, J. R. (2009). Optimization and validation of a laser ablation inductively coupled plasma mass spectrometry method for the routine analysis of soils and sediments. *Spectrochimica Acta Part B: Atomic Spectroscopy, 64*, 16–25.

Arruda, S. C. C., Rodriguez, A. P. M. & Arruda, M. A. Z. (2003). Ultrasound-assisted extraction of Ca, K and Mg from *In Vitro* Citrus culture. *Journal of the Brazilian Chemical Society, 14*, 470–474.

Associação Brasileira de Cimento Portland, Manual Técnico 4, Procedimentos para análises químicas de cimentos e suas matérias-primas, 2003 (134 pp) ABCP.

Atalay, Y. T., Vermeir, S., Witters, D., Vergauwe, N., Verbruggen, B., Verboven, P., Nicolaï B. M. & Lammertyn, J. (2011). Microfluidic analytical systems for food analysis. *Trends in Food Science and Technology, 22*, 386–404.

Aydin, M. E., Tor, A. & Özcan, S. (2006). Determination of selected polychlorinated biphenyls in soil by miniaturised ultrasonic solvent extraction and gas chromatography-mass-selective detection. *Analytica Chimica Acta, 577*, 232–237.

Azzouz, A. & Ballesteros, E. (2012). Combined microwave-assisted extraction and continuous solid-phase extraction prior to gas chromatography–mass spectrometry determination of pharmaceuticals, personal care products and hormones in soils, sediments and sludge. *Science of Total Environment, 419*, 208–215.

Bachmann, L., Zezell, D. M., Ribeiro, A. D., Gomes L. & Ito, A.S. (2006). Fluorescence spectroscopy of biological tissues - A review. *Applied Spectroscopy Reviews, 41*, 575–590.

Bagherian, H., Ashtiani, F. Z., Fouladitajar, A. & Mohtashamy, M. (2011). Comparisons between conventional, microwave- and ultrasound-assisted methods for extraction of pectin from grapefruit. *Chemical Engineering and Processing: Process Intensification, 50*, 1237–1243.

Balarama-Krishna, M. V. & Arunachalam, J. (2004). Ultrasound-assisted extraction procedure for the fast estimation of major, minor and trace elements in lichen and mussel samples by ICP-MS and ICP-AES. *Analytica Chimica Acta, 522*, 179–187.

Balarama-Krishna, M. V., Ranjit, M., Karunasagar, D. & Arunachalam, J. (2005). A rapid ultrasound-assisted thiourea extraction method for the determination of inorganic and methyl mercury in biological and environmental samples by CVAAS. *Talanta, 67*, 70–80.

Balarama-Krishna, M. V., Venkateswarlu, G., Thangavel, S. & Karunasagar, D. (2013) Oxidative pyrolysis combined with the microwave-assisted extraction method for the multi-elemental analysis of boron carbide powders by inductively coupled plasma optical emission spectrometry (ICP-OES). *Analytical Methods, 5*, 1515–1523.

Baldwin, S., Deaker, M. & Maher, W. (1994). Low-volume microwave digestion of marine biological tissues for the measurement of trace elements. *The Analyst, 119*, 1701–1704.

Bali Prasad, B., Prasad Tiwari, M., Madhuri, R. & Sindhu, P. (2010). Enantioselective quantitative separation of d- and l-thyroxine by molecularly imprinted micro-solid phase extraction silver fiber coupled with complementary molecularly imprinted polymer-sensor. *Journal of Chromatography A, 1217*, 4255–4266.

Bangroo, P. N., Jagga, C. R., Arora, H. C. & Rao, G. N. (1995). Spectrometric determination of wear metals and additive elements in lubricating oils by AAS and ICP-AES. *Atomic Spectrosccopy, 16*, 118–120.

Barbosa, A. F., Segatelli, M. G., Pereira, A. C., Santos, A. D., Kubota, L. T., Luccas, P. O. & Teixeira-Tarley, C. R. (2007). Solid-phase extraction system for Pb (II) ions enrichment based on multiwall carbon nanotubes coupled on-line to flame atomic absorption spectrometry. *Talanta, 71*, 1512–1519.

Barfi, V., Asghari, A., Rajabi, M., Barfi, A. & Saeidi, I. (2013). Simplified miniaturized ultrasound-assisted matrix solid phase dispersion extraction and high performance liquid chromatographic determination of seven flavonoids in citrus fruit juice and human fluid samples: Hesperetin and naringenin as biomarkers. *Journal of Chromatography A, 1311*, 30–40.

Barker, S. A., (2000). Matrix solid-phase dispersion, *Journal of Chromatography A, 885*, 115–127.

Barker, S. A., (2007). Matrix solid-phase dispersion (MSPD), *Journal of Biochemical and Biophysical Methods, 70*, 151–162.

Barrera-Vázquez, M. F., Comini, L. R., Martini, R. E., Núñez-Montoya, S. C., Bottini, S., Cabrera, J. L. (2014). Comparisons between conventional, ultrasound-assisted and microwave-assisted methods for extraction of anthraquinones from Heterophyllaea pustulata Hook f. (Rubiaceae). *Ultrasonics Sonochemistry, 21*, 478–484.

Barták, P., Bednar, P., Čáp L., Ondráková, L. & Stránsky, Z. (2003). SPME – A valuable tool for investigation of flower scent. *Journal of Separation Science, 26*, 715–721.

Basheer, C., Narasimhan, K., Yin, M., Zhao, C., Choolani, M. & Lee, H. K. (2008). Application of micro-solid-phase extraction for the determination of persistent organic pollutants in tissue samples. *Journal of Chromatography A, 1186*, 358–364.

Batista, B. L., Rodrigues, J. L., Souza, V. C. D. & Barbosa Jr., F. (2009). A fast ultrasound-assisted extraction procedure for trace elements determination in hair samples by ICP-MS for forensic analysis. *Forensic Science International, 192*, 88–93.

Beck, J. J., Merrill, G. B., Palumbo, J. D. & O'Keeffe, T. L. (2008). Strain of *Fusariumoxysporum* isolated from almond hulls produces styrene and 7-methyl-1,3,5-cyclooctatriene as the principal volatile components. *Journal of Agricultural Food Chemistry, 56*, 11392–11398.

Belarra, M. A., Resano, M., Vanhaecke, F. & Moens, L. (2002). Direct solid sampling with electrothermal vaporization/atomization: what for and how? *TrAC Trends in Analytical Chemistry, 21*, 828–839.

Bellido-Milla, D., Ordaz-Garcia, S. M., Guerrero-Valiente, J. L. & Hernández-Artiga, M. P. (2002). Rapid procedure to determine wear metals in lubricating oils and the analysis of variance in the evaluation of sample preparation procedures. *Microchimica Acta, 138*, 59–64.

Bendicho, C, Lavilla, I., Pena, F. & Costas, M., (2011). Green sample preparation methods. In M. de la Guardia & S. Garrigues (Eds.), *Challenges in Green Analytical Chemistry* (pp. 63–106). Cambridge: Royal Society of Chemistry.

Bendicho, C. & de Loos-Vollebregt, M.T.C. (1991). Solid sampling in electrothermal atomic absorption spectrometry using commercial atomizers: A review. *Journal of Analytical Atomic Spectrometry, 6*, 353–374.

Bendicho, C., De La Calle, I., Pena, F., Costas, M., Cabaleiro, N. & Lavilla, I. (2012a). Ultrasound-assisted pretreatment of solid samples in the context of green analytical chemistry. *TrAC Trends in Analytical Chemistry, 31*, 50–60.

Bendicho, C., Lavilla, I., Pena, F. & Romero, V. (2012b). Green chemistry in analytical atomic spectrometry: a review. *Journal Analytical Atomic Spectroscopy*, 2012, 27, 1831–1857.

Bengtson, B. & Lundholm M., (1988). Surface analysis with the glow discharge lamp. State of the art and prospects for future development. *Journal of Analytical Atomic Spectroscopy, 3*, 879–882.

Benson, W. H. Francis, P. C., Birge, W. J., Black, J. A. (1983). A simple method for acid extraction of Cd from fish eggs or fish tissues. *Atomic Spectroscopy, 4*, 212–213.

Bermejo, P., Capelo, J. L., Mota, A., Madrid, Y. & Cámara, C. (2004). Enzymatic digestion and ultrasonication: A powerful combination in analytical chemistry. *TrAC Trends in Analytical Chemistry, 23*, 654–663.

Bermejo-Barrera, P., Fernández-Nocelo, S., Moreda-Piñeiro, A. & Bermejo-Barrera, A. (1999). Usefulness of enzymatic hydrolysis procedures based on the use of pronase E as sample pre-treatment for multi-element determination in biological materials. *Journal of Analytical Atomic Spectroscopy, 14*, 1893–1900.

Bermejo-Barrera, P., Muñiz-Naveiro, O., Moreda-Piñeiro, A. & Bermejo-Barrera, A. (2000). Experimental designs in the optimisation of ultrasonic bath–acid-leaching procedures for the determination of trace elements in human hair samples by atomic absorption spectrometry. *Forensic Science International, 107*, 105–120.

Bettiol, C., Stievano, L., Bertelle, M., Delfino, F. & Argese, E., (2008). Evaluation of microwave-assisted acid extraction procedures for the determination of metal content and potential bioavailability in sediments. *Applied Geochemistry, 23*, 1140–1151.

Bi, W., Tian, M. & Row, K. H. (2013). Evaluation of alcohol-based deep eutectic solvent in extraction and determination of flavonoids with response surface methodology optimization. *Journal of Chromatography A, 1285*, 22–30.

Bicchi, C., Drigo, S. & Rubiolo, P. (2000). Influence of fibre coating in headspace solid-phase microextraction–gas chromatographic analysis of aromatic and medicinal plants. *Journal of Chromatography A, 892*, 469–485.

Bizkarguenaga, E., Iparragirre, A., Zabaleta, I., Vallejo, A., Fernández, L. A., Prieto, A. & Zuloaga, O. (2014). Focused ultrasound assisted extraction for the determination of PBDEs in vegetables and amended soil. *Talanta, 119*, 53–59.

Bocca, B., Lamazza, A., Pino, A., De Masi, E., Iacomino, M., Mattei, D., Rahimi, S., Fiori, E., Schillaci, A., Alimonti, A. & Forte, G. (2007). Determination of 30 elements in colorectal biopsies by sector field inductively coupled plasma mass spectrometry: method development and preliminary baseline levels. *Rapid Communication in Mass Spectrometry, 21*, 1776–1782.

Borges, A. R., Becker, E. M., Lequeux, C., Vale, M. G. R., Ferreira, S. L. C. & Welz, B. (2011). Method development for the determination of cadmium in fertilizer samples using high-resolution continuum source graphite furnace atomic absorption spectrometry and slurry sampling. *Spectrochimica Acta Part B: Atomic Spectroscopy, 66*, 529–535.

Borgese, L., Zacco, A., Bontempi, E., Colombi, P., Bertuzzi, R., Ferretti, E., Tenini, S. & Depero, L. E. (2009). Total reflection of X-ray fluorescence (TXRF): a mature technique for environmental chemical nanoscale metrology. *Measurement Science and Technology, 20*, 1–7.

Borgese, L., Zacco, A., Bontempi, E., Pellegatta, M., Vigna, L., Patrini, L., Riboldi, L., Rubino, F. M. & Depero, L. E. (2010). Use of total reflection X-ray fluorescence (TXRF) for the evaluation of heavy metal poisoning due to the improper use of a traditional ayurvedic drug. *Journal of Pharmaceutical and Biomedical Analysis, 52*, 787–790.

Borkowska-Burnecka, J. (2000). Microwave assisted extraction for trace element analysis of plant materials by ICP-AES. *Fresenius Journal of Anaytical Chemistry, 368*, 633–637.

Brady, D. V., Montalvo Jr., J. G., Glowacki, G. & Pisciotta, A. (1974). Direct determination of zinc in sea-bottom sediments by carbon tube atomic absorption spectrometry. *Analytica Chimica Acta, 70*, 448–452.

Brancalion, M. L., Aurelio, M. & Arruda, Z. (2005). Evaluation of medicinal plant decomposition efficiency using microwave ovens and mini-vials for Cd determination by TS-FF-AAS. *Microchimica Acta, 150*, 283–290.

Bredendiek-Kämper, S., von Bohlen, A., Klockenkämper, R., Quentmeier, A. & Klockow, D. (1996). Microanalysis of solid samples by laser ablation and total reflection X-ray fluorescence. *Journal of Analytical Atomic Spectroscopy, 11*, 537–541.

Brunori, C., Ipolyi, I., Macaluso, L. & Morabito, R. (2004). Evaluation of an ultrasonic digestion procedure for total metal determination in sediment reference materials, *Analytica Chimica Acta, 510*, 101–107.

Burguera, J. L. & Burguera, M. (2012). Analytical applications of emulsions and miroemulsions. *Talanta, 96*, 11–20.

Cabaleiro, N., Pena-Pereira, F., De La Calle, I., Bendicho, C., & Lavilla, I. (2011). Determination of triclosan by cuvetteless UV–vis micro-spectrophotometry following simultaneous ultrasound assisted emulsification–microextraction with derivatization: Use of a micellar-ionic liquid as extractant. *Microchemical Journal, 99*, 246–251

Caballo, C., Sicilia, M. D. & Rubio, S. (2014). Fast, simple and efficient supramolecular solvent-based microextraction of mecoprop and dichlorprop in soils prior to their enantioselective determination by liquid chromatography–tandem mass spectrometry. *Talanta, 119*, 46–52.

Caballo-López, A. & Luque de Castro, M. D. (2007). Determination of cadmium in leaves by ultrasound-assisted extraction prior to hydride generation, pervaporation and atomic absorption detection. *Talanta, 71*, 2074–2079.

Cabañero, A. I., Madrid, Y. & Cámara, C. (2005). Enzymatic probe sonication extraction of Se in animal-based food samples: a new perspective on sample preparation for total and Se speciation analysis. *Analytical and Bioanalytical Chemistry, 381*, 373–379.

Cal-Prieto, M. J., Felipe-Sotelo, M., Carlosena, A., Andrade, J. M., López-Mahía, P., Muniategui, S. & Prada, D. (2002). Slurry sampling for direct analysis of solid materials by electrothermal atomic absorption spectrometry (ETAAS). A literature review from 1990 to 2000. *Talanta, 56*, 1–51.

Camel, V. (2000). Microwave-assisted solvent extraction of environmental samples. *TrAC Trends in Analytical Chemistry. 19*, 229–248.

Camel, V. (2003). Solid phase extraction of trace elements. *Spectrochimica Acta Part B: Atomic Spectroscopy, 58*, 1177–1233.

Campillo, N., Viñas, P., Martínez-Castillo, N. & Hernández-Córdoba, M. (2011). Determination of volatile nitrosamines in meat products by microwave-assisted extraction and dispersive liquid–liquid microextraction coupled to gas chromatography–mass spectrometry. *Journal of Chromatography A, 1218*, 1815–1821.

Campíns-Falcó, P., Verdú-Andrés, J., Sevillano-Cabeza, A., Molins-Legua, C., Herráez-Hernández, R. (2008). New micromethod combining miniaturized matrix solid-phase dispersion and in-tube in-valve solid-phase microextraction for estimating polycyclic aromatic hydrocarbons in bivalves. *Journal of Chromatography A, 1211*, 13–21.

Canepari, S., Cardarelli, E., Ghighi, S. & Scimonelli, L. (2005). Ultrasound and microwave-assisted extraction of metals from sediment: a comparison with the BCR procedure. *Talanta, 66*, 1122–1130.

Capelo, J. L., Lavilla, I. & Bendicho, C. (1998). Ultrasound-assisted extraction of cadmium from slurried biological samples for electrothermal atomic absorption spectrometry. *Journal of Analytical Atomic Spectrometry, 13*, 1285–1290.

Capelo, J. L., Lavilla, I. & Bendicho, C. (1999). Solid-liquid extraction of copper from slurried samples using high intensity probe sonication for electrothermal atomic absorption spectrometry. *Talanta, 50*, 905–911.

Capelo, J. L., Maduro, C. & Vilhena, C. (2005). Discussion of parameters associated with the ultrasonic solid–liquid extraction for elemental analysis (total content) by electrothermal atomic absorption spectrometry. An overview. *Ultrasonics Sonochemistry, 12*, 225–232.

Capelo, J. L., Ximénez-Embún, P., Madrid-Albarrán, Y. & Cámara, C. (2004). Enzymatic probe sonication: Enhancement of pProtease-catalyzed hydrolysis of selenium bound to proteins in yeast. *Analytical Chemistry, 76*, 233–237.

Capitelli, F., Colao, F., Provenzano, M. R., Fantoni, R., Brunetti G. & Senesi. N. (2002). Determination of heavy metals in soils by Laser induced breakdown spectroscopy. *Geoderma, 106*, 45–62.

Capriotti, A. L., Cavaliere, C., Giansanti, P., Gubbiotti, R., Samperi, R. & Lagana, A. (2010). Recent developments in matrix solid-phase dispersion extraction. *Journal of Chromatography A, 1217*, 2521–2532.

Capriotti, A. L., Cavaliere, C., Lagana, A., Piovesana, S. & Samperi, R. (2013). Recent trends in matrix solid-phase dispersion. *TrAC Trends in Analytical Chemistry, 43*, 53–66.

Carballo, S., Terán, J., Soto, R. M., Carlosena, A., Andrade, J. M. & Prada, D. (2013). Green approaches to determine metals in lubricating oils by electrothermal atomic absorption spectrometry (ETAAS). *Microchemical Journal, 108*, 74–80.

Carballo-Paradelo, S., Soto-Ferreiro, R. M., Carlosena-Zubieta, A., Terán-Baamonde, J., Andrade-Garda, J. M. & Prada-Rodríguez, D. (2012). Pressurized liquid extraction to determine Mg, Al, Ti, Cu, Sn and Pb in lubricating oils by inductively coupled plasma mass spectrometry. *Journal of Analytical Atomic Spectrometry, 27*, 1694–1700.

Carpenter, R. C., (1981) The determination of cadmium, copper, lead and thallium in human liver and kidney tissue by flame atomic absorption spectrometry after enzymatic digestion. *Analytica Chimica Acta, 125,* 209–213.

Carpinteiro, J., Rodríguez, I. & Cela, R. (2004). Applicability of solid-phase microextraction combined with gas chromatography atomic emission detection (GC-MIP AED) for the determination of butyltin compounds in sediment samples. *Analytical and Bioanalytical Chemistry, 380,* 853–857.

Carrillo, J. D. & Tena, M. T. (2006). Determination of volatile compounds in antioxidant rosemary extracts by multiple headspace solid-phase microextraction and gas chromatography. *Flavour and Fragrance Journal, 21,* 626–633.

Carvalho-Santos, W. P., Teixeira-Castro, J., Almeida-Bezerra, M., Pires-Fernandes, A., Costa-Ferreira, S. L. & Andrade-Korn, M. G. (2009). Application of multivariate optimization in the development of an ultrasound-assisted extraction procedure for multielemental determination in bean seeds samples using ICP OES. *Microchemical Journal, 91,* 153–158.

Chailapakul, O., Korsrisakul, S., Siangproh, W. & Grudpan, K. (2008). Fast and simultaneous detection of heavy metals using a simple and reliable microchip-electrochemistry route: An alternative approach to food analysis. *Talanta, 74,* 683–689.

Chang, I.-H., Tulock, J. J., Liu, J., Kim, W.–S., Cannon Jr., D. M., Lu, Y., Bohn, P. W., Sweedler , J. V. & Cropek, D. M. (2005). Miniaturized lead sensor based on lead-specific DNAzyme in a nanocapillary interconnected microfluidic device. *Environmental Science & Technology, 39,* 3756–3761.

Chao, Y., Jian, Z., Tu, Y. & Huang, Y. (2013). Rapid on-line microextraction of neutral analytes in plastic-bottled beverages through ultrasound-assisted push/pull perfusion hollow-fiber liquid–liquid–liquid microextraction. *Microchemical Journal, 110,* 386–394.

Chen, S., Liu, C., Yang, M., Lu, D., Zhu, L. & Wang, Z. (2009). Solid-phase extraction of Cu, Co and Pb on oxidized single-walled carbon nanotubes and their determination by inductively coupled plasma mass spectrometry. *Journal of Hazardous Materials, 170,* 247–251.

Chen, X. & Marshall, W. D. (1999). Enzymatic digestion-high-pressure homogenization prior to slurry introduction graphite furnace atomic absorption spectrometry for selenium determination in plant and animal tissues. *Journal of Agricultural and Food Chemistry, 47,* 3727–3732.

Chen, X., Zhou, L., Zhao, Y., Pan, S. & Jin, M. (2014). Application of nanoring amino-functionalized magnetic polymer dispersive micro-solid-phase extraction and ultrafast liquid chromatography–tandem mass spectrometry in dicyandiamide residue analysis of powdered milk. *Talanta, 119,* 187–192.

Chen, X.-W. & Wang, J.-H. (2007). The miniaturization of bioanalytical assays and sample pretreatments by exploiting meso-fluidic lab-on-valve configurations: A review. *Analytica Chimica Acta, 602,* 173–180.

Choi, J. H., Mamun, M. I. R., Abd El-Aty, A. M., Kim, K. T., Koh, H. B., Shin, H. C., Kim, J., Lee, K. B. & Shim, J. H. (2009). Inert matrix and Na_4EDTA improve the supercritical fluid extraction efficiency of fluoroqulnolones for HPLC determination in pig tissues. *Talanta, 78,* 348–357.

Claisse F. (2003). Fusion and fluxes In Z. Mester & R. Sturgeon (Eds.) *Sample preparationn for trace element analysis*, (pp. 301–312). The Netherlands: Elsevier.

Clement, R.-E. & Hao, C. (2010). Sample preparation techniques for environmental organic pollutant analysis. In J. Pawliszyn & H. L. Lord (Eds.) *Handbook of sample preparation* (pp. 249–265), Hoboken, N.J.: John Wiley &Sons, Inc.

Colombini, V., Bancon-Montigny, C., Yang, L., Maxwell, P., Sturgeon, R. E. & Mester, Z. (2004). Headspace single-drop microextraction for the detection of organotin compounds. *Talanta, 63,* 555–560.

Costas, M., Lavilla, I., Gil, S., Pena, F., de la Calle, I., Cabaleiro, N. & Bendicho, C. (2010). Evaluation of ultrasound-assisted extraction as sample pre-treatment for quantitative determination of rare earth elements in marine biological tissues by inductively coupled plasma-mass spectrometry. *Analytica Chimica Acta, 679*, 49–55.

Covaci, A., Mashiya, J. G., Molope, I., Tshiame, I., Molatlhegi, R., Ngobeni, P. (2010). Miniaturized method based on matrix solid-phase dispersion for the rapid screening of 36 pesticides in agricultural food commodities. *Analytical Letters, 43*, 1400–1410.

Crescenzi, C., Di Corcia, A., Nazzari, M. & Samperi, R. (2000). Hot phosphate-buffered water extraction coupled on-line with liquid chromatography/mass spectrometry for analyzing contaminants in soil. *Analytical Chemistry, 72*, 3050–3055.

Cucheval, A. & Chow, R.C.Y. (2008). A study on the emulsification of oil by power ultrasound. *Ultrasonics Sonochemistry, 15*, 916–920.

Czégény, Z., Berente, B., Óvári, M., García-Tapia, M. & Záray, G. (1998). Microwave-assisted vapor-phase acid digestion of cellulose nitrate filters for elemental analysis of airborne dust samples. *Microchemical Journal, 59*, 100–106.

Damm, M. & Kappe, C. O. (2011). A high-throughput platform for low-volume high-temperature/pressure sealed vessel solvent extractions. *Analytica Chimica Acta, 707*, 76–83.

Dantas, A. N. S., Matos, W. O., Gouveia, S. T. & Lopes, G. S. (2013). The combination of infrared and microwave radiation to quantify trace elements inorganic samples by ICP-OES. *Talanta, 107*, 292–296.

Davidson, C. M. & Delevoye, G. (2001). Effect of ultrasonic agitation on the release of copper, iron, manganese and zinc from soil and sediment using the BCR three-stage sequential extraction. *Journal of Environmental Monitoring, 3*, 398–403.

De Jesus, A., Silva, M. M. & Vale, M. G. R. (2008). The use of microemulsion for determination of sodium and potassium in biodiesel by flame atomic absorption spectrometry. *Talanta, 74*, 1378–1384.

De Jesus, R. M., Silva, L. O. B., Castro, J. T., de Azevedo-Neto, A. D., De Jesus, R. M. & Ferreira, S. L. C. (2013). Determination of mercury in phosphate fertilizers by cold vapor atomic absorption spectrometry. *Talanta, 106*, 293–297.

De la Calle, I., Cabaleiro, N., Costas, M., Pena, F., Gil, S., Lavilla, I. & Bendicho, C. (2011a) Ultrasound-assisted extraction of gold and silver from environmental samples using different extractants followed by electrothermal-atomic absorption spectrometry. *Microchemical Journal, 97*, 93–100.

De la Calle, I., Cabaleiro, N., Lavilla, I. & Bendicho, C. (2009). Analytical evaluation of a cup-horn sonoreactor used for ultrasound-assisted extraction of trace metals from troublesome matrices. *Spectrochimica Acta Part B: Atomic Spectroscopy, 64*, 874–883.

De la Calle, I., Cabaleiro, N., Lavilla, I. & Bendicho, C. (2013a). Ultrasound-assisted single extraction tests for rapid assessment of metal extractability from soils by total reflection X-ray fluorescence. *Journal of Hazardous Materials, 260*, 202– 209.

De la Calle, I., Cabaleiro, N., Romero, V., Lavilla, I. & Bendicho, C. (2013b). Sample pretreatment strategies for total reflection X-ray fluorescence analysis: A tutorial review. *Spectrochimica Acta Part B: Atomic Spectroscopy, 90*, 23–54.

De la Calle, I., Costas, M., Cabaleiro, N., Lavilla, I. & Bendicho, C. (2012). Use of high-intensity sonication for pre-treatment of biological tissues prior to multielemental analysis by total reflection X-ray fluorescence spectrometry. *Spectrochimica Acta Part B: Atomic Spectroscopy, 67*, 43–49.

De la Calle, I., Costas, M., Cabaleiro, N., Lavilla, I. & Bendicho, C. (2013c). Fast method for multielemental analysis of plants and discrimination according to the anatomical part by total reflection X-ray fluorescence spectrometry. *Food Chemistry 138*, 234–241.

De la Calle, I., Pena, F., Cabaleiro, N., Lavilla, I. & Bendicho, C. (2011b). Ion pair-based dispersive liquid–liquid microextraction for gold determination at ppb level in solid samples after ultrasound-assisted extraction and in waters by electrothermal-atomic absorption spectrometry. *Talanta, 84,* 109–115.

De la Guardia, M. & Armenta, S. (2011). Avoiding Sample treatments. In D. Barceló (Ed.), *Green Analytical Chemistry: Theory & Practice* (pp. 59–86). Oxford: Elsevier.

De Oliveira, E. (2003). Sample Preparation for Atomic Spectroscopy: Evolution and Future Trends, *Journal of Brazilian Chemical Society, 14,* 174–182.

De Paula, C. E. R., Caldas, L. F. S., Brum, D. M. & Cassella, R. J. (2012). Development of an ultrasonic slurry sampling method for the determination of Cu and Mn in antibiotic tablets by electrothermal atomic absorption Spectrometry. *Journal of Pharmaceutical and Biomedical Analysis, 66,* 197– 203.

De Vevey, E., Bitton, G., Rossel, D., Ramos, L. D., Munguia-Gerrero, L. & Tarradellas, J. (1993). Concentration and bioavailability of heavy metals in sediments in lake yojoa (honduras). *Bulletin of Environmental Contamination and Toxicology, 50,* 253–259.

Deaker, M. & Maher, W. (1997). Low volume microwave digestion for the determination of selenium in marine biological tissues by graphite furnace atomic absorption spectroscopy. *Analytica Chimica Acta, 350,* 287–294.

Deaker, M. & Maher, W. (1999). Determination of arsenic in arsenic compounds and marine biological tissues using low volume microwave digestion and electrothermal atomic absorption spectrometry. *Journal of Anaytical Atomic Spectrometry, 14,* 1193–1207.

Del Castilho, P. & Breder, R. (1997). Soils and soil solutions. In M. Stoeppler (Ed.) *Sampling and sample preparation* (pp. 43–56). Germany: Springer-Verlag Berlin Heidelberg.

Delgado, A., Usobiaga, A., Prieto, A., Zuloaga, O., de-Diego, A. & Madariaga, J. M., (2008). Optimisation of the headspace-solid phase microextraction for organomercury and organotin compound determination in sediment and biota. *Journal of Separation Science, 31,* 768–774.

Deng, G. & Collins, G. E. (2003). Nonaqueous based microchip separation of toxic metal ions using 2-(5-bromo-2-pyridylazo)-5-(*N*-propyl-*N*-sulfopropylamino)phenol. *Journal of Chromatography A, 989,* 311–316.

Dobrowolski, R. & Otto, M. (2012). Determination of nickel and cobalt in reference plant materials by carbon slurry sampling GFAAS technique after their simultaneous preconcentration onto modified activated carbon. *Journal of Food Composition and Analysis, 26,* 58–65.

Dobrowolski, R. & Otto, M. (2013). Preparation and evaluation of Fe-loaded activated carbon for enrichment of selenium for analytical and environmental purposes. *Chemosphere, 90,* 683–690.

Dobrowolski, R., Adamczyk, A. & Otto, M. (2013). Determination of vanadium in soils and sediments by the slurry sampling graphite furnace atomic absorption spectrometry using permanent modifiers. *Talanta, 113,* 19–25.

Dobrowolski, R., Kurylo, M., Otto, M. & Mróz, A. (2012). Determination of gold in geological materials by carbon slurry sampling graphite furnace atomic absorption spectrometry. *Talanta, 99,* 750–757.

Doménech-Carbó, A., Labuda, J., Scholz, F. (2013). Electroanalytical chemistry for the analysis of solids: Characterization and classification (IUPAC Technical Report). *Pure and Applied Chemistry, 85,* 609–631.

Domeño, C., Blasco, M., Sánchez, C. & Nerín, C. (2006). A fast extraction technique for extracting polycyclic aromatic hydrocarbons (PAHs) from lichens samples used as biomonitors of air pollution: Dynamic sonication versus other methods. *Analytica Chimica Acta, 569,* 103–112.

Domeño, C., Munizza, G. & Nerín, C. (2005). Development of a solid-phase microextraction method for direct determination of pentachlorophenol in paper and board samples: Comparison with conventional extraction method. *Journal of Chromatography A, 1095,* 8–15.

Domínguez-González, R., Moreda-Piñeiro, A., Bermejo-Barrera, A. & Bermejo-Barrera, P. (2005). Application of ultrasound-assisted acid leaching procedures for major and trace elements determination in edible seaweed by inductively coupled plasma-optical emission spectrometry. *Talanta, 66*, 937–942.

Donard, O. F. X., Lalere, B., Martin, F. & Lobinski, R., (1995). Microwave-assisted leaching of organotin compounds from sediments for speciation analysis. *Analytical Chemistry, 67,* 4250–4254.

Du, W., Lei, C., Zhang, S., Bai, G., Zhou, H., Sun, M., Fu, Q. & Chang, C. (2014). Determination of clenbuterol from pork samples using surface molecularly imprinted polymers as the selective sorbents for microextraction in packed syringe. *Journal of Pharmaceutical and Biomedical Analysis, 91,* 160–168.

Duarte, F. A., Pereira, J. S. F., Barin, J. S., Mesko, M. F., Dressler, V. L., Flores, E. M. D. & Knapp, G. (2009). Seafood digestion by microwave-induced combustion for total arsenic determination by atomic spectrometry techniques with hydride generation. *Journal of Analytical Atomic Spectrometry, 24*, 224–227.

Ducree, J., Haeberle, S., Lutz, S., Pausch, S., von Stetten, F. & Zengerle, R. (2007). The centrifugal microfluidic bio-disk platform. *Journal of Micromechanics and Microengineering, 17*, S103–S115.

Duford, D. A., Peng, D. D. & Salin, E. D. (2009). Magnetically driven solid sample preparation for centrifugal microfluidic devices. *Analytical Chemistry, 81,* 4581–4584.

Ebdon, L., Fisher, A.S., Betti, M. & Leroy, M. (2003). Detection methods for the quantitation of trace elements. In Z. Mester & R. Sturgeon (Eds.), *Sample preparation for trace element analysis* (pp. 117–190). The Netherlands: Elsevier.

Eilola, K. & Perämäki., P. (2001). Microwave heated vapor-phase digestion method for biological sample materials. *Fresenius Journal Analytical Chemistry, 369,* 107–112.

Eilola, K. & Perämäki., P. (2003). Development of a modified medium pressure microwave vapor-phase digestion method for difficult to digest organic samples. *The Analyst, 128,* 194–197.

Eilola, K. & Perämäki., P. (2009). Microwave-assisted double insert vapour-phase digestion of organic samples. *Analytica Chimica Acta, 634*, 205–208.

El Azouzi, H., Cervera, M. L. & de la Guardia, M. (1998). Multi-elemental analysis of mussel samples by atomic absorption spectrometry after room temperature sonication. *Journal of Analytical Atomic Spectrometry, 13*, 533–538.

El-Amrani, S., Pena-Abaurrea, M., Sanz-Landaluze, J., Ramos, L., Guinea, J. & Cámara, C. (2012). Bioconcentration of pesticides in Zebrafish eleutheroembryos (Daniorerio). *Science of Total Environment, 425,* 184–190.

Elik, A. (2005). Ultrasound assisted pseudo-digestion of street dust samples prior to determination by atomic absorption spectrometry. *Talanta, 66*, 882–888.

Esslemont, G., Maher, W., Ford, P. & Krikowa, F. (2000). The determination of phosphorus and other elements in plant leaves by ICP-MS after low-volume microwave digestion with nitric acid. *Atomic Spectroscopy, 21*, 42–45.

Felton, M. J. (2003). Lab on a chip: poised on the brink. *Analytical Chemistry, 75,* 505A-508A.

Ferreira, S. L. C. , Miró, M., da Silva, E. G. P., Matos, G. D., dos Reis, P. S., Cardoso-Brandao, G. C., dos Santos, W. N. L., Tavares-Duarte, A. T., Rodrigues-Vale, M. G. R. & Araujo, R. G. A. (2010). Slurry sampling-an analytical strategy for the determination of metals and metalloids by spectroanalytical techniques. *Applied Spectroscopy Reviews, 45*, 44–62.

Ferreira-Damin, I. C., Zmozinski, A. V., Borges, A. R., Vale, M. G. R. & da Silva, M. M. (2011) Determination of cadmium and lead in fresh meat by slurry sampling graphite furnace atomic absorption spectrometry. *Analytical Methods, 3*, 1379–1385.

Figueredo-Rego, J., Virgilio, Nóbrega, J. A. & Gomes-Neto, J. A. (2012). Determination of lead in medicinal plants by high-resolution continuum source graphite furnace atomic absorption spectrometry using direct solid sampling. *Talanta, 100*, 21–26.

Filgueiras, A. V., Capelo, J. L., Lavilla, I. & Bendicho, C. (2000). Comparison of ultrasound-assisted extraction and microwave-assisted digestion for determination of magnesium, manganese and zinc in plant samples by flame atomic absorption spectrometry. *Talanta, 53*, 433–441.

Filgueiras, A. V., Lavilla, I. & Bendicho, C. (2001). Ultrasound-assisted solubilization of trace and minor metals from plant tissue using ethylenediaminetetraacetic acid in alkaline medium. *Fresenius Journal Analytical Chemistry, 369*, 451–456.

Filgueiras, A. V., Lavilla, I. & Bendicho, C. (2002a). Chemical sequential extraction for metal partitioning in environmental solid samples. *Journal of Environmental Monitoring, 4*, 823–857.

Filgueiras, A. V., Lavilla, I. & Bendicho, C. (2002b). Comparison of the standard SM&T sequential extraction method with small-scale ultrasound-assisted single extractions for metal partitioning in sediments. *Analytical and Bioanalytical Chemistry, 374*, 103–108.

Flores, E. M. M., Barin, J. S., Mesko, M. F. & Knapp, G. (2007). Sample preparation techniques based on combustion reactions in closed vessels — A brief overview and recent applications. *Spectrochimica Acta Part B: Atomic Spectroscopy, 62*, 1051–1064.

Flores, E.M. (1997); *Ph.D Thesis*, Universidade Federal do Rio Grande do Sul, Porto Alegre.

Gallagher, P. A., Shoemaker, J. A., Wei, X., Brockhoff-Schwegel, C. A. & Creed, J. T. (2001). Extraction and detection of arsenicals in seaweed *via* accelerated solvent extraction with ion chromatographic separation and ICP-MS detection. *Fresenius Journal Analytical Chemistry, 369*, 71–80.

Garrigues, S. & de la Guardia, M. (2013). Non-invasive analysis of solid samples. *TrAC Trends in Analytical Chemistry, 43*, 161–173.

Gil, S., Fragueiro, S., Lavilla, I. & Bendicho, C. (2005). Determination of methylmercury by electrothermal atomic absorption with *in situ* hydride generation. *Spectrochimica Acta Part B: Atomic Spectroscopy, 60*, 145–150.

Gleit, C. E. & Holland, W. D. (1962). Use of electrically excited oxygen for the low temperature decomposition of organic substances. *Analytical Chemistry, 34*, 1454–1457.

Goenaga-Infante, H., O'Connor, G., Rayman, M., Wahlen, R., Entwisle, J., Norris, P., Hearn, R. & Catterick, T. (2004). Selenium speciation analysis of selenium-enriched supplements by HPLC with ultrasonic nebulisation ICP-MS and electrospray MS/MS detection. *Journal of Analytical Atomic Spectrometry, 19*, 1529–1538.

Goenaga-Infante, H., O'Connor, G., Rayman, M., Wahlen, R., Spallholz, J. E., Hearn, R. & Catterick, T. (2005). Identification of water-soluble gamma-glutamyl-Se-methylselenocysteine in yeast-based selenium supplements by reversed-phase HPLC with ICP-MS and electrospray tandem MS detection. *Journal of Analytical Atomic Spectrometry, 20*, 864–870.

González-Curbelo, M. A., Hernández-Borges, J., Borges-Miquel, T. M. & Rodríguez-Delgado, M. A. (2013). Determination of organophosphorus pesticides and metabolites in cereal-based baby foods and wheat flour by means of ultrasound-assisted extraction and hollow-fiber liquid-phase microextraction prior to gas chromatography with nitrogen phosphorus detection. *Journal of Chromatography A, 1313*, 166–174.

Greenway, G. M. & Song, Q. J. (2002). An ultrasound accelerated sequential extraction method and its application for element partitioning studies in compost from mixed waste streams. *Journal of Environmental Monitoring, 4*, 950–955.

Grinberg, P., Campos, R. C., Mester, Z. & Sturgeon, R. E. (2003). A comparison of alkyl derivatization methods for speciation of mercury based on solid phase microextraction gas chromatography with furnace atomization plasma emission spectrometry detection. *Journal of Analytical Atomic Spectrometry, 18*, 902–909.

Guan, H., Brewer, W. E., Garris, S. T. & Morgan. S. L. (2010). Disposable pipette extraction for the analysis of pesticides in fruit and vegetables using gas chromatography/mass spectrometry. *Journal of Chromatography A, 1217*, 1867–1874.

Günther, D. & Hattendorf, B. (2005). Solid sample analysis using laser ablation inductively coupled plasma mass spectrometry. *TrAC Trends in Analytical Chemistry, 24*, 255- 265.

Guo, L. & Kee Lee, H. (2013). Microwave assisted extraction combined with solvent bar microextraction for one-step solvent-minimized extraction, cleanup and preconcentration of polycyclic aromatic hydrocarbons in soil samples. *Journal of Chromatography A, 1286,* 9–15,

Gutiérrez Valencia, T. M. & García de Llasera, M. P. (2011). Determination of organophosphorus pesticides in bovine tissue by an on-line coupled matrix solid-phase dispersion-solid phase extraction-high performance liquid chromatography with diode array detection method. *Journal of Chromatography A, 1218,* 6869- 6877.

Halls, D. J. (1995). Analytical minimalism applied to the determination of trace elements by atomic spectrometry. *Journal of Analytical Atomic Spectrometry, 10*, 169–175.

Hansson, E. & Hakkarainen, M. (2006). Multiple headspace single-drop microextraction—a new technique for quantitative determination of styrene in polystyrene. *Journal of Chromatography A, 1102,* 91–95.

Hawthorne, S. B., Trembley, S., Moniot, C. L., Grabanski, C. B. & Miller, D. J. (2000). Static subcritical water extraction with simultaneous solid-phase extraction for determining polycyclic aromatic hydrocarbons on environmental solids. *Journal of Chromatography A, 886,* 237–244.

He, D., Zhang, Z., Huang, Y. & Hu, Y. (2007). Chemiluminescence microflow injection analysis system on a chip for the determination of nitrite in food. *Food Chemistry, 10*, 667–672.

Herrero-Latorre, C., Alvarez-Mendez, J., Barciela-Garcia, J., Garcia-Martin, S. & Peña-Crecente, R. M. (2012). Carbon nanotubes as solid-phase extraction sorbents prior to atomic spectrometric determination of metal species: A review. *Analytica Chimica Acta, 749*, 16– 35.

Hirata, S. & Toshimitsu, H. (2005). Determination of arsenic species and arsenosugars in marine samples by HPLC–ICP–MS. *Analytical and Bioanalytical Chemistry, 383*, 454–460.

Hodson, M. J., White, P.J., Mead, A. & Broadley, M. R. (2005). Phylogenetic variation in the silicon composition of plants. *Annals of Botany, 96*, 1027–1046.

Hoenig, M. & de Kersabiec, A.-M. (1996). Review Sample preparation steps for analysis by atomic spectroscopy methods: present status. *Spectrochimica Acta Part B: Atomic Spectroscopy, 51,* 1297–1307.

Hoenig, M. (2001). Preparation steps in environmental trace element analysis-facts and traps. *Talanta, 54*, 1021–1038.

Hoenig, M. (2003). Dry ashing. In Z. Mester & R. Sturgeon (Eds.), *Sample preparation of geological samples, soils and sediment* (pp. 235–255). The Netherlands: Elsevier.

Hoenig, M., Baeten, H., Vanhentenrijk, S., Vassileva, E. & Quevauviller, P. (1998). Critical discussion on the need for an efficient mineralization procedure for the analysis of plant material by atomic spectrometric methods. *Analytica Chimica Acta, 358*, 85–94.

Hong, Y. & Chen, L. (2013). Extraction of quercetin from Herba Lysimachiae by molecularly imprinted-matrix solid phase dispersion, *Journal of Chromatography B: Analytical Technologies in the Biomedical and Life Sciences, 941,* 38–44.

Horváth, M., Halász, G., Kucanová, E., Kuciková, B., Fekete, I., Remeteiová, D., Heltai, G. & Flórián, K. (2013). Sequential extraction studies on aquatic sediment and biofilm samples for the assessment of heavy metal mobility. *Microchemical Journal, 107*, 121–125.

Hristozov, D., Domini, C. E., Kmetov, V., Stefanova, V., Georgieva, D. & Canals, A. (2004). Direct ultrasound-assisted extraction of heavy metals from sewage sludge samples for ICP-OES analysis. *Analytica Chimica Acta, 516*, 187–196.

Hsu, W.-H., Jiang, S.-J., Sahayam A. C. (2013). Determination of Pd, Rh, Pt, Au in road dust by electrothermal vaporization inductively coupled plasma mass spectrometry with slurry sampling. *Analytica Chimica Acta, 794*, 15– 19.

Huang, J., Liu, J., Zhang, C., Wei, J., Mei, L., Yu, S., Li, G. & Xu, L. (2012). Determination of sulfonamides in food samples by membrane-protected micro-solid phase extraction coupled with high performance liquid chromatography. *Journal of Chromatography A, 1219,* 66–74.

Huang, M., Chen, H., Fu, S. & Ding, W. (2013). Determination of volatile N-nitrosamines in meat products by microwave-assisted extraction coupled with dispersive micro solid-phase extraction and gas chromatography -chemical ionisation mass spectrometry. *Food Chemistry, 138,* 227–233.

Huerga, A., Lavilla, I. & Bendicho, C. (2005). Speciation of the immediately mobilisable As(III), As(V), MMA and DMA in river sediments by high performance liquid chromatography–hydride generation–atomic fluorescence spectrometry following ultrasonic extraction. *Analytica Chimica Acta, 534,* 121–128.

Hwang, S.–S., Park, J.-S. & Namkoong, W. (2007). Ultrasonic-assisted extraction to release heavy metals from contaminated soil. *Journal of Industrial and Engineering Chemistry, 13,* 650–656.

Hyötyläinen, T. (2009). Critical evaluation of sample pretreatment techniques. *Analytical and Bioanalytical Chemistry, 394,* 743–758.

Hyötyläinen, T., Andersson, T, Hartonen, K., Kuosmanen, K. & Riekkola, M. L. (2000). Pressurized hot water extraction coupled on-line with LC-GC: Determination of polyaromatic hydrocarbons in sediment. *Analytical Chemistry, 72,* 3070–3076.

Iavicoli, I., Castellino, N., Carelli, G. & Schlemmer, G. (2001). Lead determination in ultra micro samples of animal diet using graphite furnace atomic absorption spectrometry. *Atomic Spectroscopy, 22,* 319–323.

Ibarlucea, B., Díez-Gil, C., Ratera, I., Veciana, J., Caballero, A., Zapata, F., Tárraga, A., Molina, P., Demming, S., Büttgenbach, S., Fernández-Sánchez, C. & Llobera, A. (2013). PDMS based photonic lab-on-a-chip for the selective optical detection of heavy metal ions. *The Analyst, 138,* 839–844.

Ieggli, C. V. S., Bohrer D., do Nascimento, P. C., de Carvalho, L. M. & Garcia, S. C. (2010). Determination of sodium, potassium, calcium, magnesium, zinc, and iron in emulsified egg samples by flame atomic absorption spectrometry. *Talanta, 80,* 1282–1286.

Ihnat M. (2003). Sample preparation for food analysis In Z. Mester & R. Sturgeon (Eds.), *Sample preparation for trace element analysis* (pp. 765–855). The Netherlands: Elsevier.

Ilander, A. & Väisänen, A. (2007). An ultrasound-assisted digestion method for the determination of toxic element concentrations in ash samples by inductively coupled plasma optical emission spectrometry. *Analytica Chimica Acta, 602,* 195–201.

Ilander, A. & Väisänen, A. (2009). The determination of trace element concentrations in fly ash samples using ultrasound-assisted digestion followed with inductively coupled plasma optical emission spectrometry. *Ultrasonics Sonochemistry, 16,* 763–768.

Intima, D. P., de Oliveira, E. & Oliveira, P. V. (2009). Feasibility of using *in situ* fusion for the determination of Co, Cr and Mn in Portland cement by direct solid sampling graphite furnace atomic absorption spectrometry. *Spectrochimica Acta Part B: Atomic Spectroscopy, 64,* 610–614.

Ipolyi, I., Brunori, C., Cremisini, C., Fodor, P., Macaluso, L. & Morabito, R. (2002). Evaluation of performance of time-saving extraction devices in the BCR three-step sequential extraction procedure. *Journal of Environmental Monitoring, 4,* 541–548.

Jamali, M. K., Kazi, T. G., Arain, M. B., Afridi, H. I., Jalbani, N., Kandhro, G. A., Shah, A. Q. & Baig, J.A. (2009). Speciation of heavy metals in untreated sewage sludge by using microwave assisted sequential extraction procedure. *Journal of Hazardous Materials, 163,* 1157–1164.

Jiao, J., Ma, D., Gai, Q., Wang, W., Luo, M., Fu, Y. & Ma, W. (2013). Rapid analysis of Fructusforsythiae essential oil by ionic liquids-assisted microwave distillation coupled with headspace single-drop microextraction followed by gas chromatography–mass spectrometry. *Analytica Chimica Acta, 804,* 143–150.

Kadenkin, A., Amberger, M., Fittschen, U. E. A. & Broekaert, J. A. C. (2011). Micro-homogeneity studies of boron carbide powders. *Microchimica Acta, 173,* 183–188.

Karageorgou, E. & Samanidou, V. (2010). Application of ultrasound-assisted matrix solid-phase dispersion extraction to the HPLC confirmatory determination of cephalosporin residues in milk. *Journal of Separation Science, 33,* 2862–2871.

Karageorgou, E. & Samanidou, V. (2011). Development and validation according to European Union Decision 2002/657/EC of an HPLC–DAD method for milk multiresidue analysis of penicillins and amphenicols based on dispersive extraction by QuEChERS in MSPD format. *Journal of Separation Science, 34,* 1893–1901.

Karageorgou, E., Armeni, M., Moschou, I. & Samanidou, V. (2014). Ultrasound-assisted dispersive extraction for the high pressure liquid chromatographic determination of tetracyclines residues in milk with diode array detection, *Food Chemistry, 150,* 328–334.

Karageorgou, E., Myridakis, A., Stephanou, E. & Samanidou, V. (2013). Multi-residue LC-MS/MS analysis of cephalosporins and quinolones in milk following ultrasound assisted matrix solid phase dispersive extraction combined with QuEChERS methodology. *Journal of Separation Science, 36,* 2020–2027.

Karageorgou, E., Samanidou, V. & Papadoyannis, I. (2012). Ultrasound-assisted matrix solid phase dispersive extraction for the simultaneous analysis of blactams (four penicillins and eight cephalosporins) in milk by high performance liquid chromatography with photodiode array detection. *Journal of Separation Science, 35,* 2599–2607.

Kazi, T. G., Jalbani, N., Arain, M. B., Jamali, M. K., Afridi, H. I. & Shah, A. Q. (2009a). Determination of toxic elements in different brands of cigarette by atomic absorption spectrometry using ultrasonic assisted acid digestion. *Environmental Monitoring and Assessment, 154,* 155–167.

Kazi, T. G., Jalbani, N., Baig, J. A., Arain, M. B., Afridi, H. I., Jamali, M. K., Shah A. Q. & Memon, A. N. (2010). Evaluation of toxic elements in baby foods commercially available in Pakistan. *Food Chemistry, 119,* 1313–1317.

Kazi, T. G., Jamali, M. K., Arain, M. B., Afridi, H. I., Jalbani, N., Sarfraz, R. A. & Ansari, R. (2009b). Evaluation of an ultrasonic acid digestion procedure for total heavy metals determination in environmental and biological samples. *Journal of Hazardous Materials, 161,* 1391–1398.

Kazi, T. G., Jamali, M. K., Siddiqui, A., Kazi, G. H., Arain, M. B. & Afridi, H. I. (2006). An ultrasonic assisted extraction method to release heavy metals from untreated sewage sludge samples. *Chemosphere, 63,* 411–420.

Ke, Y., Zhua, F., Jiang, R., Wang, Y., Yue, L., Liu, H., Zeng, F. & Ouyang, G. (2014). Determination of polycyclic aromatic hydrocarbons in leather products using solid-phase microextraction coupled with gas chromatography–mass spectrometry. *Microchemical Journal, 112,* 159–163.

Khuder, A., Sawan, M. K., Karjou, J., & Razouk, A. K. (2009). Determination of trace elements in Syrian medicinal plants and their infusions by energy dispersive X-ray fluorescence and total reflection X-ray fluorescence spectrometry. *Spectrochimica Acta Part B: Atomic Spectroscopy, 64,* 721–725.

Kim, N., Jung, M., Yoo, Z., Lee, S. & Lee, D. (2005). Headspace hanging drop liquid phase microextraction and GC-MS for the determination of linalool from evening primrose flowers. *Bulletin of the Korean Chemistry Society, 26,* 1996–2000.

Klitenick, M. A., Frederickson, C. J. & Manton, W. I. (1983). Acid-vapor decomposition for determination of zinc in brain tissue by isotope dilution mass spectrometry. *Analytical Chemistry, 55,* 921–923.

Kokosa, J., Przyjazny, A. & Jeannot, M. A. (2009). *Solvent microextraction; Theory and practice*, New Jersey: John Wiley & Sons Inc.

Kolmogorov, Y. P., Mezentsev, N. A., Mironov, A. G., Parkhomenko,V. S., Spiridonov, A. M., Shaporenko, A. D., Yusupov, T. S., Zhmodik, S. M., Zolotarev, K. V. & Anoshin, G. N. (2009). Development of a complex of instrumental nuclear-physical methods to detect PGE, Re, Au, and Ag in hard-to-analyze rocks and complex ores. *Nuclear Instruments and Methods in Physics Research Section A, 603,* 131–133.

Krasnodebska-Ostrega, B., Kaczorowska, M. & Golimowski, J. (2006). Ultrasound-assisted extraction for the evaluation of element mobility in bottom sediment collected at mining and smelting Pb–Zn ores area in Poland. *Microchimica Acta, 154,* 39–43.

Kristenson, E. M., Haverkate, E. G. J, Slooten, C. J., Ramos, L., Vreuls, R. J. J. & Brinkman, U. A. Th, (2001). Miniaturized automated matrix solid-phase dispersion extraction of pesticides in fruit followed by gas chromatographic–mass spectrometric analysis. *Journal of Chromatography A, 917,* 277–286.

Kristenson, E. M., Ramos, L. & Brinkman, U. A. T. (2006). Recent advances in matrix solid-phase dispersion. *TrAC Trends in Analytical Chemistry, 25,* 96–111.

Kulkarni, P., Chellam, S. & Mittlefehldt, D. W. (2007). Microwave-assisted extraction of rare earth elements from petroleum refining catalysts and ambient fine aerosols prior to inductively coupled plasma-mass spectrometry. *Analytica Chimica Acta, 581,* 247–259.

Lana, N. B., Berton, P., Covaci, A., Atencio, A. G., Ciocco, N. F. & Altamirano, J. C. (2013). Ultrasound leaching–dispersive liquid–liquid microextraction based on solidification of floating organic droplet for determination of polybrominateddiphenyl ethers in sediment samples by gas chromatography–tandem mass spectrometry. *Journal of Chromatography A, 1285,* 15–21.

Landajo, A., Arana, G., de Diego, A., Etxebarria, N., Zuloaga, O. & Amouroux, D. (2004). Analysis of heavy metal distribution in superficial estuarine sediments (estuary of Bilbao, Basque Country) by open-focused microwave-assisted extraction and ICP-OES. *Chemosphere, 56,* 1033–1041.

Lásztity, A., Krushevska, A., Kotrebai, M., Barnes, R. M. & Amarasiriwardena, D. (1995). Arsenic determination in environmental, biological and food samples by inductively coupled plasma mass spectrometry. *Journal of Analytical Atomic Spectrometry, 10,* 505–510.

Lavilla, I., Cabaleiro, N., Costas, M., de la Calle, I. & Bendicho C. (2009). Ultrasound-assisted emulsification of cosmetic samples prior to elemental analysis by different atomic spectrometric techniques. *Talanta, 80,* 109–116.

Lavilla, I., Cabaleiro, N., Pena, F., De La Calle, I. & Bendicho, C. (2010). Ultrasound-assisted emulsification microextraction with simultaneous derivatization coupled to fibre optics-based cuvetteless UV–vis micro-spectrophotometry for formaldehyde determination in cosmetic samples. *Analytica Chimica Acta, 674,* 59–63.

Lavilla, I., Costas, M., Gil, S., Corderi, S., Sanchez, G. & Bendicho, C. (2012). Simplified and miniaturized procedure based on ultrasound-assisted cytosol preparation for the determination of Cd and Cu bound to metallothioneins in mussel tissue by ICP-MS. *Talanta, 93,* 111–116.

Lavilla, I., Mosquera, A., Millos, J., Cameselle, J. & Bendicho, C. (2006a). Ultrasound-assisted extraction technique for establishing selenium contents in breast cancer biopsies by Zeeman-electrothermal atomic absorption spectrometry using multi-injection. *Analytica Chimica Acta, 566,* 29–36.

Lavilla, I., Vilas, P. & Bendicho, C. (2008). Fast determination of arsenic, selenium, nickel and vanadium in fish and shellfish by electrothermal atomic absorption spectrometry following ultrasound-assisted extraction. *Food Chemistry, 106,* 403–409.

Lavilla, I., Vilas, P., Millos, J. & Bendicho, C. (2006b). Development of an ultrasound-assisted extraction method for biomonitoring of vanadium and nickel in the coastal environment under the influence of the *Prestige* fuel spill (North east Atlantic ocean), *Analytica Chimica Acta, 577,* 119–125.

Lee, K. S., Shiddiky, M. J. A., Park, S. H., Park, D. S. & Shim, Y. B. (2008). Electrophoretic analysis of food dyes using a miniaturized microfluidic system. *Electrophoresis, 29,* 1910–1917.

Lee, T. P., Saad, B., Poh Ng, E. & Salleh, B. (2012). Zeolite Linde Type L as micro-solid phase extraction sorbent for the high performance liquid chromatography determination of ochratoxin A in coffee and cereal. *Journal of Chromatography A, 1237,* 46–54.

Lehmann, E. L., Fostier, A. H. & Arruda, M. A. Z. (2013). Hydride generation using a metallic atomizer after microwave-assisted extraction for inorganic arsenic speciation in biological samples. *Talanta, 104,* 187–192.

Lesniewicz, A. & Zyrnicki, W. (2003). Comparison of Conventional, Ultrasound- and microwave-assisted extraction techniques applied for the determination of metals in spruce needles. *International Journal of Environmental Analytical Chemistry, 83*, 735–748.

Ligor, T. & Buszewski, B. (2008). Single-drop microextraction and gas chromatography–mass spectrometry for the determination of volatile aldehydes in fresh cucumbers. *Analytical and Bioanalytical Chemistry, 391*, 2283–2289.

Lima, E. C., Barbosa Jr., F., Krug, F. J., Silva M. M. & Vale, M. G. R. (2000). Comparison of ultrasound-assisted extraction, slurry sampling and microwave-assisted digestion for cadmium, copper and lead determination in biological and sediment samples by electrothermal atomic absorption spectrometry. *Journal of Analytical Atomic Spectrometry, 15*, 995–1000.

Lin, F., Liu, D., Das, S. M., Prempeh, N., Hua, Y. & Lu, J. (2014). Recent progress in heavy metal extraction by supercritical CO_2 fluids. *Industrial & Engineering Chemistry Research, 53*, 1866–1877.

Lin, M.-L. & Jiang, S.-H. (2013). Determination of As, Cd, Hg and Pb in herbs using slurry sampling electrothermal vaporisation inductively coupled plasma mass spectrometry. *Food Chemistry, 141*, 2158–2162.

Liu, Q., Shi, J., Sun, J., Wang, T., Zeng, L., Zhu, N. & Jiang, G. (2011). Graphene-assisted matrix solid-phase dispersion for extraction of polybrominated diphenyl ethers and their methoxylated and hydroxylatedanalogs from environmental samples. *Analytica Chimica Acta, 708*, 61--68.

Liu, W., Hwang, B., Li, Z., Jen, J. & Lee, M. (2011). Headspace solid phase microextraction *in-situ* supercritical fluid extraction coupled to gas chromatography–tandem mass spectrometry for simultaneous determination of perfluorocarboxylic acids in sediments. *Journal of Chromatography A, 1218*, 7857–7863.

Liu, Y., Li, Y. & Yang, X.-P. (2008) Preparation, characterization, and application of L-cysteine functionalized multiwalled carbon nanotubes as a selective sorbent for separation and preconcentration of heavy metals. *Advanced Functional Materials, 18*, 1536–1543.

Liu, Y., Li, Y., Wu, Z.-Q. & Yan, X.-P. (2009). Fabrication and characterization of hexahistidine-tagged protein functionalized multi-walled carbon nanotubes for selective solid-phase extraction of Cu^{2+} and Ni^{2+}. *Talanta, 79*, 1464–1471.

Liva, M., Muñoz-Olivas, R. & Cámara C. (2000). Determination of Cd in sonicate slurries and leachates of biological and environmental materials by FI-CV-AAS. *Talanta, 51*, 381–387.

Long, X., Miró, M. & Hansen, E. H. (2005). Universal approach for selective trace metal determinations *via* sequential injection-bead injection-lab-on-valve sing renewable hydrophobic bead surfaces as reagent carriers. *Analytical Chemistry, 77*, 6032–6040.

López-García, I., Campillo, N., Arnau-Jerez, I. & Hernández-Córdoba, M. (2005). Electrothermal atomic absorption spectrometric determination of germanium in soils using ultrasound-assisted leaching. *Analytica Chimica Acta, 531*, 125–129.

Lu, Y. & Zhu, Y. (2013). Determination of chlorobenzenes in textiles by pressurized hot water extraction followed by vortex-assisted liquid–liquid microextraction and gas chromatography–mass spectrometry. *Journal of Chromatography A, 1319*, 27–34.

Lu, Y. & Zhu, Y. (2014). Combination of accelerated solvent extraction and vortex-assisted liquid–liquid microextraction for the determination of dimethyl fumarate in textiles and leathers by gas chromatography–mass spectrometry. *Talanta, 119*, 430–434.

Lu, Y., Shen, Q., Dai, Z., Zhang, H. & Wang, J. (2011). Development of an on-line matrix solid-phase dispersion/fast liquid chromatography/tandem mass spectrometry system for the rapid and simultaneous determination of 13 sulfonamides in grass carp tissues. *Journal of Chromatography A, 1218*, 929–937.

Lukkari, T., Taavitsainen, M., Vaäisänen, A. & Haimi, J. (2004). Effects of heavy metals on earthworms along contamination gradients in organic rich soils. *Ecotoxicology and Environmental Safety, 59*, 340–348.

Luque de Castro, M. D. & Luque-García, J. L. (2002). *Acceleration and automation of solid sample pretreatment*. Amsterdam: Elsevier.
Luque de Castro, M. D. & Priego-Capote, F. (2007). Ultrasound-assisted preparation of liquid samples. *Talanta, 72*, 321–334.
Luque-García, J. L. & Luque de Castro, M. D. (2003). Where is microwave-based analytical equipement for solid sample pre-treatment going? *TrAC Trends in Analytical Chemistry, 22*, 90–98.
MacDonald, A. M. G. (1961). The oxygen flask method. *The Analyst, 86*, 3–12.
Maduro, C., Vale, G., Alves, S., Galesio, M., Gomes da Silva, M. D. R., Fernandez, C., Catarino, S., Rivas, M. G., Mota, A. M. & Capelo, J. L. (2006). Determination of Cd and Pb in biological reference materials by electrothermal atomic absorption spectrometry: A comparison of three ultrasonic-based sample treatment procedures. *Talanta, 68*, 1156–1161.
Magalhães, T., Carvalho, M. L., von Bohlen, A. & Becker, M. (2010). Study on trace elements behaviour in cancerous and healthy tissues of colon, breast and stomach: total reflection X-ray fluorescence applications, *Spectrochimica Acta Part B: Atomic Spectroscopy, 65*, 493–498.
Mages, M., Bandow, N., Küster, E., Brack, W. & von Tümpling, W. (2008). Zinc and cadmium accumulation in single zebrafish (Danio rerio) embryos — a total reflection X-ray fluorescence spectrometry application. *Spectrochimica Acta Part B: Atomic Spectroscopy, 63*, 1443–1449.
Mages, M., Woelfl, S., Ovari, M., von Tümpling Jr., W. & Encina, F. (2004.) The use of a portable total reflection X-ray fluorescence spectrometer for trace element determination in freshwater microcrustaceans (Daphnia). *Spectrochimica Acta Part B: Atomic Spectroscopy, 59*, 1265–1272.
Maher, W., Forster, S., Krikowa, F., Snitch, P., Chapple, G. & Craig, P. (2001). Measurement of trace elements and phosphorus in marine animal and plant tissues by low-volume microwave digestion and ICP-MS. *Atomic Spectroscopy, 22*, 361–370.
Marcó, L. M., Capote, T., Hernández, E. A. & Greaves, E. D. (2001). Feasibility study on *in situ* microwave digestion prior to analysis of biological samples by total reflection X-ray fluorescence. *Spectrochimica Acta Part B: Atomic Spectroscopy, 56*, 2187–2193.
Marín, A., López-Gonzálvez, A. & Barbas, C. (2001). Development and validation of extraction methods for determination of zinc and arsenic speciation in soils using focused ultrasound: Application to heavy metal study in mud and soils. *Analytica Chimica Acta, 442*, 305–318.
Mark, D., Haeberle, S., Roth, G., von Stetten, F. & Zengerle, R. (2010). Microfluidic lab-on-a-chip platforms: requirements, characteristics and applications. *Chemical Society Reviews, 39*, 1153–1182.
Mark, J. J. P., Kumar, A., Demattio, H., Hoffmann, W., Malik, A. & Matysik, F. M. (2011). Combination of headspace single-drop microextraction, microchip electrophoresis and contactless conductivity detection for the determination of aliphatic amines In the biodegradation process of seafood samples. *Electroanalysis, 23*, 61–168.
Márquez-Sillero, I., Aguilera-Herrador, E., Cárdenas, S. & Válcarcel, M. (2011). Ion-mobility spectrometry for environmental analysis. *TrAC Trends in Analytical Chemistry, 30*, 677–690.
Martínez-Fernández, M., Barciela-Alonso, M. C., Moreda-Piñeiro, A. & Bermejo-Barrera, P. (2011). Matrix solid phase dispersion-assisted BCR sequential extraction method for metal partitioning in surface estuarine sediments. *Talanta, 83*, 840–849.
Mason, T. & Lorimer, J. (2002). *Applied Sonochemistry. The Uses of Power Ultrasound in Chemistry and Processing*. Germany: Wiley-VCH.
Matin, A. A., Biparva, P., Gheshlaghi, M., Farhadi, K. & Gheshlaghi, A. (2013). Environmental monitoring of complex hydrocarbon mixtures in water and soil samples after solid phase microextraction using PVC/MWCNTs nanocomposite fiber. *Chemosphere, 93*, 1920–1926.
Matos, W. O., Menezes, E. A., Gonzalez, M. H., Costa, L. M., Trevizan, L. C. & Nogueira, A. R. A. (2009). Partial microwave-assisted wet digestion of animal tissue using a baby-bottle sterilizer

for analyte determination by inductively coupled plasma optical emission spectrometry. *Spectrochimica Acta Part B: Atomic Spectroscopy, 64*, 615–618.

Matschat, R., Haßler, J., Traub, H. & Dette, A. (2005). Multielement trace determination in SiC powders: assessment of interlaboratory comparisons aimed at the validation and standardization of analytical procedures with direct solid sampling based on ETV ICP OES and DC arc OES. *Analytical and Bioanalytical Chemistry, 383*, 1060–1074.

Matsubara, A., Harada, K., Hirata, K., Fukusaki, E. & Bamba, T. (2012). High-accuracy analysis system for the redox status of coenzyme Q10 by online supercritical fluid extraction–supercritical fluid chromatography/mass spectrometry. *Journal of Chromatography A, 1250*, 7679.

Matusiewicz, H. & Golik, B. (2004). Determination of major and trace elements in biological materials by microwave induced plasma optical emission spectrometry (MIP-OES) following tetramethylammonium hydroxide (TMAH) solubilisation. *Microchemical Journal, 76*, 23–29.

Matusiewicz, H. (1989). Acid vapour-phase pressure decomposition for the determination of elements in biological materials by flame atomic emission, spectrometry. *Journal of Analytical Atomic Spectrometry. 4*, 265–269.

Matusiewicz, H. (1991). A review of acid vapor-phase sample digestion of inorganic and organic matrices for elemental analysis, *International Journal of Spectroscopy, 3*, 22–28.

Matusiewicz, H. (2003). Wet digestion methods. In Z. Mester, R. Sturgeon (Eds.) *Sample preparation for trace element analysis* (pp. 193–233), Amsterdam, Elsevier.

Matusiewicz, H., Sturgeon, R. E. & Bernan, S. S. (1991). Vapor-phase acid digestion of inorganic and organic matrixes for trace element analysis using a microwave heated bomb. *Journal of Analytical Atomic Spectrometry, 6*, 283–287.

McKiernan, J. W., Creed, J. T., Brockhoff, C. A., Caruso, J. A. & Lorenzana, R. M. (1999). A comparison of automated and traditional methods for the extraction of arsenicals from fish. *Journal of Analytical Atomic Spectrometry, 14*, 607–613.

Mehdinia, A. & Aziz-Zanjani, M. O. (2013). Advances for sensitive, rapid and selective extraction in different configurations of solid-phase microextraction. *TrAC Trends in Analytical Chemistry, 51*, 13–22.

Meloche, L. M., de Bruyn, A. M. H., Otton, S. V., Ikonomou, M. G. & Gobas, F. A. P. C. (2009). Assessing exposure of sediment biota to organic contaminants by thin-film solid phase extraction. *Environmental Toxicology andl Chemistry, 28*, 247–253.

Mesko, M. F., Hartwig, C. A., Bizzi, C.A., Pereira, J.S.F., Mello, P.A. & Flores, E. M. M. (2011). Sample preparation strategies for bioinorganic analysis by inductively coupled plasma mass spectrometry, *International Journal of Mass Spectrometry, 307*, 123–136.

Mesko, M. F., Mello, P. A., Bizzi, C. A., Dressler, V. L., Knapp, G & Flores, E. M. M. (2010). Iodine determination in food by inductively coupled plasma mass spectrometry after digestion by microwave-induced combustion. *Analytical and Bioanalytical Chemistry, 398*, 1125–1131.

Mester, Z. & Sturgeon, R. E. (2002). Detection of volatile organometal chloride species in model atmosphere above seawater and sediment. *Environment Science and Technology, 36*, 1198–1201.

Mester, Z. (2002). Gas phase sampling of volatile (organo)metallic compounds above solid samples. *Journal of Analytical Atomic Spectrometry, 17*, 868–871.

Mester, Z., Lam, J., Sturgeon, R. & Pawliszyn, J. (2000). Determination of methylmercury by solid-phase microextraction inductively coupled plasma mass spectrometry: a new sample introduction method for volatile metal species. *Journal of Analytical Atomic Spectrometry, 15*, 837–842.

Mierzwa, J., Adeloju, S. B. & Dhindsa, H. S. (1997). Ultrasound accelerated solid-liquid extraction for the determination of selenium in biological samples by electrothermal atomization atomic absorption spectrometry. *Analytical Sciences, 13*, 189–193.

Miller-Ihli, N. J. (1989). Slurry sample preparation for simultaneous multi-element graphite furnace atomic absorption spectrometry, *Journal of Analytical Atomic Spectrometry, 3*, 73–81.
Miller-Ihli, N. J. (1993). Advances in ultrasonic slurry graphite furnace atomic absorption spectrometry. *Fresenius Journal Analytical Chemistry, 345*, 482–489.
Millos, J., Costas-Rodríguez, M., Lavilla, I. & Bendicho, C. (2009). Multiple small volume microwave-assisted digestions using conventional equipment for multielemental analysis of human breast biopsies by inductively coupled plasma optical emission spectrometry. *Talanta, 77*, 1490–1496.
Millos, J., Costas-Rodríguez, M., Lavilla, I. & Bendicho, C. (2008). Multielemental determination in breast cancerous and non-cancerous biopsies by inductively coupled plasma-mass spectrometry following small volume microwave-assisted digestion. *Analytica Chimica Acta, 622*, 77–84.
Miró, M. & Hansen, E. H. (2007). Miniaturization of environmental chemical assays in flowing systems: The lab-on-a-valve approach vis-à-vis lab-on-a-chip microfluidic devices. *Analytica Chimica Acta, 600*, 46–57.
Miró, M., Hansen, E. H. & Buanuam, J. (2006). The potentials of the third generation of flow injection analysis for nutrient monitoring and fractionation analysis. *Environmental Chemistry, 3*, 26–30.
Mishra, S., Tripathi, R. M., Bhalke, S., Shukla, V. K. & Puranik, V. D. (2005). Determination of methylmercury and mercury (II) in a marine ecosystem using solid-phase microextraction gas chromatography-mass spectrometry. *Analytica Chimica Acta, 551*, 192–198.
Mitra, S. & Brukh, R. (2003). Sample preparation: an analytical perspective. In S. Mitra (Ed.) *Sample preparation techniques in analytical chemistry* (pp. 1–36), Hoboken, N.J.: John Wiley & Sons, Inc.
Moliner-Martinez, Y., Campíns-Falcó, P., Molins-Legua, C., Segovia-Martínez, L. & Seco-Torrecillas, A. (2009). Miniaturized matrix solid phase dispersion procedure and solid phase microextraction for the analysis of organochlorinated pesticides and polybrominated diphenylethers in biota samples by gas chromatography electron capture detection. *Journal of Chromatography A, 1216*, 6741–6745.
Montes-Bayón, M., Díaz-Molet, M. J., Blanco-González, E. & Sanz-Medel, A. (2006). Evaluation of different sample extraction strategies for selenium determination in selenium-enriched plants (*Allium sativum* and *Brassica juncea*) and Se speciation by HPLC-ICP-MS. *Talanta, 68*, 1287–1293.
Morales-Muñoz, S. & Luque de Castro, M. D. (2005). Dynamic ultrasound-assisted extraction of colistin from feeds with on-line pre-column derivatization and liquid chromatography-fluorimetric detection. *Journal of Chromatography A, 1066*, 1–7.
Moreda-Piñeiro, A., Peña-Vázquez, E., Hermelo-Herbello, P., Bermejo-Barrera, P., Moreda-Piñeiro, J., Alonso-Rodríguez, E., Muniategui-Lorenzo, S., López-Mahía, P. & Prada-Rodríguez, D. (2008). Matrix solid-phase dispersion as a sample pretreatment for the speciation of arsenic in seafood products. *Analytical Chemistry, 80*, 9272–9278.
Moreda-Piñeiro, J., Alonso-Rodríguez, E., López-Mahía, P., Muniategui-Lorenzo, S., Prada-Rodríguez, D., Romarís-Hortas, V., Míguez-Framil, M., Moreda-Piñeiro, A. & Bermejo-Barrera, P. (2009). Matrix solid-phase dispersión of organic compounds and its feasibility for extracting inorganic and organometallic compounds. *TrAC Trends in Analytical Chemistry, 28*, 110–116.
Moreda–Piñeiro, J., Alonso-Rodriguez, E., Romaris-Hortas, V., Moreda-Pineiro, A., Lopez-Mahia, P., Muniategui-Lorenzo, S., Prada-Rodriguez, D. & Bermejo-Barrera, P. (2012). Assessment of the bioavailability of toxic and non-toxic arsenic species in seafood samples. *Food Chemistry, 130*, 552–560.
Moreno, P., Quijano, M. A., Gutiérrez, A. M., Pérez-Conde, M. C. & Cámara, C. (2001). Fractionation studies of selenium compounds from oysters, and their determination by high-performance liquid chromatography coupled to inductively coupled plasma mass spectrometry. *Journal of Analytical Atomic Spectrometry, 16*, 1044–1050.
Müller, I. (1998). Digestion and preparation of organic and biological microsamples for ultratrace elemental analysis. *Atomic Spectroscopy, 19*, 45–47.

Müller, L. (1999). Field Analysis by SPME, in Pawliszyn, J. (Ed.), *Applications of solid phase microextraction* (pp. 269–283). Cambridge: Royal Society of Chemistry.
Müller-Esterl, W. (2008). *Bioquímica. Fundamentos para medicina y ciencias de la vida.* Editorial Reverté, S. A.: Barcelona.
Muñoz, A. H. S., Kubachka, K., Wrobel, K., Gutierrez-Corona, F., Yathavakill, S. K. V., Caruso, J. A. & Wrobel, K. (2005). Metallomics approach to trace element analysis in Ustilago maydis using cellular fractionation, atomic absorption spectrometry, and size exclusion chromatography with ICP-MS detection. *Journal of Agricultural and Food Chemistry, 53,* 5138–5143.
Muñoz, J., Gallego, M. & Valcárcel, M. (2005). Speciation of organometallic compounds in environmental samples by gas chromatography after flow preconcentration on fullerenes and nanotubes, *Analytical Chemistry, 77,* 5389–5395.
Muñoz-Ortuño, M., Moliner-Martínez, Y., Cogollos-Costa, S., Herráez-Hernández, R. & Campíns-Falcó, P. (2012). A miniaturized method for estimating di(2-ethylhexyl) phthalate in bivalves as bioindicators. *Journal of Chromatography A, 1260,* 169–173.
Муránszky, G., Óvári, M., Virág, I., Csiba, P., Dobai, R. & Záray, G. (2011). Chemical characterization of PM_{10} fractions of urban aerosol. *Microchemical Journal, 98,* 1–10.
Naeeni, M. H., Yamini, Y. & Rezaee, M. (2011). Combination of supercritical fluid extraction with dispersive liquid–liquid microextraction for extraction of organophosphorus pesticides from soil and marine sediment samples. *Journal of Supercritical Fluids, 57,* 219–226.
Nakazato, T., Akasaka, M. & Tao, H. (2006). A rapid fractionation method for heavy metals in soil by continuous-flow sequential extraction assisted by focused microwaves. *Analytical and Bioanalytical Chemistry, 386,* 1515–1523.
Neves, R. C. F., Moraes, P. M., Saleh, M. A.D., Loureiro, V. R., Silva, F. A., Barros, M. M., Padilha, C.C. F., Jorge, S. M. A. & Padilha, P. M. (2009). FAAS determination of metal nutrients in fish feed after ultrasound extraction. *Food Chemistry, 113,* 679–683.
Niemelä, M., Kola, H., Eilola, K. & Perämäki, P. (2004). Development of analytical methods for the determination of sub-ppm concentrations of palladium and iron in methotrexate. *Journal of Pharmaceutical and Biomedical Analysis, 35,* 433–439.
Nieuwenhuize, J., Poley-Vos, C. H., van den Akker, A. H. & van Delft, W. (1991). Comparison of microwave and conventional extraction techniques for the determination of metals in soil, sediment and sludge samples by atomic spectrometry. *The Analyst, 116,* 347–351.
Nóbrega, J. A., Santos, M. C., de Sousa, R. A., Cadore, S., Barnes, R. M. & Tatro, M. (2006). Sample preparation in alkaline media. *Spectrochimica Acta Part B: Atomic Spectroscopy, 61,* 465–495.
Osberghaus, U. & Helmers, E. (1997). Waste. In M. Stoeppler (Ed.) *Sampling and sample preparation* (pp. 57–74). Germany, Springer-Verlag Berlin Heidelberg.
Ozbek, N. & Akman, S. (2012). A slurry sampling method for the determination of iron and zinc in baby food by flame atomic absorption spectrometry. *Food Additives and Contaminants, 29,* 208–216.
Öztan, S., Düring, R.- A. (2012). Microwave assisted EDTA extraction-determination of pseudototal contents of distinct trace elements in solid environmental matrices, *Talanta, 99,* 594–602.
Palacios, M. A., Varga, A., Gómez, M., Cámara C. & Gavilanes, F. (1999). Evaluation of acid hydrolysis of proteins on Se-amino acids and trimethylselenonium species by liquid chromatography-microwave digestion - hydride generation-atomic absorption spectrometry. *Quimica Analítica, 18,* 163–168.
Pawliszyn, J. & Lord, H. L (2010). *Handbook of Sample Preparation.* New Jersey: Wiley.
Pawliszyn, J. (2002). *Sampling and sample preparation for field and laboratory.* Netherlands: Elsevier.
Pawliszyn, J. (2003). Sample Preparation: Quo Vadis? *Analytical Chemistry, 75,* 2543–2558.
Pedrero, Z., Ruiz-Encinar, J., Madrid, Y. & Cámara, C. (2007). Identification of selenium species in selenium-enriched *Lens esculenta* plants by using two-dimensional liquid chromatography-

inductively coupled plasma mass spectrometry and [^{77}Se]selenomethionine selenium oxide spikes. *Journal of Chromatography A, 1139*, 247–253.
Pellati, F., Pio Prencipe P., Bertelli, D. & Benvenuti, S. (2013). An efficient chemical analysis of phenolic acids and flavonoids in raw propolis by microwave-assisted extraction combined with high-performance liquid chromatography using the fused-core technology. *Journal of Pharmaceutical and Biomedical Analysis, 81–82*, 126–132.
Pena-Abaurrea, M., García de la Torre, V. S. & Ramos, L. (2013a). Ultrasound-assisted extraction followed by disposable pipette purification for the determination of polychlorinated biphenyls in small-size biological tissue samples. *Journal of Chromatography A, 1317*, 223–229.
Pena-Abaurrea, M., Ramos, J. J., Gonzalez, M. J. & Ramos, L. (2013b). Miniaturized selective pressurized liquid extraction of polychlorinated biphenyls and polybrominateddiphenyl ethers from feedstuffs. *Journal of Chromatography A, 1273*, 18–25.
Peña-Farfal, C., Moreda-Piñeiro, A., Bermejo-Barrera, A., Bermejo-Barrera, P., Pinochet-Cancino, H. & de Gregori-Henríquez, I. (2004). Ultrasound bath-assisted enzymatic hydrolysis procedures as sample pretreatment for the multielement determination in mussels by inductively coupled plasma atomic emission spectrometry. *Analytical Chemistry, 76*, 3541–3547.
Pena-Pereira, F., Lavilla, I. & Bendicho, C. (2009). Miniaturized preconcentration methods based on liquid–liquid extraction and their application in inorganic ultratrace analysis and speciation: A review. *Spectrochimica Acta Part B: Atomic Spectroscopy, 64*, 1–15.
Pena-Pereira, F., Lavilla, I. & Bendicho, C. (2010a). Liquid-phase microextraction techniques within the framework of green chemistry. *TrAC Trends in Analytical Chemistry, 29*, 617–628.
Pena-Pereira, F., Lavilla, I. & Bendicho, C. (2010b). Liquid-phase microextraction approaches combined with atomic detection: A critical review. *Analytica Chimica Acta, 669*, 1–16.
Pena-Pereira, F., Lavilla, I. & Bendicho, C. (2010c). Colorimetric assay for determination of trimethylamine-nitrogen (TMA-N) in fish by combining headspace-single-drop microextraction and microvolume UV–vis spectrophotometry. *Food Chemistry, 119*, 402–407.
Pérez, G., Valiente, M. & Bendicho, C. (2008). A comparative study of metal readsorption in the application of a three-stage sequential extraction scheme and two accelerated versions (ultrasonic and single extractions). *The Open Analytical Chemistry Journal, 2*, 40–46.
Pérez-Cid, B. & Boia, C. (2001). Comparison between total determination and extractable heavy metals from river sediments using conventional and microwave accelerated leaching tests. *International Journal of Environmental Analytical Chemistry, 81*, 101–115.
Pérez-Cid, B., Lavilla, I. & Bendicho, C. (1998). Speeding up of a three-stage sequential extraction method for metal speciation using focused ultrasound. *Analytica Chimica Acta, 360*, 35–41.
Pérez-Cid, B., Lavilla, I. & Bendicho, C. (1999). Application of microwave extraction for partitioning of heavy metals in sewage sludge. *Analytica Chimica Acta, 378*, 201–210.
Petrovic, M. & Barceló, D. (2004) Environmental organic analytes. In J. M. Rosenfeld (Ed.) *Sample preparation for hyphenated analytical techniques* (pp. 150–172). U.S. and Canada, Oxford: Blackwell.
Pino, V., Ayala, J. H., González, V. & Afonso, A. M. (2013). Focused microwave-assisted micellar extraction combined with solid-phase microextraction-gas chromatography/mass spectrometry to determine chlorophenols in wood samples. *Analytica Chimica Acta, 582*, 10–18.
Pirsaheb, M., Fattahi, N. & Shamsipur, M. (2013). Determination of organophosphorous pesticides in summer crops using ultrasound-assisted solvent extraction followed by dispersive liquid–liquid microextraction based on the solidification of floating organic drop. *Food Control, 34*, 378–385.
Pisonero, J., Fernández, B., Pereiro, R., Bordel, N. & Sanz-Medel, A. (2006). Glow-discharge spectrometry for direct analysis of thin and ultra-thin solid films. *TrAC Trends in Analytical Chemistry, 25*, 11–18.

Plata, M. R., Hernando, J., Zougagh, M., Contento, A. M., Villaseñor, M. J., Sánchez-Rojas & J. L., Ríos, A. (2008). Characterization and analytical validation of a microcantilever-based sensor for the determination of total carbonate in soil samples, *Sensors and Actuators B, 134,* 245–251.

Pontes, F. V. M., Mendes, B. A. D., de Souza, E. M. F., Ferreira, F. N., da Silva, L.I. D., Carneiro, M. C., Monteiro, M. I. C., de Almeida, M. D., Neto, A. A. & Vaitsman, D. S. (2010). Determination of metals in coal fly ashes using ultrasound-assisted digestion followed by inductively coupled plasma optical emission spectrometry. *Analytica Chimica Acta, 659,* 55–59.

Portet-Koltalo, F., Oukebdane, K., Dionner, F. & Desbène, P. L. (2009). Optimisation of supercritical fluid extraction of polycyclic aromatic hydrocarbons and their nitrated derivatives adsorbed on highly sorptive diesel particulate matter. *Analytica Chimica Acta, 28,* 48–56.

Potts, P. J. & Robinson, P. (2003). Sample preparation of geological samples, soils and sediments. In Z. Mester, R. Sturgeon (Eds.) *Sample preparation for trace element analysis* (pp. 723–764), Amsterdam, Elsevier.

Pourmortazavi, S. M., Rahimi-Nasrabadi & M., Hajimirsadeghi, S. S. (2014). Supercritical fluid technology in analytical chemistry- Review. *Current Analytical Chemistry, 10,* 3–28.

Priego-Capote, F. & Luque de Castro, M, (2007). Ultrasound in analytical chemistry. *Analytical and Bioanalytical Chemistry, 387*, 249–257.

Pu, Q., Sun, Q., Hu, Z. & Su, Z. (1998). Application of 2-mercaptobenzothiazole-modified silica gel to on-line preconcentration and separation of silver for its atomic absorption spectrometric determination. *The Analyst, 123,* 239–243.

Pumure, I., Renton, J. J. & Smart R. B. (2009). Accelerated aqueous leaching of selenium and arsenic from coal associated rock samples with selenium speciation using ultrasound extraction. *Environmental Geology, 56,* 985–991.

Puttemans, F., van den Winkel, P. & Massart D. L. (1983). The determination of arsenic by electro-thermal atomic absorption spectrometry after liquid-liqiud extraction. *Analytical Chimica Acta, 149*, 123–128.

Qiao, F. & Sun, H. (2010). Simultaneous extraction of enrofloxacin and ciprofloxacin from chicken tissue bymolecularly imprinted matrix solid-phase dispersion. *Journal of Pharmaceutical and Biomedical Analysis, 53,* 795–798.

Qiao, F., & Yan, H. (2010). Simultaneous analysis of fluoroquinolones and xanthine derivatives in serum by molecularly imprinted matrix solid-phase dispersion coupled with liquid chromatography, *Journal of Chromatography B: Analytical Technologies in the Biomedical and Life Sciences, 879,* 3551–3555.

Quevauviller, P., Rauret, G. & Griepink, B. (1993). Single and sequential extraction in sediments and soils. *International Journal of Environmental Analytical Chemistry, 51*, 231–235.

Quevauviller, P., Rauret, G., Rubio, R., López-Sánchez, J.-F., Ure, A., Bacon, J. & Muntau, H. (1997). Certified reference materials for the quality control of EDTA- and acetic acid-extractable contents of trace elements in sewage sludge amended soils (CRMs 483 and 484). *Fresenius Journal of Analytical Chemistry, 357*, 611–618.

Quijano, M. A., Moreno, P., Gutiérrez, A. M., Pérez-Conde, M. C. & Cámara, C. (2000). Selenium speciation in animal tissues after enzymatic digestion by high-performance liquid chromatography coupled to inductively coupled plasma mass spectrometry. *Journal of Mass Spectrometry, 35*, 878–884.

Quintana, J. B., Boonjob, W., Miró, M., Cerda, V. (2009). Online coupling of bead injection lab-on-valve analysis to gas chromatography: application to the determination of trace levels of polychlorinated biphenyls in solid waste leachates. *Analytical Chemistry, 81,* 4822–4830.

Ramalhosa, E., Río-Segade, S., Pereira, E., Vale, C. & Duarte, A. (2001). Microwave-assisted extraction for methylmercury determination in sediments by high performance liquid chromatography-cold vapour atomic fluorescence spectrometry. *Journal of Analytical Atomic Spectrometry, 16*, 643–647.

Ramalhosa, M. J., Paíga, P., Morais, S., Sousa, A. M. M., Gonçalves, M. P., Delerue-Matos, C. & Prior Pinto Oliveira, M. B. (2012). Analysis of polycyclic aromatic hydrocarbons in fish: Optimisation and validation of microwave-assisted extraction. *Food Chemistry, 135*, 234–242.

Ramil-Criado, M., Rodríguez-Pereiro, I. & Cela-Torrijos, R. (2004). Determination of polychlorinated biphenyls in ash using dimethylsulfoxide microwave assisted extraction followed by solid-phase microextraction. *Talanta, 63*, 533–540.

Ramos, J. J., Dietz, C., Gonzalez, M. J. & Ramos, L. (2000). Miniaturised selective pressurised liquid extraction of polychlorinated biphenyls from foodstuffs. *Journal of Chromatography A, 1152*, 254–261.

Ramos, J. J., Rial-Otero, R., Ramos, L. & Capelo, J. L. (2008). Ultrasonic-assisted matrix solid-phase dispersion as an improved methodology for the determination of pesticides in fruits. *Journal of Chromatography A, 1212*, 145–149.

Ramos, L. (2012). Critical overview of selected contemporary sample preparation techniques. *Journal of Chromatography A, 1221*, 84– 98.

Ramos, L., Ramos, J. J. & Brinkman, U. A. T. (2005). Miniaturization in sample treatment for environmental analysis. *Analytical and Bioanalytical Chemistry, 381*, 119–140.

Ramos, L., Vreuls, J. J. & Brinkman, U. A. Th. (2007). Miniaturised pressurised liquid extraction of polycyclic aromatic hydrocarbons from soil and sediment with subsequent large-volume injection–gas chromatography. *Journal of Chromatography A, 891*, 275–286.

Rasmussen, R. R., Hedegaard, R. V., Larsen, E. H. & Sloth, J. J. (2012). Development and validation of an SPE HG-AAS method for determination of inorganic arsenic in samples of marine origin. *Analytical and Bioanalytical Chemistry, 403*, 2825–2834.

Ravelo-Pérez, L. M., Herrera-Herrera, A. V., Hernández-Borges, J. & Rodríguez-Delgado, M. A. (2010). Carbon nanotubes: Solid-phase extraction. *Journal of Chromatography A, 1217*, 2618–2641.

Rechenberg, W. (1997). Decomposition [of materials] for trace analysis in the cement industry. In M. Stoeppler (Ed.) *Sampling and sample preparation* (pp. 183–198). Germany: Springer-Verlag Berlin Heidelberg.

Reid, M. K., Spencer, K. L. & Shotbolt, L. (2011). An appraisal of microwave-assisted Tessier and BCR sequential extraction methods for the analysis of metals in sediments and soils. *Journal of Soils and Sediments, 11*, 518–528.

Remeteiová, D., Ružičková, S. & Rusnák, R. (2008). Ultrasound-assisted extraction in the fractionation analysis, of gravitation dust sediments. *Microchimica Acta, 163*, 257–261.

Reus, U. (1991). Determination of trace elements in oils and greases with total reflection X-ray fluorescence: sample preparation methods. *Spectrochimica Acta Part B: Atomic Spectroscopy, 46*, 1403–1411.

Reyes, L. H., Rahman, G. M. M., Fahrenholz, T. & Kingston, H. M. S. (2008). Comparison of methods with respect to efficiencies, recoveries, and quantitation of mercury species interconversions in food demonstrated using tuna fish. *Analytical and Bioanalytical Chemistry, 390*, 2123–2132.

Reyes-Garcés, N., Gómez-Ríos, G. A., Souza Silva, É. A., & Pawliszyn, J. (2013). Coupling needle trap devices with gas chromatography–ion mobility spectrometry detection as a simple approach for on-site quantitative analysis. *Journal of Chromatography A, 1300* 193–198.

Rezaei, F. & Hosseini, M. (2011). New method based on combining ultrasonic assisted miniaturized matrix solid-phase dispersion and homogeneous liquid–liquid extraction for the determination of some organochlorinated pesticides in fish. *Analytica Chimica Acta, 702*, 274–279.

Ribeiro, A. S., Curtius, A. J. & Pozebon, D. (2000). Determination of As, Cd, Ni and Pb in human hair by electrothermal atomic absorption spectrometry after sample treatment with tetramethylammonium hydroxide. *Microchemical Journal, 64*, 105–110.

Richter, B. R., Jones, B. A., Ezzell, J. L., Porter, N. L., Avdalovic, N. & Pohl, C. (1996.) Accelerated solvent extraction: a technique for sample preparation. *Analytical Chemistry, 68*, 1033–1039.

Ríos, A. & Zougagh, M. (2013). Sample preparation for micro total analytical systems (μ-TASs). *TrAC Trends in Analytical Chemistry, 43,* 174–188.

Ríos, A., Escarpa, A., Simonet, B. (2009). *Miniaturization Systems: Principles, designs and applications.* United Kingdom: John Wiley & Sons.

Ríos, A., Zougagh, M. & Avila, M. (2012). Miniaturization through lab-on-a-chip: Utopia or reality for routine laboratories? A review. *Analytica Chimica Acta, 740,* 1–11.

Rodríguez, I., Santamarina, M., Bollaín, M. H., Mejuto, M. C. & Cela, R. (1997). Speciation of organotin compounds in marine biomaterials after basic leaching in a non-focused microwave extractor equipped with pressurized vessels. *Journal of Chromatography A, 774,* 379–387.

Romero, R., Costas-Mora, I., Lavilla, I. & Bendicho, C. (2011). Cold vapor-solid phase microextraction using amalgamation in different Pd-based substrates combined with direct thermal desorption in a modified absorption cell for the determination of Hg by atomic absorption spectrometry. *Spectrochimica Acta Part B: Atomic Spectroscopy, 66,* 156–162.

Russo R. E. & Baldwin D. P. (2003). Laser ablation sampling. In Z. Mester & R. Sturgeon (Eds.), *Sample preparation for trace element analysis* (pp. 593–609). The Netherlands: Elsevier.

Salvador, A., Pascual-Martí, M. C., Adell, J. R., Requeni, A. & March, J. G. (2000). Analytical methodologies for atomic spectrometric determination of metallic oxides in UV sunscreen creams. *Journal of Pharmaceutical and Biomedical Analysis, 22,* 301–306.

Sánchez-Moreno, R. A., Gismera, M. J., Sevilla, M. T. & Procopio, J. R. (2010). Direct and rapid determination of ultratrace heavy metals in solid plant materials by ET-AAS ultrasonic-assisted slurry sampling. *Phytochemical Analysis, 21,* 340–347.

Sansoni, B. & Panday V. K. (1994) Trace element analysis in biological specimens. In R. F. M. Herber & M. Stoeppler (Ed.) *Sample treatment of human biological materials,* (pp. 21–52). The Netherlands: Elsevier.

Santos, C., Alava-Moreno, F., Lavilla, I. & Bendicho, C. (2000). Total As in seafood as determined by transverse heated electrothermal atomic absorption spectrometry-longitudinal Zeeman background correction: An evaluation of automated ultrasonic slurry sampling, ultrasound-assisted extraction and microwave-assisted digestion methods. *Journal of Anaytical Atomic Spectrometry, 15,* 987–994.

Santos-Júnior, D., Krug, F. J., de Godoi-Pereira, M. & Korn, M. (2006). Currents on ultrasound-assisted extraction for sample preparation and spectroscopic analytes determination. *Applied Spectroscopy Reviews, 41,* 305–321.

Sanz, E., Muñoz-Olivas, R. & Cámara, C. (2005a). Evaluation of a focused sonication probe for arsenic speciation in environmental and biological samples. *Journal of Chromatography A, 1097,* 1–8.

Sanz, E., Muñoz-Olivas, R. & Cámara, C. (2005b). A rapid and novel alternative to conventional sample treatment for arsenic speciation in rice using enzymatic ultrasonic probe. *Analytica Chimica Acta, 535,* 227–235.

Sanz, E., Muñoz-Olivas, R., Dietz, C., Sanz, J. & Cámara, C. (2007) Alternative extraction methods for arsenic speciation in hair using ultrasound probe sonication and pressurised liquid extraction. *Journal of Anaytical Atomic Spectrometry, 22,* 131–139.

Sanz-Segundo, C., Hernández-Artiga, M. P., Hidalgo-Hidalgo de Cisneros, J. L., Bellido-Milla, D. & Naranjo-Rodriguez, I. (1999). Determination of wear metals in marine lubricating oils by microwave digestion and atomic absorption spectrometry. *Microchimica Acta, 132,* 89–94.

Sávoly, Z., Nagy, P., Havancsák, K. & Záray, G. (2012). Microanalytical investigation of nematodes. *Microchemical Journal, 105,* 83–87.

Scaccia, S. & Mecozzi, R. (2012). Trace Cd, Co, and Pb elements distribution during Sulcis coal pyrolysis: GFAAS determination with slurry sampling technique. *Microchemical Journal, 100,* 48–54.

Schladot, J.-D. & Backhaus, F. (1997). Collection, preparation and long-term storage of marine samples. In M. Stoeppler (Ed.) *Sampling and sample preparation* (pp. 74–87). Germany: Springer-Verlag Berlin Heidelberg.

Schmeling, M., Klockenkämper, R. & Klockow, D. (1997). Application of total-reflection X-ray fluorescence spectrometry to the analysis of airborne particulate matter. *Spectrochimica Acta Part B: Atomic Spectroscopy, 52*, 985–994.

Schöniger, W. (1955). Rapid microanalytical determination of halogen in organic compounds. *Microchimica Acta, 1*, 123–129.

Schöniger, W. (1956). The rapid microanalytical determination of halogens and sulfur in organic substances *Microchimica Acta, 1*, 869–876.

Seidi, S., Yamini, Y., Heydari, A., Moradi, M., Esrafili, A. & Rezazadeh, M. (2011). Determination of the baine in water samples, biological fluids, poppy capsule, and narcotic drugs, using electro-membrane extraction followed by high-performance liquid chromatography analysis. *Analytica Chimica Acta, 701*, 181–188.

Self, R. (2005). *Extraction of organic analytes from foods*. London: Royal Society of Chemistry.

Sereshti, H., Bakhtiari, S. & Khojeh, V. (2011). Simultaneous determination of aluminum, copper, iron and zinc in Oscillatoria, Juncus littoralis tissues and wetland water samples by ultrasound-assisted emulsification-microextraction combined with ICP-OES. *Analytical Methods, 3*, 2936–2942.

Sereshti, H., Heidari, R. & Samadi, S. (2014). Determination of volatile components of saffron by optimised ultrasound-assisted extraction in tandem with dispersive liquid–liquid microext-raction followed by gas chromatography–mass spectrometry. *Food Chemistry, 143*, 499–505.

Shamsipur, M., Ramezani, M. & Sadeghi, M. (2009). Preconcentration and determination of ultra trace amounts of palladium in water samples by dispersive liquid-liquid microextraction and graphite furnace atomic absorption spectrometry. *Microchimica Acta, 166*, 235–242.

Shen, Q., Dong, W., Wang, Y., Gong, L., Dai, Z. & Cheung, H. (2013a). Pipette tip solid-phase extraction and ultra-performance liquid chromatography/mass spectrometry based rapid analysis of picrosides from Picrorhizascrophularii flora. *Journal of Pharmaceutical and Biomedical Analysis*, 136–140.

Shen, Q., Gong, L., Baibado, J. T., Dong, W., Wang, Y., Dai, Z. & Cheung, H. (2013b). Graphene based pipette tip solid phase extraction of marine toxins in shellfish muscle followed by UPLC-MS/MS analysis. *Talanta, 116*, 770–775.

Shiddiky, M. J. A., Lee, K.-S., Son, J., Park, D.-S. & Shim, Y.-B. (2009). Development of extraction and analytical methods of nitrite ion from food samples: Microchip electrophoresis with a modified electrode. *Journal of Agricultural Food Chemistry, 57*, 4051–4057.

Sides, S. L., Polowy, K. B., Thornquest Jr., A. D., & Burinsky, D. J. (2001). Identification of a pharmaceutical packaging off-odor using solid phase microextraction gas chromatography/ mass spectrometry. *Journal of Pharmaceutical and Biomedical Analysis, 25*, 379–386.

Silva, R. G. L., Willie, S. N., Sturgeon, R. E., Santelli, R. E. & Sella, S. M. (1999). Alkaline solubilization of biological materials for trace element analysis by electrothermal atomic absorption spectrometry. *The Analyst, 124*, 1843–1846.

Siwek, M., Noubar, A. B., Bergmann, J., Niemeyer, B. & Galunsky, B. (2006). Enhancement of enzymatic digestion of Antarctic krill and successive extraction of selenium organic compounds by ultrasound treatment. *Analytical and Bioanalytical. Chemistry, 384*, 244–249.

Skorek, R., Turek, E., Zawisza, B., Marguí, E., Queralt, I., Stempin, M., Kucharski, P. & Sitko, R. (2012). Determination of selenium by X-ray fluorescence spectrometry using dispersive solid-phase microextraction with multiwalled carbon nanotubes as solid sorbent. *Journal of Analytical Atomic Spectrometry, 27*, 1688–1693.

Smith, F. E. & Arsenault, E. A. (1996). Microwave-assisted sample preparation in analytical chemistry. *Talanta, 43*, 1207–1268.

Solchaga, M., Montoro, R. & de la Guardia, M. (1986). Flame absorption spectroscopic determination of cadmium, copper, iron, lead, and zinc in mussels. *Journal of the Association of Official Analytical Chemistry, 69*, 874–876.
Song, W., Zhang, Y., Li, G., Chen, H., Wang, H., Zhao, Q., He, D., Zhao, C. & Ding, L. (2014). A fast, simple and green method for the extraction of carbamate pesticides from rice by microwave assisted steam extraction coupled with solid phase extraction. *Food Chemistry, 143*, 192–198.
Song, X., Shi, Y. & Chen, J. (2013). Carbon nanotubes-reinforced hollow fibre solid-phase microextraction coupled with high performance liquid chromatography for the determination of carbamate pesticides in apples. *Food Chemistry, 139*, 246–252,
Souza Silva, E. A., Risticevic, S. & Pawliszyn, J. (2013). Recent trends in SPME concerning sorbent materials, configurations and *in vivo* applications. *TrAC Trends in Analytical Chemistry, 43*, 24–36.
Soylak, M., Tuzen, M., Narin, I. & Sari, H. (2004). Comparison of microwave, dry and wet digestion procedures for the determination of trace metal contents in spice samples produced in Turkey. *Journal of Food and Drug Analysis, 12*, 254–258.
Spanke, J., von Bohlen, A., Klockenkamper, R., Quentmeier, A. & Klockow, D. (2000). Total reflection X-ray fluorescence analysis of laser-deposited solid sample material. *Journal of Analytical Atomic Spectrometry, 15*, 673–679.
Starr, J. M. & Selim, M. I. (2008). Supercritical fluid extraction of aflatoxin B from soil. *Journal of Chromatography A, 1209*, 37–43.
Stashenko, E.E., Ordoñez, S. A., Marin, N. A. & Martinez, J. R. (2009). Determination of the volatile and semi-volatile secondary metabolites, and aristolochic acids in Aristolochiaringens Vahl. *Journal of Chromatographic Science, 47*, 817–821.
Steiner J. W., Moy, D. C., Kramer, H. L. (1987). Rapid determination of chromium in bovine liver using an atomic absorption spectrometer with a modified carbon rod atomizer. *The Analyst, 112*, 1113–1115.
Stoeppler, M. (1997). Sampling: an introduction. In M. Stoeppler (Ed.) *Sampling and sample preparation* (pp. 1–6). Germany: Springer-Verlag Berlin Heidelberg.
Su, R., Wang, X., Xu, X., Wang, Z., Li, D., Zhao, X., Li, X., Zhang, H. & Yu, A. (2011). Application of multiwall carbon nanotubes-based matrix solid phase dispersion extraction for determination of hormones in butter by gas chromatography mass spectrometry. *Journal of Chromatography A, 1218*, 5047–5054.
Subramanian, K. S. (1996). Determination of metals in biofluids and tissues: sample preparation methods for atomic spectroscopic techniques. *Spectrochimica Acta Part B: Atomic Spectroscopy, 51*, 291–319.
Sun, Y. & Li, H. (2000) Determination of trace selenium in human plasma and hair with ternary inclusion compound–fluorescent spectrophotometry. *The Analyst, 125*, 2326–2329.
Sunarso, J. & Ismadji, S. (2009). Decontamination of hazardous substances from solid matrices and liquids using supercritical fluids extraction: A review. *Journal of Hazardous Materials, 161*, 1–20.
Sutherland, R. A. (2010). BCR®-701: A review of 10-years of sequential extraction analyses, *Analytica Chimica Acta, 680*, 10–20.
Szoboszlai, N., Réti, A., Budai, B., Szabó, Z., Kralovánszky, J. & Záray, G. (2008). Direct elemental analysis of cancer cell lines by total reflection X-ray fluorescence. *Spectrochimica Acta Part B: Atomic Spectroscopy, 63*, 1480–1484.
Szpunar, J., Bouyssiere, B. & Lobinski, R. (2003). Sample preparation for speciation analysis for metallobiomolecules, In Z. Mester & R. Sturgeon (Eds.), *Sample preparation for trace element analysis* (pp. 1185–1210). The Netherlands: Elsevier.
Szpunar, J., Schmitt, V. O., Donard, O. F. X. & Lobinski, R. (1996). Low-power focused microwave technology as a new tool for rapid preparation of solid samples for speciation analysis. *TrAC Trends in Analytical Chemistry, 15*, 181–187.

Talukder, M. M. R., Zaman, M. M., Hayashi, Y., Wu, J. C. & Kawanishi, T. (2006). Ultrasonication enhanced hydrolytic activity of lipase in water/isooctane two-phase systems. *Biocatalysis and Biotransformation, 24*, 189–194.

Tan, C.S., Oliveira, E., Oliveira, P.V., *Abstracts of 11 Encontro Nacional de Química Analítica*, Campinas, SP, 2001.

Tan, Y. & Marshall, W. D. (1997). Enzymic Digestion–High-pressure Homogenization Prior to Slurry Introduction Electrothermal Atomic Absorption Spectrometry for the Determination of Selenium in Plant and Animal Tissues. *The Analyst, 122*, 13–18.

Teixeira-Tarley, C. R., Barbosa, A. F., Segatelli, M. G., Figueiredo, E. C. & Luccas, P. O. (2006). Highly improved sensitivity of TS-FF-AAS for Cd(II) determination at ng L^{-1} levels using a simple flow injection minicolumn preconcentration system with multiwall carbon nanotubes. *Journal of Analytical Atomic Spectrometry, 21*, 1305–1313.

Tessier, A., Campbell, P. G. C. & Bisson, M. (1979). Sequential extraction procedure for the speciation of particulate trace metals. *Analytical Chemistry, 51*, 844–851.

Theisen, M. & Niessner, R. (1999). Elemental analysis of airborne dust samples with TXRF: Comparison of oxygen-plasma ashing on sapphire carriers and acid digestion for sample preparation. *Fresenius Journal of Anaytical Chemistry, 365*, 332–337.

Thomas, A. D. & Smythe, L. E. (1973) Rapid destruction of plant material with concentrated nitric acid vapour (vapour phase oxidation). *Talanta, 20*, 469–475.

Togunde, O. P., Oakes, K. D., Servos, M. R. & Pawliszyn, J. (2012). Optimization of solid phase microextraction for non-lethal *in vivo* determination of selected pharmaceuticals in fish muscle using liquid chromatography–mass spectrometry, *Journal of Chromatography A, 1261*, 99–106.

Tran, M. Q. T., Nygren, Y., Lundin, C., Naredi, P. & Björn, E. (2010). Evaluation of cell lysis methods for platinum metallomic studies of human malignant cells. *Analytical Biochemistry, 396*, 76–82.

Trevizan, L. C., Nogueira, A. R. A. & Nóbrega, J. A. (2003). Single vessel procedure for acid vapor partial digestion of bovine liver in a focused microwave: multielement determination by ICP-OES. *Talanta, 61*, 81–86.

Turner, A. & Olsen, Y. S. (2000). Chemical versus enzymatic digestion of contaminated estuarine sediment: Relative importance of iron and manganese oxides in controlling trace metal bioavailability. *Estuarine, Coastal and Shelf Science, 51*, 717–728.

Turner, A. (2006). Enzymatic mobilisation of trace metals from estuarine sediment. *Marine Chemistry, 98*, 140–147.

Turner, A., Henon, D. N. & Dale, J. L. L. (2001) Pepsin-digestibility of contaminated estuarine sediments. *Estuarine, Coastal and Shelf Science, 53*, 671–681.

Tuzen, M., Saygi, K. O. & Soylak, M. (2008). Solid phase extraction of heavy metal ions in environmental samples on multiwalled carbon nanotubes. *Journal of Hazardous Materials, 152*, 632–639.

Umale, S. & Mahanwar, P. A. (2010). Supercritical fluid extraction of heavy metals from solid matrices. *Chemical Technology: An Indian Journal, 5*, 25–44.

Väisänen, A. & Kiljunen, A. (2005). Ultrasound-assisted sequential extraction method for the evaluation of mobility of toxic elements in contaminated soils. *International Journal of Environmental Analytical Chemistry, 85*, 1037–1049.

Väisänen, A., Suontamo, R., Silvonen, J. & Rintala, J. (2002). Ultrasound-assisted extraction in the determination of arsenic, cadmium, copper, lead, and silver in contaminated soil samples by inductively coupled plasma atomic emission spectrometry. *Analytical and Bioanalytical Chemistry, 373*, 93–97.

Valcárcel, M., & Cárdenas, M. S. (2000). *Automatización y miniaturización en química analítica*. Barcelona: Springer.

Vale, G., Pereira, S., Mota, A., Fonseca, L. & Capelo, J. L. (2007) Enzymatic probe sonication as a tool for solid–liquid extraction for total selenium determination by electrothermal-atomic absorption spectrometry. *Talanta, 74*, 198–205.

Vale, G., Rial-Otero, R., Mota, A., Fonseca, L. & Capelo, J. L. (2008). Ultrasonic-assisted enzymatic digestion (USAED) for total elemental determination and elemental speciation: A tutorial. *Talanta, 75*, 872–884.

Vanhaecke, F., Resano, M. & Moens, L. (2002). Electrothermal vaporisation ICP-mass spectrometry (ETV-ICP-MS) for the determination and speciation of trace elements in solid samples – A review of real-life applications from the author's lab. *Analytical and Bioanalytical Chemistry, 374*, 188–195.

Varga, I., Rierpl, E. & Tusai, A. (1999). Comparison of slurry preparation and microwave digestion of freshwater algae for multi-element analysis by total reflection X-ray fluorescence spectrometry. *Journal of Analytical Atomic Spectrometry, 14*, 881–883.

Varga, I., Szebeni, A., Szoboszlai, N. & Kovács, B. (2005). Determination of trace elements in human liver biopsy samples by ICP–MS and TXRF: hepatic steatosis and nickel accumulation. *Analytical and Bioanalytical Chemistry, 383*, 476–482.

Vázquez, C., Custo, G., Barrio, N., Burucúa, J., Boeykens, S. & Marte, F. (2010). Inorganic pigment study of the San Pedro Gonzalez Telmo Sibyls using total reflection X-ray fluorescence. *Spectrochimica Acta Part B: Atomic Spectroscopy, 65*, 852–858.

Vazquez, M. J., Carro, A. M., Lorenzo, R. A. & Cela, R. (1997). Optimization of methylmercury microwave-assisted extraction from aquatic sediments. *Analytical Chemistry, 69*, 221–225.

Veggi, P. C., Martinez, J. & Meireles, M. A. A. (2013). Fundamentals of microwave extraction, in: Chemat, F. & Cravotto, G. (Ed.), *Microwave-assisted extraction for bioactive compounds, theory and practice* (pp. 15–53). Netherlands: Springer.

Vignola, F., Borges, D. L. G., Curtius, A. J., Welz, B. & Becker-Ross, H. (2010). Simultaneous determination of Cd and Fe in sewage sludge by high-resolution continuum source electrothermal atomic absorption spectrometry with slurry sampling. *Microchemical Journal, 95*, 333–336.

Vilanó, M. & Rubio, R. (2001). Determination of arsenic species in oyster tissue by microwave-assisted extraction and liquid chromatography-atomic fluorescence detection. *Applied Organometallic Chemistry, 15*, 658–666.

Viñas, P., Bravo-Bravo, M., López-García, I. & Hernández-Córdoba, M. (2013). Dispersive liquid–liquid microextraction for the determination of vitamins D and K in foods by liquid chromatography with diode-array and atmospheric pressure chemical ionization-mass spectrometry detection. *Talanta, 115*, 806–813.

Wagner, A. & Mages, M. (2010). Total-Reflection X-ray fluorescence analysis of elements in size-fractionated particulate matter sampled on polycarbonate filters — composition and sources of aerosol particles in Göteborg, Sweden. *Spectrochimica Acta Part B: Atomic Spectroscopy, 65*, 471–477.

Wang, H., Li, G., Zhang. Y., Chen, H., Zhao, Q., Song, W., Xu, Y., Jin, H. & Ding, L. (2012). Determination of triazine herbicides in cereals using dynamic microwave-assisted extraction with solidification of floating organic drop followed by high-performance liquid chromatography. *Journal of Chromatography A, 1233*, 36–43.

Wang, H., So, P. & Yao, Z. (2014). Direct analysis of herbal powders by pipette-tip electrospray ionization mass spectrometry. *Analytica Chimica Acta, 809*, 109–116.

Wang, H., Wu, Z., Zhang, Y., Chen, B., He, M. & Hu, B. (2013a). Chip-based liquid phase microextraction combined with electrothermal vaporization-inductively coupled plasma mass spectrometry for trace metal determination in cell samples. *Journal of Analytical Atomic Spectrometry, 28*, 1660–1665.

Wang, J. & Hansen, E. H. (2001a). Coupling sequential injection on-line preconcentration by means of a renewable microcolumn with ion-exchange beads with detection by electrothermal atomic absorption spectrometry comparing the performance of eluting the loaded beads with transporting them directly into the graphite tube, as demonstrated for the determination of nickel in environmental and biological samples. *Analytica Chimica Acta, 435,* 331–342.

Wang, J. & Hansen, E. H. (2001b). Interfacing sequential injection on-line preconcentration using a renewable micro-column incorporated in a 'lab-on-valve' system with direct injection nebulization inductively coupled plasma mass spectrometry. *Journal of Analytical Atomic Spectrometry, 16,* 1349–1355.

Wang, J., Hansen, E. H. & Miró, M. (2003). Sequential injection–bead injection-lab-on-valve schemes for on-line solid phase extraction and preconcentration of ultra-trace levels of heavy metals with determination by electrothermal atomic absorption spectrometry and inductively coupled plasma mass spectrometry. *Analytica Chimica Acta, 499,* 139–147.

Wang, S., Lin, Y. & Wai, C. M. (2003). Supercritical fluid extraction of toxic heavy metals from solid and aqueous matrices. *Separation Science and Technology 38,* 2279–2289.

Wang, X., Ding, L., Zhang, H., Cheng, J., Yu, A., Zhang, H., Liu, L., Liu, Z. & Li, Y. (2006). Development of an analytical method for organotin compounds in fortified flour samples using microwave-assisted extraction and normal-phase HPLC with UV detection. *Journal of Chromatography B: Analytical Technologies in the Biomedical and Life Sciences, 843,* 268–274.

Wang, X., Jin, H., Ding, L., Zhang, H., Zhang, H., Qu, C. & Yu, A. (2008). Organotin speciation in textile and plastics by microwave-assisted extraction HPLC–ESI-MS. *Talanta, 75,* 556–563.

Wang, Z., Zhao, X., Xu, X., Wu, L., Su. R., Zhao, Y., Jiang, C., Zhang, H., Ma, Q., Lu, C. & Dong, D. (2013b). An absorbing microwave micro-solid-phase extraction device used in non-polar solvent microwave-assisted extraction for the determination of organophosphorus pesticides. *Analyica Chimica Acta, 760,* 60–68.

Welz, B., Sperling, M. (1999). *Atomic absorption spectrometry*, Germany: Wiley-VCH.

Welz, B., Vale, M. G. R., Borges, D. L. G. & Heitmann, U. (2007). Progress in direct solid sampling analysis using line source and high-resolution continuum source electrothermal atomic absorption spectrometry. *Analytical and Bioanalytical Chemistry, 389,* 2085–2095.

Wieteska, E., Zióek, A. & Drzewihska, A. (1996). Extraction as a method for preparation of vegetable samples for the determination of trace metals by atomic absorption spectrometry, *Analytica Chimica Acta, 330,* 25 l-257.

Woelfl, S., Mages, M. & Encina, F. (2003). Cold plasma ashing improves the trace element detection of single Daphnia specimens by total reflection X-ray fluorescence spectrometry. *Spectrochimica Acta Part B: Atomic Spectroscopy, 58,* 2157–2168.

Woelfl, S., Mages, M., Mercado, S., Villalobos, L. & Encina, M. O. F. (2004). Determination of trace elements in planktonic microcrustaceans using total reflection X-ray fluorescence (TXRF): first results from two Chilean lakes. *Analytica Bioanalytical Chemistry, 378,* 1088–1094.

Wu, S. & Ding, W. (2010). Fast determination of synthetic polycyclic musks in sewage sludge and sediments by microwave-assisted headspace solid-phase microextraction and gas chromatography–mass spectrometry. *Journal of Chromatography A, 1217,* 2776–2781.

Wu, S., Liu, L. & Ding, W. (2012). One-step microwave-assisted headspace solid-phase microextraction for the rapid determination of synthetic polycyclic musks in oyster by gas chromatography–mass spectrometry. *Food Chemistry, 133,* 513–517.

Xi, Y., Duford, D. A. & Salin, E. D. (2010). Automated liquid-solid extraction of pyrene from soil on centrifugal microfluidic devices. *Talanta, 82,* 1072–1076.

Xia, L., Wu, Y. & Hu, B. (2007). Hollow-fiber liquid-phase microextraction prior to low-temperature electrothermal vaporization ICP-MS for trace element analysis in environmental and biological samples. *Journal of Mass Spectrometry, 42,* 803–810.

Xu, Z. G., Hu, Y.F., Hu, Y. L. & Li, G. K. (2010). Investigation of ractopamine molecularly imprinted stir bar sorptive extraction and its application for trace analysis of beta (2)-agonists in complex samples. *Journal of Chromatography A, 1217,* 3612–3618.
Xue, S., Liu, Y., Li, H.-F., Uchiyama, K. & Lin, J. M. (2013). A microscale solid-phase extraction poly(dimethylsiloxane) chip for enrichment and fluorescent detection of metal ions. *Talanta, 116,* 1005–1009.
Yabalak, E. & Gizir, A. M. (2013). Subcritical and supercritical fluid extraction of heavy metals from sand and sewage sludge. *Journal of the Serbian Chemical Society, 78,* 1013–1022.
Yamamoto, K. Y., Cremers, D. A., Ferris, M. J. & Foster, L. E. (1996). Detection of metals in the environment using a portable laser-induced breakdown spectroscopy instrument. *Applied Spectroscopy, 50,* 222–233.
Yan, H., Cheng, X., Sun, N., Cai, T., Wu, R. & Han, K. (2012a). Rapid and selective screening of melamine in bovine milk using molecularly imprinted matrix solid-phase dispersion coupled with liquid chromatography-ultraviolet detection. *Journal of Chromatography B: Analytical Technologies in the Biomedical and Life Sciences, 908,* 137–142.
Yan, H., Wang, F., Wang, H. & Yang, G. (2012b). Miniaturized molecularly imprinted matrix solid-phase dispersion coupled with high performance liquid chromatography for rapid determination of auxins in orange samples. *Journal of Chromatography A, 1256,* 1–8.
Yan, H., Wang, H., Qiao, J. & Yang, G. (2011). Molecularly imprinted matrix solid-phase dispersion combined with dispersive liquid–liquid microextraction for the determination of four Sudan dyes in egg yolk, *Journal of Chromatography A, 1218,* 2182–2188.
Yang, G., Xu, J., Xu, L., Chen, G. & Fu, F. (2010). Analysis of ultratrace triorganotin compounds in aquatic organisms by using capillary electrophoresis–inductively coupled plasma mass spectrometry. *Talanta, 80,* 1913–1918.
Yi, Y.-Z., Jiang, S.-J. & Sahayam, A. C. (2012). Palladium nanoparticles as the modifier for the determination of Zn, As, Cd, Sb, Hg and Pb in biological samples by ultrasonic slurry sampling electrothermal vaporization inductively coupled plasma mass spectrometry. *Journal of Analytical Atomic Spectrometry, 27,* 426–431.
Yin, X., Frech, W., Hoffmann, E., Lüdke, C. & Skole, J. (1998). Mercury speciation by coupling cold vapour atomic absorption spectrometry with flow injection on-line preconcentration and liquid chromatographic separation. *Fresenius Journal of Analytical Chemistry, 361,* 761–766.
Yu, Y., Chen, B., Shen, C., Cai, Y., Xie, M., Zhou, W., Chen, Y., Li, Y. & Duan, G. (2010). Multiple headspace single-drop microextraction coupled with gas chromatography for direct determination of residual solvents in solid drug product. *Journal of Chromatography A, 1217,* 5158–5164.
Yu, Y.-L., Jiang, Y., Chen, M.-L. & Wang, J.-H. (2011). Lab-on-valve in the miniaturization of analytical systems and sample processing for metal analysis. *TrAC Trends in Analytical Chemistry, 30,* 1649–1658.
Yu, Z., Robinson, P. & McGoldrick, P. (2001). An evaluation of methods for the chemical decomposition of geological materials for trace element determination using ICP-MS. *Geostandard Newsletter, 25,* 199–217.
Zang, Z., Hu, Z., Li, Z., He, Q. & Chang, X. (2009). Synthesis, characterization and application of ethylenediamine-modified multiwalled carbon nanotubes for selective solid-phase extraction and preconcentration of metal ions. *Journal of Hazardous Materials, 172,* 958–963.
Zhang, H. & Shi, Y. (2010). Temperature-assisted ionic liquid dispersive liquid–liquid microextraction combined with high performance liquid chromatography for the determination of anthraquinones in Radix and Rhizoma Rhei samples. *Talanta, 82,* 1010–1016.
Zhang, W., Hu, Z., Liu, Y., Chen, H., Gao, S. & Gaschnig, R. M. (2012). Total rock dissolution using ammonium bifluoride (NH_4HF_2) in screw-top teflon vials: A new development in open-vessel digestion. *Analytical Chemistry, 84,* 10686–10693.

Zhang, X, Oakes, K. D., Wang, S., Cui, S., Pawliszyn, J., Metcalfe, C. D. & Servos, M. R. (2012). *In vivo* sampling of environmental organic contaminants in fish by solid-phase microextraction. *TrAC Trends in Analytical Chemistry, 32,* 31–39.

Zhao, Q. X., Chen, Y. W., Belzile, N. & Wang, M. (2010). Low volume microwave digestion and direct determination of selenium in biological samples by hydride generation-atomic fluorescence spectrometry. *Analytica Chimica Acta, 665,* 123–128.

Zhou, C.-Y., Wong, M.-K. & Koh, L.-L. (1996a). Microwave-assisted dilute acid extraction of trace metals from biological samples for atomic absorption spectrometric determination. *Journal of Analytical Atomic Spectrometry, 11,* 585–590.

Zhou, C.-Y., Wong, M.-K., Koh, L.-L. & Wee, Y.-C. (1996b). Microwave digestion of biological samples with tetramethylammonium hydroxide and ethylenediaminetetraacetic acid for elemental determination. *Talanta, 43,* 1061–1068.

Zhou, Y., Zhang. C., Yan, Z., Li, K., Wang, L., Xie, Y., Li, F., Liu, Z. & Yang, J. (2012). The use of copper (II) isonicotinate-based micro-solid-phase extraction for the analysis of polybrominated diphenyl ethers in soils. *Analytica Chimica Acta, 747,* 36–41.

Zhu, F., Xu, J., Ke, Y., Huang, S., Zeng, F., Luan, T. & Ouyang, G. (2013). Applications of *in vivo* and in vitro solid-phase microextraction techniques in plant analysis: A review. *Analytica Chimica Acta, 794,* 1–14.

Zou, Z., Jang, A., MacKnight, E., Wu, P.-M., Do, J., Bishop, P. L. & Ahn, C. H. (2008). Environmentally friendly disposable sensors with microfabricated on-chip planar bismuth electrode for *in situ* heavy metal ions measurement. *Sensors and Actuators B, 134,* 18–24.

8 Green Aspects of Miniaturized Sample Preparation Techniques

Adam Kloskowski[a*], Łukasz Marcinkowski[a] and Jacek Namieśnik[b]
[a] *Gdansk University of Technology, Faculty of Chemistry, Department of Physical Chemistry, Narutowicza Str.11/12, Gdansk 80–233, Poland.*
[b] *Gdansk University of Technology, Faculty of Chemistry, Department of Analytical Chemistry, Narutowicza Str.11/12, Gdansk 80–233, Poland.*
*e-mail: adam.kloskowski@pg.gda.pl

8.1 Introduction

Correct monitoring and control of the state of the environment and scientific research on the subject of environmental protection often require that determinations of compounds present at very low concentrations be made in samples characterized by a matrix of complex composition (*e.g.*, biological samples, water and soil samples). Moreover, due to the limited number of analytical techniques which are sufficiently sensitive to conduct direct determinations of trace components, it is necessary to employ analytical procedures that involve an analyte isolation and enrichment step prior to final analyte quantification. Technical and methodological solutions described in previous chapters allow for reduced dimensions of equipment and the amount of material used at the sample preparation stage prior to analysis. Therein, the reader will find a detailed description of specific techniques and their associated analytical methods, including their range of application for isolating certain types of analytes from various kinds of samples. The aim of this chapter is to evaluate pro-ecological effects resulting from the miniaturization of sample preparation techniques. Although the elaboration of most of the aforementioned solutions was dictated by very different demands, *e.g.*, cost reduction, ease of application and the ability to create hyphenated analytical systems, the environmental effects are also considered to be a key benefit to their development.

A discussion of the relationship between the miniaturization of analyte isolation and enrichment techniques and the requirements posed by green analytical techniques should begin by reminding the reader of the fundamental targets that guide green analytical chemistry (GAC). At present, it is commonly accepted that the development of analytical techniques should proceed by following the priority issues:

1. discontinuation of or significant reduction in the use of environmentally-harmful chemical compounds in analytical procedures, with a special emphasis on organic solvents
2. limiting the emission of vapors and gases

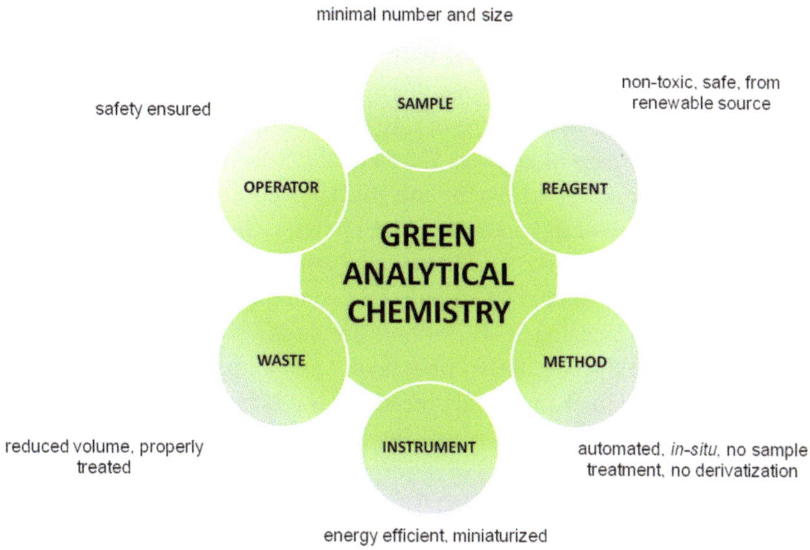

Figure 8.1 Important components of analytical procedures in relation to the principles of GAC.

3. limiting the amount of liquid and solid wastes generated by analytical laboratories
4. introduction of solutions into analytical practice which would allow the application of environmentally-friendly solvents such as water and supercritical CO_2
5. reduction of energy- and labor-intensive analytical procedures
6. possible automation

Important components of analytical procedures in relation to the principles of GAC are shown in **Figure 8.1**.

First of all, it should be mentioned that the fulfillment of some of the aforementioned priorities is an obvious consequence of the miniaturization of sample preparation techniques. Undoubtedly, a reduction in the use of organic solvents, vapor emissions, and the amount of generated wastes will accompany miniaturization. A similar relationship can be found between the miniaturization and lowered energy intensity, for example, *via* decreasing energy used per unit when thermostating a sample or by reducing the amount of solvent being evaporated in the process of analyte condensation. On the other hand, the same aims can be and often are achieved not by scaling down, but rather by choosing appropriate materials used in microextraction techniques as well as the procedures for their application. As an example might serve the application of polydimethylsiloxane (PDMS). This polymer is applied as a sorbent, solely, in microextraction techniques. Its properties allow multiple use and thermal desorption of analytes, thus eliminating the need of using organic solvents. It should also be mentioned that the sample preparation techniques are not easily classified according to their green features because in many cases the application of one technological solution allows for achieving multiple aims. However, other goals, such as automation of the procedure often require additional efforts. Solid-phase microextraction (SPME) is a flagship example of GAC because the use of polymers as a sorption phase and the release of analytes *via* thermal desorption allows for a complete elimination of organic solvents from the analytical procedure, and therefore points 1–3 are fulfilled. In the case of thermally sensitive analytes, the same technique requires that analytes are extracted from the fiber with solvents, thus reducing the green character of the procedure. A similar situation is related to the possible automation of the analyte isolation step by using SPME. Technically, such a solution was achieved only a few years ago when the specialized fiber holder and automated sampler were constructed. The automation of analytical procedures is relatively simple only in the case of thermal desorption of analytes where the SPME fiber is introduced directly into the GC injector, whereas in the case of desorption with solvents, more sophisticated devices are used (Hutchinson et al., 2007).

Considering the above information, we will discuss microextraction techniques for sample preparation with regard to the materials used, design solutions and possible automation. Finally, the application of chemometric tools in microextraction techniques is briefly described.

8.2 Reduction of the Amount of Organic Solvents Used

To begin the process of reducing the amount of organic solvent used in an analysis, a reference point has to be chosen. This is usually the technique that is most commonly used in practice, which for many processes is the sample preparation step based on liquid-liquid extraction (LLE) with organic solvents. LLE is one of the oldest extraction techniques, mostly applied to aqueous samples that are characterized by a matrix of simple to complex composition. In general, the technique involves mixing the sample with a fresh aliquot of solvent, leading to the separation of analytes depending on their affinity for the two components of the system (sample phase and extractant). The use of solvents with increasing polarity creates an opportunity to obtain different fractions which are enriched in various analyte groups. Moreover, another step in the procedure is solvent evaporation aimed at condensing the dissolved analytes. Multi-step procedures are highly time- and labor-intensive and, as a consequence, result in longer exposure of the laboratory personnel to harmful vapors of chemical reagents, particularly of organic solvents. Also, a relatively large number of operations performed on the sample increases the risk of analyte loss and sample contamination. From the point of view of green chemistry priorities, in this type of procedure the only requirement which can be fulfilled is the application of environmentally friendly solvents, *e.g.*, water and supercritical CO_2. All design solutions lead, in a natural way, to the miniaturization of the dimensions of equipment used for isolating analytes and, at the same time, to decreased use of organic solvents. As a consequence, the amount of waste produced and the emissions of vapors and gases are diminished. Nevertheless, it should be mentioned that the same phenomenon is used for isolating analytes (separation based on partitioning between phases) by liquid-phase microextraction (LPME) techniques and classical LLE techniques. Therefore, LPME is not always a practical alternative to typical LLE techniques, for example, in situations when high recovery instead of high enrichment of analytes is required. In such a scenario, due to the limited volume of extractant, the competitiveness of LPME techniques is based on the extractant's physical properties, in particular on its affinity for target analytes. In this context, miniaturization of extraction techniques allows a search for alternative extractants among rare and/or expensive materials.

Single-drop microextraction (SDME) is one of the most popular techniques in which the use of solvent has been significantly reduced to only one drop with a volume of 1–8 μL compared to classical LLE. SDME was developed in 1996 under the name Solvent Extraction in a Microdrop (Liu & Dasgupta, 1996) or Solvent Microextraction into a Single Drop (Jeannot & Cantwell, 1996). In this technique, the extraction process occurs *via* dissolution of analytes in a drop of solvent suspended at the end of microsyringe needle. The needle is immersed in the sample or sample headspace (HS-SDME mode) (Tankeviciute et al., 2001). In the case of HS-SDME, the choice of solvent is wider because its solubility in the sample is of no significance. In order to improve the stability of the suspended drop, needles with specifically shaped ends and made of various materials are used (Ye et al., 2007; Batlle et al., 2008; Sharma et al., 2011).

In some cases, it is also possible to reduce the sample volume. The drop-to-drop SDME technique (DD-SDME) enables fast extraction and eliminates the sample mixing because the analytes are extracted into a drop of solvent from the sample of a very small volume, usually no more than 10 μL (Wu et al., 2006). After the extraction of analytes, the drop of liquid sorbent is drawn into the syringe with the plunger and then injected into the dosing device of measuring instrument. The extraction of analytes by SDME techniques is characterized by short extraction time, low cost, simplicity of operation, and no need for complicated equipment.

A further reduction of the amount of solvents used has been achieved in the lab-on-a-chip version known as droplet-membrane-droplet-LPME, in which the analytes are extracted through a liquid membrane into a few microliters of acceptor phase (Sikanen et al., 2010). The chip used in droplet-membrane-droplet-LPME technique is made of appropriately molded aluminum foil. The sample phase and the extractant phase (10 μL) are separated by microporous polypropylene membrane that had been impregnated with a solvent. A small amount of sample (10–15 μL) is placed onto the membrane. After the extraction, the acceptor phase is collected with a microsyringe and then introduced into an analytical instrument.

To make SDME compatible with gas chromatography (GC), mainly organic solvents (1-octanol, toluene, dodecanol, undecanol) or chloroorganic solvents (dichloromethane, chloroform, carbon tetrachloride, trichloroethylene, dichloroethylene) are used as extractants (Romero et al., 2007). However, the application of the latter is being discontinued due to their toxicity. Information about the application of other less traditional solvents, such as aqueous solutions of β-cyclodextrin, can be found in the literature (Wu et al., 2008).

The modifications of classical SDME are also worth mentioning. The directly suspended droplet microextraction technique (DSDME) involves the placement of a drop of water-immiscible solvent directly on the sample surface. After the extraction, solvent is collected from the sample surface with a microsyringe (Yangcheng et al., 2006; Pena-Pereira et al., 2012). From the point of view of green principles, the main disadvantage of this solution is that the transfer of extract to the next steps of analytical procedure cannot be automated. Solidified floating organic drop microextraction technique (SFODME) is also an interesting option (Khalili Zanjani et al., 2007). In this technique, a drop of extractant immiscible with the sample matrix is placed on the sample surface and then stirred at constant temperature. After the extraction, the vessel containing the sample and extractant is cooled until the extraction medium solidifies. Next, the extractant phase with the absorbed analytes is transferred to another vessel where it melts. Then the extraction medium is dosed into a measuring device (Sobhi et al., 2008). As in the case of DSDME, the GAC limitation of SFODME is that the transfer of sample into a measuring device cannot be automated.

Hollow fiber liquid-phase microextraction techniques (HF-LPME) are another solution that allows a reduction in the amount of solvents used (Pedersen-Bjergaard & Rasmmussen, 1999). In this group of techniques, a small volume of liquid extraction medium (a few μL) is contained within the lumen of porous polypropylene fiber which is attached to the end of microsyringe needle and immersed in the sample (Zhu et al., 2001). Extraction by HF-LPME technique can be conducted either in a two-phase system, where the wall and lumen of the fiber are filled with the same solvent, or in a three-phase system in which the fiber wall is impregnated with different solvent than that present in the fiber lumen. A wide choice of available fibers (Kosaraju & Sirkar, 2007) and solvents enables the achievement of high extraction selectivity for different analytes, and allows the application of HF-LPME for sampling analytes from samples which are contaminated or have complex matrices.

As mentioned previously, the limitation of microextraction techniques based on the partitioning mechanism is the small amount of extractant used. As a consequence, the high affinity of the extractant for the analytes is much more significant for the quality of the obtained results in comparison to classical LLE. For example, in order to increase the extraction efficiency, multiwall carbon nanotubes (MWCNTs) dispersed in a specific extractant can be used. The hollow fiber solid–liquid phase microextraction (HF-SLPME) technique ensures high selectivity and good efficiency of analyte extraction even when the volume of sorption phase is very small (Es'haghi et al., 2010). Extraction can also take place in a two-phase system in which carbon nanotubes suspended in organic solvent fill the lumen of the hollow fiber (Es'haghi et al., 2011). A similar system has been designed for solvent bar microextraction (SBME) technique (Jiang & Lee, 2004), where the extractant is immobilized inside the pores of a polypropylene fiber. Both ends of the fiber are closed and the fiber is either filled with the extractant (liquid-liquid system), or with an acceptor phase that is different than the extractant (liquid-liquid-liquid system) (Yu et al., 2008). After the extraction, the acceptor phase is collected from the fiber lumen with a microsyringe and injected into the GC or high performance liquid chromatography (HPLC) device. In relation to HF-LPME, it is noteworthy that automation of analytical procedure (Ouyang et al., 2006) and concurrent sampling of analytes from multiple samples is possible, which is important with regard to the green character of this technique. In the case of solvent cooling assisted dynamic hollow-fiber-supported headspace liquid phase microextraction (SC-DHF-HS-LPME), the extractant is pumped through the porous polymeric fiber which has been cooled to -1 °C, to lower the solvent vapor pressure and therefore the solvent loss (Huang et al., 2007).

Another group of solvent microextration techniques is based on the dispersion of a small amount of solvent (up to 200 µL) in the investigated sample. In the basic variant of dispersive liquid-liquid microextraction (DLLME), the dispersed droplets of extractant are generated by using the third liquid phase, which does not mix with the sample or the extractant and acts as a dispersant (Rezaee et al., 2006). After the extraction, samples are shaken and centrifuged to separate the extraction phases, which is later injected into the appropriate analytical instrument. The presence of a dispersant in the system (usually methanol, acetone or acetonitrile in the amount of 0.5–2µL) may lower the value of the analyte-specific partition coefficient during the partitioning of analytes between the sample phase and extraction phase. This is undesirable from the point of view of green chemistry, therefore, dispersion by means of ultrasound (ultrasound-assisted emulsification-microextraction (USAEME) or ultrasound-assisted dispersive liquid-liquid microextraction (US-DLLME) (Regueiro et al., 2008; Fontana et al., 2009) or vortex mixing (vortex-assisted liquid–liquid microextraction (VALLME) is often applied (Jia et al., 2010; Yiantzi et al., 2010). Moreover, generation of very small droplets of extractant results in a significant increase in the interphase surface of both immiscible liquids, which speeds up the transfer of analytes to the extractant and consequently decreases the time necessary for isolating the analytes. Another solution which eliminates the use of dispersing solvents from the procedure is the application of surfactant solutions, often in combination with ultrasound treatment. In this way, techniques such as surfactant-assisted dispersive liquid–liquid microextraction (SA-DLLME) (Saraji & Bidgoli, 2010), ion pair-based surfactant-assisted microextraction (IP-SAME) (Moradi et al., 2011), ultrasound-assisted surfactant-enhanced emulsification microextraction (UASEME) (Wu et al., 2010) and vortex-assisted surfactant-enhanced-emulsification liquid–liquid microextraction (VSLLME) (Yang et al., 2011) were developed. Another aspect of modified DLLME techniques demonstrates that the principles of GAC are often realized by making a compromise on selected targets. It turns out that the previously criticized use of dispersant may result, under specific conditions, in the demulsification of the extractant. Guo and Lee (2011b) described a procedure where a mixture of hexane and acetone was added, leading to the creation of an emulsion. After addition of the next portion of acetone, demulsification occurred, causing separation of the extractant from the sample phase. Separated extractant could easily be retrieved with a syringe, so omission of certain steps of the analytical procedure, such as mixing, emulsification in an ultrasonic bath or freezing out the extraction phase was possible. As a consequence, the number of operations that involve sample handling decreases, which usually improves the quality of the results. It is an individual decision which option is considered more valuable in a given situation with regard to how green the whole analytical procedure should be. Demulsification is conducted after the extraction step *via* the addition of another portion of dispersant, which acts as a demulsifying agent. The techniques based on demulsification can be found in literature under the name of solvent-terminated dispersive liquid–liquid microextraction (ST-DLLME) (H. Chen et al., 2010), and solvent demulsification dispersive liquid–liquid microextraction (SD-DLLME) (Guo & Lee, 2011b). In both these techniques, solvents less dense than water are used which enables the collection of extract from the sample surface with a syringe.

Fully automated variants of DLLME techniques have also been developed, such as sequential injection dispersive liquid–liquid microextraction (SI-DLLME) which allows on-line injection of extracts into flame (or electrothermal) atomic absorption spectrometry atomizers (Anthemidis & Ioannou, 2010, 2011).

A further reduction in the amount of solvents used was achieved by forcing the transport of analytes through a semipermeable membrane in an electric field. In on-chip electro membrane extraction (EME), the isolation process was conducted by using a 25 μm thick porous polypropylene membrane whose pores had been impregnated with the acceptor phase. The membrane separates the microchannel (50 μm deep and 2 mm wide) bored in one of the two connected plates made of poly(methyl) methacrylate (PMMA) from a well containing a small amount (a few μL) of acceptor phase (2-nitrophenyl octyl ether (NPOE) or dodecyl acetate). The platinum wire anode is attached inside the microchannel into which sample is pumped, while the cathode is placed inside the well containing the acceptor phase. The transfer of extract to the next steps of analytical procedure is performed with a micropipette (Petersen et al., 2010).

8.3 Green Extraction Phases for Microextraction Techniques

In order to increase the green character of microextraction techniques, extractants which are less harmful to the environment are introduced into analytical practices. Initially, solvents such as chlorinated hydrocarbons were mainly used in DLLME techniques due to practical reasons. Because they are denser than water, it is possible to simply separate the phases by sample centrifugation. Recently, applications of solvents that are less dense than water as extractants in DLLME has been published (Hashemi et al., 2009). The use of this type of solvent creates a possibility to perform the extraction without a dispersant and, at times, allows for eliminating the sample centrifugation step. In most cases, the application of extractants which have lower density than water necessitated the elaboration of novel designs, for example, a narrow-necked vessel for collecting the extractant (Farajzadeh et al., 2009; Hashemi et al., 2009), a special vessel for centrifuging the emulsion in the USAEME technique (Saleh et al., 2009), a vessel which speeds up the extraction *via* mixing with a magnetic stirrer (Zhang et al., 2011) and a membrane system which facilitates the dosing and collection of extractant from the sample (Farajzadeh et al., 2012). Long-chain alcohols such as dodecanol and hexadecanol are often used as solvents in dispersive techniques. These solvents are commonly recognized as not very harmful to the environment. In addition, since their melting points are close to room temperature, it is possible to solidify the extractants which facilitates their quantitative collection from the sample mixture. In dispersive liquid-liquid microextraction based on the solidification of a floating organic drop (DLLME-SFOD), the extraction proceeds as in the typical DLLME. The extractant is added to and then dispersed in the analyzed sample, followed by sample centrifugation and cooling to solidify the extractant present on the top of the sample vessel. The solidified extractant is transferred to another vessel where it melts. Next, the extractant is analyzed using chromatographic techniques (Leong & Huang, 2008; Xu et al., 2009).

Another example of the implementation of GAC principles in LPME techniques is the use of extractants which are not harmful to the environment, fo example, aqueous solutions of surfactants in cloud point extraction (CPE) (Paleologos et al., 2005). In supramolecular-based dispersive liquid–liquid microextraction (SM-DLLME), coacervates and supramolecular systems serve as extractants (Jafarvand & Shemirani, 2011). Coacervates, which are colloidal micelles (microscopic structures in the shape of droplets), self-assemble in colloidal systems while their external layer is semi-permeable and allows the analytes to permeate to the interior. Contrary to basic dispersive techniques, SM-DLLME is characterized by a short extraction time due to the dispersion of the sorption phase, and it enables the isolation of analytes with widely ranging polarity while eliminating toxic solvents and sample mixing. In the case of surfactants present in the ionized form (anionic and cationic forms), coacervates are obtained by cooling the solution below the cloud point. For non-ionic surfactants, the solution should be warmed up above the cloud point. During the isolation process, poorly soluble and non-polar analytes are dissolved in the hydrophobic inner space of micelles, while polar compounds become incorporated into the external micellar structure (Yu, 2005; Madej, 2009). Because the surfactant can be adsorbed onto the capillary wall or the packing of chromatographic column, in some applications the extract is diluted before quantitative analysis (Giokas et al.,

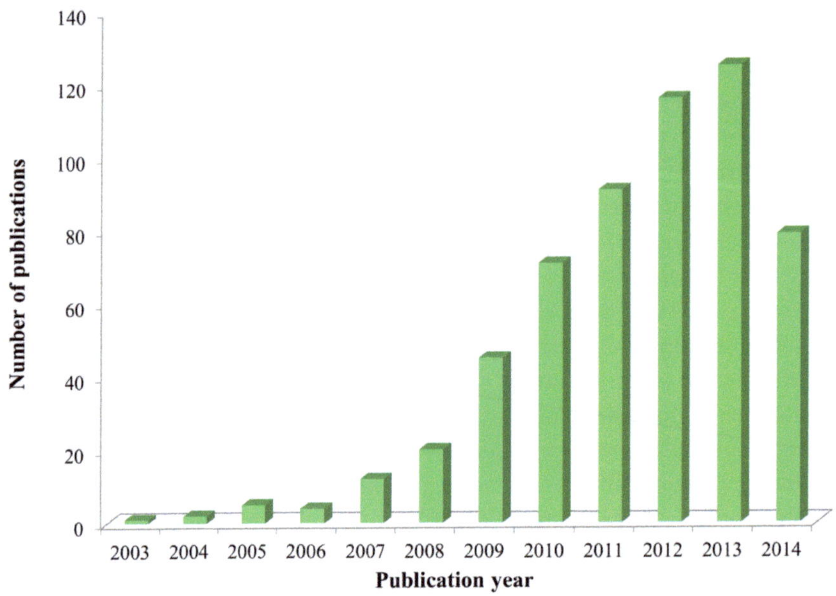

Figure 8.2 Number of publications regarding ionic liquids in microextraction techniques in the past decade.

2002), although this limits the green character of the technique. Therefore, the dual cloud point extraction (dCPE) technique was developed with the use of different surfactant (Yin, 2007). In the first step of this procedure, a surfactant solution is added to the sample containing hydrophobic analytes, which results in the formation of hydrophobic complexes with the analyte. Next, the sample is heated and centrifuged, and the complexes are extracted into the newly created phase which is rich in another surfactant. In the final step, an aqueous solution of L-cysteine is used to form a stable nd hydrophilic water soluble complexes of mercury for their subsequent transfer to aqueous phase, which is later analyzed by capillary electrophoresis (CE).

Another example of using greener substitutes is the application of ionic liquids (ILs) in microextraction techniques. The main property of ILs, which allows this group of compounds to be called green, is their very low vapor pressure and high thermal stability. Due to their high viscosity and surface tension, ILs can form a stable phase boundary which enables the extraction of analytes from the sample headspace and *via* direct contact with the investigated sample. As a result, one can notice a continuous growth in the the number of publications describing application of ILs in microextraction techniques, as shown in **Figure 8.2**. As in the case of previously described coacervates, it is noteworthy that the application of ILs results, to a certain degree, from the miniaturization of devices used in the sample preparation step. A decrease in the amount of extractants allowed the use of rare and expensive materials. Considering the use of ILs in microextraction techniques, one should also bear in mind that the physicochemical properties of these compounds can be controlled by properly selecting the component anions and cations.

As a result of the aforementioned properties, ILs have been used for coating fibers in SPME (Ho et al., 2011). The high viscosity of ILs allows one to obtain coatings of relatively high thickness and good stability. The first attempts at coating fibers with ILs were undertaken in 2005 (Liu et al., 2005). Unfortunately, the IL coatings produced could be used only once, and the fibers had to be coated with an IL after each use. The application of Nafion as a pre-coating substratum increased the surface wettability and therefore enabled the formation of a thicker layer of IL. The increased amount of sorption phase allowed the application of IL-coated fibers for isolating polycyclic aromatic hydrocarbons (PAHs) from aqueous samples. However, the use of Nafion did not solve the problem of repeated coating, and it was still necessary to renew the sorbent layer after each extraction (Hsieh et al., 2006). The immobilization of ILs on the fiber of the SPME device can be achieved *via* impregnation of silicone elastomer coatings on the fiber surface (He et al., 2009). To this end, sol-gel techniques (López-Darias et al., 2010) or *in situ* cross-linking on stainless steel fibers coated with a layer of silver is used (Feng et al., 2011).

The use of ILs in SDME techniques has produced better results and the first application of an IL in SDME was reported in 2003. It was demonstrated that the extraction of PAH analytes with an IL gave better results compared to the extraction with 1-octanol for both direct sample extraction and sample headspace extraction (Liu et al., 2003). In consecutive publications, the applicability of ILs for isolating a wide spectrum of compounds including pesticides (Ye et al., 2006), phenols (Wang et al., 2010), BTEX (Aguilera-Herrador et al., 2008) and PAHs (Yao et al., 2009) was demonstrated. ILs were also successfully used to isolate inorganic compounds from food samples (Pena-Pereira et al., 2009). The instability of an IL drop and the small drop volume still remain the main problems in SDME techniques that employ ILs but new technical solutions continue to appear. In 2009, the use of an IL in dynamic liquid-phase microextraction (dLPME) was described (Cruz-Vera et al., 2008). The constructed device designed for controlling the flow rate allows precise dosing of a given drop volume and enables increasing the drop volume to 50 μL which translates into improved measurement sensitivity. Due to the thermal resistance of ILs, it was possible to couple SDME with GC (Aguilera-Herrador et al., 2007). After the extraction, the same syringe is used to directly introduce the droplet of ionic liquid into the GC injector where the analytes are released *via* thermal desorption. Only a slight modification of standard injector is necessary which involve using an inlet liner with a larger diameter to avoid the transfer of IL into the column (Chisvert et al., 2009).

The use of ILs in dispersive techniques is the natural consequence of their properties, i.e. viscosity, surface tension, and hydrophilicity/hydrophobicity. In the case of DLLME, ILs were applied in the extraction of organic compounds (aromatic amines,

PAHs) (Pena et al., 2009) as well as metals (cadmium, lanthanides) (Martinis et al., 2008). The DLLME techniques employing ILs mainly differ in the way the extractant is dispersed. Some techniques are based on a complexing agent and surfactant, which lowers adhesive interactions between the IL and the vessel wall. On the other hand, in ionic liquid-based ultrasound-assisted dispersive liquid–liquid microextraction (IL-USA-DLLME) the dispersion effect is achieved by means of ultrasound (Arvand et al., 2012). In vortex-assisted liquid–liquid microextraction (VALLME), the IL is dispersed throughout the sample by vortex mixing (Chamsaz et al., 2013).

In *in situ* solvent-formation microextraction (ISFME) a hydrophobic IL and a reagent that aids in the formation of ion pairs are added simultaneously to the sample. As a result, the solution becomes cloudy due to the formation of microdroplets consisting of the IL and the aiding agent. Finally, the extract is separated by centrifugation (Baghdadi & Shemirani, 2008). Another interesting solution, based on the broad hysteresis of ILs and their propensity to occur in the supercritical state, is cold-induced aggregation microextraction (CIAME). At first, the IL is dissolved in a warm sample. Next, the mixture is cooled in an ice bath in order to separate the extractant phase. After reaching the cloud point, the IL and the retained analytes are centrifuged. In both CIAME and temperature-controlled ionic liquid exhaustively dispersive liquid phase microextraction (TILDLME), the dispersion effect can be achieved without a dispersant (Zhou et al., 2008; Bai et al., 2009).

The application of ILs in previously mentioned techniques such as SBME (Guo & Lee, 2011a) or a three-phase liquid extraction system such as hollow fiber membrane liquid–liquid–liquid microextraction (HFM-LLLME) (Basheer et al., 2008) has also been reported.

Microextraction techniques that use water as an analyte-retaining medium have the greenest characteristics. HF-LPME is one of such techniques. The extraction can be conducted either in a two-phase system, where the aqueous solution of acceptor is separated from the sample by a semipermeable membrane, or in a three-phase system in which the membrane pores are filled with an appropriate organic solvent. In the latter case, the technique is practically solvent-free because the organic solvent is not consumed during the extraction. The isolation of analytes proceeds *via* appropriate selection of pH of the acceptor phase and sample (donor). It must be noted that this technique can be applied to ionized analytes only. It is possible to control the selectivity and efficiency of the extraction by appropriately selecting the liquid membrane and the pH of acceptor and donor phases.

The limited dimensions of the HF-LPME device allowed for the use of electric fields to accelerate the transport of analytes. In this way, an electrochemically-enhanced hollow fiber microextraction technique has been developed (Pedersen-Bjergaard & Rasmussen, 2006; Jamt et al., 2012). The extraction of analytes takes place in a three-phase system. The analytes migrate *via* a liquid membrane (immobilized in the pores of a polypropylene fiber) from the aqueous sample, in which the electrode is immersed, to the aqueous acceptor phase inside the fiber where the other electrode is present. After the extraction, the acceptor phase is introduced into a CE instrument by means of a syringe. In the case of electrochemically-aided extraction, the transport of analytes is forced by the difference in the potentials of the acceptor and donor phases. Electrochemically-enhanced hollow fiber microextraction enables a significant decrease in extraction time compared to typical HF-LPME. It is noteworthy that a miniaturized version of the technique exists which is analogous to the previously described technique of EME in which the difference in pH of the acceptor phase and sample was the driving force behind the analyte extraction into the acceptor phase (Ramos Payán et al., 2012). Electrochemically-enhanced extraction allows for very efficient and selective isolation of analytes from samples with complex matrices, including biological and environmental samples (Nojavan & Fakhari, 2010; Seidi et al., 2011).

Another group of microextraction techniques is distinguished by the use of solid and pseudo-liquid materials in the process of analyte extraction. These techniques constitute the green version of solid phase extraction (SPE) techniques. In the classical version of SPE the analytes are retained on the sorbent bed and then washed out with significant amounts of organic solvents. The additional disadvantage of this type of procedure is the necessary preconcentration of the extract prior to its use in the later steps of the analytical procedure. Preconcentration is usually performed *via* solvent evaporation. In microextraction techniques, due to the small amount of sorption phase used (even less than 1 μL), the amount of solvents necessary for washing out the analytes is significantly reduced and is a step towards implementing GAC. However, it has to be stressed that the biggest advantages of SPME techniques become apparent when thermal desorption methods are employed to release the analytes. It should be kept in mind that the solvent-free character of a given technique is not directly connected to a concrete technical design because the majority of the presented techniques can also be applied as a variant with solvent extraction. In the two dominant microextraction techniques, i.e., SPME and stir bar sorptive extraction (SBSE) there is a plethora of methodological, technical and material solutions that allow for conducting the analyte isolation with some use of solvents (a step involving washing out of analytes with solvents) as well as completely solvent-free procedures. The analysis of available literature indicates that the above-mentioned division of techniques is mainly based on the type of analytes and sample matrix, and only to the limited extent on miniaturization. However, it is worth mentioning that certain schemes fundamentally result from the fact that small amounts of sorbents are used. The need to achieve low quantification limits forces the use of sorbents which enable obtaining high extraction efficiencies. First of all, this favors the application of adsorbents, and secondly, drives the development of new adsorbent types. On the other hand, the use of adsorbents, which strongly interact with analytes, significantly limits the application of thermal desorption as the method for releasing the analytes. The use of quasi-liquid polymers as sorption phase, for example, PDMS, does not mean at all that the desorption of analytes will be exclusively limited to thermal methods.

Considering the above information, the application of the most popular SPME techniques is discussed in the next part of the chapter, with a particular emphasis placed on the materials used and without the methodology for analyte release.

As already mentioned, SPME is one of the most popular microextraction techniques employing adsorbents. This technique has numerous practical applications which encompass a wide spectrum of analytes and samples in different states of matter and in complex matrices. SPME is one of rare techniques which offers the concurrent fulfillment of a number of requirements posed by GAC, such as complete elimination of organic solvents, simple automation, low labor- and time-intensity procedures for isolating the analytes, and the possibility of *in situ* and *in vivo* sample collection (Vuckovic et al., 2010; Zhang et al., 2012). Nevertheless, it should be kept in mind that not all analytical procedures employing this technique fulfill the aforementioned conditions.

Therefore, despite many advantages of SPME, there is still a need to find novel technical, methodological and material solutions which would allow for improving the efficiency of analyte extraction and a further reduction in the sample preparation time. To this end, new designs have been presented, such as internally cooled fibers (Zhang & Pawliszyn, 1995) and in-tube SPME (Eisert & Pawliszyn, 1997). Also, information about electrosorption-enhanced solid-phase microextraction (EE-SPME) can be found in the literature (Chai et al., 2007). In the latter technique, a fiber coated with an electrical conductor such as Nafion/carbon nanotubes (Zeng et al., 2010) allows for fiber polarization such that the fiber plays the role of a working electrode. The application of this technique enables a significant decrease in the extraction time and creates a possibility to extract ionizable analytes from samples with polar matrices (Zeng et al., 2011).

However, the greatest progress in SPME techniques has been in the field of novel materials, which aside from increased extraction efficiency should also have better mechanical resistance. These improved features translate into lower material use and higher thermal resistance, which allows for the application of thermal desorption. The most frequently reported sorption materials synthesized for SPME purposes are polymeric sorbents such as conductive polymers (Wu & Pawliszyn, 2001), molecularly imprinted polymers (MIPs) (Turiel & Martín-Esteban, 2009), and the materials obtained *via* sol-gel processes (Kumar et al., 2008). Information about the materials that are more technologically advanced can also be found in the literature, *e.g.*, polymeric ILs whose development has solved the aforementioned problem with low durability of microextraction fibers coated with ILs (Zhao et al., 2008). Polymeric ILs are characterized by high thermal resistance which extend the lifetime of microextraction fibers (up to 150 extractions) (Wanigasekara et al., 2010).

SPME is also suitable for applying nanotechnology products. Nanomaterials are attractive because they possess high surface area and regular and reproducible structure *e.g.*, graphene lattice), while the properties of their adsorption centers can be modified. The system which is most frequently described in literature consists of microextraction fibers coated with carbon nanotubes (single-walled carbon nanotubes and MWCNTs). These materials display unusual mechanical resistance and chemical stability as well as unique electrical properties (Wang et al., 2006). Other forms of carbon such as carbon nanocones/disks (Jiménez-Soto et al., 2010), hydroxyfullerenes (Yu et al., 2002), carbon aerogels (CAs) and wormhole-like mesoporous carbons (WMCs) (Zhu et al., 2010) are also used as sorbents. Moreover, graphene has been applied in this context because it has high thermal resistance (above 300°C) (J. Chen et al., 2010). Meso- and nanoporous silica materials have even higher thermal resistance reaching 900 °C, for example, commercially available MCM-41 and SBA-15 (Du et al., 2005). In the literature, there are also reports about silica-based composite materials such as nanocomposite polypyrrole and SBA-15 (Gholivand et al., 2011), silica-carbon composites (Zeng et al., 2008) and the mechanically-resistant composite of aniline and silica nanoparticles (Bagheri & Roostaie, 2012). An interesting solution is the techniques in which a metal wire is coated with silica microstructures called flower-like. The created highly porous coating serves as a support for liquid organic extractant (typically propyl benzoate). After extraction, both the extractant and retained analytes are evaporated in the GC injection port due to their high volatility. Coatings of gold and silver nanoparticles are other nanosorbents described in the literature. The obtained adsorption layer is characterized by very high thermal and chemical resistance (Feng et al., 2012).

The second-most reported microextraction technique which fulfills the principles of GAC is SBSE. It was described for the first time in 1999 as an alternative to SPME and has a similar range of applications. In principle, SBSE was developed to fulfill the need for large sorbent volumes (compared to SPME) and to retain the simplicity of operation. This technique allows for partial or even total elimination of organic solvents from the analytical procedure, which has been achieved by the application of sorbents that are identical or similar to the ones used in SPME (Baltussen et al., 1997). Extraction can be conducted directly from the sample or from the sample headspace. The analytes can be released *via* thermal desorption as well as desorption with a solvent (Kawaguchi et al., 2006). As in the case of SPME, significant progress in the field of sorbent development has been made in SBSE techniques although its scope is more limited since SBSE requires high mechanical resistance (*e.g.*, to friction on the walls of the sample container). Nevertheless, besides the typical sorbents such as PDMS, a number of new materials can be used such as polyurethane foams (PU) (Portugal et al., 2008; Silva et al., 2008), poly(phthalazine ether sulfone ketone) (PPESK) (Guan et al., 2008), alkyl-diol-silica (ADS) restricted access material (RAM) (Lambert et al., 2005), monolithic adsorbents (Huang et al., 2009a, 2009b, 2009c), and composite sorbents such as PDMS/polypyrrole (PPY) (Melo et al., 2009). The descriptions of coating a stirrer with MIP-type materials have also been published (Zhu et al., 2006).

Aside from the two leading techniques described above, many novel designs can be found in the literature. This indicates that a continuous need for technical solutions still exists in order to decrease the amount of solvents that are used. For example, in micro-SPE (μ-SPE) a solid sorbent is placed inside a bag made of a porous membrane (Basheer et al., 2006). The use of the membrane allows one to extract the analyte from contaminated samples and suspensions. Moreover, in comparison to classical SPE, μ-SPE is simpler and it decreases the time needed for analyte extraction (Basheer et al., 2007). A comparatively simple way of introducing sorbent into the sample was presented in adsorptive microextraction techniques, such as gluing the sorbent to a carrier made of polypropylene fibers or polystyrene spheres (Neng et al., 2010). Stir cake sorptive extraction (SCSE) is an interesting solution which extends the life span of sorptive elements because the monolithic adsorbent has been synthesized inside the holder that protects its content against mechanical damage. In this way, the life span of a sorptive element can reach up to 1000 h of extraction (Huang et al., 2011). Solid phase nanoextraction (SPNE) techniques are also worth mentioning because they are similar, to a certain degree, to dispersive techniques although in this case a solution of gold nanoparticles is used as suspension (Wang & Campiglia, 2008).

Unfortunately, in the techniques just described it is necessary to desorb the analytes with solvents, which is a disadvantage. However, it has to be highlighted that the use of solvents has been significantly reduced due to the very small dimensions of the extraction devices.

8.4 Automation in Microextraction Techniques

Possible automation of sample preparation is a significant aspect used to evaluate whether a procedure fulfills the principles of GAC. When making such an evaluation, it is critical to consider both the possible automation of sample preparation and the possible automation of the entire analytical procedure, including the introduction of analytes into a measuring instrument. Based on the analysis of potential cost reduction of materials and energy used, it seems that a possible automation of the entire analytical procedure has much lower impact on the overall costs. On the other hand, the automation of sample preparation translates into a direct decrease of materials used, including organic solvents when applicable. This mainly results from the reduction of human error, which increases the reproducibility of the results and therefore reduces the number of necessary analyses. Another obvious consequence of automation is the reduced generation of waste, lower energy use for obtaining a sufficient amount of data, and lower costs of data analysis. Considering the miniaturization of microextraction devices, it seems that the automation of microextraction techniques should not pose big problems. However, analysis of published literature shows that miniaturization may help with automation, but it is not always the case. In some techniques, *e.g.*, SPME, the automation happened almost naturally because it has been based on the previously automated devices. In the case of other techniques, it was necessary to develop specialized technological solutions. The examples of different technical solutions used in the aforementioned microextraction techniques are presented below.

The automation of SPME (not always implemented) has been achieved by designing an appropriate holder for the SPME device which is compatible with typical autosamplers. In this way, it is possible to conduct all steps of the SPME procedure in fully automated mode, i.e., sample incubation, analyte extraction, desorption and fiber cleaning. Moreover, thanks to the application of computer-controlled integrated systems, the critical parameters such as extraction time and sample temperature can be controlled (O'Reilly et al., 2005; Vuckovic et al., 2008). At present, a number of commercial devices are available, including TriPlus (Thermo Fisher Scientific; Milan, Italy), Combi-PAL (CTC Analytics; Zwingen, Switzerland), MPS 2 (Gerstel Inc.; Mulheim and der Ruhr, Germany), and Concept 96 (PAS Technology; Magdala, Germany). Modifications allowing for multiple concurrent extractions are also known, for example, multi-well plate format SPME. These have helped to reduce the time necessary for obtaining the final quantification, and in the improvement of technique precision and extraction repeatability (Vatinno et al., 2008). Some variants of SPME have also been automated, like in the case of internally-cooled coated fiber (CCF) device. Here the automation enables a better control of fiber temperature as well as the application of the CCF-SPME device in routine analyses (Chen & Pawliszyn, 2006).

Procedure automation is also possible in the case of SBSE techniques. Moreover, devices for analyte desorption have been introduced to analytical practices which allow for concurrent determination of analytes released from a number of sorption elements (Kawaguchi et al., 2004).

In general, the attempts to automate liquid phase extraction techniques have encountered bigger difficulties. Nevertheless, examples of successful automation of the sample preparation step exist.

The development of an automated version of SDME was relatively simple. The variant is based on a typical microsyringe which is coupled with an automatic sample feeder (Ouyang et al., 2007). A successful attempt at automating the analytical procedure was also undertaken in the case of HF-LPME. In addition, in the dynamic hollow-fiber liquid-phase microextraction (dynamic-HF-LPME) the possibility of concurrent analyte sampling from a number of samples has been achieved (Pezo et al., 2007). Information about the fully automated procedures based on DLLME techniques can also be found in the published literature. SI-DLLME enables on-line dispensing of the extraction phase samples into FAAS (Anthemidis & Ioannou, 2009) and ETAAS atomizers. In this technique, a dispersing agent, extractant and complexing compound is introduced into the stream of aqueous sample which flows through a microcolumn filled with adsorbent. Next, the retained complexes of analytes are washed out with isobutyl methyl ketone directly into the atomizer of an atomic absorption spectrometer (Anthemidis & Ioannou, 2011). In another, more complicated technical solution, the extraction is conducted in a conical vial, followed by spontaneous phase separation. The extract is transferred into a micro-volume Z-flow cell and then analyzed spectrophotometrically (Andruch et al., 2012).

8.5 Chemometric Approaches for Optimization and Evaluation of Microextraction Techniques

Basically, all techniques of analyte isolation apply a similar scheme, i.e., contact between the sample and the sorption phase, the latter displaying higher affinity for the analytes than the sample matrix. As a result, the analytes are collected on the sorbent in accordance with the system-specific partition coefficient and the ratio of sorption phase volume to sample volume. The kinetics of analyte transport depend on the values of the diffusion coefficient and the geometry of the system. The extraction efficiency in microextraction techniques is fundamentally disadvantageous because of the volume ratio of phases. As a consequence, exhaustive extraction occurs only in very rare cases. In most situations, the amount of analyte retained during the extraction is sufficient to deplete their content in the sample to the level which can be assumed to be constant. Under these circumstances, the precise determination of optimal extraction conditions has a fundamental role. In-depth knowledge of physicochemical processes taking place in the system during extraction is necessary for selecting critical parameters which should be considered as independent variables in the optimization process. Nevertheless, the quantitative effect of these parameters, particularly in the case of interactions among variables, is difficult to predict. In such situations, the classical methodology for establishing optimal conditions of analyte isolation fails completely because it is based on the one-factor-at-time approach in which only one parameter is optimized and the others are treated as constant. At this point, the application of chemometrics becomes necessary because it offers more advanced tools based on mathematical and statistical methods. In the optimally designed set of experiments, chemometrics can fulfill the need for finding the maximum amount of information which will allow for determining optimal conditions of chemical processes and extending the knowledge about the system operation. The aforementioned aspects of chemometric applications implement the principles of GAC in several ways:

1. highlighting the parameters which are significant for the process of analyte extraction and therefore creating possibilities to reduce the number of variables being optimized, and gaining the full control over the sample preparation process
2. a significant reduction in the number of experiments that are necessary to determine the optimal conditions of analyte isolation
3. a possibility to determine optimal conditions irrespective of interactions, which guarantees the most effective procedure for analyte isolation.

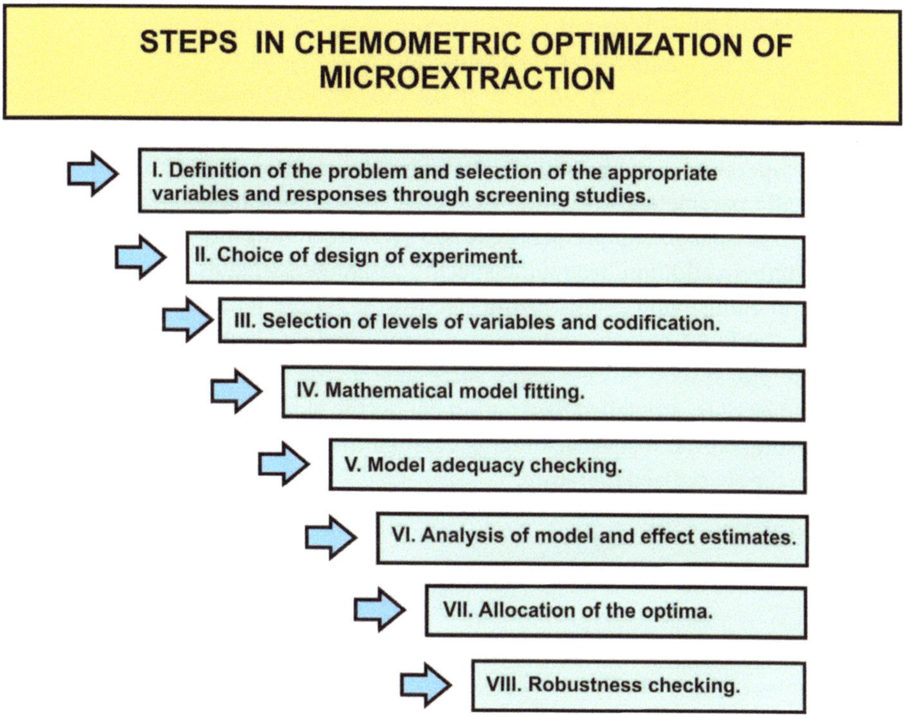

Figure 8.3 Steps in chemometric optimization of microextraction.

As mentioned before, a key issue from the point of view of microextraction techniques is the inclusion of interactions among the variables. To this end, an appropriate mathematical model (equation 8.1) is used, which is fitted to the input data describing the process parameters (x_i, x_j) and the system response (Y), the latter usually being the parameter related to the amount of analytes retained (*e.g.*, the area of a chromatographic peak).

$$Y = \beta_0 + \sum \beta_i x_i + \sum \beta_{ii} x_i^2 + \sum \beta_{ij} x_i x_j + \varepsilon \tag{8.1}$$

where β_0 is constant, β_i is a regression coefficient of linear terms, β_{ii} is a regression coefficient of quadratic terms, β_{ij} is a regression coefficient of interactions between the parameters x_i and x_j, and ε is the residual associated with the experiment.

Equation 8.1 is presented in the most generalized form which allows for determining the linear effects, maximal values or saddle points, and the interactions between the parameters. In practice, the number of terms in equation 8.1 is determined from the goodness of fit analysis (analysis of variance, ANOVA) using the observed values. The optimization of microextraction process (with respect to the number of variables, **Figure 8.3**) is conducted in a number of key steps. Most of the steps are realized by utilizing chemometric tools, while the experimenter has the fundamental influence on the final results with regard to the appropriate choice of optimized parameters (step 1) and the determination of their variation level (step 3).

As mentioned before, a number of parameters should be considered when dealing with microextraction techniques. These range from the most general ones, such as temperature, sample volume (or headspace volume) and extraction time, to more specific parameters, including extractant type and volume, extraction mode (direct or headspace), sample pH, agitation type (including the stirring rate) and ionic strength. In addition, technique-specific parameters such as the volume of dispersant and emulsifying agents or the electrical potential must be considered and optimized. **Table 8.1** contains examples of literature data about the optimization of the sample preparation process using some microextraction techniques, as well as the applied models. A detailed description of specific models goes beyond the scope of this chapter so the reader wishing to broaden his/her knowledge in this field is referred to the specialized literature.

8.6 Conclusions

The miniaturization of sample preparation techniques fits directly into the trend set by the principles of GAC. The resulting reduction in the amount of organic solvents used during the isolation of analytes and their subsequent release is practically unquestionable. Miniaturization also produces less waste than in conventional systems. Nevertheless, the evaluation based exclusively on this one indicator leads to superficial assessment which does not reflect the whole picture. Therefore, we would like to turn the reader's attention to a couple of additional aspects of microextraction techniques which should be developed further in order to implement the majority of GAC principles. Doubts arise when we consider not only the final application of some materials, but also the process of their production. MIPs and ILs can serve as an example here. In both cases, the amount of chemical reagents used in the synthesis and refining of these compounds (ILs in particular) can outweigh the amount of reagents saved due to the application of new materials in the isolation of analytes. The possibility of automating the sample preparation technique and coupling it directly to the final measurement step is another important factor. The operator-independent procedures have better reproducibility of results and throughput; therefore, they are characterized by a significantly lower use of materials and energy per analytical process. In this context, the shortcomings are noticeable, particularly in the liquid phase extraction techniques. In this case, the sample preparation step is quite often automated while the introduction of the sample into a measuring instrument is, with rare exceptions, a manual process. It seems that despite huge creativity of analytical chemists with respect to novel technical solutions, during the designing process more attention still should be paid to the possibility of future automation of both the sample preparation step alone and the entire analytical procedure. Also, the analysis of published litera-

Table 8.1 Examples of simultaneous microextraction techniques and chemometric treatment.

Compounds	Sample matrices	Microextraction technique	Screening step	Variables	Optimization step	Ref.
methyl *t*-butyl ether	water samples	HS-SPME	fractional factorial design	concentration of NaCl, extraction temperature, extraction time, agitation, headspace volume	central composite design (CCD)	(Dron et al., 2002)
dialkyl ethers and BTEX	water samples	HS-SPME	Plackett–Burman design	type of fiber, pH, ionic strength, headspace volume, extraction temperature, extraction time, desorption time	CCD – 2^3 augmented with (2×3 + 1)	(Arambarri et al., 2004)
organochlorine pesticides and PCBs	serum	HS-SPME	Plackett–Burman design	absorption time, salt concentration	CCD	(Aguilar et al., 1999)
chlorobenzenes	water samples	microwave enhanced HS-SDME	Plackett–Burman design	drop volume, sample volume, stirring speed, ionic strength, extraction time, IL type, microwave power, length of the Y-type glass tube	mixed-level factorial design	(Vidal et al., 2007b)
benzophenone-3	human urine	SDME	Plackett–Burman design	ionic strength, extraction time, stirring speed, pH, IL type	CCD	(Vidal et al., 2007a)
2,4,6-trichloroanisole; 2,4,6-tribromoanisole	water samples, wine samples	HS-SDME	fractional factorial experimental	extraction time, pH, salt concentration, extraction time, sample time	Box-Behnken design	(Martendal et al., 2007)
nonsteroidal anti-inflammatory drugs	wastewater samples	LLLME	orthogonal array design (4(1) x 2(12))	type of organic solvent, composition of donor and acceptor phase, stirring speed, extraction time, salt concentration	orthogonal array design (4 x 4)	(Wu & Lee, 2005)
Cu(II)	water samples	DLLME		sample size, volume of dispersive solvent, type of dispersive solvent, type of extracting solvent, volume of extracting solvent, concentration of 8-hydroxyquinolinium, salt concentration	CCD	(Farajzadeh et al., 2008)

ture indicates that the potential presented by chemometric tools is underutilized. At present, the one-factor-at-a-time method of optimization still dominates, while this approach should be used only for the simplest systems, particularly with regard to microextraction techniques.

Obviously, the aforementioned remarks do not discredit the enormous impact of miniaturization as one of many ways of fulfilling the requirements posed by GAC. They rather point to the fact that the full implementation of GAC principles demands additional activities with respect to miniaturization in order to fully live up to the GAC designation.

Abbreviations

μ-SPE	micro-solid phase extraction
ADS	alkyl-diol-silica
ANOVA	analysis of variance
AμE	adsorptive μ-extraction
CAs	carbon aerogels
CCD	Central Composite Design
CE	capillary electrophoresis
CIAME	cold-induced aggregation microextraction
CPE	cloud point extraction
dCPE	dual cloud point extraction
DD-SDME	drop-to-drop single-drop microextraction
DLLME	dispersive liquid-liquid microextraction
DLLME-SFO	dispersive liquid-liquid microextraction based on the solidification of a floating organic drop
dLPME	dynamic liquid-phase microextraction
DSDME	directly suspended droplet microextraction technique
dynamic-HF-LPME	dynamic hollow-fiber liquid-phase microextraction
EE-SPME	electrosorption enhanced solid-phase microextraction
EME	electro membrane extraction
ETAAS	electrothermal atomic absorption spectrometry
FAAS	flame atomic absorption spectrometry
GC	gas chromatography
HF-LPME	hollow fiber liquid phase microextraction
HF-SLPME	Hollow fiber solid–liquid phase microextraction
HPLC	high performance liquid chromatography
HS-SDME	headspace single-drop microextraction
IL	ionic liquid
IP-SAME	ion pair based surfactant assisted microextraction
ISFME	*in situ* solvent-formation microextraction

LLE	liquid-liquid extraction
LPME	liquid-phase microextraction
MIPs	molecularly imprinted polymers
MWCNT	multiwall carbon nanotubes
NPOE	2-nitrophenyl octyl ether
PAHs	polycyclic aromatic hydrocarbons
PDMS	polydimethylsiloxane
PMMA	poly(*methyl)*methacrylate
PPESK	poly(phthalazine ether sulfone ketone)
PPY	polypyrrole
PU	polyurethane foam
RAM	restricted access material
SA-DLLME	surfactant-assisted dispersive liquid–liquid microextraction
SBME	solvent bar microextraction
SBSE	stir bar sorptive extraction
SC-DHF-HS-LPME	solvent cooling assisted dynamic hollow-fiber-supported head-space liquid phase microextraction
SCSE	stir cake sorptive extraction
SD-DLLME	solvent demulsification dispersive liquid–liquid microextraction
SDME	single-drop microextraction
SFODME	Solidified floating organic drop microextraction
SI-DLLME	sequential injection dispersive liquid–liquid microextraction
SM-DLLME	supramolecular-based dispersive liquid–liquid microextraction
SPE	solid phase extraction
SPME	solid-phase microextraction
SPNE	solid phase nanoextraction
ST-DLLME	solvent terminated dispersive liquid–liquid microextraction
TILDLME	temperature-controlled ionic liquid exhaustively dispersive liquid phase micro-extraction
UASEME	ultrasound-assisted surfactant-enhanced emulsification microex-traction
USA-DLLME	ultrasound-assisted dispersive liquid–liquid microextraction
USAEME	ultrasound-assisted emulsification-microextraction
US-DLLME	ultrasound-assisted dispersive liquid-liquid microextraction
VALLME	vortex-assisted liquid–liquid microextraction
VSLLME	vortex-assisted surfactant-enhanced-emulsification liquid–liquid microextraction

References

Aguilar, C., Peñalver, A., Pocurull, E., Ferré, J., Borrull, F., & Marcé, R. M. (1999). Optimization of solid-phase microextraction conditions using a response surface methodology to determine organochlorine pesticides in water by gas chromatography and electron-capture detection. *Journal of Chromatography A, 844*, 425–432.

Aguilera-Herrador, E., Lucena, R., Cárdenas, S., & Valcárcel, M. (2007). Direct coupling of ionic liquid based single-drop microextraction and GC/MS. *Analytical Chemistry, 80*, 793–800.

Aguilera-Herrador, E., Lucena, R., Cárdenas, S., & Valcárcel, M. (2008). Ionic liquid-based single-drop microextraction/gas chromatographic/mass spectrometric determination of benzene, toluene, ethylbenzene and xylene isomers in waters. *Journal of Chromatography A, 1201*, 106–111.

Andruch, V., Acebal, C. C., Škrlíková, J., Sklenářová, H., Solich, P., Balogh, I. S., et al. (2012). Automated on-line dispersive liquid–liquid microextraction based on a sequential injection system. *Microchemical Journal, 100*, 77–82.

Anthemidis, A. N., & Ioannou, K.-I. G. (2009). On-line sequential injection dispersive liquid–liquid microextraction system for flame atomic absorption spectrometric determination of copper and lead in water samples. *Talanta, 79*, 86–91.

Anthemidis, A. N., & Ioannou, K.-I. G. (2010). Development of a sequential injection dispersive liquid–liquid microextraction system for electrothermal atomic absorption spectrometry by using a hydrophobic sorbent material: Determination of lead and cadmium in natural waters. *Analytica Chimica Acta, 668*, 35–40.

Anthemidis, A. N., & Ioannou, K.-I. G. (2011). Sequential injection dispersive liquid–liquid microextraction based on fatty alcohols and poly(etheretherketone)-turnings for metal determination by flame atomic absorption spectrometry. *Talanta, 84*, 1215–1220.

Arambarri, I., Lasa, M., Garcia, R., & Millán, E. (2004). Determination of fuel dialkyl ethers and BTEX in water using headspace solid-phase microextraction and gas chromatography–flame ionization detection. *Journal of Chromatography A, 1033*, 193–203.

Arvand, M., Bozorgzadeh, E., Shariati, S., & Zanjanchi, M. A. (2012). Trace determination of linear alkylbenzene sulfonates using ionic liquid based ultrasound-assisted dispersive liquid-liquid microextraction and response surface methodology. *Analytical Methods, 4*, 2272–2277.

Baghdadi, M., & Shemirani, F. (2008). Cold-induced aggregation microextraction: A novel sample preparation technique based on ionic liquids. *Analytica Chimica Acta, 613*, 56–63.

Bagheri, H., & Roostaie, A. (2012). Aniline–silica nanocomposite as a novel solid phase microextraction fiber coating. *Journal of Chromatography A, 1238*, 22–29.

Bai, H., Zhou, Q., Xie, G., & Xiao, J. (2009). Enrichment and sensitive determination of dichlorodiphenyltrichloroethane and its metabolites with temperature controlled ionic liquid dispersive liquid phase microextraction prior to high performance liquid phase chromatography. *Analytica Chimica Acta, 651*, 64–68.

Baltussen, E., Janssen, H.-G., Sandra, P., & Cramers, C. A. (1997). A new method for sorptive enrichment of gaseous samples: Application in air analysis and natural gas characterization. *Journal of High Resolution Chromatography, 20*, 385–393.

Basheer, C., Alnedhary, A. A., Madhava Rao, B. S., Balasubramanian, R., & Lee, H. K. (2008). Ionic liquid supported three-phase liquid–liquid–liquid microextraction as a sample preparation technique for aliphatic and aromatic hydrocarbons prior to gas chromatography-mass spectrometry. *Journal of Chromatography A, 1210*, 19–24.

Basheer, C., Alnedhary, A. A., Rao, B. S. M., Valliyaveettil, S., & Lee, H. K. (2006). Development and application of porous membrane-protected carbon nanotube micro-solid-phase extraction combined with gas chromatography/mass spectrometry. *Analytical Chemistry, 78*, 2853–2858.

Basheer, C., Chong, H. G., Hii, T. M., & Lee, H. K. (2007). Application of porous membrane-protected micro-solid-phase extraction combined with HPLC for the analysis of acidic drugs in wastewater. *Analytical Chemistry, 79*, 6845–6850.

Batlle, R., López, P., Nerín, C., & Crescenzi, C. (2008). Active single-drop microextraction for the determination of gaseous diisocyanates. *Journal of Chromatography A, 1185*, 155–160.

Chai, X., He, Y., Ying, D., Jia, J., & Sun, T. (2007). Electrosorption-enhanced solid-phase microextraction using activated carbon fiber for determination of aniline in water. *Journal of Chromatography A, 1165*, 26–31.

Chamsaz, M., Atarodi, A., Eftekhari, M., Asadpour, S., & Adibi, M. (2013). Vortex-assisted ionic liquid microextraction coupled to flame atomic absorption spectrometry for determination of trace levels of cadmium in real samples. *Journal of Advanced Research, 4*, 35–41.

Chen, H., Chen, R., & Li, S. (2010). Low-density extraction solvent-based solvent terminated dispersive liquid–liquid microextraction combined with gas chromatography-tandem mass spectrometry for the determination of carbamate pesticides in water samples. *Journal of Chromatography A, 1217*, 1244–1248.

Chen, J., Zou, J., Zeng, J., Song, X., Ji, J., Wang, Y., et al. (2010). Preparation and evaluation of graphene-coated solid-phase microextraction fiber. *Analytica Chimica Acta, 678*, 44–49.

Chen, Y., & Pawliszyn, J. (2006). Miniaturization and automation of an internally cooled coated fiber device. *Analytical Chemistry, 78*, 5222–5226.

Chisvert, A., Román, I. P., Vidal, L., & Canals, A. (2009). Simple and commercial readily-available approach for the direct use of ionic liquid-based single-drop microextraction prior to gas chromatography: Determination of chlorobenzenes in real water samples as model analytical application. *Journal of Chromatography A, 1216*, 1290–1295.

Cruz-Vera, M., Lucena, R., Cárdenas, S., & Valcárcel, M. (2008). Ionic liquid-based dynamic liquid-phase microextraction: Application to the determination of anti-inflammatory drugs in urine samples. *Journal of Chromatography A, 1202*, 1–7.

Dron, J., Garcia, R., & Millán, E. (2002). Optimization of headspace solid-phase microextraction by means of an experimental design for the determination of methyl tert.-butyl ether in water by gas chromatography–flame ionization detection. *Journal of Chromatography A, 963*, 259–264.

Du, X.-Z., Wang, Y.-R., Tao, X.-J., & Deng, H.-L. (2005). An approach to application of mesoporous hybrid as a fiber coating of solid-phase microextraction. *Analytica Chimica Acta, 543*, 9–16.

Eisert, R., & Pawliszyn, J. (1997). New trends in solid phase microextraction. *Critical Reviews in Analytical Chemistry, 27*, 103–135.

Es'haghi, Z., Golsefidi, M. A., Saify, A., Tanha, A. A., Rezaeifar, Z., & Alian-Nezhadi, Z. (2010). Carbon nanotube reinforced hollow fiber solid/liquid phase microextraction: A novel extraction technique for the measurement of caffeic acid in *Echinacea purpurea* herbal extracts combined with high-performance liquid chromatography. *Journal of Chromatography A, 1217*, 2768–2775.

Es'haghi, Z., Khooni, M. A.-K., & Heidari, T. (2011). Determination of brilliant green from fish pond water using carbon nanotube assisted pseudo-stir bar solid/liquid microextraction combined with UV–vis spectroscopy–diode array detection. *Spectrochimica Acta Part A: Molecular and Biomolecular Spectroscopy, 79*, 603–607.

Farajzadeh, M. A., Bahram, M., Mehr, B. G., & Jönsson, J. Å. (2008). Optimization of dispersive liquid–liquid microextraction of copper (II) by atomic absorption spectrometry as its oxinate chelate: Application to determination of copper in different water samples. *Talanta, 75*, 832–840.

Farajzadeh, M. A., Djozan, D., & Khorram, P. (2012). Development of a new dispersive liquid–liquid microextraction method in a narrow-bore tube for preconcentration of triazole pesticides from aqueous samples. *Analytica Chimica Acta, 713*, 70–78.

Farajzadeh, M. A., Seyedi, S. E., Shalamzari, M. S., & Bamorowat, M. (2009). Dispersive liquid–liquid microextraction using extraction solvent lighter than water. *Journal of Separation Science, 32*, 3191–3200.
Feng, J., Sun, M., Li, J., Liu, X., & Jiang, S. (2011). A novel silver-coated solid-phase microextraction metal fiber based on electroless plating technique. *Analytica Chimica Acta, 701*, 174–180.
Feng, J., Sun, M., Li, J., Liu, X., & Jiang, S. (2012). A novel aromatically functional polymeric ionic liquid as sorbent material for solid-phase microextraction. *Journal of Chromatography A, 1227*, 54–59.
Fontana, A. R., Wuilloud, R. G., Martínez, L. D., & Altamirano, J. C. (2009). Simple approach based on ultrasound-assisted emulsification-microextraction for determination of polibrominated flame retardants in water samples by gas chromatography–mass spectrometry. *Journal of Chromatography A, 1216*, 147–153.
Gholivand, M. B., Abolghasemi, M. M., & Fattahpour, P. (2011). Polypyrrole/hexagonally ordered silica nanocomposite as a novel fiber coating for solid-phase microextraction. *Analytica Chimica Acta, 704*, 174–179.
Giokas, D., Paleologos, E., & Karayannis, M. (2002). Speciation of Fe(II) and Fe(III) by the modified ferrozine method, FIA–spectrophotometry, and flame AAS after cloud-point extraction. *Analytical and Bioanalytical Chemistry, 373*, 237–243.
Guan, W., Wang, Y., Xu, F., & Guan, Y. (2008). Poly(phthalazine ether sulfone ketone) as novel stationary phase for stir bar sorptive extraction of organochlorine compounds and organophosphorus pesticides. *Journal of Chromatography A, 1177*, 28–35.
Guo, L., & Lee, H. K. (2011a). Ionic liquid based three-phase liquid–liquid–liquid solvent bar microextraction for the determination of phenols in seawater samples. *Journal of Chromatography A, 1218*, 4299–4306.
Guo, L., & Lee, H. K. (2011b). Low-density solvent-based solvent demulsification dispersive liquid–liquid microextraction for the fast determination of trace levels of sixteen priority polycyclic aromatic hydrocarbons in environmental water samples. *Journal of Chromatography A, 1218*, 5040–5046.
Hashemi, P., Beyranvand, S., Mansur, R. S., & Ghiasvand, A. R. (2009). Development of a simple device for dispersive liquid–liquid microextraction with lighter than water organic solvents: Isolation and enrichment of glycyrrhizic acid from licorice. *Analytica Chimica Acta, 655*, 60–65.
He, Y., Pohl, J., Engel, R., Rothman, L., & Thomas, M. (2009). Preparation of ionic liquid based solid-phase microextraction fiber and its application to forensic determination of methamphetamine and amphetamine in human urine. *Journal of Chromatography A, 1216*, 4824–4830.
Ho, T. D., Canestraro, A. J., & Anderson, J. L. (2011). Ionic liquids in solid-phase microextraction: A review. *Analytica Chimica Acta, 695*, 18–43.
Hsieh, Y.-N., Huang, P.-C., Sun, I. W., Whang, T.-J., Hsu, C.-Y., Huang, H.-H., & Kuei, C.-H. (2006). Nafion membrane-supported ionic liquid–solid phase microextraction for analyzing ultra trace PAHs in water samples. *Analytica Chimica Acta, 557*, 321–328.
Huang, S.-P., & Huang, S.-D. (2007). Determination of organochlorine pesticides in water using solvent cooling assisted dynamic hollow-fiber-supported headspace liquid-phase microextraction. *Journal of Chromatography A, 1176*, 19–25.
Huang, X., Chen, L., Lin, F., & Yuan, D. (2011). Novel extraction approach for liquid samples: Stir cake sorptive extraction using monolith. *Journal of Separation Science, 34*, 2145–2151.
Huang, X., Lin, J., Yuan, D., & Hu, R. (2009a). Determination of steroid sex hormones in wastewater by stir bar sorptive extraction based on poly(vinylpyridine-ethylene dimethacrylate) monolithic material and liquid chromatographic analysis. *Journal of Chromatography A, 1216*, 3508–3511.

Huang, X., Qiu, N., Yuan, D., & Huang, B. (2009b). A novel stir bar sorptive extraction coating based on monolithic material for apolar, polar organic compounds and heavy metal ions. *Talanta, 78*, 101–106.

Huang, X., Qiu, N., Yuan, D., & Lin, Q. (2009c). Sensitive determination of strongly polar aromatic amines in water samples by stir bar sorptive extraction based on poly(vinylimidazole-divinylbenzene) monolithic material and liquid chromatographic analysis. *Journal of Chromatography A, 1216*, 4354–4360.

Hutchinson, J. P., Setkova, L., & Pawliszyn, J. (2007). Automation of solid-phase microextraction on a 96-well plate format. *Journal of Chromatography A, 1149*, 127–137.

Jafarvand, S., & Shemirani, F. (2011). Supramolecular-based dispersive liquid–liquid microextraction: A novel sample preparation technique utilizes coacervates and reverse micelles. *Journal of Separation Science, 34*, 455–461.

Jamt, R. E. G., Gjelstad, A., Eibak, L. E. E., Øiestad, E. L., Christophersen, A. S., Rasmussen, K. E., & Pedersen-Bjergaard, S. (2012). Electromembrane extraction of stimulating drugs from undiluted whole blood. *Journal of Chromatography A, 1232*, 27–36.

Jeannot, M. A., & Cantwell, F. F. (1996). Solvent microextraction into a single drop. *Analytical Chemistry, 68*, 2236–2240.

Jia, C., Zhu, X., Wang, J., Zhao, E., He, M., Chen, L., & Yu, P. (2010). Extraction of pesticides in water samples using vortex-assisted liquid–liquid microextraction. *Journal of Chromatography A, 1217*, 5868–5871.

Jiang, X., & Lee, H. K. (2004). Solvent bar microextraction. *Analytical Chemistry, 76*, 5591–5596.

Jiménez-Soto, J. M., Cárdenas, S., & Valcárcel, M. (2010). Carbon nanocones/disks as new coating for solid-phase microextraction. *Journal of Chromatography A, 1217*, 3341–3347.

Kawaguchi, M., Ishii, Y., Sakui, N., Okanouchi, N., Ito, R., Inoue, K., et al. (2004). Stir bar sorptive extraction with *in situ* derivatization and thermal desorption–gas chromatography–mass spectrometry in the multi-shot mode for determination of estrogens in river water samples. *Journal of Chromatography A, 1049*, 1–8.

Kawaguchi, M., Ito, R., Saito, K., & Nakazawa, H. (2006). Novel stir bar sorptive extraction methods for environmental and biomedical analysis. *Journal of Pharmaceutical and Biomedical Analysis, 40*, 500–508.

Khalili Zanjani, M. R., Yamini, Y., Shariati, S., & Jönsson, J. Å. (2007). A new liquid-phase microextraction method based on solidification of floating organic drop. *Analytica Chimica Acta, 585*, 286–293.

Kosaraju, P. B., & Sirkar, K. K. (2007). Novel solvent-resistant hydrophilic hollow fiber membranes for efficient membrane solvent back extraction. *Journal of Membrane Science, 288*, 41–50.

Kumar, A., Gaurav, Malik, A. K., Tewary, D. K., & Singh, B. (2008). A review on development of solid phase microextraction fibers by sol-gel methods and their applications. *Analytica Chimica Acta, 610*, 1–14.

Lambert, J.-P., Mullett, W. M., Kwong, E., & Lubda, D. (2005). Stir bar sorptive extraction based on restricted access material for the direct extraction of caffeine and metabolites in biological fluids. *Journal of Chromatography A, 1075*, 43–49.

Leong, M.-I., & Huang, S.-D. (2008). Dispersive liquid–liquid microextraction method based on solidification of floating organic drop combined with gas chromatography with electron-capture or mass spectrometry detection. *Journal of Chromatography A, 1211*, 8–12.

Liu, H., & Dasgupta, P. K. (1996). Analytical chemistry in a drop. Solvent extraction in a microdrop. *Analytical Chemistry, 68*, 1817–1821.

Liu, J.-f., Jiang, G.-b., Chi, Y.-g., Cai, Y.-q., Zhou, Q.-x., & Hu, J.-T. (2003). Use of ionic liquids for liquid-phase microextraction of polycyclic aromatic hydrocarbons. *Analytical Chemistry, 75*, 5870–5876.

Liu, J.-f., Li, N., Jiang, G.-b., Liu, J.-m., Jönsson, J. Å., & Wen, M.-j. (2005). Disposable ionic liquid coating for headspace solid-phase microextraction of benzene, toluene, ethylbenzene, and xylenes in paints followed by gas chromatography–flame ionization detection. *Journal of Chromatography A, 1066*, 27–32.

López-Darias, J., Pino, V., Anderson, J. L., Graham, C. M., & Afonso, A. M. (2010). Determination of water pollutants by direct-immersion solid-phase microextraction using polymeric ionic liquid coatings. *Journal of Chromatography A, 1217*, 1236–1243.

Madej, K. (2009). Microwave-assisted and cloud-point extraction in determination of drugs and other bioactive compounds. *TrAC Trends in Analytical Chemistry, 28*, 436–446.

Martendal, E., Budziak, D., & Carasek, E. (2007). Application of fractional factorial experimental and Box-Behnken designs for optimization of single-drop microextraction of 2,4,6-trichloroanisole and 2,4,6-tribromoanisole from wine samples. *Journal of Chromatography A, 1148*, 131–136.

Martinis, E. M., Olsina, R. A., Altamirano, J. C., & Wuilloud, R. G. (2008). Sensitive determination of cadmium in water samples by room temperature ionic liquid-based preconcentration and electrothermal atomic absorption spectrometry. *Analytica Chimica Acta, 628*, 41–48.

Melo, L. P., Nogueira, A. M., Lanças, F. M., & Queiroz, M. E. C. (2009). Polydimethylsiloxane/polypyrrole stir bar sorptive extraction and liquid chromatography (SBSE/LC-UV) analysis of antidepressants in plasma samples. *Analytica Chimica Acta, 633*, 57–64.

Moradi, M., Yamini, Y., Kakehmam, J., Esrafili, A., & Ghambarian, M. (2011). A new strategy to simultaneous microextraction of acidic and basic compounds. *Journal of Chromatography A, 1218*, 3945–3951.

Neng, N. R., Silva, A. R. M., & Nogueira, J. M. F. (2010). Adsorptive micro-extraction techniques—Novel analytical tools for trace levels of polar solutes in aqueous media. *Journal of Chromatography A, 1217*, 7303–7310.

Nojavan, S., & Fakhari, A. R. (2010). Electro membrane extraction combined with capillary electrophoresis for the determination of amlodipine enantiomers in biological samples. *Journal of Separation Science, 33*, 3231–3238.

O'Reilly, J., Wang, Q., Setkova, L., Hutchinson, J. P., Chen, Y., Lord, H. L., et al. (2005). Automation of solid-phase microextraction. *Journal of Separation Science, 28*, 2010–2022.

Ouyang, G., & Pawliszyn, J. (2006). Kinetic calibration for automated hollow fiber-protected liquid-phase microextraction. *Analytical Chemistry, 78*, 5783–5788.

Ouyang, G., Zhao, W., & Pawliszyn, J. (2007). Automation and optimization of liquid-phase microextraction by gas chromatography. *Journal of Chromatography A, 1138*, 47–54.

Paleologos, E. K., Giokas, D. L., & Karayannis, M. I. (2005). Micelle-mediated separation and cloud-point extraction. *TrAC Trends in Analytical Chemistry, 24*, 426–436.

Pedersen-Bjergaard, S., & Rasmussen, K. E. (1999). Liquid–liquid–liquid microextraction for sample preparation of biological fluids prior to capillary electrophoresis. *Analytical Chemistry, 71*, 2650–2656.

Pedersen-Bjergaard, S., & Rasmussen, K. E. (2006). Electrokinetic migration across artificial liquid membranes: New concept for rapid sample preparation of biological fluids. *Journal of Chromatography A, 1109*, 183–190.

Pena-Pereira, F., Costas-Mora, I., Lavilla, I., & Bendicho, C. (2012). Rapid screening of polycyclic aromatic hydrocarbons (PAHs) in waters by directly suspended droplet microextraction-microvolume fluorospectrometry. *Talanta, 89*, 217–222.

Pena-Pereira, F., Lavilla, I., Bendicho, C., Vidal, L., & Canals, A. (2009). Speciation of mercury by ionic liquid-based single-drop microextraction combined with high-performance liquid chromatography-photodiode array detection. *Talanta, 78*, 537–541.

Pena, M. T., Casais, M. C., Mejuto, M. C., & Cela, R. (2009). Development of an ionic liquid based dispersive liquid–liquid microextraction method for the analysis of polycyclic aromatic hydrocarbons in water samples. *Journal of Chromatography A, 1216*, 6356–6364.

Petersen, N. J., Foss, S. T., Jensen, H., Hansen, S. H., Skonberg, C., Snakenborg, D., et al. (2010). On-chip electro membrane extraction with online ultraviolet and mass spectrometric detection. *Analytical Chemistry, 83*, 44–51.
Pezo, D., Salafranca, J., & Nerín, C. (2007). Development of an automatic multiple dynamic hollow fibre liquid-phase microextraction procedure for specific migration analysis of new active food packagings containing essential oils. *Journal of Chromatography A, 1174*, 85–94.
Portugal, F. C. M., Pinto, M. L., & Nogueira, J. M. F. (2008). Optimization of polyurethane foams for enhanced stir bar sorptive extraction of triazinic herbicides in water matrices. *Talanta, 77*, 765–773.
Ramos Payán, M. D., Jensen, H., Petersen, N. J., Hansen, S. H., & Pedersen-Bjergaard, S. (2012). Liquid-phase microextraction in a microfluidic-chip – High enrichment and sample clean-up from small sample volumes based on three-phase extraction. *Analytica Chimica Acta, 735*, 46–53.
Regueiro, J., Llompart, M., Garcia-Jares, C., Garcia-Monteagudo, J. C., & Cela, R. (2008). Ultrasound-assisted emulsification–microextraction of emergent contaminants and pesticides in environmental waters. *Journal of Chromatography A, 1190*, 27–38.
Rezaee, M., Assadi, Y., Milani Hosseini, M.-R., Aghaee, E., Ahmadi, F., & Berijani, S. (2006). Determination of organic compounds in water using dispersive liquid–liquid microextraction. *Journal of Chromatography A, 1116*, 1–9.
Romero, J., López, P., Rubio, C., Batlle, R., & Nerín, C. (2007). Strategies for single-drop microextraction optimisation and validation: Application to the detection of potential antimicrobial agents. *Journal of Chromatography A, 1166*, 24–29.
Saleh, A., Yamini, Y., Faraji, M., Rezaee, M., & Ghambarian, M. (2009). Ultrasound-assisted emulsification microextraction method based on applying low density organic solvents followed by gas chromatography analysis for the determination of polycyclic aromatic hydrocarbons in water samples. *Journal of Chromatography A, 1216*, 6673–6679.
Saraji, M., & Bidgoli, A. (2010). Dispersive liquid–liquid microextraction using a surfactant as disperser agent. *Analytical and Bioanalytical Chemistry, 397*, 3107–3115.
Seidi, S., Yamini, Y., Saleh, A., & Moradi, M. (2011). Electromembrane extraction of levamisole from human biological fluids. *Journal of Separation Science, 34*, 585–593.
Sharma, N., Jain, A., Singh, V. K., & Verma, K. K. (2011). Solid-phase extraction combined with headspace single-drop microextraction of chlorophenols as their methyl ethers and analysis by high-performance liquid chromatography-diode array detection. *Talanta, 83*, 994–999.
Sikanen, T., Pedersen-Bjergaard, S., Jensen, H., Kostiainen, R., Rasmussen, K. E., & Kotiaho, T. (2010). Implementation of droplet-membrane-droplet liquid-phase microextraction under stagnant conditions for lab-on-a-chip applications. *Analytica Chimica Acta, 658*, 133–140.
Silva, A. R. M., Portugal, F. C. M., & Nogueira, J. M. F. (2008). Advances in stir bar sorptive extraction for the determination of acidic pharmaceuticals in environmental water matrices: Comparison between polyurethane and polydimethylsiloxane polymeric phases. *Journal of Chromatography A, 1209*, 10–16.
Sobhi, H. R., Yamini, Y., Esrafili, A., & Abadi, R. H. H. B. (2008). Suitable conditions for liquid-phase microextraction using solidification of a floating drop for extraction of fat-soluble vitamins established using an orthogonal array experimental design. *Journal of Chromatography A, 1196–1197*, 28–32.
Tankeviciute, A., Kazlauskas, R., & Vickackaite, V. (2001). Headspace extraction of alcohols into a single drop. *The Analyst, 126*, 1674–1677.
Turiel, E., & Martín-Esteban, A. (2009). Molecularly imprinted polymers for solid-phase microextraction. *Journal of Separation Science, 32*, 3278–3284.
Vatinno, R., Vuckovic, D., Zambonin, C. G., & Pawliszyn, J. (2008). Automated high-throughput method using solid-phase microextraction–liquid chromatography–tandem mass spectrometry

for the determination of ochratoxin A in human urine. *Journal of Chromatography A, 1201*, 215–221.

Vidal, L., Chisvert, A., Canals, A., & Salvador, A. (2007a). Sensitive determination of free benzophenone-3 in human urine samples based on an ionic liquid as extractant phase in single-drop microextraction prior to liquid chromatography analysis. *Journal of Chromatography A, 1174*, 95–103.

Vidal, L., Domini, C. E., Grané, N., Psillakis, E., & Canals, A. (2007b). Microwave-assisted headspace single-drop microextration of chlorobenzenes from water samples. *Analytica Chimica Acta, 592*, 9–15.

Vuckovic, D., Cudjoe, E., Hein, D., & Pawliszyn, J. (2008). Automation of solid-phase microextraction in high-throughput format and applications to drug analysis. *Analytical Chemistry, 80*, 6870–6880.

Vuckovic, D., Zhang, X., Cudjoe, E., & Pawliszyn, J. (2010). Solid-phase microextraction in bioanalysis: New devices and directions. *Journal of Chromatography A, 1217*, 4041–4060.

Wang, H., & Campiglia, A. D. (2008). Determination of polycyclic aromatic hydrocarbons in drinking water samples by solid-phase nanoextraction and high-performance liquid chromatography. *Analytical Chemistry, 80*, 8202–8209.

Wang, J.-X., Jiang, D.-Q., Gu, Z.-Y., & Yan, X.-P. (2006). Multiwalled carbon nanotubes coated fibers for solid-phase microextraction of polybrominated diphenyl ethers in water and milk samples before gas chromatography with electron-capture detection. *Journal of Chromatography A, 1137*, 8–14.

Wang, Q., Qiu, H., Li, J., Liu, X., & Jiang, S. (2010). On-line coupling of ionic liquid-based single-drop microextraction with capillary electrophoresis for sensitive detection of phenols. *Journal of Chromatography A, 1217*, 5434–5439.

Wanigasekara, E., Perera, S., Crank, J., Sidisky, L., Shirey, R., Berthod, A., & Armstrong, D. (2010). Bonded ionic liquid polymeric material for solid-phase microextraction GC analysis. *Analytical and Bioanalytical Chemistry, 396*, 511–524.

Wu, H.-F., Yen, J.-H., & Chin, C.-C. (2006). Combining drop-to-drop solvent microextraction with gas chromatography/mass spectrometry using electronic ionization and self-ion/molecule reaction method to determine methoxyacetophenone isomers in one drop of water. *Analytical Chemistry, 78*, 1707–1712.

Wu, J., & Lee, H. K. (2005). Orthogonal array designs for the optimization of liquid-liquid-liquid microextraction of nonsteroidal anti-inflammatory drugs combined with high-performance liquid chromatography-ultraviolet detection. *Journal of Chromatography A, 1092*, 182–190.

Wu, J., & Pawliszyn, J. (2001). Preparation and applications of polypyrrole films in solid-phase microextraction. *Journal of Chromatography A, 909*, 37–52.

Wu, Q., Chang, Q., Wu, C., Rao, H., Zeng, X., Wang, C., & Wang, Z. (2010). Ultrasound-assisted surfactant-enhanced emulsification microextraction for the determination of carbamate pesticides in water samples by high performance liquid chromatography. *Journal of Chromatography A, 1217*, 1773–1778.

Wu, Y., Xia, L., Chen, R., & Hu, B. (2008). Headspace single drop microextraction combined with HPLC for the determination of trace polycyclic aromatic hydrocarbons in environmental samples. *Talanta, 74*, 470–477.

Xu, H., Ding, Z., Lv, L., Song, D., & Feng, Y.-Q. (2009). A novel dispersive liquid–liquid microextraction based on solidification of floating organic droplet method for determination of polycyclic aromatic hydrocarbons in aqueous samples. *Analytica Chimica Acta, 636*, 28–33.

Yang, Z.-H., Lu, Y.-L., Liu, Y., Wu, T., Zhou, Z.-Q., & Liu, D.-H. (2011). Vortex-assisted surfactant-enhanced-emulsification liquid–liquid microextraction. *Journal of Chromatography A, 1218*, 7071–7077.

Yangcheng, L., Quan, L., Guangsheng, L., & Youyuan, D. (2006). Directly suspended droplet microextraction. *Analytica Chimica Acta, 566*, 259–264.

Yao, C., Pitner, W. R., & Anderson, J. L. (2009). Ionic liquids containing the tris(pentafluoroethyl) trifluorophosphate anion: A new class of highly selective and ultra hydrophobic solvents for the extraction of polycyclic aromatic hydrocarbons using single drop microextraction. *Analytical Chemistry, 81*, 5054–5063.

Ye, C.-L., Zhou, Q.-X., & Wang, X.-M. (2006). Headspace liquid-phase microextraction using ionic liquid as extractant for the preconcentration of dichlorodiphenyltrichloroethane and its metabolites at trace levels in water samples. *Analytica Chimica Acta, 572*, 165–171.

Ye, C., Zhou, Q., & Wang, X. (2007). Improved single-drop microextraction for high sensitive analysis. *Journal of Chromatography A, 1139*, 7–13.

Yiantzi, E., Psillakis, E., Tyrovola, K., & Kalogerakis, N. (2010). Vortex-assisted liquid–liquid microextraction of octylphenol, nonylphenol and bisphenol-A. *Talanta, 80*, 2057–2062.

Yin, X.-B. (2007). Dual-cloud point extraction as a preconcentration and clean-up technique for capillary electrophoresis speciation analysis of mercury. *Journal of Chromatography A, 1154*, 437–443.

Yu, C., Liu, Q., Lan, L., & Hu, B. (2008). Comparison of dual solvent-stir bars microextraction and U-shaped hollow fiber–liquid phase microextraction for the analysis of Sudan dyes in food samples by high-performance liquid chromatography–ultraviolet/mass spectrometry. *Journal of Chromatography A, 1188*, 124–131.

Yu, J., Dong, L., Wu, C., Wu, L., & Xing, J. (2002). Hydroxyfullerene as a novel coating for solid-phase microextraction fiber with sol-gel technology. *Journal of Chromatography A, 978*, 37–48.

Yu, L.-P. (2005). Cloud point extraction preconcentration prior to high-performance liquid chromatography coupled with cold vapor generation atomic fluorescence spectrometry for speciation analysis of mercury in fish samples. *Journal of Agricultural and Food Chemistry, 53*, 9656–9662.

Zeng, J., Chen, J., Song, X., Wang, Y., Ha, J., Chen, X., & Wang, X. (2010). An electrochemically enhanced solid-phase microextraction approach based on a multi-walled carbon nanotubes/ Nafion composite coating. *Journal of Chromatography A, 1217*, 1735–1741.

Zeng, J., Yu, B., Chen, W., Lin, Z., Zhang, L., Lin, Z., et al. (2008). Application of ceramic/carbon composite as a novel coating for solid-phase microextraction. *Journal of Chromatography A, 1188*, 26–33.

Zeng, J., Zou, J., Song, X., Chen, J., Ji, J., Wang, B., et al. (2011). A new strategy for basic drug extraction in aqueous medium using electrochemically enhanced solid-phase microextraction. *Journal of Chromatography A, 1218*, 191–196.

Zhang, P.-P., Shi, Z.-G., Yu, Q.-W., & Feng, Y.-Q. (2011). A new device for magnetic stirring-assisted dispersive liquid–liquid microextraction of UV filters in environmental water samples. *Talanta, 83*, 1711–1715.

Zhang, X., Oakes, K. D., Wang, S., Servos, M. R., Cui, S., Pawliszyn, J., et al. (2012). *In vivo* sampling of environmental organic contaminants in fish by solid-phase microextraction. *TrAC Trends in Analytical Chemistry, 32*, 31–39.

Zhang, Z., & Pawliszyn, J. (1995). Quantitative extraction using an internally cooled solid phase microextraction device. *Analytical Chemistry, 67*, 34–43.

Zhao, F., Meng, Y., & Anderson, J. L. (2008). Polymeric ionic liquids as selective coatings for the extraction of esters using solid-phase microextraction. *Journal of Chromatography A, 1208*, 1–9.

Zhou, Q., Bai, H., Xie, G., & Xiao, J. (2008). Temperature-controlled ionic liquid dispersive liquid phase micro-extraction. *Journal of Chromatography A, 1177*, 43–49.

Zhu, F., Guo, J., Zeng, F., Fu, R., Wu, D., Luan, T., et al. (2010). Preparation and characterization of porous carbon material-coated solid-phase microextraction metal fibers. *Journal of Chromatography A, 1217*, 7848–7854.

Zhu, L., Zhu, L., & Lee, H. K. (2001). Liquid–liquid–liquid microextraction of nitrophenols with a hollow fiber membrane prior to capillary liquid chromatography. *Journal of Chromatography A, 924*, 407–414.

Zhu, X., Cai, J., Yang, J., Su, Q., & Gao, Y. (2006). Films coated with molecular imprinted polymers for the selective stir bar sorption extraction of monocrotophos. *Journal of Chromatography A, 1131*, 37–44.

Index

A

accelerated solvent extraction 348, 366
aerogel 38, 429
alkanethiols 43
analytical process 1, 2, 3, 4, 7, 9, 17, 22, 311, 315, 326, 378, 381, 434
analytical separation 1, 2, 3, 7, 11, 12, 14, 16, 19, 29, 197, 381
anodization 67, 90
antibodies 105, 125, 131, 376, 381
antigens 64, 125
aptamers 131, 376
Artificial Neural Networks 297
atomic spectrometry 18, 19, 261, 331, 333, 359
automation 1, 3, 33, 34, 75, 76, 88, 192, 194, 203, 204, 205, 206, 207, 210, 214, 215, 219, 221, 229, 235, 270, 276, 299, 308, 310, 329, 330, 349, 357, 359, 360, 369, 378, 379, 380, 381, 382, 418, 421, 422, 428, 430, 431, 434

B

BCR protocol 344, 346, 347, 348, 350, 352
Box Behnken Design 293, 297

C

calibration 298, 299, 330, 367, 370
calixarenes 41, 42
capillary electrophoresis 11, 16, 261, 278, 367
– chip-based 16, 17, 358
carbon nanomaterials 95, 98
carbon nanotubes 41, 91, 95, 96, 220, 223, 351, 354, 368, 420, 429
Central Composite Design 293, 294, 297
certified reference materials 299, 315
chemical vapor deposition 92
chemometrics 277, 290, 300, 418, 432, 433, 435, 436
cloud point extraction 423
– dual 424
coacervates 196, 232, 269, 377, 423, 424
cold-induced aggregation microextraction 11, 231, 269, 287, 426
combustion
– microwave-induced 332
complex-imprinted polymer 114
composites 58, 59, 429
– microcomposites 59
– multifunctional composites 139, 157
– nanocomposites 37, 59, 60, 61, 90, 381
conductive polymers 37, 44, 47, 49, 90, 428
continuous-flow microextraction 193, 209, 261
crosslinkers 54, 55, 64, 66, 91, 105
crown ethers 41
cycle-flow microextraction 210
cyclodextrins 41, 377

D

data processing 3
deep eutectic solvents 369, 381
demulsification 229, 422
dendrimers 95
derivatization 11, 97, 144, 191, 225, 278, 287, 288, 289, 298, 300, 355
desorption 31, 33, 71, 124, 278, 289, 316, 427, 430
– laser 290
– liquid 122, 217, 290, 429
– thermal 10, 74, 198, 289, 353, 418, 425, 427, 429
detection 1, 2, 3, 17, 18, 29, 194, 197, 380
diazonium salts 43, 44
digestion 229
– acid 309, 331, 335
– enzymatic 340, 341
– microwave-assisted 335, 336
– small-volume microwave-assisted 336
– ultrasound-assisted 339
directly suspended droplet microextraction 11, 194, 201, 212, 214, 264, 280, 298, 299, 300, 420
direct solid analysis 308, 326, 366
dispersive liquid-liquid microextraction 10, 194, 225, 228, 255, 256, 257, 268, 279, 356, 372, 421, 422, 425, 431
– air-assisted 269
– ultrasound-assisted 200, 269, 421
dispersive micro-solid-phase extraction 354
dispersive solvent 268, 269, 421
distribution coefficient 74, 254, 266, 283

Doehlert Design 293, 294, 297
drop-in-drop 192, 201, 207
drop-to-drop 201, 208, 224, 419
dry ashing 309, 331, 332
dynamic liquid-phase microextraction 192, 206, 207, 425

E

electrochemical techniques 21
electrodeposition 90
electrokinetic migration 221
electroless plating 101
electromembrane extraction 193, 216, 217, 220, 221, 222, 223, 224, 422, 427
electrophoretic deposition 70, 90, 91
electropolymerization 49
electrospinning 37, 63, 70
electrospun nanofibers 63, 71
emulsification 213, 226, 228, 326, 330
– ultrasound-assisted 331
emulsification microextraction
– ultrasound-assisted 200, 357, 421
enrichment factor 6, 9, 10, 191, 192, 194, 208, 225, 234, 254, 265, 268, 271, 276, 277, 280, 298
experimental design 200, 201, 277, 290, 291, 292, 294, 295, 297
extraction
– acid 343
– microwave-assisted 347, 366, 370
– ultrasound-assisted 344, 366
extraction efficiency 10, 36, 64, 93, 96, 122, 261, 265, 268, 277, 280, 368, 374, 427, 428, 432

F

factorial design 291, 292
flow injection 3, 4, 358, 359
fractionation 343, 344, 346, 347, 350
fullerenes 41, 95, 96

G

gas chromatography 11, 12, 31, 33, 70, 261, 278
– chip-based 13
– micro 13, 14
graphene 62, 95, 97, 368, 429
graphite oxides 98
green analytical chemistry 192, 201, 234, 378, 416, 427, 434
green chemistry 289, 418, 421, 430

H

Hildebrand solubility parameter 254, 255, 279
hollow fiber 10, 193, 215, 217, 266, 426
hollow fiber liquid-phase microextraction 10, 193, 194, 215, 255, 256, 257, 258, 266, 357, 420, 431
– three phase 216, 266, 267
– two phase 215, 266, 267

I

immunosorbents 89, 125, 157
initiator 54, 60, 91, 124
inside needle capillary adsorption trap 75
in situ solvent-formation microextraction 269, 426
integration 1, 11, 18, 19, 310, 357, 359, 378, 379, 381, 382
interferences 191, 198, 287, 289, 299, 300, 308, 311, 326, 334, 359, 372, 382
in-tube solid-phase microextraction 31, 57, 65, 88, 142, 144, 428
ionic liquids 89, 122, 123, 126, 127, 128, 129, 130, 157, 194, 195, 197, 220, 226, 228, 230, 232, 260, 262, 267, 268, 279, 377, 381, 424, 425
– micellar 197
– polymeric 68, 89, 124, 126, 127, 128, 129, 130, 157, 381, 428
ion imprinted polymers 94
ion-imprinted polymers 115

L

lab-on-a-chip 1, 20, 21, 378, 419
lab-on-a-robot 17
lab-on-a-valve 3, 4, 359, 380, 381
lab-on-a-valve bead-injection 359
lab-on-disk 357
levitation 8
limit of detection 75, 298, 366
limit of quantification 298, 366

liquid chromatography 7, 11, 14, 31, 33, 34, 261, 278
– chip-based 14, 15, 16
liquid-liquid extraction 9, 10, 191, 192, 253, 309, 350, 356, 418
– chip-based 207
liquid-liquid-liquid microextraction 193, 201, 210, 278
liquid-liquid microextraction
– air-assisted 234, 270
– vortex-assisted 421, 426
liquid phase deposition 92
liquid-phase microextraction 10, 192, 253, 356, 366, 372, 374, 377, 419
– chip-based 358
– three phase 300
– two phase 300

M

mass analyzers 20
mass spectrometry 18, 20
mass transfer 10, 51, 58, 96, 101, 141, 203, 212, 213, 215, 218, 219, 223, 230, 270, 277, 283, 316
matrix effects 286, 299, 356, 368, 381
matrix solid-phase dispersion 366, 367
membranes 10, 11, 215, 219, 223, 267, 278, 374, 426, 430
mesoporous materials 101, 102, 104
metal-organic frameworks 94, 131, 135, 136, 139, 140, 157
metathesis reactions 228, 232
microanalysis 326, 332
microelectrodes 22
micro-electromechanical systems 11, 17
microextraction in packed syringe 10, 70, 73, 88, 144, 154, 375
microplasmas 19, 20
micro-solid-phase extraction 57, 374, 380, 430
micro-total analysis 1, 3, 374, 378
microwave energy 200, 205, 276, 330, 331, 333, 335, 343, 347, 352
miniaturization 1, 3, 4, 5, 7, 9, 11, 12, 14, 16, 17, 18, 19, 20, 21, 22, 23, 192, 198, 308, 310, 317, 331, 333, 335, 336, 349, 356, 357, 359, 360, 369, 378, 379, 381, 382, 416, 417, 419, 424, 430, 434, 436
molecularly-imprinted materials 105, 157
molecularly imprinted polymers 37, 64, 65, 89, 91, 94, 276, 351, 428
molecularly imprinted xerogels 65
molecular spectrometry 18
monolithic materials 94, 139
monolithic polymers 54, 56, 57, 58
monomers 49, 54, 55, 64, 66, 91, 105, 113, 124

N

nanocomposites 59
nanoelectrodes 22
nanoelectromembrane extraction 224
nanomaterials 37, 38, 61, 63, 89, 94, 157, 428
– metal oxide 99
nanoparticles 20, 60, 61, 91, 276, 277, 376, 429
needle trap device 75, 76, 88, 144
nonimprinted polymers 115

O

organically modified silica 39, 113
organic polymer monoliths 141

P

permanent modifiers 329, 330
Plackett-Burman design 293
polymerization 48, 54, 60, 91
– chemical 48
– electrochemical 49
porogen 54, 55, 91, 124
porosity 41, 54, 65
– mesopores 54
– micropores 54
portability 1, 3, 7, 16, 19, 21, 23, 29, 199, 380, 381
power consumption 7, 14, 17, 19, 20
preconcentration 9, 15, 47, 76, 94, 115, 191, 206, 265, 266, 278, 282, 288, 329, 353, 366

Q

quantum cascade lasers 19
quantum dots 95, 197

R

recovery studies 299
refractive index 253, 258
response surface methodologies 291
restricted access materials 94, 142, 429

S

salting-out 219, 285, 286
sample collection 3, 9, 315, 428
sample preparation 1, 2, 3, 9, 191, 308, 311, 315, 317, 434
sample preservation 3, 9, 315, 316
sample pre-treatment 7, 9, 313, 315
sample storage 7, 315, 316
sample throughput 7, 336, 338, 349, 381, 382
screen-printed electrodes 198, 223, 232
selectivity 2, 10, 11, 20, 35, 39, 65, 88, 93, 94, 125, 157, 234, 253, 254, 277, 278, 279, 280, 282, 285, 298, 299, 300, 378, 381, 420, 426
self-assembled monolayers 42, 43
sensitivity 2, 6, 7, 15, 19, 20, 21, 75, 88, 94, 144, 199, 223, 234, 253, 298, 378
sequential extraction schemes 346, 348, 352
sequential injection 206, 229, 422
single-drop microextraction 10, 192, 193, 194, 195, 201, 202, 225, 255, 256, 257, 258, 261, 262, 419, 431
– direct 201, 205, 261, 264, 284
– headspace 193, 200, 201, 203, 261, 278, 356, 372, 419
slurry sampling 309, 326, 327, 360
sol-gel 37, 38, 39, 41, 42, 61, 65, 89, 96, 113, 123, 124, 144, 425, 428
solidification of floating organic drop microextraction 11, 194, 201, 214, 264, 265, 287, 420, 423
solid-liquid extraction 343
solid-phase dynamic extraction 75
solid-phase extraction 9, 54, 88, 191, 309, 353, 427
solid-phase microextraction 9, 29, 75, 76, 88, 352, 366, 372, 374, 375, 418, 425, 427, 428
– coating preparation techniques 88
– conventional coatings 35, 92
– headspace 278, 355
– modes 31
– novel materials 94
– off-line coupling 31
– on-line coupling 31
– theoretical aspects 29
solvent bar microextraction 193, 218, 421, 426
solvents 1, 5, 7, 195, 214, 220, 226, 253, 260, 261, 267, 268, 279, 343, 377, 417, 419, 420
– boiling point 254, 259, 261
– chemical reactivity 259
– corrosiveness 259
– density 258
– dielectric constant 254
– flammability 260
– freezing point 260
– immiscibility 258
– interfacial tension 259
– physicochemical properties 253, 255, 256, 257, 281
– solubility 253
– surface tension 254, 261
– toxicity 260
– vapor pressure 254, 259
– viscosity 259
sonication devices 230, 329, 344
Soxhlet extraction 9, 309, 350
speciation 340, 341, 342, 343, 344, 347
stir bar sorptive extraction 10, 74, 75, 88, 144, 427, 429
supercritical fluid extraction 9, 343, 349, 366
supercritical fluids 343, 371, 417, 418
supported liquid membrane 193, 215, 216, 221, 223
supramolecular systems 226, 232, 269, 279, 377, 381, 423
surfactants 72, 194, 196, 197, 228, 230, 263, 269, 342, 377, 421, 423, 424, 426
sustainability 5, 22, 234, 235
syneresis 38, 39

T

Tessier protocol 346, 348
thin film microextraction 10, 88, 144, 150, 151

U

ultrasound irradiation 194, 200, 205, 219, 228, 230, 270, 276, 329, 330, 331, 341, 343, 344, 345, 351, 352, 356, 421, 426

V

validation 298, 309
vapor emissions 416, 417
variables 291, 432, 434
– optimization 200, 201, 277, 280, 288, 290, 291, 293, 294, 297, 432, 433, 434, 436
– screening 291, 292, 294
vortex agitation 200, 228, 230

W

wastes 5, 9, 11, 74, 191, 192, 289, 314, 326, 341, 350, 417, 419, 430, 434

Bei Fragen zur Produktsicherheit wenden Sie sich bitte an:
If you have any questions regarding product safety,
please contact:

Walter de Gruyter GmbH
Genthiner Straße 13
10785 Berlin
productsafety@degruyterbrill.com